T0186022

LONDON MATHEMATICAL SOCIETY LECTURE NOTE SERIES

Managing Editor: Professor N. J. Hitchin, Mathematical Institute, University of Oxford, 24-29 St Giles, Oxford OX1 3LB, United Kingdom

The titles below are available from booksellers, or from Cambridge University Press at www.cambridge.org/mathematics

London Mathematical Society Lecture Note Series. 337

Methods in Banach Space Theory

Proceedings of the V Conference on Banach Spaces, Cáceres, Spain, 13–18 September 2004

Edited by

JESÚS M. F. CASTILLO
Universidad de Extremadura

WILLIAM B. JOHNSON
Texas A&M University

CAMBRIDGE
UNIVERSITY PRESS

CAMBRIDGE
UNIVERSITY PRESS

Shaftesbury Road, Cambridge CB2 8EA, United Kingdom

One Liberty Plaza, 20th Floor, New York, NY 10006, USA

477 Williamstown Road, Port Melbourne, VIC 3207, Australia

314–321, 3rd Floor, Plot 3, Splendor Forum, Jasola District Centre, New Delhi – 110025, India

103 Penang Road, #05–06/07, Visioncrest Commercial, Singapore 238467

Cambridge University Press is part of Cambridge University Press & Assessment, a department of the University of Cambridge.

We share the University's mission to contribute to society through the pursuit of education, learning and research at the highest international levels of excellence.

www.cambridge.org
Information on this title: www.cambridge.org/9780521685689

© Cambridge University Press & Assessment 2006

First published 2006

A catalogue record for this publication is available from the British Library

ISBN   978-0-521-68568-9   Paperback

Contents

Preface

In the end, mathematics is what matters. In addition to the contributors to this volume of proceedings, the Conference included talks by C. Benítez, F. Cobos, P. Domanski, G. Godefroy, W. B. Johnson, A. Molto, J. P. Moreno, P. L. Papini and R. Phelps among the main speakers, and Julio Flores, Yves Raynaud, F. Naranjo, José Orihuela, Pedro Martín, I. J. Cabrera, V. Montesinos, P. Hájek, Luis Oncina, G. Gruenhage, Fernando García, Vassilis Kanellopoulos, I. Villanueva, P. Bandyopadhyay, Sebastian Lajara, Diego Yañez, David Pérez-García, C. Michels, Bernardo Cascales, Pandelis Dodos, Antonio Pulgarín, M. Muñoz, José Rodríguez, Luz M. Fernández-Cabrera, Carmen Calvo, Juan Casado, Belmensnaoui Aqzzouz, Redouane Nouira, Antonio Martínez, Fernando Mayoral, Maria del Carmen Calderón Moreno, Luiza Amália de Moraes, J. A. Prado Bassas, My Hachem Lalaoui Rhali, Houcine Benabdellah, Rachid El Harti, Michael Stessin among the shorter talks.

The conference was organized, like this proceedings volume, in five sections,

- Geometrical methods
- Homological methods
- Topological methods
- Operator theory methods
- Function space methods

The idea was to organize Banach space theory (and related categories) according to the methods used to approach problems. Of course, the methods can overlap, and (fortunately) usually do, but typically one of the methods sticks out. This proceedings volume has been organized accordingly; so, in each of its five sections the reader will encounter survey and research papers describing the state-of-the-art in each of those areas. All together the sixteen papers depict a rather attractive panorama of modern infinite dimensional Banach space theory.

Acknowledgements

From September 13th to September 18th, 2004, the Banach space theory group of the Mathematics Department of the University of Extremadura played host to the V Conference on Banach spaces and related topics. The conference was sponsored by:

- Departamento de Matemáticas de la Uex
- Universidad de Extremadura
- Diputación de Cáceres
- Consejería de Educación de la Junta de Extremadura
- Caja de Extremadura

The editors of this proceedings volume express their gratitude to those organizations for financial support. The Junta de Extremadura must moreover be acknowledged for putting at our disposal the beautiful XVIth century palace "Complejo San Francisco". We thank Cambridge University Press for making this volume proceedings a reality, and are grateful for the coordination efforts of Roger Astley and Catherine Appleton.

The conference united 86 participants from 14 countries, including leading specialists in the field. Thanks are to due to the members of the Scientific Committee, Fernando Bombal, María Jesús Carro, Manuel González, Bob Phelps, and Angel Rodríguez Palacios, and to the main speakers of the conference for their generous participation. All related information can be found at the website http://www.banachspaces.com/banach04 Last but not least, on behalf of all the participants we thank the volunteer students for the truly outstanding effort they made throughout the conference, doing whatever had to be done with extra smiles: Sandra Anes, Patricia Arjona, Javier Cabello, María Fejo, Ruth García, José Navarro and Jesús Suárez.

Junta de
Extremadura

SATURATED EXTENSIONS, THE ATTRACTORS METHOD AND HEREDITARILY JAMES TREE SPACES

SPIROS A. ARGYROS, ALEXANDER D. ARVANITAKIS, AND ANDREAS G. TOLIAS

Dedicated to the memory of R.C. James

CONTENTS

0. INTRODUCTION

The purpose of the present work is to provide examples of HI Banach spaces with no reflexive subspace and study their structure. As is well known W.T. Gowers [G1] has constructed a Banach space $\mathfrak{X}_{gt}$ with a boundedly complete basis $(e_n)_n$, not containing ℓ_1, and such that all of its infinite dimensional subspaces have non separable dual. We shall refer to this space as the Gowers Tree space. The predual $(\mathfrak{X}_{gt})_*$, namely the space generated by the biorthogonal of the basis, also has the property that it does not contain c_0 or a reflexive subspace. It remains unknown whether $\mathfrak{X}_{gt}$ is HI and moreover the structure of $\mathcal{L}(\mathfrak{X}_{gt})$ is unclear. Notice that Gowers dichotomy [G2] yields that $\mathfrak{X}_{gt}$ and $(\mathfrak{X}_{gt})_*$ contain HI subspaces. The structure of $\mathfrak{X}^*_{gt}$ also remains unclear. The main obstacle for understanding the structure of $\mathfrak{X}_{gt}$ or $\mathcal{L}(\mathfrak{X}_{gt})$ is the use of

Research partially supported by EPEAEK program "PYTHAGORAS".

a probabilistic argument for establishing the existence of vectors with certain properties.

Our approach in constructing HI spaces with no reflexive subspace, is different from Gowers' one. In particular we avoid the use of any probabilistic argument and thus we are able to control the structure of the spaces as well as the structure of the spaces of bounded linear operators acting on them. Moreover we are able to provide examples of spaces X exhibiting a vast difference between the structures of X and X^*.

The following are the highlight of our results:

- There exists a HI Banach space X with a shrinking basis and with no reflexive subspace. Moreover every $T : X \to X$ is of the form $\lambda I + W$ with W weakly compact (and hence strictly singular).

The absence of reflexive subspaces in X in conjunction with the property that every strictly singular operator is weakly compact is evidence supporting the existence of Banach spaces such that every non Fredholm operator is compact.

- The dual X^* of the previous X is HI and reflexively saturated and the dual of every subspace Y of X is also HI.

This shows a strong divergence between the structure of X and X^*. We recall that in [AT2] a reflexive HI space X is constructed whose dual X^* is unconditionally saturated. The analogue of this in the present setting is the following one:

- There exists a HI Banach space Y with a shrinking basis and with no reflexive subspace, such that the dual space Y^* is reflexive and unconditionally saturated.

The definition of the space Y requires an adaptation of the methods of [AT2] within the present framework of building spaces with no reflexive subspace.

- There exists a partition of the basis $(e_n)_n$ of the previous X into two sets $(e_n)_{n \in L_1}$, $(e_n)_{n \in L_2}$ such that setting $X_{L_1} = \overline{\mathrm{span}}\{e_n : n \in L_1\}$, $X_{L_2} = \overline{\mathrm{span}}\{e_n : n \in L_2\}$, both $X_{L_1}^*$, $X_{L_2}^*$ are HI with no reflexive subspace.

The pairs $X_{L_i}, X_{L_i}^*$ for $i = 1, 2$ share similar properties with the pair $(\mathfrak{X}_{gt})_*$ and $\mathfrak{X}_{gt}$.

- The space X^{**} is non separable and every w^*-closed subspace of X^{**}, is either isomorphic to ℓ_2 or is non-separable and contains ℓ_2. Therefore every quotient of X^* has a further quotient isomorphic to ℓ_2. Moreover X^{**}/X is isomorphic to $\ell_2(\Gamma)$.

It seems also possible that $\mathfrak{X}_{gt}^*$ satisfies a similar to the above property although this is not easily shown. Further X^* is the first example of a HI space with the following property: X^*/Y is HI whenever Y is w^*-closed (this is equivalent to say that for every subspace Z of X, Z^* is HI) and also every quotient of X^* has a further quotient which is isomorphic to ℓ_2.

- There exists a non separable HI Banach space Z not containing a reflexive subspace and such that every bounded linear operator $T :$

$Z \to Z$ is of the form $T = \lambda I + W$ with W a weakly compact (hence strictly singular) operator with separable range.

This is an extreme construction resulting from a variation of the methods used in the construction of the space X involved in the previous results. The fact of the matter is that these methods are not stable. Thus some minor changes in the initial data could produce spaces with entirely different structure. Notice that the space Z is of the form Y^{**} with Y and Y^* sharing similar properties with the pair X, X^* appearing in the previous statements.

We shall proceed to a more detailed presentation of the results of the paper and also of the methods used for constructing the spaces, which are interesting on their own. We have divided the rest of the introduction in three subsections. The first concerns the structure of Banach spaces not containing ℓ_1, c_0 or reflexive subspace. The second is devoted to saturated extensions and in the third we explain the method of attractors which permits the construction of dual pairs X, X^* with strongly divergent structure.

0.1. Hereditarily James Tree spaces.

Separable spaces like Gowers Tree space undoubtedly have peculiar structure. Roughly speaking, in every subspace one can find a structure similar to the James tree basis. Next we shall attempt to be more precise. Thus we shall define the Hereditarily James Tree spaces, making more transparent their structure. We begin by recalling some of the fundamental characteristics of James' paradigm.

In the sequel we shall denote by $(\mathcal{D}, \prec)$ the dyadic tree and by $[\mathcal{D}]$, the set of all branches (or the body) of $\mathcal{D}$. As usual we would consider that the nodes of $\mathcal{D}$ consist of finite sequences of 0's and 1's and $a \prec b$ iff a is an initial part of b. The lexicographic order of $\mathcal{D}$, denoted by $\prec_{lex}$ defines a well ordering which is consistent with the tree order (i.e. $a \prec b$ yields that $a \prec_{lex} b$).

The space JT.

The James Tree space JT ([J]) is defined as the completion of $(c_{00}(\mathcal{D}), \| \cdot \|_{JT})$ where for $x \in c_{00}(\mathcal{D})$, $\|x\|_{JT}$ is defined as follows:

$$\|x\|_{JT} = \sup \left\{ \left(\sum_{i-1}^{n} \left(\sum_{n \subset s_i} x(n) \right)^2 \right)^{1/2} : (s_i)_{i=1}^n \text{ pairwise disjoint segments} \right\}.$$

The main properties of the space JT, is that does not contain ℓ_1 and has nonseparable dual.

Next, we list some properties of JT related to our consideration.

- The Hamel basis $(e_a)_{a \in \mathcal{D}}$ of $c_{00}(\mathcal{D})$ ordered with the lexicographic order defines a (conditional) boundedly complete basis of JT.
- For every branch b in $[\mathcal{D}]$, $b = (a_1 \prec a_2 \prec \cdots \prec a_n \cdots)$ the sequence $(e_{a_n})_n$ is non trivial weak Cauchy and moreover $b^* = w^* - \sum_{n=1}^{\infty} e_{a_n}^*$ defines a norm one functional in JT^*.
- The biorthogonal functionals of the basis $(e_a^*)_{a \in \mathcal{D}}$ generate the predual JT_* of JT and they satisfy the following property.
  For every segment s of $\mathcal{D}$ setting $s^* = \sum_{a \in s} e_a^*$ we have that $\|s^*\| = 1$.

4 SPIROS A. ARGYROS, ALEXANDER D. ARVANITAKIS, AND ANDREAS G. TOLIAS

It is worth pointing out an alternative definition of the norm of JT. Thus we consider the following subset of $c_{00}(\mathcal{D})$,

$$G_{JT} = \Big\{ \sum_{i=1}^{n} \lambda_i s_i^* : (s_i)_{i=1}^{n} \text{ are disjoint finite segments and } \sum_{i=1}^{n} \lambda_i^2 \leq 1 \Big\}$$

Here s_i^* are defined as before. It is an easy exercise to see that the norm induced by the set G_{JT} on $c_{00}(\mathcal{D})$ coincides with the norm of JT.

The James Tree properties.

Let X be a space with a Schauder basis $(e_n)_n$. A block subspace Y of X has the boundedly complete (shrinking) James tree property if there exists a seminormalized block (in the lexicographical order $\prec_{lex}$ of $\mathcal{D}$) sequence $(y_a)_{a\in\mathcal{D}}$ in Y and a $c > 0$ such that the following holds.

(1) **(boundedly complete)** There exists a bounded family $(b^*)_{b\in[\mathcal{D}]}$ in X^*, such that for each $b \in [\mathcal{D}]$, $b = (a_1, a_2, \ldots, a_n, \ldots)$ the sequence $(y_{a_n})_n$ is non trivial weakly Cauchy with $\lim b^*(y_{a_n}) > c$ and $\lim b_1^*(y_{a_n}) = 0$ for all $b_1 \neq b$.

(2) **(shrinking)** For all finite segments s of $\mathcal{D}$, $\| \sum_{a\in s} y_a \| \leq c$.

Let's observe that $(e_a)_{a\in\mathcal{D}}$ in JT satisfies the boundedly complete James Tree property while $(e_a^*)_{a\in\mathcal{D}}$ in JT_* satisfies the shrinking one. Also, if the initial space X has a boundedly complete basis only the boundedly complete James Tree property could occur. A similar result holds if X has a shrinking basis. Finally if Y has the boundedly complete James Tree property, then Y^* is non separable and if X has a shrinking basis and Y has the (shrinking) James Tree property, then Y^{**} is non separable.

For simplicity, in the sequel we shall consider that the initial space X has either a boundedly complete or a shrinking basis. Thus if a block subspace has the James Tree property, then it will be determined as either boundedly complete or shrinking according to the corresponding property of the initial basis.

Definition 0.1. Let X be a Banach space with a Schauder basis.

(a) A family $\mathcal{L}$ of block subspaces of X has the James Tree property, provided every Y in $\mathcal{L}$ has that property.

(b) The space X is said to be Hereditarily James Tree (HJT) if it does not contain c_0, ℓ_1 and every block subspace Y of X, has the James Tree property.

It follows from Gowers' construction that the Gowers Tree space $\mathfrak{X}_{gt}$, and its predual $(\mathfrak{X}_{gt})_*$ are HJT spaces.

One of the results of the present work is that HJT property is not preserved under duality. Namely, there exists a HJT space X with a shrinking basis, such that X^* is reflexively (even unconditionally) saturated. However, in the same example there exists a subspace Y of X with Y^* also an HJT space.

One of the basic ingredients in our approach to building HJT spaces is the following space:

Proposition 0.2. There exists a space $JT_{\mathcal{F}_2}$ with a boundedly complete basis $(e_n)_n$ such that the following hold:

   (i) The space $JT_{\mathcal{F}_2}$ is ℓ_2 saturated.

   (ii) The basis $(e_n)_n$ is normalized weakly null and for every $M \in [\mathbb{N}]$ the subspace $X_M = \overline{\text{span}}\{e_n : n \in M\}$ has the James tree property.

It is clear that none subsequence $(e_n)_{n \in M}$ is unconditional. Thus the basis of $JT_{\mathcal{F}_2}$ shares similar properties with the classical Maurey Rosenthal example [MR]. We shall return to this space in the sequel explaining more about its structure and its difference from Gowers' space.

Codings and tree structures. As is well known, every attempt to impose tight (or conditional) structure in some Banach space, requires the definition of the conditional elements which in turn results from the existence of special sequences defined with the use of a coding. What is less well known is that the codings induce a tree structure in the special sequences. As we shall explain shortly, the James tree structure in the subspaces of HJT spaces, like $\mathfrak{X}_{gt}$, $(\mathfrak{X}_{gt})_*$ or the spaces presented in this paper, are directly related to codings.

Let's start with a general definition of a coding, and the obtained special sequences. Consider a collection $(F_j)_j$ with each F_j a countable family of elements of $c_{00}(\mathbb{N})$. To make more transparent the meaning of our definitions, let's assume that each $F_j = \{\frac{1}{m_j^2} \sum_{k \in F} e_k^* : F \subset \mathbb{N}, \#F \le n_j\}$ where $(m_j), (n_j)$ are appropriate fast increasing sequences of natural numbers. Notice that the elements of the family $\mathcal{T} = \cup_j F_j$ and the combinations of them will play the role of functionals belonging to a norming set. This explains the use of e_k^* instead of e_k. For simplicity, we also assume that the families $(F_j)_j$ are pairwise disjoint. This happens in the aforementioned example although it is not always true. Under this additional assumption to each $\phi \in \cup_j F_j$ corresponds a unique index by the rule $\text{ind}(\phi) = j$ iff $\phi \in F_j$. Further for a finite block sequence $s = (\phi_1, \ldots, \phi_d)$ with each $\phi_i \in \cup_j F_j$, we define $\text{ind}(s) = \{\text{ind}(\phi_1), \ldots, \text{ind}(\phi_d)\}$.

The σ-coding: Let Ω_1, Ω_2 be a partition of $\mathbb{N}$ into two infinite disjoint subsets. We denote by $\mathcal{S}$ the family of all block sequences $s = (\phi_1 < \phi_2 < \cdots < \phi_d)$ such that $\phi_i \subset \cup_j F_j$, $\text{ind}(\phi_1) \in \Omega_1$, $\{\text{ind}(\phi_2) < \cdots < \text{ind}(\phi_d)\} \subset \Omega_2$. Clearly $\mathcal{S}$ is countable, hence there exists an injection

$$\sigma : \mathcal{S} \to \Omega_2$$

satisfying $\sigma(s) > \text{ind}(s)$ for every $s \in \mathcal{S}$.

The σ-special sequences: A sequence $s = (\phi_1 < \phi_2 < \cdots < \phi_n)$ in $\mathcal{S}$ is said to be a σ-special sequence iff for every $1 \le i < n$ setting $s_i = (\phi_1 < \cdots < \phi_i)$ we have that

$$\phi_{i+1} \in F_{\sigma(s_i)}.$$

The following tree-like interference holds for σ-special sequences.

Let s, t be two σ-special sequences with $s = (\phi_1, \ldots, \phi_n)$, $t = (\psi_1, \ldots, \psi_m)$. We set

$$i_{s,t} = \max\{i : \text{ind}(\phi_i) = \text{ind}(\psi_i)\}$$

if the later set is non empty. Otherwise we set $i_{s,t} = 0$. Then the following are easily checked.

(a) For every $i < i_{s,t}$ we have that $\phi_i = \psi_i$.

(b) $\{\mathrm{ind}(\phi_i) : i > i_{s,t}\} \cap \{\mathrm{ind}(\psi_j) : j > i_{s,t}\} = \emptyset$.

These two properties immediately yield that the set $\mathcal{T} \cup_j F_j$ endowed with the partial order $\phi \prec_\sigma \psi$ iff there exists a σ-special sequence $(\phi_1, \ldots, \phi_n)$ and $1 \leq i < j \leq n$ with $\phi = \phi_i$ and $\psi = \phi_j$ is a tree.

Now for the given tree structure $(\mathcal{T}, \prec_\sigma)$ we will define norms similar to the classical James tree norm mentioned above.

The space $JT_{\mathcal{F}_2}$: For the first application the family $(F_j)_j$ is the one defined above.

For a σ-special sequence $s = (\phi_1, \ldots, \phi_n)$ and an interval E of $\mathbb{N}$ we set $s^* = \sum_{k=1}^n \phi_k$ and let Es^* be the restriction of s^* on E (or the pointwise product $s^* \chi_E$). A σ-special functional x^* is any element Es^* as before.

Also, for a σ-special functional $x^* = Es^*$, $s = (\phi_1, \ldots, \phi_n)$, we let $\mathrm{ind}(x^*) = \{\mathrm{ind}(\phi_k) : \mathrm{supp}\, \phi_k \cap E \neq \emptyset\}$. We consider the following set

$$\mathcal{F}_2 = \{\pm e_n^* : n \in \mathbb{N}\} \cup \{\sum_{i=1}^d a_i x_i^* : a_i \in \mathbb{Q}, \sum_{i=1}^d a_i^2 \leq 1, (x_i^*)_{i=1}^d \text{ are}$$

σ-special functionals with $(\mathrm{ind}(x_i^*))_{i=1}^d$ pairwise disjoint, $d \in \mathbb{N}\}$

The space $JT_{\mathcal{F}_2}$ is the completion of $(c_{00}, \|.\|_{\mathcal{F}_2})$ where for $x \in c_{00}$,

$$\|x\|_{\mathcal{F}_2} = \sup\{\phi(x) : \phi \in \mathcal{F}_2\}.$$

Comparing the norming set $\mathcal{F}_2$ with the norming set G_{JT} of JT one observes that σ-special functionals in $\mathcal{F}_2$ play the role of the functionals s^* defined by the segments of the dyadic tree $\mathcal{D}$. As we have mentioned in Proposition 0.2, the space $JT_{\mathcal{F}_2}$, like JT, is ℓ_2 saturated, but for every $M \in [\mathbb{N}]$, the subspace $X_M \overline{\mathrm{span}}\{e_n : n \in M\}$ has non separable dual. The later is a consequence of the fact that the tree structure $(\mathcal{T}, \prec_\sigma)$ is richer than that of the dyadic tree basis in JT. Indeed, it is easy to check that for every $M \in [\mathbb{N}]$ we can construct a block sequence $(\phi_a)_{a \in \mathcal{D}}$ such that

(i) $\phi_a = \frac{1}{m_{j_a}^2} \sum_{k \in F_a} e_k^*$ where $\#F_a = n_{j_a}$ and $F_a \subset M$, while $F_a < F_\beta$ if $a \prec_{lex} \beta$.

(ii) For a branch $b = (a_1 \prec a_2 \prec \cdots \prec a_n \prec \cdots)$ of $\mathcal{D}$ and for every $n \in \mathbb{N}$ we have that $(\phi_{a_1}, \ldots, \phi_{a_n})$ is a σ-special sequence.

Defining now $x_a = \frac{m_{j_a}^2}{n_{j_a}} \sum_{k \in F_a} e_k$, the family $(x_a)_{a \in \mathcal{D}}$ provides the James tree structure of X_M.

The Gowers Tree space. The definition of $\mathfrak{X}_{gt}$ uses similar ingredients with the corresponding of $JT_{\mathcal{F}_2}$ although structurally the two spaces are entirely different. The norming set G_{gt} of Gowers space is saturated under the operations $(\mathcal{A}_{n_j}, \frac{1}{m_j})_j$. We recall that a subset G of c_{00} is closed (or saturated) for the operation $(\mathcal{A}_n, \frac{1}{m})$ if for every $\phi_1 < \phi_2 < \cdots < \phi_k$, $k \leq n$ with $\phi_i \in G$, $i = 1, \ldots, k$, the functional $\phi = \frac{1}{m} \sum_{i=1}^k \phi_i$ belongs to G.

The norming set G_{gt} is the minimal subset of c_{00} satisfying the following conditions:

(i) $\{\pm e_k^* : k \in \mathbb{N}\} \subset G_{gt}$, G_{gt} is symmetric and closed under the operation of restricting elements to the intervals.

(ii) G_{gt} is closed in the $(\mathcal{A}_{n_j}, \frac{1}{m_j})_j$ operations. We also set

$$K_j = \{\phi \in G_{gt} : \phi \text{ is the result of a } (\mathcal{A}_{n_j}, \frac{1}{m_j}) \text{ operation}\}$$

(iii) G_{gt} contains the set

$$\{\sum_{i=1}^{d} a_i x_i^* : a_i \in \mathbb{Q}, \sum_{i=1}^{d} a_i^2 \leq 1, (x_i^*)_{i=1}^d, \sigma\text{-special functionals}$$
$$\text{with } (\text{ind}(x_i^*))_{i=1}^d \text{ pairwise disjoint}, d \in \mathbb{N}\}$$

(iv) G_{gt} is rationally convex.

We explain briefly condition (iii). For a coding σ, the σ-special sequences $(\phi_1, \ldots, \phi_n)$ are defined as in the case of $\mathcal{F}_2$. Here the set K_j plays the role of the corresponding F_j in $\mathcal{F}_2$. The σ-special functionals x^*, are defined as in the case of $\mathcal{F}_2$.

Let's observe that G_{gt} is almost identical with $\mathcal{F}_2$, although the spaces defined by them are entirely different. The essential difference between $\mathcal{F}_2$ and G_{gt} is that in the case of $\mathcal{F}_2$ each F_j, $j \in \mathbb{N}$ does not norm any subspace of $JT_{\mathcal{F}_2}$, while in $\mathfrak{X}_{gt}$ each K_j defines an equivalent norm on $\mathfrak{X}_{gt}$. The later means that in every subspace Y of $\mathfrak{X}_{gt}$, the families K_j, $j \in \mathbb{N}$ as well as $\{x^* : x^* \text{ is a } \sigma\text{-special functional}\}$ and $\{\sum_{i=1}^{d} \lambda_i x_i^* : \sum_{i=1}^{n} \lambda_i^2 \leq 1, (\text{ind}(x_i^*))_{i=1}^d \text{ are pairwise disjoint}\}$ define equivalent norms making it difficult to distinguish the action of them on the elements of Y. Thus, while the spaces of the form $JT_{\mathcal{F}_2}$ can be studied in terms of the classical theory, the space $\mathfrak{X}_{gt}$ requires advanced tools, like Gowers probabilistic argument, which do not permit a complete understanding of its structure.

0.2. Saturated extensions. The method of HI extensions appeared in the Memoirs monograph [AT1] and was used to derive the following two results:

- Every separable Banach space Z not containing ℓ_1 is a quotient of a separable HI space X, with the additional property that Q^*Z^* is a complemented subspace of X^*. (Here Q denotes the quotient map from X to Z.)
- There exists a nonseparable HI Banach space.

Roughly speaking, the method of HI extensions provides a tool to connect a given norm, usually defined through a norming set G with a HI norm. The resulting new norm will preserve some of the ingredients of the initial norm and will also be HI. To some extent, HI extensions, have similar goals with HI interpolations ([AF]) and some of the results could be obtained with both methods. However it seems that the method of extensions is very efficient when we want to construct dual pairs X, X^* with divergent structure. This actually requires the combination of extensions with the method of attractors, which appeared in [AT2] where a reflexive HI space X is constructed with X^* unconditionally saturated.

In the sequel we shall provide a general definition of saturated extensions which include several forms of extensions which appeared elsewhere (c.f. [AT1, AT2, ArTo])

Let $\mathcal{M}$ be a compact family of finite subsets of $\mathbb{N}$. For the purposes of the present paper, $\mathcal{M}$ will be either some $\mathcal{A}_n = \{F \subset \mathbb{N} : \#F \leq n\}$, or some $\mathcal{S}_n$, the nth Schreier family. For $0 < \theta < 1$, the $(\mathcal{M}, \theta)$-operation on c_{00} is a map which assigns to each $\mathcal{M}$-admissible block sequence $(\phi_1 < \phi_2 < \cdots < \phi_n)$, the functional $\theta \sum_{i=1}^{n} \phi_i$. (We recall that $\phi_1, \phi_2, \ldots, \phi_n$ is $\mathcal{M}$-admissible if $\{\min \operatorname{supp} \phi_i : i = 1, \ldots, n\}$ belongs to $\mathcal{M}$.) A subset G of c_{00} is said to be closed in the $(\mathcal{M}, \theta)$-operation, if for every $\mathcal{M}$-admissible block sequence $\phi_1, \ldots, \phi_n$, with each $\phi_i \in G$, the functional $\theta \sum_{i=1}^{n} \phi_i$ belongs to G. When we refer to saturated norms we shall mean that there exists a norming set G which is closed under certain $(\mathcal{M}_j, \theta_j)_j$ operations.

Let G be a subset of c_{00}. The set G is said to be a ground set if it is symmetric, $\{e_n^* : n \in \mathbb{N}\}$ is contained in G, $\|\phi\|_\infty \leq 1$, $\phi(n) \in \mathbb{Q}$ for all $\phi \in G$ and G is closed under the restriction of its elements to intervals of $\mathbb{N}$. A ground norm, $\|\cdot\|_G$ is any norm induced on c_{00} by a ground set G. It turns out that for every space $(X, \|\cdot\|_X)$ with a normalized Schauder basis $(x_n)_n$ there exists a ground set G_X such that the natural map $e_n \mapsto x_n$ defines an isomorphism between $(X, \|\cdot\|_X)$ and $\overline{(c_{00}, \|\cdot\|_{G_X})}$.

Saturated extensions of a ground set G. Let G be a ground set, $(m_j)_j$ an appropriate sequence of natural numbers and $(\mathcal{M}_j)_j$ a sequence of compact families such that $(\mathcal{M}_j)_j$ is either $(\mathcal{A}_{n_j})_j$ or $(\mathcal{S}_{n_j})_j$.

Denote by E_G the minimal subset of c_{00} such that

  (i) The ground set G is a subset of E_G.
 (ii) The set E_G is closed in the $(\mathcal{M}_j, \frac{1}{m_j})$ operation.
(iii) The set E_G is rationally convex.

Definition 0.3. A subset D_G of E_G is said to be a saturated extension of the ground set G if the following hold:

  (i) The set D_G is a subset of E_G, the ground set G is contained in D_G and D_G is closed under restrictions of its elements to intervals.
 (ii) The set D_G is closed under even operations $(\mathcal{M}_{2j}, \frac{1}{m_{2j}})_j$.
(iii) The set D_G is rationally convex.
(iv) Every $\phi \in D_G$ admits a tree analysis $(f_t)_{t \in T}$ with each $f_t \in D_G$.

Denoting by $\|\cdot\|_{D_G}$ the norm on c_{00} induced by D_G and letting X_{D_G} be the space $\overline{(c_{00}, \|\cdot\|_{D_G})}$, we call X_{D_G} a *saturated extension* of the space $X_G = \overline{(c_{00}, \|\cdot\|_G)}$.

Let's point out that the basis $(e_n)_n$ of c_{00} is a bimonotone boundedly complete Schauder basis of X_{D_G} and that the identity $I : X_{D_G} \to X_G$ is a norm one operator. Observe also that we make no assumption concerning the odd operations. As we will see later making several assumptions for the odd operations, we will derive saturated extensions with different properties.

A last comment on the definition of D_G, is related to the condition (iv). The tree analysis $(f_t)_{t \in T}$ of a functional f in E_G describes an inductive procedure for obtaining f starting from elements of the ground set G and either applying operations $(\mathcal{M}_j, \frac{1}{m_j})$ or taking rational convex combinations. This tree structure is completely irrelevant to the tree structures discussed in the previous subsection. Its role is to help estimate upper bounds of the norm of vectors in X_{D_G}.

Properties and variants of Saturated extensions.

As we have mentioned, for $x \in c_{00}$, $\|x\|_G \leq \|x\|_{D_G}$. This is an immediate consequence of the fact that $G \subset D_G$. On the other hand, there are cases of ground sets G such that D_G does not add more information beyond G itself. Such a case is when G defines a norm $\| \cdot \|_G$ equivalent to the ℓ_1 norm. The measure of the fact that $\| \cdot \|_{D_G}$ is strictly greater than $\| \cdot \|_G$ on a subspace Y of X_{D_G} is that the identity operator $I : X_{D_G} \to X_G$ restricted to Y is a strictly singular one. If $I : X_{D_G} \to X_G$ is strictly singular, then we refer to strictly singular extensions. The first result we want to mention is that strictly singular extensions are reflexive ones. More precisely the following holds:

Proposition 0.4. Let Y be a closed subspace of X_{D_G} such that $I|_Y : Y \to X_G$ is strictly singular. Then Y is reflexively saturated. In particular X_{D_G} is reflexively saturated whenever it is a strictly singular extension.

Next we proceed to specify the odd operations and to derive additional information on the structure of X_{D_G} whenever X_{D_G} is a strictly singular extension.

(a) Unconditionally saturated extensions.

This is the case where $D_G = E_G = D_G^u$. In this case the following holds:

Proposition 0.5. Let Y be a closed subspace of $X_{D_G^u}$ such that $I|_Y : Y \to X_G$ is strictly singular. Then Y is unconditionally (and reflexively) saturated.

(b) Hereditarily Indecomposable extensions.

HI extensions, are the most important ones. In this case the norming set D_G^{hi} is defined as follows. D_G^{hi} is the minimal subset of c_{00} satisfying the following conditions

(i) $\{e_n^* : n \in \mathbb{N}\} \subset D_G^{hi}$, D_G^{hi} is symmetric and closed under restriction of its elements to intervals.

(ii) D_G^{hi} is closed under $(\mathcal{M}_{2j}, \frac{1}{m_{2j}})_j$ operations.

(iii) For each j, D_G^{hi} is closed under $(\mathcal{M}_{2j-1}, \frac{1}{m_{2j-1}})$ operation on $2j - 1$ special sequences.

(iv) D_G^{hi} is rationally convex.

The $2j - 1$ special sequences are defined through a coding σ and satisfy the following conditions.

(a) $(f_1, \ldots, f_d)$ is $\mathcal{M}_{2j-1}$ admissible

(b) For $i \leq i \leq d$ there exists some $j \in \mathbb{N}$ such that $f_i \in K_{2j} = \{ \frac{1}{m_{2j}} \sum_{l=1}^{k} \phi_l : \phi_1 < \cdots < \phi_k \text{ is } \mathcal{M}_{2j} \text{ admissible}, \phi_l \in D_G^{hi} \}$ and if $i > 1$ then $2j = \sigma(f_1, \ldots, f_{i-1})$.

Notice that in the definition of D_G^{hi} we do not refer to the tree analysis. The reason is that the existence of a tree analysis follows from the minimality of D_G^{hi} and the conditions involved in its definition.

The analogue of the previous results also holds for HI extensions.

Proposition 0.6. Let Y be a closed subspace of $X_{D_G^{hi}}$ such that $I|_Y : Y \to X_G$ is strictly singular. Then Y is a HI space. In particular strictly singular and HI extensions yield HI spaces.

The above three propositions indicate that if we wish to have additional structure on $X_{D_G}, X_{D_G^u}, X_{D_G^{hi}}$ we need to consider strictly singular extensions. As is shown in [AT1], this is always possible. Indeed, for every ground set G such that the corresponding space X_G does not contain ℓ_1 there exists a family $(\mathcal{M}_j, \frac{1}{m_j})_j$ such that the saturated extension of G by this family is a strictly singular one. Thus the following is proven ([AT1]).

Theorem 0.7. Let X be a Banach space with a normalized Schauder basis $(x_n)_n$ such that X contains no isomorphic copy of ℓ_1. Then there exists a HI space Z with a normalized basis $(z_n)_n$ such that the map $z_n \mapsto x_n$ has a linear extension to a bounded operator $T : Z \to X$.

This theorem in conjunction with the following one yields that every separable Banach space X not containing ℓ_1 is the quotient of a HI space.

Theorem 0.8 ([AT1]). Let X be a separable Banach space not containing ℓ_1. Then there exists a space Y not containing ℓ_1, with a normalized Schauder basis $(y_n)_n$ and a bounded linear operator $T : Y \to X$ such that $(Ty_n)_n$ is a dense subset of the unit sphere of X.

The predual $(X_{D_G^{hi}})_*$. As we have mentioned before the basis $(e_n)_{n \in \mathbb{N}}$ of $X_{D_G^{hi}}$ is boundedly complete, hence the space $(X_{D_G^{hi}})_*$, which is the subspace of $X_{D_G^{hi}}^*$ norm generated by the biorthogonal functionals $(e_n^*)_{n \in \mathbb{N}}$, is a predual of $X_{D_G^{hi}}$. In many cases it is shown that $(X_{D_G^{hi}})_*$ is also a HI space. This requires some additional information concerning the weakly null block sequences in X_G, which is stronger than the assumption that the identity map $I : X_{D_G^{hi}} \to X_G$ is strictly singular. For example in [AT1], for extensions using the operations $(\mathcal{S}_{n_j}, \frac{1}{m_j})_j$, had been assumed that the ground set G is $\mathcal{S}_2$ bounded. In the present paper for the operations $(\mathcal{A}_{n_j}, \frac{1}{m_j})_j$ we introduce the concept of strongly strictly singular extension which yields that $(X_{D_G^{hi}})_*$ is HI. It is also worth pointing out that $(X_{D_G^{hi}})_*$ is not necessarily reflexively saturated as happens for the strictly singular extensions $X_{D_G} \, X_{D_G^{hi}}$. This actually will be a key point in our approach for constructing HI spaces with no reflexive subspace.

0.3. **The attractors method.** Let's return to our initial goal, namely constructing HI spaces with no reflexive subspace. It is clear from our preceding discussion that HI extensions of ground sets G such that X_G does not contain ℓ_1 yield reflexively saturated HI spaces. Therefore there is no hope to obtain HI spaces with no reflexive subspace as a result of a HI extension of a ground set G. As mentioned in [ArTo] saturation and HI methods share common metamathematical ideas with the forcing method in set theory. In particular the fact that HI extensions are reflexively saturated is similar to the well known collapse phenomena in the extensions of models of set theory via the forcing method. An illustrating example of such phenomena in HI extensions is that $\mathfrak{X}_{gt}$ is a quotient of a HI and reflexively saturated space. In spite of all these discouraging observations we claim that HI extensions could help to yield HI spaces with no reflexive subspace, and this is closely related to the structure of $(X_{D_G^{hi}})_*$. Evidently from the initial stages of HI theory, ([GM1],[GM2],[AD]) and for many years, a norming set D was defined, using saturation methods and codings, in such a way as to impose certain properties in the space $(X, \|\cdot\|_D)$. In [AT2] the norming set D was designed to impose divergent properties in $(X, \|\cdot\|_D)$ and $(X, \|\cdot\|_D)^*$. The method used for this is the attractors method, not so named in [AT2], which will also be used in the present work.

The general principle of the attractors method is the following:

We are interested in designing a ground set G and a HI extension D_G^{hi} such that the following two divergent properties hold:

(a) $X_{D_G^{hi}}$ is a strictly singular extension of X_G. In other words every subspace Y of $X_{D_G^{hi}}$ contains a further subspace Z on which the G-norm becomes negligible.

(b) The set G is asymptotic in $(X_{D_G^{hi}})_*$. This means that there exists $c > 0$ such that for every subspace Y of $(X_{D_G^{hi}})_*$ and every $\varepsilon > 0$ there exists $\phi \in G$ with $\|\phi\|_{(X_{D_G^{hi}})_*} \geq c$ and $\mathrm{dist}(\phi, Y) < \varepsilon$.

In other words, we want G to be small, as a norming set for the space $X_{D_G^{hi}}$ and large as a subset of $(X_{D_G^{hi}})_*$. Notice that such a relation between G and D_G^{hi} requires the two sets to be built with similar materials, and moreover to impose certain special functionals in D_G^{hi} (we call these attractor functionals) which will allow us to attract in every subspace of $(X_{D_G^{hi}})_*$ part of the structure of the set G.

Let us be more precise explaining how we define the corresponding sets G and D_G^{hi} to obtain a HI extension $X_{D_G^{hi}}$, such that $(X_{D_G^{hi}})_*$ is also HI and does not contain reflexive subspaces.

The ground set $\mathcal{F}_2$ and the norming set $D_{\mathcal{F}_2}$. We start by defining the following family $(F_j)_j$. We shall use the sequence of positive integers $(m_j)_j$, $(n_j)_j$ recursively defined as follows:

- $m_1 = 2$ and $m_{j+1} = m_j^5$.
- $n_1 = 4$, and $n_{j+1} = (5n_j)^{s_j}$ where $s_j = \log_2 m_{j+1}^3$.

We set $F_0 = \{\pm e_n^* : n \in \mathbb{N}\}$ and for $j = 1, 2, \ldots$ we set

$$F_j = \{\frac{1}{m_{4j-3}^2} \sum_{i \in I} \pm e_i^* : \#(I) \leq \frac{n_{4j-3}}{2}\} \cup \{0\}.$$

Using the family $(F_j)_j$ and a coding $\sigma_{\mathcal{F}}$, we define the ground set $\mathcal{F}_2$ in the same manner as in the first subsection.

Next we define the set $D_{\mathcal{F}_2}$ which is a HI extension of $\mathcal{F}_2$ with attractors as follows:

The set $D_{\mathcal{F}_2}$ is a minimal subset of c_{00} satisfying the following properties:

(i) $\mathcal{F}_2 \subset D_{\mathcal{F}_2}$, $D_{\mathcal{F}_2}$ is symmetric (i.e. if $f \in D_{\mathcal{F}_2}$ then $-f \in D_{\mathcal{F}_2}$) and $D_{\mathcal{F}_2}$ is closed under the restriction of its elements to intervals of $\mathbb{N}$ (i.e. if $f \in D_{\mathcal{F}_2}$ and E is an interval of $\mathbb{N}$ then $Ef \in D_{\mathcal{F}_2}$).

(ii) $D_{\mathcal{F}_2}$ is closed under $(\mathcal{A}_{n_{2j}}, \frac{1}{m_{2j}})$ operations, i.e. if $f_1 < f_2 < \cdots < f_{n_{2j}}$ belong to $D_{\mathcal{F}_2}$ then the functional $f = \frac{1}{m_{2j}}(f_1 + f_2 + \cdots + f_{n_{2j}})$ belongs also to $D_{\mathcal{F}_2}$.

(iii) $D_{\mathcal{F}_2}$ is closed under $(\mathcal{A}_{n_{4j-1}}, \frac{1}{m_{4j-1}})$ operations on special sequences i.e. for every n_{4j-1} special sequence $(f_1, f_2, \ldots, f_{n_{4j-1}})$ the functional $f = \frac{1}{m_{4j-1}}(f_1 + f_2 + \cdots + f_{n_{4j-1}})$ belongs to $D_{\mathcal{F}_2}$. In this case we say that f is a **special functional**.

(iv) $D_{\mathcal{F}_2}$ is closed under $(\mathcal{A}_{n_{4j-3}}, \frac{1}{m_{4j-3}})$ operations on attractor sequences i.e. for every $4j - 3$ attractor sequence $(f_1, f_2, \ldots, f_{n_{4j-3}})$ the functional $f = \frac{1}{m_{4j-3}}(f_1 + f_2 + \cdots + f_{n_{4j-3}})$ belongs to $D_{\mathcal{F}_2}$. In this case we say that f is an **attractor**.

(v) The set $D_{\mathcal{F}_2}$ is rationally convex.

In the above definition, the special functionals and the attractors, defined in (iii) and (iv) respectively, require some more explanation. First, the n_{4j-1} special sequences $(f_1, \ldots, f_{n_{4j-1}})$ are defined through a coding σ as in the case of the aforementioned HI extensions. Thus each f_i, $1 \leq i \leq n_{4j-1}$ belongs to some

$$K_{2j} = \{\frac{1}{m_{2j}} \sum_{l=1}^{n_{2j}} \phi_l : \phi_1 < \cdots < \phi_{n_{2j}}, \phi_l \in D_{\mathcal{F}_2}\}$$

where $2j$ is equal to $\sigma(f_1, \ldots, f_{i-1})$ whenever $1 < i$.

The special functionals will determine the HI property of the extension $D_{\mathcal{F}_2}$.

Each $4j - 3$ attractor sequence $f_1 < \cdots < f_{n_{4j-3}}$ is of the following form. All the odd members of the sequence are elements of $\cup_j K_{2j}$ while the even members are $f_{2i} = e_{\ell_{2i}}^*$ and furthermore the sequence $f_1, \ldots, f_{n_{4j-3}}$ is determined by the coding σ in a similar manner to the n_{4j-1} special sequence. Let us observe that for every $j \in \mathbb{N}$ there exist many $P \subset \mathbb{N}$ with the following properties. First $\#P \frac{n_{4j-3}}{2}$, hence $\frac{1}{m_{4j-3}^2} \sum_{\ell \in P} e_\ell^* \in F_j$ and also there exists an attractor sequence $(f_1, \ldots, f_{4j-3})$ with $\{e_\ell^* : \ell \in P\}$ coinciding with the even terms of the sequence. The purpose of the attractors is to make the family $\cup_j F_j$ asymptotic in the space $(X_{D_{\mathcal{F}_2}})_*$. In particular using attractors, the following is proved.

For every subspace Y of $(X_{D_{\mathcal{F}_2}})_*$ and every $j \in \mathbb{N}$ there exist $\phi_j \in Y$ and $\psi_j = \frac{1}{m_{4j-3}^2} \sum_{\ell \in P} e_\ell^* \in F_j$, such that

(a) $\|\phi_j + \psi_j\| > \frac{1}{144}$.
(b) $\|\phi_j - \psi_j\| \leq \frac{1}{m_{4j-3}}$.

This shows that indeed $\cup_j F_j$ is asymptotic and furthermore we can copy a complete dyadic block subtree of $(\mathcal{T} = \cup_j F_j, \prec_{\sigma_{\mathcal{F}}})$ into the subspace Y which yields that Y is not reflexive.

The following summarizes the main steps in our approach to constructing HI spaces with no reflexive subspace.

(1) For two appropriately chosen sequences $(m_j)_j$, $(n_j)_j$ we set $F_j = \{\frac{1}{m_{4j-3}^2} \sum_{k \in F} e_k^* : \#F \leq \frac{n_{4j-3}}{2}\}$ and for the family $(F_j)_j$ we construct the norming set $\mathcal{F}_2$ and the James Tree space $JT_{\mathcal{F}_2}$.

(2) The space $JT_{\mathcal{F}_2}$ does not contain ℓ_1 and for every weakly null sequence $(x_n)_n$ in $JT_{\mathcal{F}_2}$ with $\|x_n\| \leq C$, $\lim \|x_n\|_\infty = 0$ and every $m \in \mathbb{N}$ there exists $L \in [\mathbb{N}]$ such that for every $y^* \in \mathcal{F}_2$

(1) $$\#\{n \in L : |y^*(x_n)| \geq \frac{1}{m}\} \leq 66m^2 C^2.$$

(3) We consider the HI extension with attractors $D_{\mathcal{F}_2}^{hi}$ of $\mathcal{F}_2$ defined by the operations $(\mathcal{A}_{n_j}, \frac{1}{m_j})$, and we denote by $\mathfrak{X}_{\mathcal{F}_2}$ the space $\overline{(c_{00}, \|\cdot\|_{D_{\mathcal{F}_2}^{hi}})}$.

(4) Inequality (1) yields that $\mathfrak{X}_{\mathcal{F}_2}$ is a strongly strictly singular extension of $JT_{\mathcal{F}_2}$. Therefore:
  (i) The space $\mathfrak{X}_{\mathcal{F}_2}$ is HI and reflexively saturated.
  (ii) The predual $(\mathfrak{X}_{\mathcal{F}_2})_*$ is HI.

(5) Using the attractor functionals, we copy into every subspace of $(\mathfrak{X}_{\mathcal{F}_2})_*$ a complete dyadic subtree of $(\mathcal{T}, \prec_{\mathcal{F}})$ which shows that $(\mathfrak{X}_{\mathcal{F}_2})_*$ is a Hereditarily James Tree space (HJT) and hence it does not contain a reflexive subspace.

Notice that $(\mathfrak{X}_{\mathcal{F}_2})_*$ shares with the space X, in the statements presented at the beginning of the introduction, most of the properties stated there. However for some of the properties a variation is required. In fact there exists a complete subtree $(\mathcal{T}', \prec_{\sigma_F})$ of $(\mathcal{T}, \prec_{\sigma_F})$ such that for the corresponding space $\mathfrak{X}_{\mathcal{F}_2'}$ we have that $(\mathfrak{X}_{\mathcal{F}_2'})^*/(\mathfrak{X}_{\mathcal{F}_2'})_* = \ell_2(\Gamma)$ with $\#\Gamma = 2^\omega$. The space $(\mathfrak{X}_{\mathcal{F}_2'})_*$ coincides with X in the aforementioned statements.

The construction of a nonseparable HI space Z not containing reflexive subspaces requires changing the framework with the operations $(\mathcal{S}_{n_j}, \frac{1}{m_j})_j$ instead of $(\mathcal{A}_{n_j}, \frac{1}{m_j})_j$. In this framework the set $\mathcal{F}_s$ and the space $JT_{\mathcal{F}_s}$ are defined. More precisely $\mathcal{F}_s$ is defined in a similar manner as $\mathcal{F}_2$ based on the families

$$F_j = \{\frac{1}{m_{4j-3}^2} \sum_{i \in I} \pm e_i^* : I \in \mathcal{S}_{n_{4j-3}}\}.$$

Using the coding $\sigma_{\mathcal{F}}$, we define the special functionals and their indices as in the $\mathcal{F}_2$ case. Finally we set

$$\mathcal{F}_s = \{\pm e_n^* : n \in \mathbb{N}\} \cup \{\sum_{i=1}^{d} x_i^* : \min \operatorname{supp} x_i^* \geq d, \ i = 1, \ldots, n,$$

$$(\operatorname{ind}(x_i^*))_{i=1}^{d} \text{ are pairwise disjoint}\}$$

The HI extension with attractors is defined similarly to the $\mathfrak{X}_{\mathcal{F}_2}$ case. Then $\mathfrak{X}_{\mathcal{F}_s}$ is an asymptotic ℓ_1 and reflexively saturated HI space and also $(\mathfrak{X}_{\mathcal{F}_s})_*$ is HI while not containing any reflexive subspace. Passing to a complete subtree $(T', \prec_{\sigma_{\mathcal{F}}})$ of $(T, \prec_{\sigma_{\mathcal{F}}})$ and to the corresponding $\mathcal{F}_s'$, $\mathfrak{X}_{\mathcal{F}'_s}$, we obtain the additional property that $\mathfrak{X}_{\mathcal{F}'_s}^*/(\mathfrak{X}_{\mathcal{F}'_s})_* \cong c_0(\Gamma)$ with $\#\Gamma = 2^\omega$. As is shown in [AT1], this yields that $\mathfrak{X}_{\mathcal{F}'_s}^*$ is HI and since it contains a subspace $((\mathfrak{X}_{\mathcal{F}'_s})_*)$ with no reflexive subspace, the space $\mathfrak{X}_{\mathcal{F}'_s}^*$ has the same property.

Let's mention also that an HI asymptotic ℓ_1 Banach space X not containing a reflexive subspace, with nonseparable dual X^* which is also HI not containing any reflexive subspace, has been constructed in [AGT]. This space is the analogue of $\mathfrak{X}_{gt}$ in the frame of the operations $(\mathcal{S}_{n_j}, \frac{1}{m_j})_j$.

The last variant we present, concerns the HI space Y with a shrinking basis, not containing a reflexive subspace, such that the dual Y^* is unconditionally and reflexively saturated.

For this, starting with the set $\mathcal{F}_2$ we pass to an extension only with attractors and additionally we subtract a large portion of the conditional structure of the attractors. This permits us to show that the extension space $\mathfrak{X}_{\mathcal{F}_2}^{us}$ is unconditionally saturated. The remaining part of the conditional structure of the attractors, forces the predual $(\mathfrak{X}_{\mathcal{F}_2}^{us})_*$ to be HI and not to contain any reflexive subspace.

The paper is organized as follows. The first two sections are devoted to the presentation of the strictly singular and strongly strictly singular (HI) extensions with attractors of a ground space Y_G. We shall denote these as X_G. For the results presented in these two sections the attractors play no role. Thus all statements remain valid whether we consider extensions with attractors or not. The strictly singular extension, as they have defined before, provide information about X_G and $\mathcal{L}(X_G)$. In particular the following is proven:

Theorem 0.9. If X_G is a strictly singular extension (with or without attractors) then the natural basis of X_G is boundedly complete, the space X_G is HI, reflexively saturated and every T in $\mathcal{L}(X_G)$ is of the form $T = \lambda I + S$ with S a strictly singular operator.

Strongly strictly singular extensions concern the HI property of $(X_G)_*$ and the structures of $\mathcal{L}(X_G)$, $\mathcal{L}((X_G)_*)$. The following theorem includes the main results of Section 2.

Theorem 0.10. If X_G is a strongly strictly singular extension (with or without attractors) then in addition to the above we have the following

(i) The predual $(X_G)_*$ is HI.
(ii) Every strictly singular S in $\mathcal{L}(X_G)$ is weakly compact.
(iii) Every T in $\mathcal{L}((X_G)_*)$ is of the form $T = \lambda I + S$ with S being a strictly singular and weakly compact operator.

The following result concerning the quotients of X_G is also proved in Section 2.

Theorem 0.11. If X_G is a strongly strictly singular extension (with or without attractors) and Z is a w^* closed subspace of X_G then the quotient X_G/Z is HI.

The dual form of the above theorem is the following. For every subspace W of $(X_G)_*$ the dual space W^* is HI. Notice also that the additional assumption that Z is w^* closed can not be dropped, as the results of Section 5 indicate.

In Section 3 we study the spaces $JT_{\mathcal{F}_2}$. We are mainly concerned with proving the aforementioned (1) yielding that the extension $\mathfrak{X}_{\mathcal{F}_2}$ is a strongly strictly singular one. In Section 4 using attractors we prove that $(\mathfrak{X}_{\mathcal{F}_2})_*$ is a Hereditarily James Tree (HJT) space and hence it does not contain any reflexive subspace. Section 5 is devoted to the study of $(\mathfrak{X}_{\mathcal{F}_2})^*$. It is shown that for every subspace Y of $(\mathfrak{X}_{\mathcal{F}_2})_*$, the space ℓ_2 is isomorphic to a subspace of the nonseparable space Y^{**}. We also describe the definition of $\mathfrak{X}_{\mathcal{F}_2'}$ which has the additional property that $(\mathfrak{X}_{\mathcal{F}_2'})^*/(\mathfrak{X}_{\mathcal{F}_2'})_*$ is isomorphic to $\ell_2(\Gamma)$. Section 6 and Section 7 contain the variants $\mathfrak{X}_{\mathcal{F}_s}$, $\mathfrak{X}_{\mathcal{F}_2}^{us}$ mentioned before. We have also included two appendices. In Appendix A we present a proof of a form of the basic inequality used for estimating upper bounds for the action of functionals on certain vectors. In Appendix B we proceed to a systematic study of the James Tree spaces $JT_{\mathcal{F}_2}$, $JT_{\mathcal{F}_s}$ and $JT_{\mathcal{F}_{2,s}}$. We actually show that $JT_{\mathcal{F}_2}$ is ℓ_2 saturated while $JT_{\mathcal{F}_s}$ and $JT_{\mathcal{F}_{2,s}}$ are c_0 saturated. The study of James Tree spaces in Section 3 and Appendix B is not related to HI techniques and uses classical Banach space theory with Ramsey's theorem also playing a key role.

1. STRICTLY SINGULAR EXTENSIONS WITH ATTRACTORS

In this section we introduce the ground sets G and then we define the extensions $X_G = T[G, (\mathcal{A}_{n_j}, \frac{1}{m_j})_j, \sigma]$ with low complexity saturation methods. Attractors are also defined. We provide conditions yielding the HI property of the extension X_G and we study the space of the operators $\mathcal{L}(X_G)$. The results and the techniques are analogue to the corresponding of [AT1] where extensions using higher complexity saturation methods are presented. We refer the reader to [ArTo] for an exposition of low complexity extensions. We also point out that the attractors in the present and next section will be completely neutralized. Their role will be revealed in Section 4 where we study the structure of $(\mathfrak{X}_{\mathcal{F}_2})_*$.

Definition 1.1. (ground sets) A set $G \subset c_{00}(\mathbb{N})$ is said to be ground if the following conditions are satisfied

(i) $e_n^* \in G$ for $n = 1, 2, \ldots$, G is symmetric (i.e. if $g \in G$ then $-g \in G$), and closed under restriction of its elements to intervals of $\mathbb{N}$ (i.e. if $g \in G$ and E is an interval of $\mathbb{N}$ then $Eg \in G$).

(ii) $\|g\|_\infty \le 1$ for every $g \in G$ and $g(n) \in \mathbb{Q}$ for every $g \in G$ and $n \in \mathbb{N}$.

(iii) Denoting by $\| \ \|_G$ the ground norm on $c_{00}(\mathbb{N})$ defined by the rule $\|x\|_G = \sup\{g(x) : g \in G\}$, the ground space Y_G, which is the completion of $(c_{00}(\mathbb{N}), \| \ \|_G)$ contains no isomorphic copy of ℓ_1.

It follows readily that the standard basis $(e_n)_n$ of Y_G is a bimonotone Schauder basis. The converse is also true. Namely if Y has a bimonotone basis $(y_n)_n$ and ℓ_1 does not embed into Y then there exists a ground set G such that Y_G is isometric to Y. Also, as is well known, every space $(Y, \| \ \|)$ with a basis $(e_n)_{n \in \mathbb{N}}$ admits an equivalent norm $\|| \cdot \||$ such that $(e_n)_{n \in \mathbb{N}}$ is a bimonotome basis for $(Y, \|| \ \||)$.

Definition 1.2. (HI extensions with attractors) We fix two strictly increasing sequences of even positive integers $(m_j)_{j \in \mathbb{N}}$ and $(n_j)_{j \in \mathbb{N}}$ defined as follows:

- $m_1 = 2$ and $m_{j+1} = m_j^5$.
- $n_1 = 4$, and $n_{j+1} = (5n_j)^{s_j}$ where $s_j = \log_2 m_{j+1}^3$.

We let D_G be the minimal subset of $c_{00}(\mathbb{N})$ satisfying the following conditions:

(i) $G \subset D_G$, D_G is symmetric (i.e. if $f \in D_G$ then $-f \in D_G$) and D_G is closed under the restriction of its elements to intervals of $\mathbb{N}$ (i.e. if $f \in D_G$ and E is an interval of $\mathbb{N}$ then $Ef \in D_G$).

(ii) D_G is closed under $(\mathcal{A}_{n_{2j}}, \frac{1}{m_{2j}})$ operations, i.e. if $f_1 < f_2 < \cdots < f_{n_{2j}}$ belong to D_G then the functional $f = \frac{1}{m_{2j}}(f_1 + f_2 + \cdots + f_{n_{2j}})$ also belongs to D_G. In this case we say that the functional f is the result of an $(\mathcal{A}_{n_{2j}}, \frac{1}{m_{2j}})$ operation.

(iii) D_G is closed under $(\mathcal{A}_{n_{4j-1}}, \frac{1}{m_{4j-1}})$ operations on special sequences, i.e. for every n_{4j-1} special sequence $(f_1, f_2, \ldots, f_{n_{4j-1}})$, the functional $f = \frac{1}{m_{4j-1}}(f_1 + f_2 + \cdots + f_{n_{4j-1}})$ belongs to D_G. In this case we say that f is a result of an $(\mathcal{A}_{n_{4j-1}}, \frac{1}{m_{4j-1}})$ operation and that f is a **special functional**.

(iv) D_G is closed under $(\mathcal{A}_{n_{4j-3}}, \frac{1}{m_{4j-3}})$ operations on attractor sequences, i.e. for every $4j - 3$ attractor sequence $(f_1, f_2, \ldots, f_{n_{4j-3}})$, the functional $f = \frac{1}{m_{4j-3}}(f_1 + f_2 + \cdots + f_{n_{4j-3}})$ belongs to D_G. In this case we say that f is a result of an $(\mathcal{A}_{n_{4j-3}}, \frac{1}{m_{4j-3}})$ operation and that f is an **attractor**.

(v) The set D_G is rationally convex.

The space $X_G = T[G, (\mathcal{A}_{n_j}, \frac{1}{m_j})_j, \sigma]$, which is the completion of the space $(c_{00}(\mathbb{N}), \| \ \|_{D_G})$, is called a **strictly singular extension with attractors** of the space Y_G, provided that the identity operator $I : X_G \to Y_G$ is strictly singular.

The norm satisfies the following implicit formula.

$$\|x\| \;=\; \max\left\{ \|x\|_G,\; \sup_j\{\sup \frac{1}{m_{2j}} \sum_{i=1}^{n_{2j}} \|E_i x\|\},\right.$$

$$\left.\sup\{\phi(x):\; \phi \text{ special functional}\},\; \sup\{\phi(x):\; \phi \text{ attractor}\}\right\}$$

where the inside supremum in the second term is taken over all choices $(E_i)_{i=1}^{n_{2j}}$ of successive intervals of $\mathbb{N}$.

We next complete the definition of the norming set D_G by giving the precise definition of special functionals and attractors.

From the minimality of D_G it follows that each $f \in D_G$ has one of the following forms.

(i) $f \in G$. We then say that f is of type 0.

(ii) $f = \pm Eh$ where h is a result of an $(\mathcal{A}_{n_j}, \frac{1}{m_j})$ operation and E is an interval. In this case we say that f is of type I. Moreover we say that the integer m_j is a weight of f and we write $w(f) = m_j$. We notice that an $f \in D_G$ may have many weights.

(iii) f is a rational convex combination of type 0 and type I functionals. In this case we say that f is of type II.

Definition 1.3. (σ coding, special sequences and attractor sequences)
Let $\mathbb{Q}_s$ denote the set of all finite sequences $(\phi_1, \phi_2, \ldots, \phi_d)$ such that $\phi_i \in c_{00}(\mathbb{N})$, $\phi_i \neq 0$ with $\phi_i(n) \in \mathbb{Q}$ for all i, n and $\phi_1 < \phi_2 < \cdots < \phi_d$. We fix a pair Ω_1, Ω_2 of disjoint infinite subsets of $\mathbb{N}$. From the fact that $\mathbb{Q}_s$ is countable we are able to define a Gowers-Maurey type injective coding function $\sigma : \mathbb{Q}_s \to \{2j :\; j \in \Omega_2\}$ such that $m_{\sigma(\phi_1, \phi_2, \ldots, \phi_d)} > \max\{\frac{1}{|\phi_i(e_l)|} :\; l \in \operatorname{supp}\phi_i,\; i = 1, \ldots, d\} \cdot \max \operatorname{supp}\phi_d$. Also, let $(\Lambda_i)_{i \in \mathbb{N}}$ be a sequence of pairwise disjoint infinite subsets of $\mathbb{N}$ with $\min \Lambda_i > m_i$.

(A) A finite sequence $(f_i)_{i=1}^{n_{4j-1}}$ is said to be a n_{4j-1} **special sequence** provided that

   (i) $(f_1, f_2, \ldots, f_{n_{4j-1}}) \in \mathbb{Q}_s$ and $f_i \in D_G$ for $i = 1, 2, \ldots, n_{4j-1}$.

   (ii) $w(f_1) = m_{2k}$ with $k \in \Omega_1$, $m_{2k}^{1/2} > n_{4j-1}$ and for each $1 \le i < n_{4j-1}$, $w(f_{i+1}) = m_{\sigma(f_1, \ldots, f_i)}$.

(B) A finite sequence $(f_i)_{i=1}^{n_{4j-3}}$ is said to be a n_{4j-3} **attractor sequence** provided that

   (i) $(f_1, f_2, \ldots, f_{n_{4j-3}}) \in \mathbb{Q}_s$ and $f_i \in D_G$ for $i = 1, 2, \ldots, n_{4j-3}$.

   (ii) $w(f_1) = m_{2k}$ with $k \in \Omega_1$, $m_{2k}^{1/2} > n_{4j-3}$ and $w(f_{2i+1}) = m_{\sigma(f_1, \ldots, f_{2i})}$ for each $1 \le i < n_{4j-3}/2$.

   (iii) $f_{2i} = e_{l_{2i}}^*$ for some $l_{2i} \in \Lambda_{\sigma(f_1, \ldots, f_{2i-1})}$, for $i = 1, \ldots, n_{4j-3}/2$.

The definition of the special functionals and the attractors completes the definition of the norming set D_G of the space X_G.

Remarks 1.4. (i) Since the sequence $(\frac{n_{2j}}{m_{2j}})_j$ increases to infinity and the norming set D_G is closed in the $(\mathcal{A}_{n_{2j}}, \frac{1}{m_{2j}})$ operations we get that the Schauder basis $(e_n)_{n\in\mathbb{N}}$ of X_G is boundedly complete.

(ii) The special sequences, as in previous constructions (see for example [GM1],[G2],[AD],[AT1]), are responsible for the absence of unconditionality in every subspace of X_G.

(iii) The attractors do not effect the results of the present section. Their role is to attract the structure of G, for certain G, in every subspace of the predual $(X_G)_* = \overline{\mathrm{span}}\{e_n^* : n \in \mathbb{N}\}$ of the space X_G. So, if we discard condition (iv) in the definition of the norming set D_G (Definition 1.2) then the corresponding space X_G, which we call a **strictly singular extension** provided that the identity operator $I : X_G \to Y_G$ is strictly singular, shares all the properties we shall prove in this section with those X_G's which are strictly singular extensions with attractors.

Definition 1.5. (Rapidly increasing sequences) A block sequence (x_k) in X_G is said to be a (C, ε) rapidly increasing sequence (R.I.S.), if $\|x_k\| \le C$, and there exists a strictly increasing sequence (j_k) of positive integers such that

(a) $(\max \mathrm{supp}\, x_k)\frac{1}{m_{j_{k+1}}} < \varepsilon.$

(b) For every $k = 1, 2, \ldots$ and every $f \in D_G$ with $w(f) = m_i$, $i < j_k$ we have that $|f(x_k)| \le \frac{C}{m_i}$.

Remark 1.6. A subsequence of a (C, ε) R.I.S. remains a (C, ε) R.I.S. while a sequence which is a (C, ε) R.I.S. is also a (C', ε') R.I.S. if $C' \ge C$ and $\varepsilon' \ge \varepsilon$.

Proposition 1.7. Let $(x_k)_{k=1}^{n_{j_0}}$ be a (C, ε) R.I.S. with $\varepsilon \le \frac{2}{m_{j_0}^2}$ such that for every $g \in G$, $\#\{k : |g(x_k)| > \varepsilon\} \le n_{j_0-1}$. Then

1) For every $f \in D_G$ with $w(f) = m_i$,

$$|f(\frac{1}{n_{j_0}}\sum_{k=1}^{n_{j_0}} x_k)| \le \begin{cases} \frac{3C}{m_{j_0}m_i}, & \text{if } i < j_0 \\ \frac{C}{n_{j_0}} + \frac{C}{m_i} + C\varepsilon, & \text{if } i \ge j_0 \end{cases}$$

In particular $\|\frac{1}{n_{j_0}}\sum_{k=1}^{n_{j_0}} x_k\| \le \frac{2C}{m_{j_0}}.$

2) If $(b_k)_{k=1}^{n_{j_0}}$ are scalars with $|b_k| \le 1$ such that

(2) $$|h(\sum_{k\in E} b_k x_k)| \le C(\max_{k\in E} |b_k| + \varepsilon \sum_{k\in E} |b_k|)$$

for every interval E of positive integers and every $h \in D_G$ with $w(h) = m_{j_0}$, then

$$\|\frac{1}{n_{j_0}}\sum_{k=1}^{n_{j_0}} b_k x_k\| \le \frac{4C}{m_{j_0}^2}.$$

The proof of the above proposition is based on what we call the basic inequality (see also [AT1], [ALT]). Its proof is presented in Appendix A.

Remark 1.8. The validity of Proposition 1.7 is independent of the assumption that the operator $I : X_G \to Y_G$ is strictly singular.

In the present section we shall prove several properties of the space X_G provided that the space X_G is a strictly singular extension of Y_G.

Definition 1.9. (exact pairs) A pair (x, ϕ) with $x \in X_G$ and $\phi \in D_G$ is said to be a (C, j, θ) exact pair (where $C \geq 1$, $j \in \mathbb{N}$, $0 \leq \theta \leq 1$) if the following conditions are satisfied:

(i) $1 \leq \|x\| \leq C$, for every $\psi \in D_G$ with $w(\psi) = m_i$, $i \neq j$ we have that $|\psi(x)| \leq \frac{2C}{m_i}$ if $i < j$, while $|\psi(x)| \leq \frac{C}{m_j^2}$ if $i > j$ and $\|x\|_\infty \leq \frac{C}{m_j^2}$.

(ii) ϕ is of type I with $w(\phi) = m_j$.

(iii) $\phi(x) = \theta$ and $\operatorname{ran} x = \operatorname{ran} \phi$.

Definition 1.10. (ℓ_1^k **averages**) Let $k \in \mathbb{N}$. A finitely supported vector $x \in X_G$ is said to be a $C - \ell_1^k$ average if $\|x\| > 1$ and there exist $x_1 < \ldots < x_k$ with $\|x_i\| \leq C$ such that $x = \frac{1}{k} \sum_{i=1}^{k} x_i$.

Lemma 1.11. Let $j \in \mathbb{N}$ and $\varepsilon > 0$. Then every block subspace of X_G contains a vector x which is a $2 - \ell_1^{n_{2j}}$ average. If X_G is a strictly singular extension (with or without attractors) then we may select x to additionally satisfy $\|x\|_G < \varepsilon$.

Proof. If the identity operator $I : X_G \to Y_G$ is strictly singular we may pass to a further block subspace on which the restriction of I has norm less than $\frac{\varepsilon}{2}$. For the remainder of the proof in this case, and the proof in the general case, we refer to [GM1] (Lemma 3) or to [AM]. □

Lemma 1.12. Let x be a $C - \ell_1^k$ average. Then for every $n \leq k$ and every sequence of intervals $E_1 < \ldots < E_n$, we have that $\sum_{i=1}^{n} \|E_i x\| \leq C(1 + \frac{2n}{k})$. In particular if x is a $C - \ell_1^{n_j}$ average then for every $f \in D_G$ with $w(f) = m_i$, $i < j$ then $|f(x)| \leq \frac{1}{m_i} C(1 + \frac{2n_{j-1}}{n_j}) \leq \frac{3C}{2} \frac{1}{m_i}$.

We refer to [S] or to [GM1] (Lemma 4) for a proof.

Remark 1.13. Let $(x_k)_k$ be a block sequence in X_G such that each x_k is a $\frac{2C}{3} - \ell_1^{n_{j_k}}$ average and let $\varepsilon > 0$ be such that $\#(\operatorname{ran}(x_k)) \frac{1}{m_{j_{k+1}}} < \varepsilon$. Then Lemma 1.12 yields that condition (b) in the definition of R.I.S. (Definition 1.5) is also satisfied hence $(x_k)_k$ is a (C, ε) R.I.S. In this case we shall call $(x_k)_k$ a (C, ε) R.I.S. of ℓ_1 averages. From this observation and Lemma 1.11 it follows that if X_G is a strictly singular extension of Y_G then for every $\varepsilon > 0$, every block subspace of X_G contains a $(3, \varepsilon)$ R.I.S. of ℓ_1 averages $(x_k)_{k \in \mathbb{N}}$ with $\|x_k\|_G < \varepsilon$.

Proposition 1.14. Suppose that X_G is a strictly singular extension of Y_G (with or without attractors). Let Z be a block subspace of X_G, let $j \in \mathbb{N}$ and let $\varepsilon > 0$. Then there exists a $(6, 2j, 1)$ exact pair (x, ϕ) with $x \in Z$ and $\|x\|_G < \varepsilon$.

Proof. From the fact that the identity operator $I : X_G \to Y_G$ is strictly singular we may assume, passing to a block subspace of Z, that $\|z\|_G < \frac{\varepsilon}{6}\|z\|$ for every $z \in Z$. We choose a $(3, \frac{1}{n_{2j}})$ R.I.S. of ℓ_1 averages in Z, $(x_k)_{k=1}^{n_{2j}}$. For $k = 1, 2, \ldots, n_{2j}$ we choose $\phi_k \in D_G$ with $\operatorname{ran}\phi_k = \operatorname{ran} x_k$ such that $\phi_k(x_k) > 1$. We set $\phi = \frac{1}{m_{2j}} \sum_{k=1}^{n_{2j}} \phi_k$. We have that $\eta = \phi(\frac{m_{2j}}{n_{2j}} \sum_{k=1}^{n_{2j}} x_k) > 1$.

On the other hand Proposition 1.7 yields that $\|\frac{m_{2j}}{n_{2j}} \sum_{k=1}^{n_{2j}} x_k\| \leq 6$. We set

$$x = \frac{1}{\eta} \frac{m_{2j}}{n_{2j}} \sum_{k=1}^{n_{2j}} x_k.$$

We have that $1 = \phi(x) \leq \|x\| \leq 6$, hence also $\|x\|_G \leq \varepsilon$, while $\operatorname{ran}\phi = \operatorname{ran} x$. From Proposition 1.7 it follows that for every $\psi \in D_G$ with $w(\psi) = m_i$, $i \neq 2j$ we have that $|\psi(x)| \leq \frac{9}{m_i}$ if $i < 2j$ while $|\psi(x)| \leq m_{2j}(\frac{3}{n_{2j}} + \frac{3}{m_i} + \frac{3}{n_{2j}}) \leq \frac{1}{m_{2j}^2}$ if $i > m_{2j}$. Finally $\|x\|_\infty \leq \frac{m_{2j}}{n_{2j}} \max_k \|x_k\|_\infty \leq \frac{3m_{2j}}{n_{2j}} < \frac{1}{m_{2j}^2}$.

Therefore (x, ϕ) is a $(6, 2j, 1)$ exact pair with $x \in Z$ and $\|x\|_G < \varepsilon$. □

Definition 1.15. (dependent sequences and attracting sequences)

(A) A double sequence $(x_k, x_k^*)_{k=1}^{n_{4j-1}}$ is said to be a $(C, 4j - 1, \theta)$ dependent sequence (for $C > 1$, $j \in \mathbb{N}$, and $0 \leq \theta \leq 1$) if there exists a sequence $(2j_k)_{k=1}^{n_{4j-1}}$ of even integers such that the following conditions are fulfilled:

   (i) $(x_k^*)_{k=1}^{n_{4j-1}}$ is a $4j - 1$ special sequence with $w(x_k^*) = m_{2j_k}$ for each k.

   (ii) Each (x_k, x_k^*) is a $(C, 2j_k, \theta)$ exact pair.

(B) A double sequence $(x_k, x_k^*)_{k=1}^{n_{4j-3}}$ is said to be a $(C, 4j - 3, \theta)$ attracting sequence (for $C > 1$, $j \in \mathbb{N}$, and $0 \leq \theta \leq 1$) if there exists a sequence $(2j_k)_{k=1}^{n_{4j-3}}$ of even integers such that the following conditions are fulfilled:

   (i) $(x_k^*)_{k=1}^{n_{4j-3}}$ is a $4j - 3$ attractor sequence with $w(x_{2k-1}^*) = m_{2j_{2k-1}}$ and $x_{2k}^* = e_{l_{2k}}^*$ where $l_{2k} \in \Lambda_{2j_{2k}}$ for all $k \leq n_{4j-3}/2$.

   (ii) $x_{2k} = e_{l_{2k}}$.

   (iii) Each (x_{2k-1}, x_{2k-1}^*) is a $(C, 2j_{2k-1}, \theta)$ exact pair.

Remark 1.16. If $(x_k, x_k^*)_{k=1}^{n_{4j-1}}$ is a $(C, 4j - 1, \theta)$ dependent sequence (resp. $(x_k, x_k^*)_{k=1}^{n_{4j-3}}$ is a $(C, 4j - 1, \theta)$ attracting sequence) then the sequence $(x_k)_k$ is a $(2C, \frac{1}{n_{4j-1}^2})$ R.I.S. (resp. a $(2C, \frac{1}{n_{4j-3}^2})$ R.I.S.). Let examine this for a $(C, 4j - 3, \theta)$ attracting sequence (the proof for a dependent sequence is similar). First $\|x_k\| = \|e_{l_k}\| = 1$ if k is even while $\|x_k\| \leq C$ if k is odd, as follows from the fact that (x_k, x_k^*) is a $(C, 2j_k, \theta)$ exact pair.

Second, the growth condition of the coding function σ in Definition 1.3 and condition (ii) in the same definition yield that for each k we have that $(\max \operatorname{supp} x_k)\frac{1}{m_{2j_{k+1}}} = \max \operatorname{supp} x_k^* \cdot \frac{1}{m_{\sigma(x_1^*, \ldots, x_k^*)}}$

$< \min\{|x_i^*(e_l)| : l \in \operatorname{supp} x_i^*, i = 1, \ldots, k\} \leq \frac{1}{m_{2j_1}} < \frac{1}{n_{4j-3}^2}$.

Finally, if $f \in D_G$ with $w(f) = m_i$, $i < 2j_k$ then $|f(x_k)| = |f(e_{l_k})| \leq \|f\|_\infty \leq \frac{1}{m_i}$ if k is even, while $|f(x_k)| \leq \frac{2C}{m_i}$ if k is odd, since in this case (x_k, x_k^*) is a $(C, 2j_k, \theta)$ exact pair.

Proposition 1.17. (i) Let $(x_k, x_k^*)_{k=1}^{n_{4j}-1}$ be a $(C, 4j-1, \theta)$ dependent sequence such that $\|x_k\|_G \leq \frac{2}{m_{4j-1}^2}$ for $1 \leq k \leq n_{4j}-1$. Then we have that

$$\|\frac{1}{n_{4j-1}} \sum_{k=1}^{n_{4j}-1} (-1)^{k+1} x_k\| \leq \frac{8C}{m_{4j-1}^2}.$$

(ii) If $(x_k, x_k^*)_{k=1}^{n_{4j}-3}$ is a $(C, 4j-3, \theta)$ attracting sequence with $\|x_{2k-1}\|_G \leq \frac{2}{m_{4j-3}^2}$ for $1 \leq k \leq n_{4j-3}/2$ and for every $g \in G$ we have that $\#\{k : |g(x_{2k})| > \frac{2}{m_{4j-3}^2}\} \leq n_{4j-4}$ then

$$\|\frac{1}{n_{4j-3}} \sum_{k=1}^{n_{4j}-3} (-1)^{k+1} x_k\| \leq \frac{8C}{m_{4j-3}^2}.$$

(iii) If $(x_k, x_k^*)_{k=1}^{n_{4j}-1}$ is a $(C, 4j-1, 0)$ dependent sequence with $\|x_k\|_G \leq \frac{2}{m_{4j-1}^2}$ for $1 \leq k \leq n_{4j}-1$ then we have that

$$\|\frac{1}{n_{4j-1}} \sum_{k=1}^{n_{4j}-1} x_k\| \leq \frac{8C}{m_{4j-1}^2}.$$

Proof. The conclusion will follow from Proposition 1.7 2) after showing that the required conditions are fulfilled. We shall only show (i); the proof of (ii) and (iii) is similar.

From the previous remark the sequence $(x_k)_{k=1}^{n_{4j}-1}$ is a $(2C, \frac{1}{n_{4j-1}^2})$ R.I.S. hence it is a $(2C, \frac{2}{m_{4j-1}^2})$ R.I.S. (see Remark 1.6). It remains to show that for $f \in D_G$ with $w(f) = m_{4j-1}$ and for every interval E of positive integers we have that

$$|f(\sum_{k \in E} (-1)^{k+1} x_k)| \leq 2C(1 + \frac{2}{m_{4j-1}^2} \#(E)).$$

Such an f is of the form $f = \frac{1}{m_{4j-1}}(F x_{t-1}^* + x_t^* + \cdots + x_r^* + f_{r+1} + \cdots + f_d)$ for some $4j-1$ special sequence $(x_1^*, x_2^*, \ldots, x_r^*, f_{r+1}, \ldots, f_{n_{4j-1}})$ where $x_{r+1}^* \neq f_{r+1}$ with $w(x_{r+1}^*) = w(f_{r+1})$, $d \leq n_{4j-1}$ and F is an interval of the form $[m, \max \operatorname{supp} x_{t-1}^*]$.

We estimate the value $f(x_k)$ for each k.

- If $k < t - 1$ we have that $f(x_k) = 0$.
- If $k = t - 1$ we get $|f(x_{t-1})| = \frac{1}{m_{4j-1}}|F x_{t-1}^*(x_{t-1})| \leq \frac{1}{m_{4j-1}}\|x_{t-1}\| \leq \frac{C}{m_{4j-1}}$.
- If $k \in \{t, \ldots, r\}$ we have that $f(x_k) = \frac{1}{m_{4j-1}} x_k^*(x_k) = \frac{\theta}{m_{4j-1}}$.
- If $k > r + 1$, then the injectivity of the coding function σ and the definition of special functionals yield that $w(f_i) \neq m_{2j_k}$ for all $i \geq$

$r + 1$. Using the fact that (x_k, x_k^*) is a $'(C, 2j_k, \theta)$ exact pair and taking into account that $n_{4j-1}^2 < m_{2j_1} \leq \sqrt{m_{2j_k}}$ we get that

$$
\begin{aligned}
|f(x_k)| &= \frac{1}{m_{4j-1}}|(f_{r+1} + \ldots + f_d)(x_k)| \\
&\leq \frac{1}{m_{4j-1}}\Big(\sum_{w(f_i)<m_{2j_k}} |f_i(x_k)| + \sum_{w(f_i)>m_{2j_k}} |f_i(x_k)| \Big) \\
&\leq \frac{1}{m_{4j-1}}\Big(\sum_{4j-1<l<2j_k} \frac{2C}{m_l} + n_{4j-1}\frac{C}{m_{2j_k}^2} \Big) \\
&\leq \frac{C}{m_{4j-1}^2}
\end{aligned}
$$

- For $k = r + 1$, using a similar argument to the previous case we get that $|f(x_{r+1})| \leq \frac{C}{m_{4j-1}} + \frac{C}{m_{4j-1}^2} < \frac{2C}{m_{4j-1}}$.

Let E be an interval. From the previous estimates we get that

$$
\begin{aligned}
|f\big(\sum_{k\in E}(-1)^{k+1}x_k\big)| &\leq |f(x_{t-1})| + \Big| \sum_{k\in E\cap[t,r]} \frac{\theta}{m_{4j-1}}(-1)^{k+1} \Big| \\
&\quad + |f(x_{r+1})| + \sum_{k\in E\cap(r+1,n_{4j-1}]} |f(x_k)| \\
&\leq \frac{C}{m_{4j-1}} + \frac{1}{m_{4j-1}} + \frac{C+1}{m_{4j-1}} + \frac{C}{m_{4j-1}^2}\#(E) \\
&< 2C(1 + \frac{2}{m_{4j-1}^2}\#(E)).
\end{aligned}
$$

The proof of the proposition is complete. $\qquad\square$

Theorem 1.18. If the space X_G is a strictly singular extension, (with or without attractors) then it is Hereditarily Indecomposable.

Proof. Let Y and Z be a pair of block subspaces of X_G and let $\delta > 0$. We choose $j \in \mathbb{N}$ with $m_{4j-1} > \frac{48}{\delta}$. Using Proposition 1.14 we inductively construct a $(6, 4j - 1, 1)$ dependent sequence $(x_k, x_k^*)_{k=1}^{n_{4j-1}}$ with $x_{2k-1} \in Y$, $x_{2k} \in Z$ and $\|x_k\|_G < \frac{1}{m_{4j-1}^2}$ for all k. From Proposition 1.17 (i) we get

that $\|\frac{1}{n_{4j-1}} \sum_{k=1}^{n_{4j-1}}(-1)^{k+1}x_k\| \leq \frac{48}{m_{4j-1}^2}$. On the other hand the functional

$x^* = \frac{1}{m_{4j-1}} \sum_{k=1}^{n_{4j-1}} x_k^*$ belongs to D_G and the estimate of x^* on the vector

$\frac{1}{n_{4j-1}} \sum_{k=1}^{n_{4j-1}} x_k$ gives that $\|\frac{1}{n_{4j-1}} \sum_{k=1}^{n_{4j-1}} x_k\| \geq \frac{1}{m_{4j-1}}$.

Setting $y = \sum_{k=1}^{n_{4j-1}/2} x_{2k-1}$ and $z = \sum_{k=1}^{n_{4j-1}/4} x_{2k-1}$ we get that $y \in Y$, $z \in Z$

and $\|y - z\| < \delta\|y + z\|$. Therefore the space X_G is Hereditarily Indecomposable. $\qquad\square$

Proposition 1.19. If X_G is a strictly singular extension (with or without attractors) then the dual X_G^* of the space $X_G = T[G, (\mathcal{A}_{n_j}, \frac{1}{m_j})_j, \sigma]$ is the norm closed linear span of the w^* closure of G.

$$X_G^* = \overline{\text{span}}(\overline{G}^{w^*}).$$

Proof. Assume the contrary. Then setting $Z = \overline{\text{span}}(\overline{G}^{w^*})$ there exist $x^* \in X_G^* \setminus Z$ with $\|x^*\| = 1$ and $x^{**} \in X_G^{**}$ such that $Z \subset \ker x^{**}$, $\|x^{**}\| = 2$ and $x^{**}(x^*) = 2$. The space X_G contains no isomorphic copy of ℓ_1, since X_G is a HI space, thus from the Odell-Rosenthal theorem there exist a sequence $(x_k)_{k \in \mathbb{N}}$ in X_G with $\|x_k\| \leq 2$ such that $x_k \xrightarrow{w^*} x^{**}$. Since each e_n^* belongs to Z we get that $\lim_k e_n^*(x_k) = 0$ for all n, thus, using a sliding hump argument, we may assume that $(x_k)_{k \in \mathbb{N}}$ is a block sequence. Since also $x^*(x_k) \to x^{**}(x^*) = 2$ we may also assume that $1 < x^*(x_k)$ for all k. Let's observe that every convex combination of $(x_k)_{k \in \mathbb{N}}$ has norm greater than 1.

Considering each x_k as a continuous function $x_k : \overline{G}^{w^*} \to \mathbb{R}$ we have that the sequence $(x_k)_{k \in \mathbb{N}}$ is uniformly bounded and tends pointwise to 0, hence it is a weakly null sequence in $C(\overline{G}^{w^*})$. Since Y_G is isometric to a subspace of $C(\overline{G}^{w^*})$ we get that $x_k \xrightarrow{w} 0$ in Y_G thus there exists a convex block sequence $(y_k)_{k \in \mathbb{N}}$ of $(x_k)_{k \in \mathbb{N}}$ with $\|y_k\|_G \to 0$. We may thus assume that $\|y_k\|_G < \frac{\varepsilon}{2}$ for all k, where $\varepsilon = \frac{1}{n_4}$. We may construct a block sequence $(z_k)_{k \in \mathbb{N}}$ of $(y_k)_{k \in \mathbb{N}}$ such that $(z_k)_{k \in \mathbb{N}}$ is a $(3, \varepsilon)$ R.I.S. of ℓ_1 averages while each z_k is an average of $(y_k)_{k \in \mathbb{N}}$ with $\|z_k\|_G < \varepsilon$ (see Remark 1.13). Proposition 1.7 yields that the vector $z = \frac{1}{n_4} \sum_{k=1}^{n_4} z_k$ satisfies $\|z\| \leq \frac{2 \cdot 3}{m_4} < 1$. On the other hand, the vector z, being a convex combination of $(x_k)_{k \in \mathbb{N}}$, satisfies $\|z\| > 1$. This contradiction completes the proof of the proposition. □

Remark 1.20. The content of the above proposition is that the strictly singular extension (with or without attractors) $X_G = T[G, (\mathcal{A}_{n_j}, \frac{1}{m_j})_j, \sigma]$ of the space Y_G is actually a **reflexive extension**. Namely if $\overline{G}^{w^*}$ is a subset of $c_{00}(\mathbb{N})$ then a consequence of Proposition 1.19 is that the space X_G is reflexive. Furthermore, if X_G is nonreflexive then the quotient space $X_G^*/(X_G)_*$ is norm generated by the classes of the elements of the set $\overline{G}^{w^*}$. Related to this is also the next.

Proposition 1.21. The strictly singular extension (with or without attractors) X_G is reflexively saturated (or somewhat reflexive).

Proof. Let Z be a block subspace of X_G. From the fact that the identity operator $I : X_G \to Y_G$ is strictly singular we may choose a normalized block sequence $(z_n)_{n \in \mathbb{N}}$ in Z, with $\sum_{n=1}^{\infty} \|z_n\|_G < \frac{1}{2}$. We claim that the space $Z' = \overline{\text{span}}\{z_n : n \in \mathbb{N}\}$ is a reflexive subspace of Z.

It is enough to show that the Schauder basis $(z_n)_{n \in \mathbb{N}}$ of Z' is boundedly complete and shrinking. The first follows from the fact that $(z_n)_{n \in \mathbb{N}}$ is a block sequence of the boundedly complete basis $(e_n)_{n \in \mathbb{N}}$ of X_G. To see that $(z_n)_{n \in \mathbb{N}}$ is shrinking it is enough to show that $\|f|_{\overline{\text{span}}\{z_i : i \geq n\}}\| \xrightarrow{n \to \infty} 0$ for every $f \in X_G^*$. From Proposition 1.19 it is enough to prove it only for $f \in \overline{G}^{w^*}$. Since $\sum\limits_{n=1}^{\infty} \|z_n\|_G < \frac{1}{2}$ the conclusion follows. $\qquad\square$

Proposition 1.22. Let Y be an infinite dimensional closed subspace of X_G. Every bounded linear operator $T : Y \to X_G$ takes the form $T = \lambda I_Y + S$ with $\lambda \in \mathbb{R}$ and S a strictly singular operator (I_Y denotes the inclusion map from Y to X_G).

The proof of Proposition 1.22 is similar to the corresponding result for the space of Gowers and Maurey (Lemmas 22 and 23 of [GM1]) and is based on the following lemma.

Lemma 1.23. Let Y be a subspace of X_G and let $T : Y \to X_G$ be a bounded linear operator. Let $(y_l)_{l \in \mathbb{N}}$ be a block sequence of $2 - \ell_1^{n_l}$ averages with increasing lengths in Y such that $(Ty_l)_{l \in \mathbb{N}}$ is also a block sequence and $\lim\limits_{l} \|y_l\|_G = 0$. Then $\lim\limits_{l} \text{dist}(Ty_l, \mathbb{R}y_l) = 0$.

Proof of Proposition 1.22. Assume that T is not strictly singular. We shall determine a $\lambda \neq 0$ such that $T - \lambda I_Y$ is strictly singular.

Let Y' be an infinite dimensional closed subspace of Y such that $T : Y' \to T(Y')$ is an isomorphism. By standard perturbation arguments and using the fact that X_G is a strictly singular extension of Y_G, we may assume, passing to a subspace, that Y' is a block subspace of X_G spanned by a normalized block sequence $(y_n')_{n \in \mathbb{N}}$ such that $(Ty_n')_{n \in \mathbb{N}}$ is also a block sequence and $\sum\limits_{n=1}^{\infty} \|y_n'\|_G < 1$. From Lemma 1.11 we may choose a block sequence $(y_n)_{n \in \mathbb{N}}$ of $2 - \ell_1^{n_i}$ averages of increasing lengths in $\text{span}\{y_n' : n \in \mathbb{N}\}$ with $\|y_n\|_G \to 0$. Lemma 1.23 yields that $\lim\limits_{n} \text{dist}(Ty_n, \mathbb{R}y_n) = 0$. Thus there exists a $\lambda \neq 0$ such that $\lim\limits_{n} \|Ty_n - \lambda y_n\| = 0$.

Since the restriction of $T - \lambda I_Y$ to any finite codimensional subspace of $\overline{\text{span}}\{y_n : n \in \mathbb{N}\}$ is clearly not an isomorphism and since also Y is a HI space, it follows from Proposition 1.2 of [AT1] that the operator $T - \lambda I_Y$ is strictly singular. $\qquad\square$

2. STRONGLY STRICTLY SINGULAR EXTENSIONS

It is not known whether the predual $(X_G)_* = \overline{\text{span}}\{e_n^* : n \in \mathbb{N}\}$ of the strictly singular extension X_G is in general a Hereditarily Indecomposable space. In this section we introduce the concept of strongly strictly singular extensions which permit us to ensure the HI property for the space $(X_G)_*$ and to obtain additional information for this space as well as for the spaces $\mathcal{L}(X_G)$, $\mathcal{L}((X_G)_*)$. We also study the quotients of X_G with w^* closed subspaces Z and we show that these quotients are HI.

Definition 2.1. Let G be a ground set and X_G be an extension of the space Y_G. The space X_G is said to be a strongly strictly singular extension provided the following property holds:

For every $C > 0$ there exists $j(C) \in \mathbb{N}$ such that for every $j \geq j(C)$ and every C-bounded block sequence $(x_n)_{n \in \mathbb{N}}$ in X_G with $\|x_n\|_\infty \to 0$ and $(x_n)_{n \in \mathbb{N}}$ a weakly null sequence in Y_G, there exists $L \in [\mathbb{N}]$ such that for every $g \in G$

$$\#\{n \in L : |g(x_n)| > \frac{2}{m_{2j}^2}\} \leq n_{2j-1}.$$

Remark 2.2. Let's observe, for later use, that if $(x_n)_{n \in \mathbb{N}}$ is a R.I.S. of ℓ_1 averages (Remark 1.13), then $\|x_n\|_\infty \to 0$. Therefore if X_G is a strongly singular extension of Y_G, there exists a subsequence $(x_{l_n})_{n \in \mathbb{N}}$ such that the sequence $y_n = x_{l_{2n-1}} - x_{l_{2n}}$ is weakly null and satisfies the above stated property.

Proposition 2.3. If X_G is a strongly strictly singular extension (with or without attractors) of Y_G, then the identity map $I : X_G \to Y_G$ is a strictly singular operator.

Proof. In any block subspace of X_G we may consider a block sequence $(x_n)_{n \in \mathbb{N}}$ with $1 \leq \|x_n\|_{X_G} \leq 2$, $\|x_n\|_\infty \to 0$ and $(x_n)_{n \in \mathbb{N}}$ being weakly null. Passing to a subsequence $(x_{l_n})_{n \in \mathbb{N}}$ for $j \geq j(2)$ we obtain that

$$\|\frac{1}{n_{2j}} \sum_{i=1}^{n_{2j}} x_{l_i}\|_{X_G} \geq \frac{1}{m_{2j}}$$

and on the other hand

$$\|\frac{1}{n_{2j}} \sum_{i=1}^{n_{2j}} x_{l_i}\|_{Y_G} \leq \frac{2}{m_{2j}^2} + \frac{2n_{2j-1}}{n_{2j}} < \frac{3}{m_{2j}^2}$$

which yields that I is not an isomorphism in any block subspace of X_G. $\square$

Definition 2.1 is the analogue of the definition of $\mathcal{S}_2$ bounded or $\mathcal{S}_\xi$ bounded sets (see [AT1] where the norming sets are defined with the use of saturation methods of the form $(\mathcal{S}_{\xi_j}, \frac{1}{m_j})_j$) in the context of saturation methods of low complexity, i.e. of the form $(\mathcal{A}_{n_j}, \frac{1}{m_j})_j$. As we have noticed earlier the assumption of a strongly strictly singular extension (with or without attractors) is required in order to prove that the predual space $(X_G)_*$ is Hereditarily Indecomposable.

The HI property of the dual space X_G^* essentially depends on the internal structure of the set G. Thus we shall see examples of strongly strictly singular extensions (with or without attractors) such that X_G^* is either HI or contains $\ell_2(\mathbb{N})$.

Definition 2.4. (c_0^k vectors) Let $k \in \mathbb{N}$. A finitely supported vector $x^* \in (X_G)_*$ is said to be a $C - c_0^k$ vector if there exist $x_1^* < \cdots < x_k^*$ such that $\|x_i^*\| > C^{-1}$, $x^* = x_1^* + \cdots + x_k^*$ and $\|x^*\| \leq 1$.

Remark 2.5. The fact that the norming set D_G is rationally convex yields that D_G is pointwise dense in the unit ball of the space $B_{X_G^*}$. Since also the norming set D_G is closed in $(\mathcal{A}_{n_{2j}}, \frac{1}{m_{2j}})$ operations we get that for every j, if $f_1, f_2, \ldots, f_{n_{2j}}$ is a block sequence in X_G^* with $\|f_i\| \leq 1$ then $\|\frac{1}{m_{2j}} \sum_{i=1}^{n_{2j}} f_i\| \leq 1$.

Lemma 2.6. Let $(x_\ell^*)_{\ell \in \mathbb{N}}$ be a block sequence in $(X_G)_*$. Then for every k there exists a $x^* \in \text{span}\{x_\ell^* : \ell \in \mathbb{N}\}$ which is a $2 - c_0^k$ vector.

The proof is based on Remark 2.5 and can be found in [AT2] (Lemma 5.4).

Lemma 2.7. For every $2 - c_0^k$ vector x^* and every $\varepsilon > 0$ there exists a $2 - c_0^k$ vector f with $f \in D_G$, $\text{ran} f = \text{ran} x^*$ and $\|x^* - f\| < \varepsilon$.

Proof. This follows from the fact that the norming set D_G is pointwise dense in $B_{X_G^*}$. $\square$

Lemma 2.8. If x^* is a $C - c_0^k$ vector then there exists a $C - \ell_1^k$ average x with $\text{ran}(x) = \text{ran}(x^*)$ and $x^*(x) > 1$.

Proof. Let $x^* = \sum_{i=1}^{k} x_i^*$ where $x_1^* < \cdots < x_k^*$, $\|x_i^*\| > C^{-1}$ and $\|x^*\| \leq 1$. For $i = 1, \ldots, k$ we choose $x_i \in X_G$ with $\|x_i\| \leq 1$, $x_i^*(x_i) > C^{-1}$ and $\text{ran}(x_i) = \text{ran}(x_i^*)$. We set $x = \frac{1}{k} \sum_{i=1}^{k} (Cx_i)$. Then $\|Cx_i\| \leq C$ for $i = 1, \ldots, k$, while $\|x\| \geq x^*(x) > 1$. Also, since $\text{ran}(x) = \text{ran}(x^*)$, x is the desired $C - \ell_1^k$ average. $\square$

Proposition 2.9. Let Z be a block subspace of $(X_G)_*$ and let $k \in \mathbb{N}$, $\varepsilon > 0$. Then there exists a $2 - \ell_1^k$ vector y and $y^* \in D_G$ such that $y^*(y) > 1$, $\text{ran}(y^*) = \text{ran}(y)$ and $\text{dist}(y^*, Z) < \varepsilon$.

Proof. From Lemma 2.6 we can choose a $2 - c_0^k$ vector $x^* = \sum_{i=1}^{k} x_i^*$ in Z. Lemma 2.8 yields the existence of a $2 - \ell_1^k$ average y with $\text{ran}(y) = \text{ran}(x^*)$ and $x^*(y) > 1$. Applying Lemma 2.7 we can find $y^* \in D_G$ with $\text{ran}(y^*) = \text{ran}(x^*)$ and $\|y^* - x^*\| < \min\{\varepsilon, \frac{x^*(y)-1}{2}\}$. It is clear that y and y^* satisfy the desired conditions. $\square$

Lemma 2.10. Suppose that X_G is a strongly strictly singular extension (with or without attractors) and let Z be a block subspace of $(X_G)_*$. Then for every $j > 1$ and $\varepsilon > 0$ there exists a $(18, 2j, 1)$ exact pair (z, z^*) with $\text{dist}(z^*, Z) < \varepsilon$ and $\|z\|_G \leq \frac{3}{m_{2j}}$.

Proof. Using Proposition 2.9 we may select a block sequence $(y_l)_{l \in \mathbb{N}}$ in X_G and a sequence $(y_l^*)_{l \in \mathbb{N}}$ such that

(i) Each y_l is a $2 - \ell_1^{n_{i_l}}$ average where $(i_l)_{l \in \mathbb{N}}$ is an increasing sequence of integers.

(ii) $y_l^* \in D_G$ for all l and $\sum_{l=1}^{\infty} \text{dist}(y_l^*, Z) < \varepsilon$.

(iii) $y_l^*(y_l) > 1$ and $\operatorname{ran} y_l^* = \operatorname{ran} y_l$.

From Remark 1.13 we may assume (passing, if necessary, to a subsequence) that $(y_l)_{l \in \mathbb{N}}$ is a $(3, \varepsilon)$ R.I.S. Since Y_G contains no isomorphic copy of ℓ_1 we may assume, passing again to a subsequence, that $(y_l)_{l \in \mathbb{N}}$ is a weakly Cauchy sequence in Y_G. Setting $x_l = \frac{1}{2}(y_{2l-1} - y_{2l})$ it is clear that $\|x_l\|_\infty \to 0$ while $(x_l)_{l \in \mathbb{N}}$ is a weakly null sequence in Y_G. From the fact that X_G is a strongly strictly singular extension (with or without attractors) it follows that there exists $M \in [\mathbb{N}]$ such that for every $g \in G$ the set $\{l \in M : |g(x_l)| > \frac{2}{m_{2j}^2}\}$ has at most n_{2j-1} elements (notice that $66m_{2j}^4 < n_{2j-1}$). We may assume that $M = \mathbb{N}$. Also $(x_l)_{l \in \mathbb{N}}$ is a $(3, \varepsilon)$ R.I.S. We set

$$z^* = \frac{1}{m_{2j}} \Big(\sum_{l=1}^{n_{2j}-1} y_{2l-1}^* - y_{2l}^* \Big).$$

From Proposition 1.7 we get that $\|\frac{1}{n_{2j}} \sum_{l=1}^{n_{2j}} x_l\| \le \frac{6}{m_{2j}}$ while $z^*\big(\frac{m_{2j}}{n_{2j}} \sum_{l=1}^{n_{2j}} x_l\big) > 1$ hence there exists η with $\frac{1}{6} \le \eta < 1$ such that $z^*\big(\eta \frac{m_{2j}}{n_{2j}} \sum_{l=1}^{n_{2j}} x_l\big) = 1$. We set

$$z = \eta \frac{m_{2j}}{n_{2j}} \sum_{l=1}^{n_{2j}} x_l.$$

It follows easily from Proposition 1.7 that (z, z^*) is a $(18, 2j, 1)$ exact pair. From condition (ii) we get that $\operatorname{dist}(z^*, Z) < \varepsilon$. Finally we have that $\|z\|_G \le \frac{3}{m_{2j}}$. Indeed, let $g \in G$. Since $\#\{l : |g(x_l)| > \frac{2}{m_{2j}^2}\} \le n_{2j-1}$ and $|g(x_l)| \le \|x_l\| \le 2$ for all l we have that

$$|g(z)| \le \frac{m_{2j}}{n_{2j}} \sum_{l=1}^{n_{2j}} |g(x_l)| \le \frac{m_{2j}}{n_{2j}} \Big(\frac{2}{m_{2j}^2} n_{2j} + 2n_{2j-1} \Big) < \frac{3}{m_{2j}}.$$

Therefore $\|z\|_G \le \frac{3}{m_{2j}}$. $\qquad \square$

Lemma 2.11. Let X_G be a strongly strictly singular extension (with or without attractors) and let Y, Z be a pair of block subspaces of $(X_G)_*$. Then for every $\varepsilon > 0$ and $j > 1$ there exists a $(18, 4j - 1, 1)$ dependent sequence $(x_k, x_k^*)_{k=1}^{n_{4j-1}}$ with $\sum \operatorname{dist}(x_{2k-1}^*, Y) < \varepsilon$, $\sum \operatorname{dist}(x_{2k}^*, Z) < \varepsilon$ and $\|x_k\|_G \le \frac{2}{n_{4j-1}^2}$ for all k.

Proof. This is an immediate consequence of Lemma 2.10. $\qquad \square$

Theorem 2.12. If X_G is a strongly strictly singular extension (with or without attractors), then the predual space $(X_G)_*$ is Hereditarily Indecomposable.

Proof. Let Y, Z be a pair of block subspaces of $(X_G)_*$. For every $j > 1$ using Lemma 2.11 we may select a $(18, 4j - 1, 1)$ dependent sequence $(x_k, x_k^*)_{k=1}^{n_{4j-1}}$ with $\sum \operatorname{dist}(x_{2k-1}^*, Y) < 1$ and $\sum \operatorname{dist}(x_{2k}^*, Z) < 1$ and $\|x_k\|_G \le \frac{2}{m_{4j-1}^2}$ for all k.

The functional $x^* = \frac{1}{m_{4j-1}} \sum_{k=1}^{n_{4j-1}} x_k^*$ belongs to the norming set D_G hence $\|x^*\| \leq 1$. From Proposition 1.17 we get that $\|\frac{1}{n_{4j-1}} \sum_{k=1}^{n_{4j-1}} (-1)^{k+1} x_k\| \leq \frac{144}{m_{4j-1}^2}$.

We set

$$h_Y = \frac{1}{m_{4j-1}} \sum_{k=1}^{n_{4j-1}/2} x_{2k-1}^* \text{ and } h_Z = \frac{1}{m_{4j-1}} \sum_{k=1}^{n_{4j-1}/2} x_{2k}^*.$$

Estimating $h_Y - h_Z$ on the vector $\frac{1}{n_{4j-1}} \sum_{k=1}^{n_{4j-1}} (-1)^{k+1} x_k$ yields that $\|h_Y - h_Z\| \geq \frac{m_{4j-1}}{144}$ while we obviously have that $\|h_Y + h_Z\| = \|x^*\| \leq 1$.

From the fact that $\text{dist}(h_Y, Y) < 1$ and $\text{dist}(h_Z, Z) < 1$ we may select $f_Y \in Y$ and $f_Z \in Z$ with $\|h_Y - f_Y\| < 1$ and $\|h_Z - f_Z\| < 1$. From the above estimates we conclude that $\|f_Y - f_Z\| \geq (\frac{m_{4j-1}}{432} - \frac{2}{3}) \|f_Y + f_Z\|$. Since we can find such f_Y and f_Z for arbitrary large j it follows that $(X_G)_*$ is Hereditarily Indecomposable. $\qquad\square$

The next two theorems concern the structure of $\mathcal{L}(\mathfrak{X}_G)$, $\mathcal{L}((\mathfrak{X}_G)_*)$. We start with the following lemmas. The first is the analogue of Lemma 1.23 for strongly strictly singular extensions.

Lemma 2.13. Assume that X_G is a strongly strictly singular extension. Let Y be a subspace of X_G and let $T : Y \to X_G$ be a bounded linear operator. Let $(y_\ell)_{\ell \in \mathbb{N}}$ be a block sequence in Y of C-$\ell_1^{j_k}$ averages with $\lim j_k = \infty$. Furthermore assume that $(Ty_\ell)_{\ell \in \mathbb{N}}$ is also a block sequence. Then

$$\lim \text{dist}(Ty_\ell, \mathbb{R}y_\ell) = 0.$$

Proof. Assume that the conclusion fails. We may assume, passing to a subsequence that there exists $\delta > 0$ such that for every $\ell \in \mathbb{N}$, $\text{dist}(Ty_\ell, \mathbb{R}y_\ell) > \delta$ and moreover that $(y_\ell)_\ell$ is a R.I.S. Next for each $\ell \in \mathbb{N}$, we choose ϕ_ℓ such that $\text{supp}\,\phi_\ell \subset \text{ran}(y_\ell \cup Ty_\ell)$, $\phi_\ell \in D_G$, $\phi_\ell(Ty_\ell) > \frac{\delta}{2}$ and $\phi_\ell(y_\ell) = 0$. From Remark 2.2 and since X_G is a strictly singular extension of Y_G, for every $j \in \mathbb{N}$, $j > j(C)$ we can find a subsequence $(y_{\ell_k})_{k \in \mathbb{N}}$ such that the sequence $w_k = (y_{\ell_{2k-1}} - y_{\ell_{2k}})/2$ is weakly null and for every $g \in G$,

$$\#\left\{ k \in \mathbb{N} : |g(w_k)| > \frac{2}{m_{2j}^2} \right\} \leq n_{2j-1}.$$

This yields that for every $j > j(C)$ there exists $w_{k_1} < w_{k_2} < \cdots < w_{k_{n_{2j}}}$ and $\phi_{k_1} < \phi_{k_2} < \cdots < \phi_{k_{n_{2j}}}$ such that setting $w \frac{m_{2j}}{n_{2j}} \sum_{i=1}^{n_{2j}} w_{k_i}$ and $\phi \frac{1}{m_{2j}} \sum_{i=1}^{n_{2j}} \phi_i$, we have that

$$\|w\| \leq 6C, \quad \phi \in D_G, \quad \phi(Tw) > \frac{\delta}{2} \quad \phi(w) = 0, \quad \text{and} \quad \|w\|_G < \frac{3}{m_{2j}^2}.$$

In particular (w, ϕ) is $(6C, 2j, 0)$ exact pair with $\|w\|_G < \frac{3}{m_{2j}^2}$. The remaining part of the proof follows the arguments of Lemmas 22 and 23 of [GM1] using Proposition 1.17, (iii). $\qquad \square$

The next lemma is easy and its proof is included in the proof of Theorem 9.4 of [AT1].

Lemma 2.14. Let X be a Banach space with a boundedly complete basis $(e_n)_{n\in\mathbb{N}}$ not containing ℓ_1. Assume that $T : X \to X$ is a bounded linear non weakly compact operator. Then there exist two block sequences $(x_n)_{n\in\mathbb{N}}$ $(y_n)_{n\in\mathbb{N}}$ and y in X such that the following hold:

(i) $(x_n)_{n\in\mathbb{N}}$ is normalized, $x_n \xrightarrow{w^*} x^{**} \in X^{**} \setminus X$.

(ii) $(y_n)_{n\in\mathbb{N}}$ is bounded, $y_n \xrightarrow{w^*} y^{**} \in X^{**} \setminus X$.

(iii) $\|Tx_n - (y + y_n)\| \to 0$.

Theorem 2.15. If X_G is a strongly strictly singular extension (with or without attractors), then every bounded linear operator $T : X_G \to X_G$ takes the form $T = \lambda I + S$ with S a strictly singular and weakly compact operator.

Proof. We already know from Proposition 1.22 that every bounded linear operator $T : X_G \to X_G$ is of the form $T = \lambda I + S$ with S a strictly singular operator so it remains to show that every strictly singular operator $S : X_G \to X_G$ is weakly compact.

Assume now that there exists a strictly singular $T \in \mathcal{L}(X_G)$ which is not weakly compact. Then from Lemma 2.14, there exist $(x_n)_{n\in\mathbb{N}}$, $(y_n)_{n\in\mathbb{N}}$, y in X_G satisfying the conclusions of Lemma 2.14. It follows that there exists a subsequence $(x_n)_{n\in L}$ such that setting $Z = \overline{\text{span}}\{x_n : n \in L\}$ there exists a compact perturbation of $T|Z$ denoted by $\tilde{T}$ such that for $n \in L$, we have that $\tilde{T}(x_n) = y + z_n$. For simplicity of notation assume that $L = \mathbb{N}$. Since $x_n \xrightarrow{w^*} x^{**} \in X_G^{**} \setminus X_G$ and $y_n \xrightarrow{w^*} y^{**} \in X_G^{**} \setminus X_G$ we may assume that every convex combination of $(x_n)_{n\in\mathbb{N}}$ has norm greater than $\delta > 0$.

Choose $(z_k)_{k\in\mathbb{N}}$, $z_k = \frac{1}{n_{j_k}} \sum_{i\in F_k} \frac{1}{\delta} x_i$ with $\#F_k = n_{j_k}$ and $F_k < F_{k+1}$. Then

setting $w_k = \frac{z_{2k-1} - z_{2k}}{\|z_{2k-1} - z_{2k}\|}$, Lemma 2.13 (actually its proof) yields that

$$\lim \text{dist}(\tilde{T}w_k, \mathbb{R}w_k) = 0.$$

From this we conclude that for some subsequence $(w_k)_{k\in L}$, $\tilde{T}|\overline{\text{span}}\{w_k : k \in L\}$ is an isomorphism contradicting our assumption that T is strictly singular. $\qquad \square$

Theorem 2.16. Let X_G be a strongly singular extension (with or without attractors) of Y_G. Then every bounded linear operator $T : (X_G)_* \to (X_G)_*$ is of the form $T = \lambda I + S$ with S strictly singular.

The proof of this result follows the lines of Proposition 7.1 in [AT2]. We first state two auxiliary lemmas.

Lemma 2.17. Let X be a HI space with a Schauder basis $(e_n)_{n \in \mathbb{N}}$. Assume that $T : X \rightarrow X$ is a bounded linear operator which is not of the form $T = \lambda I + S$ with S strictly singular. Then there exists n_0 and $\delta > 0$ such that for every $z \in X_{n_0} = \overline{\text{span}}\{e_n : n \geq n_0\}$, $\text{dist}(Tz, \mathbb{R}z) \geq \delta \|z\|$.

Proof. If not, then there exists a normalized block sequence $(z_n)_{n \in \mathbb{N}}$ such that $\text{dist}(Tz_n, \mathbb{R}z_n) \leq \frac{1}{n}$. Choose $\lambda \in \mathbb{R}$ such that $\|Tz_n - \lambda z_n\|_{n \in L} \rightarrow 0$ for a subsequence $(z_n)_{n \in L}$. Then for a further subsequence $(z_n)_{n \in M}$ we have that $T - \lambda I|_{\overline{\text{span}}\{z_n : n \in M\}}$ is a compact operator. The HI property of X easily yields that $T - \lambda I$ is a strictly singular operator, contradicting our assumption. $\square$

Lemma 2.18. Let $T : (X_G)_* \rightarrow (X_G)_*$ be a bounded linear operator with $\|T\| = 1$. Assume that for some $\delta > 0$, and $n_0 \in \mathbb{N}$, $\text{dist}(Tf, \mathbb{R}f) \geq \delta \|f\|$ for all $f \in (X_G)_*$ with $n_0 < \text{supp } f$. Then for every $k \in \mathbb{N}$ and every block subspace Z of $(X_G)_*$ there exist a $z^* \in Z$ with $\|z^*\| \leq 1$ and a $\frac{2}{\delta}$-ℓ_1^k average z such that $z^*(z) = 0$, $Tz^*(z) > 1$ and $\text{ran } z \subset \text{ran } z^* \cup \text{ran } Tz^*$.

Proof. By Lemma 2.6 there exists a 2-c_0^k vector $z^* = \sum_{i=1}^{k} z_i^*$ in Z with $n_0 < \min \text{supp } z^*$. Since $\text{dist}(Tz_i^*, \mathbb{R}z_i^*) \geq \delta \|z_i^*\| > \frac{\delta}{2}$ we may choose for each $i = 1, \ldots, k$ a vector $z_i \in X_G$ with $\|z_i\| < \frac{2}{\delta}$ and $\text{supp } z_i \subset \text{ran } z_i^* \cup \text{ran } Tz_i^*$ satisfying $Tz_i^*(z_i) > 1$ and $z_i^*(z_i) = 0$. We set $z = \frac{1}{k} \sum_{i=1}^{k} z_i$. It is easy to check that z is the desired vector. $\square$

Proof of Theorem 2.16. On the contrary assume that there exists $T \in \mathcal{L}((X_G)_*)$ which is not of the desired form. Assume further that $\|T\| = 1$ and Te_n^* is finitely supported with $\lim \min \text{supp } Te_n^* = \infty$. (We may assume the later conditions from the fact that the basis $(e_n^*)_{n \in \mathbb{N}}$ of $(X_G)_*$ is weakly null.) In particular for every block sequence $(z_n^*)_{n \in \mathbb{N}}$ in $(X_G)_*$ there exists a subsequence $(z_n^*)_{n \in L}$ such that $(\text{ran } z_n^* \cup \text{ran } Tz_n^*)_n$ is a sequence of successive subsets of $\mathbb{N}$.

Let $\delta > 0$ and $n_0 \in \mathbb{N}$ be as in Lemma 2.18 and let $j(\frac{2}{\delta})$ be the corresponding index such that for all $j \geq j(\frac{2}{\delta})$ the conclusion of Definition 2.1 holds for $\frac{2}{\delta}$-bounded block sequences of X_G. Using arguments similar to those of Lemma 2.10 for $j \geq j(\frac{2}{\delta})$ we can find an $(\frac{18}{\delta}, 2j, 0)$ exact pair (z, z^*), with $Tz^*(z) > 1$ and $\|z\|_G \leq \frac{3}{m_{2j}}$. Then for every $j \in \mathbb{N}$ there exists a $(\frac{18}{\delta}, 4j - 1, 0)$ dependent sequence $(z_k, z_k^*)_{k=1}^{n_{4j-1}}$, such that $z_k^*(z_k) = 0$, $Tz_k^*(z_k) > 1$, $\|z_k\|_G \leq \frac{1}{m_{4j-1}^2}$, $(\text{ran } z_k^* \cup \text{ran } Tz_k^*)_{k=1}^{n_{4j-1}}$ are successive subsets of $\mathbb{N}$ and $\text{ran } z_k \subset I_k$ where I_k is the minimal interval of $\mathbb{N}$ containing $\text{ran } z_k^* \cup \text{ran } Tz_k^*$.

Proposition 1.17 yields that

$$\left\| \frac{1}{n_{4j-1}} \sum_{k=1}^{n_{4j-1}} z_k \right\| \leq \frac{144}{m_{4j-1}^2 \delta}.$$

Finally $\|\frac{1}{m_{4j-1}} \sum\limits_{k=1}^{n_{4j-1}} T z_k^*\| \leq 1$ (since $\|T\| \leq 1$) and also

$$1 \geq \|\frac{1}{m_{4j-1}} \sum_{k=1}^{n_{4j-1}} T z_k^*\| \geq \frac{m_{4j-1}^2 \delta}{144 m_{4j-1}} \frac{1}{n_{4j-1}} \sum_{k=1}^{n_{4j-1}} T z_k^*(z_k) \geq \frac{m_{4j-1} \delta}{144}.$$

This yields a contradiction for sufficiently large $j \in \mathbb{N}$.   □

The following lemma is similar to a corresponding result used by V. Ferenczi [Fe] in order to show that every quotient of the space constructed by W.T Gowers and B. Maurey remains Hereditarily Indecomposable.

Lemma 2.19. Suppose that X_G is a strictly singular extension of Y_G (with or without attractors). Let Z be w^* closed subspace of X_G and let Y be a closed subspace of X_G with $Z \subset Y$ such that the quotient space Y/Z is infinite dimensional. Then for every $m, N \in \mathbb{N}$ and $\varepsilon > 0$ there exists $x \in \mathrm{span}\{e_i : i \geq m\}$ which is a $2 - \ell_1^N$ average with $\mathrm{dist}(x, Y) < \varepsilon$ and there exists $f \in B_{(X_G)_*}$ with $\mathrm{dist}(f, Z_\perp) < \varepsilon$ such that $f(x) > 1$.

Proof. We recall that from the fact that Z is w^* closed the quotient space X_G/Z may be identified with the dual of the annihilator $Z_\perp = \{f \in (X_G)_* : f(z) = 0 \ \forall z \in Z\}$, i.e. $X_G/Z = (Z_\perp)^*$. Pick a normalized sequence $(\hat{y}_n')_{n \in \mathbb{N}}$ in Y/Z with $\hat{y}_n' \xrightarrow{w^*} 0$. From W.B. Johnson's and H.P. Rosenthal's work on w^*- basic sequences ([JR]) and their w^* analogue of the classical Bessaga - Pelczynski theorem, we may assume, passing to a subsequence, that $(\hat{y}_n')_{n \in \mathbb{N}}$ is a w^* basic sequence. Hence there exists a bounded sequence $(z_n^*)_{n \in \mathbb{N}}$ in $Z_\perp$ such that $(z_n^*, \hat{y}_n')_{n \in \mathbb{N}}$ are biorthogonal $(z_n^*(\hat{y}_m') = \delta_{nm})$ and $\sum\limits_{i=1}^{n} \hat{y}(z_n^*)\hat{y}_i' \to \hat{y}$ for every $\hat{y}$ in the weak* closure of the linear span of the sequence $(\hat{y}_n')_{n \in \mathbb{N}}$.

Since $(X_G)_*$ contains no isomorphic copy of ℓ_1 (as it a space with separable dual) we may assume, passing to a subsequence, that $(z_n^*)_{n \in \mathbb{N}}$ is weakly Cauchy, hence $(z_{2n-1}^* - z_{2n}^*)_{n \in \mathbb{N}}$ is weakly null. Using a sliding hump argument and passing to a subsequence we may assume that with an error up to ε this sequence is a block sequence with respect to the standard basis of $(X_G)_*$.

We set $y_n^* = z_{2n-1}^* - z_{2n}^*$ and $\hat{y}_n = \hat{y}_{2n-1}'$ for $n = 1, 2, \ldots$. Then $(y_n^*, \hat{y}_n)_{n \in \mathbb{N}}$ are biorthogonal, $(y_n^*)_{n \in \mathbb{N}}$ is a weakly null block sequence in $(X_G)_*$ with $y_n^* \in Z_\perp$, while $(\hat{y}_n)$ is a normalized w^*- basic sequence in Y/Z.

We choose $k, j \in \mathbb{N}$ such that $2^k > m_{2j}$ and $(2N)^k \leq n_{2j}$. We set

$$\mathcal{A}_1 = \{L \in [\mathbb{N}], \ L = \{l_i, \ i \in \mathbb{N}\} : \ \|\frac{1}{2N} \sum_{i=1}^{2N} (-1)^{i+1} \hat{y}_{l_i}\| > \frac{1}{2}\}$$

and $\mathcal{B}_1 = [\mathbb{N}] \setminus \mathcal{A}_1$

From Ramsey's theorem we may find a homogenous set L either in $\mathcal{A}_1$ or in $\mathcal{B}_1$. We may assume that $L = \mathbb{N}$.

Suppose first that the homogenous set is in $\mathcal{A}_1$, i.e. $\|\frac{1}{2N} \sum\limits_{i=1}^{2N} (-1)^{i+1} \hat{y}_{l_i}\| > \frac{1}{2}$ for every $l_1 < l_2 < \ldots < l_{2N}$ in $\mathbb{N}$. For each n we may choose $y_n \in Y \subset X_G$

with $\|y_n\| = 1$ and $Q(y_n) = \hat{y}_n$. Passing to a subsequence we may assume (again with an error up to ε) that the sequence $x_n = y_{2n-1} - y_{2n}$ is a weakly null block sequence in X_G with $\min \operatorname{supp} x_i \geq m$. We set $x = \frac{1}{N} \sum_{i=1}^{N} x_i$. It is clear that x is a 2-ℓ_1^N average while since $Qx \in Y/Z \subset X_G/Z = (Z_\perp)^*$ and $\|Qx\| > 1$ there exists $f \in Z_\perp$ with $\|f\| \leq 1$ such that $f(x) > 1$.

On the other hand if the homogenous set is in $\mathcal{B}_1$ then we may assume, passing again to a subsequence that there exists $a_1 \geq 2$ such that setting
$$\hat{y}_{2,n} = a_1 \cdot \frac{1}{2N} \sum_{i=(n-1)(2N)+1}^{n(2N)} (-1)^{i+1} \hat{y}_i \text{ for } i = 1, 2, \ldots, (\hat{y}_{2,n})_{n \in \mathbb{N}} \text{ is a normalized}$$
sequence in Y/Z. We may again apply Ramsey's theorem defining $\mathcal{A}_2$, $\mathcal{B}_2$ as before, using the sequence $(\hat{y}_{2,n})_{n \in \mathbb{N}}$ instead of $(\hat{y}_n)_{n \in \mathbb{N}}$. If the homogenous set is in $\mathcal{A}_2$ the proof finishes as before while if it is in $\mathcal{B}_2$ we continue defining $(\hat{y}_{3,n})_{n \in \mathbb{N}}$ $\mathcal{A}_3$, $\mathcal{B}_3$ and so on.

If in none of the first k steps we arrived at a homogenous set in some $\mathcal{A}_i$ then there exist $a_1, a_2, \ldots, a_k \geq 2$ and $l_1 < l_2 < \cdots < l_{(2N)^k}$ in $\mathbb{N}$ such that the vector
$$\hat{y} = a_1 a_2 \cdot \ldots \cdot a_k \frac{1}{(2N)^k} \sum_{i=1}^{(2N)^k} (-1)^{i+1} \hat{y}_{l_i}$$
satisfies $\|\hat{y}\| = 1$.

But then the functional $y^* = \frac{1}{m_{2j}} \sum_{i=1}^{(2N)^k} (-1)^{i+1} y_{l_i}^*$ belongs to $Z_\perp$ and satisfies $\|y^*\| \leq 1$ (as $(y_{l_i}^*)_i$ is a block sequence with $\|y_{l_i}^*\| \leq 1$ and $(2N)^k \leq n_{2j}$). Therefore, taking into account the biorthogonality, we ge that
$$1 = \|\hat{y}\| \leq y^*(\hat{y}) = \frac{1}{m_{2j}} \frac{2^k}{(2N)^k} (2N)^k = \frac{2^k}{m_{2j}}$$
which contradicts our choice of k and j. $\qquad \square$

Lemma 2.20. Suppose that X_G is a strongly strictly singular extension of Y_G (with or without attractors) and let Y and Z be as in Lemma 2.19. Then for every $j > 1$ and every $\varepsilon > 0$ there exists a $(18, 2j, 1)$ exact pair (y, f) with $\operatorname{dist}(y, Y) < \varepsilon$, $\|y\|_G < \frac{3}{m_{2j}}$ and $\operatorname{dist}(f, Z_\perp) < \varepsilon$.

Proof. Let $(\varepsilon_i)_{i \in \mathbb{N}}$ be a sequence of positive reals with $\sum_{i=1}^{\infty} \varepsilon_i < \varepsilon$. Using Lemma 2.19 we may inductively construct a block sequence $(x_i)_{i \in \mathbb{N}}$ in X_G, a sequence $(\phi_i)_{i \in \mathbb{N}}$ in $B_{(X_G)_*}$ and a sequence of integers $t_1 < t_2 < \cdots$ such that the following are satisfied:

(i) $\operatorname{dist}(x_i, Y) < \varepsilon_i$, $\operatorname{dist}(\phi_i, Z_\perp) < \varepsilon_i$ and $\phi_i(x_i) > 1$.

(ii) The sequence $(x_i)_{i \in \mathbb{N}}$ is a sequence of $2 - \ell_1$ averages of increasing length and $\min \operatorname{supp} x_i \geq t_i$.

(iii) The restriction of the functional ϕ_i to the space $\overline{\operatorname{span}}\{e_n : n \geq t_{i+1}\}$ has norm at most ε_i.

Passing to a subsequence we may assume that the sequences $(\phi_i)_{i \in \mathbb{N}}$ and $(x_i)_{i \in \mathbb{N}}$ are weakly Cauchy. Thus the sequence $(-\phi_{2n-1} + \phi_{2n})_{n \in \mathbb{N}}$ is weakly null; so we may assume, passing again to a subsequence, that it is a block sequence and that, since $(x_{2n})_{n \in \mathbb{N}}$ is a weakly Cauchy sequence, the sequence $y_n = x_{4n-2} - x_{4n}$ is a weakly null block sequence in X_G and thus also in Y_G, therefore, from the fact that X_G is a strongly strictly singular extension of Y_G we may assume, passing to a subsequence, that for every $g \in G$ the set $\{n \in \mathbb{N} : |g(y_n)| > \frac{2}{m_{2j}^2}\}$ contains at most n_{2j-1} elements and also that $(y_n)_{n \in \mathbb{N}}$ is a $(3, \varepsilon)$ R.I.S. of ℓ_1 averages.

We set $f_n = \frac{1}{2}(-\phi_{4n-3} + \phi_{4n-2})$ for $n = 1, 2, \ldots$, and we may assume that $\max(\operatorname{supp} f_n \cup \operatorname{supp} y_n) < \min(\operatorname{supp} f_{n+1} \cup \operatorname{supp} y_{n+1})$ for all n. Finally we set

$$ y' = \frac{m_{2j}}{n_{2j}} \sum_{i=1}^{n_{2j}} y_i \text{ and } f = \frac{1}{m_{2j}} \sum_{i=1}^{n_{2j}} f_i. $$

As in the proof of Lemma 2.10 we obtain that (y, f), where y is a suitable scalar multiple of y', is the desired exact pair. $\qquad \square$

Using Lemma 2.20 we prove the following:

Theorem 2.21. If X_G is a strongly strictly singular extension of Y_G (with or without attractors) and Z is a w^* closed subspace of X_G of infinite codimension then the quotient space X_G/Z is Hereditarily Indecomposable.

Proof. Let Y_1 and Y_2 be subspaces of X_G with $Z \hookrightarrow Y_1 \cap Y_2$ such that Z is of infinite codimension in each Y_i, $i = 1, 2$. Then for every $\varepsilon > 0$ and $j \in \mathbb{N}$ we may select an $(18, 4j - 1, 1)$ dependent sequence $\chi = (x_k, x_k^*)_{k=1}^{n_{4j-1}}$ such that

(i) $\|x_k\|_G < \frac{1}{m_{4j-1}^2}$, $k = 1, \ldots, n_{4j-1}$.

(ii) $\operatorname{dist}(x_{2k-1}, Y_1) < \varepsilon$, $\operatorname{dist}(x_{2k}, Y_2) < \varepsilon$.

(iii) $\operatorname{dist}(x_k^*, Z_\perp) < \varepsilon$.

Let $Q : X_G \to X_G/Z$ be the quotient map. If $\varepsilon > 0$ is sufficiently small Proposition 1.17 easily yields that

$$ \operatorname{dist}(S_{Q(Y_1)}, S_{Q(Y_2)}) < \frac{C}{m_{4j-1}} $$

where C is a constant independent of j. The proof is complete. $\qquad \square$

3. The James Tree Space $JT_{\mathcal{F}_2}$.

In this section we define a class of James Tree-like spaces. These spaces share some of the main properties of the classical JT space. Namely they do not contain an isomorphic copy of the space ℓ_1. Furthermore they have a bimonotone basis. In particular their norming set is a ground set and a specific example of this form will be the ground set for our final constructions. The principal goal is to prove the inequality in Proposition 3.14 yielding that that the ground set $\mathcal{F}_2$ defined in the next section admits a strongly strictly singular extension. In Appendix B we present a systematic study of $JT_{\mathcal{F}_2}$ spaces and of some variants of them.

Definition 3.1. (JTG families) A family $\mathcal{F} = (F_j)_{j=0}^{\infty}$ of subsets of $c_{00}(\mathbb{N})$ is said to be a **James Tree Generating family** (JTG family) provided it satisfies the following conditions:

(A) $F_0 = \{\pm e_n^* : n \in \mathbb{N}\}$ and each F_j is nonempty, countable, symmetric, compact in the topology of pointwise convergence and closed under restrictions to intervals of $\mathbb{N}$.

(B) Setting $\tau_j = \sup\{\|f\|_\infty : f \in F_j\}$, the sequence $(\tau_j)_{j\in\mathbb{N}}$ is strictly decreasing and $\sum_{j=1}^{\infty} \tau_j \le 1$.

Definition 3.2. (The $\sigma_{\mathcal{F}}$ coding) Let $(F_j)_{j=0}^{\infty}$ be a JTG family. We fix a pair Ξ_1, Ξ_2 of disjoint infinite subsets of $\mathbb{N}$. Let $W = \{(f_1,\ldots,f_d) : f_i \in \cup_{j=1}^{\infty} F_j,\ f_1 < \cdots < f_d,\ d \in \mathbb{N}\}$. The set W is countable so we may select an 1–1 coding function $\sigma_{\mathcal{F}} : W \to \Xi_2$ such that for every $(f_1,\ldots,f_d) \in W$,

$$\sigma_{\mathcal{F}}(f_1,\ldots,f_d) > \max\{k : \exists i \in \{1,\ldots,d\}\ \text{with}\ f_i \in F_k\}.$$

A finite or infinite block sequence $(f_i)_i$ in $\bigcup_{j=1}^{\infty} F_j \setminus \{0\}$ is said to be a $\sigma_{\mathcal{F}}$ special sequence provided $f_1 \in \bigcup_{l\in\Xi_1} F_l$ and $f_{i+1} \in F_{\sigma_{\mathcal{F}}(f_1,\ldots,f_i)}$ for all i. A $\sigma_{\mathcal{F}}$ special functional x^* is any functional of the form $x^* = E\sum_i f_i$ with $(f_i)_i$ a $\sigma_{\mathcal{F}}$ special sequence (when the sum $\sum_i f_i$ is infinite it is considered in the pointwise topology) and E an interval of $\mathbb{N}$. If the interval E is finite then x^* is said to be a finite $\sigma_{\mathcal{F}}$ special functional. We denote by $\mathcal{S}$ the set of all finite $\sigma_{\mathcal{F}}$ special functionals. Let's observe that $\overline{\mathcal{S}}^{w^*}$ is the set of all $\sigma_{\mathcal{F}}$ special functionals.

Definition 3.3. (A) Let $s = (f_i)_i$ be a $\sigma_{\mathcal{F}}$ special sequence. Then for for each i we define the $\mathrm{ind}_s(f_i)$ as follows. $\mathrm{ind}_s(f_1) = \min\{j : f_1 \in F_j\}$ while for $i = 2, 3, \ldots$ $\mathrm{ind}_s(f_i) = \sigma_{\mathcal{F}}(f_1,\ldots,f_{i-1})$.

(B) Let $s = (f_i)_i$ be a $\sigma_{\mathcal{F}}$ special sequence and let E be an interval. The set of indices of the $\sigma_{\mathcal{F}}$ special functional $x^* = E\sum_i f_i$ is the set

$$\mathrm{ind}_s(x^*) = \{\mathrm{ind}_s(f_i) : Ef_i \ne 0\}.$$

(C) A (finite or infinite) family of $\sigma_{\mathcal{F}}$ special functionals $(x_k^*)_k$ is said to be disjoint if for each k there exists a $\sigma_{\mathcal{F}}$ special sequence $s_k = (f_i^k)_i$ and interval E_k such that $x_k^* = E_k \sum_i f_i^k$ and $(\mathrm{ind}_{s_k}(x_k^*))_k$ are pairwise disjoint.

Remark 3.4. (a) Our definition of $\mathrm{ind}_s(f_i)$ and $\mathrm{ind}_s(x^*)$, which is rather technical, is required by the fact that we did not assume $(F_i \setminus \{0\})_i$ to be pairwise disjoint, hence the same f could occur in several different $\sigma_{\mathcal{F}}$ special sequences.

(b) Let's observe that for every family $(x_i^*)_{i=1}^{d}$ of disjoint $\sigma_{\mathcal{F}}$ special functionals, $\|\sum_{i=1}^{d} x_i^*\|_\infty \le 1$ (recall that $\sum_{j=1}^{\infty} \tau_j \le 1$).

(c) Let $s_1 = (f_i)_i$, $s_2 = (h_i)_i$ be two distinct $\sigma_{\mathcal{F}}$ special sequences. Then $\text{ind}_{s_1}(f_i) \neq \text{ind}_{s_2}(h_j)$ for $i \neq j$ while there exists i_0 such that $f_i = h_i$ for all $i < i_0$ and $\text{ind}_{s_1}(f_i) \neq \text{ind}_{s_2}(h_i)$ for $i > i_0$.

(d) For every family $(s_i)_{i=1}^d$ of infinite $\sigma_{\mathcal{F}}$ special sequences there exists n_0 such that $(Es_i^*)_{i=1}^d$ are disjoint, where $E = [n_0, \infty)$ and s_i^* denotes the $\sigma_{\mathcal{F}}$ special functional defined by the $\sigma_{\mathcal{F}}$ special sequence s_i.

Definition 3.5. (The norming set $\mathcal{F}_2$). Let $(F_j)_{j=0}^{\infty}$ be a JTG family. We set

$$\mathcal{F}_2 = F_0 \cup \Big\{ \sum_{k=1}^d a_k x_k^* : a_k \in \mathbb{Q}, \ \sum_{k=1}^d a_k^2 \leq 1, \ \text{and}$$

$$(x_k^*)_{k=1}^d \text{ is a family of disjoint finite } \sigma_{\mathcal{F}} \text{ special functionals} \Big\}$$

The space $JT_{\mathcal{F}_2}$ is defined as the completion of the space $(c_{00}(\mathbb{N}), \|\ \|_{\mathcal{F}_2})$ where $\|x\|_{\mathcal{F}_2} = \sup\{f(x) : f \in \mathcal{F}_2\}$ for $x \in c_{00}(\mathbb{N})$.

Remark 3.6. Let's observe that the standard basis $(e_n)_{n \in \mathbb{N}}$ of $c_{00}(\mathbb{N})$ is a normalized bimonotone Schauder basis of the space $JT_{\mathcal{F}_2}$.

Theorem 3.7. (i) The space $JT_{\mathcal{F}_2}$ does not contain ℓ_1.

(ii) $JT_{\mathcal{F}_2}^* = \overline{\text{span}}(\{e_n^* : n \in \mathbb{N}\} \cup \{b^* : b \in \mathcal{B}\})$ where $\mathcal{B}$ is the set of all infinite $\sigma_{\mathcal{F}}$ special sequences.

The proof of the above theorem is almost identical with the proofs of Propositions 10.4 and 10.11 of [AT1]. We proceed to a short description of the basic steps.

Let's start by observing the following.

$$\overline{\mathcal{F}_2}^{w^*} = F_0 \cup \Big\{ \sum_{k=1}^{\infty} a_k x_k^* : a_k \in \mathbb{Q}, \ \sum_{k=1}^{\infty} a_k^2 \leq 1, \ \text{and}$$

$$(x_k^*)_{k=1}^{\infty} \text{ is a family of disjoint } \sigma_{\mathcal{F}} \text{ special functionals} \Big\}$$

Also for a disjoint family $(x_i^*)_{i=1}^{\infty}$ of special functionals and $(a_i)_{i=1}^{\infty}$ in $\mathbb{R}$, we have that $\|\sum_{i=1}^{\infty} a_i x_i^*\|_{JT_{\mathcal{F}_2}^*} \leq (\sum_{i=1}^{\infty} a_i^2)^{1/2}$. The above observations yield the following:

Lemma 3.8. $\overline{\mathcal{F}_2}^{w^*} \subset \overline{\text{span}}(\{e_n^* : n \in \mathbb{N}\} \cup \{b^* : b \in \mathcal{B}\})$ where $\mathcal{B}$ is the set of all infinite $\sigma_{\mathcal{F}}$ special sequences.

Observe also that $\overline{\mathcal{F}_2}^{w^*}$ is w^* compact and 1-norming hence contains the set $\text{Ext}(B_{JT_{\mathcal{F}_2}^*})$. Rainwater's theorem and the above results yield that a bounded sequence $(x_k)_{k \in \mathbb{N}}$ is weakly Cauchy if and only if $\lim_k e_n^*(x_k)$ and $\lim_k b^*(x_k)$ exist for all n and infinite special sequences b. This is established by the following.

Lemma 3.9. Let $(x_k)_{k \in \mathbb{N}}$ be a bounded sequence in $JT_{\mathcal{F}_2}$ and let $\varepsilon > 0$. Then there exists a finite family $x_1^*, \ldots, x_d^*$ of disjoint special functionals and an $L \in [\mathbb{N}]$ such that

$$\limsup_{k \in L} |x^*(x_k)| \leq \varepsilon$$

for every special functional x^* such that the family $x^*, x_1^*, \ldots, x_d^*$ is disjoint.

For a proof we refer the reader to the proof of Lemma 10.5 [AT1].

Lemma 3.10. Let $(x_k)_{k \in \mathbb{N}}$ be a bounded sequence in $JT_{\mathcal{F}_2}$. There exists an $M \in [\mathbb{N}]$ such that for every special functional x^* the sequence $(x^*(x_k))_{k \in M}$ converges.

Also for the proof of this we refer the reader to the proof of Lemma 10.6 of [AT1]

Proof of Theorem 3.7. (i) Let $(x_k)_{k \in \mathbb{N}}$ be a bounded sequence in $JT_{\mathcal{F}_2}$. By an easy diagonal argument we may assume that for every $n \in \mathbb{N}$, $\lim_n e_n^*(x_k)$ exists. Lemma 3.10 also yields that there exists a subsequence $(x_{l_k})_{k \in \mathbb{N}}$ such that for every special sequence b, $\lim_k b^*(x_{l_k})$ also exists. As we have mentioned above Lemma 3.8 yields that $(x_{l_k})_{k \in \mathbb{N}}$ is weakly Cauchy.

(ii) Since $\text{Ext}(B_{JT_{\mathcal{F}_2}^*}) \subset \overline{\mathcal{F}_2}^{w^*}$ and ℓ_1 does not embed into $JT_{\mathcal{F}_2}$ Haydon's theorem [Ha] yields that $\overline{\mathcal{F}_2}^{w^*}$ norm generates $JT_{\mathcal{F}_2}^*$. Lemma 3.8 yields the desired result. $\qquad \square$

The remaining part of this section concerns the proof of Proposition 3.14, stated below. This will be used in the next section to show that a specific ground set $\mathcal{F}_2$ admits a strongly strictly singular extension.

Definition 3.11. Let $(x_n)_n$ be a bounded block sequence in $JT_{\mathcal{F}_2}$ and $\varepsilon > 0$. We say that $(x_n)_n$ is ε-*separated* if for every $\phi \in \cup_{j \in \mathbb{N}} F_j$

$$\#\{n : |\phi(x_n)| \geq \varepsilon\} \leq 1.$$

In addition, we say that $(x_n)_n$ is *separated* if for every $L \in [\mathbb{N}]$ and $\varepsilon > 0$ there exists an $M \in [L]$ such that $(x_n)_{n \in M}$ is ε-separated.

Lemma 3.12. Let $(x_n)_n$ be a bounded separated sequence in $JT_{\mathcal{F}_2}$ such that for every infinite $\sigma_{\mathcal{F}}$ special functional b^* we have that $\lim_n b^*(x_n) = 0$. Then for every $\varepsilon > 0$, there exists an $L \in [\mathbb{N}]$ such that for all $y^* \in \overline{\mathcal{S}}^{w^*}$,

$$\#\{n \in L : |y^*(x_n)| \geq \varepsilon\} \leq 2.$$

Proof. Assume the contrary and fix an $\varepsilon > 0$ such that the statement of the lemma is false. Define

$$A = \left\{ (n_1 < n_2 < n_3) \in [\mathbb{N}]^3 : \exists y^* \in \overline{\mathcal{S}}^{w^*}, |y^*(x_{n_1})|, |y^*(x_{n_2})|, |y^*(x_{n_2})| \geq \varepsilon \right\}$$

and $B = [\mathbb{N}]^3 \setminus A$. Then Ramsey's Theorem yields that there exists an $L \in [\mathbb{N}]$ such that either $[L]^3 \subset A$ or $[L]^3 \subset B$. Our assumption rejects the second case, so we conclude that for all $n_1 < n_2 < n_3 \in L$, there is a $y_{n_1,n_2,n_3}^* \in \overline{\mathcal{S}}^{w^*}$ such that $|y_{n_1,n_2,n_3}^*(x_{n_i})| \geq \varepsilon$, $i = 1, 2, 3$.

Since $(x_n)_{n \in L}$ is separated, we may assume by passing to a subsequence that for $\varepsilon' = \frac{\varepsilon}{8}$, $(x_n)_{n \in L}$ is ε'-separated. For reasons of simplicity in the notation we may moreover and do assume that $(x_n)_{n \in \mathbb{N}}$ has both properties.

For all triples $(1 < n < k)$, let $y_{n,k}^*$ denote an element in $\overline{S}^{w^*}$ such that $|y_{n,k}^*(x_i)| \geq \varepsilon$, $i = 1, n, k$. Moreover, let $y_{n,k}^* = E_{n,k} \sum_{i=1}^{\infty} \phi_{n,k}^i$ where $(\phi_{n,k}^i)_{i \in \mathbb{N}}$ is a $\sigma_{\mathcal{F}}$ special sequence and $E_{n,k} \subset \mathbb{N}$ is an interval. For $1 < n < k$ we define the number $[n, k]$ as follows:

$$[n, k] = \min\{i \in \mathbb{N} : \ \max \operatorname{supp} \phi_{n,k}^i \geq \min \operatorname{supp} x_k\}.$$

Also, let $A = \{(n < k) \in [\mathbb{N} \setminus \{1\}]^2 : |\phi_{n,k}^{[n,k]}(x_n)| \leq \varepsilon'\}$ and $B = [\mathbb{N} \setminus \{1\}]^2 \setminus A$.

Again, using Ramsey's theorem and passing to a subsequence, we may and do assume that $[\mathbb{N} \setminus \{1\}]^2 \subset A$ or $[\mathbb{N} \setminus \{1\}]^2 \subset B$. Notice in the second case, that since $(x_n)_n$ is ε'-separated, we have that for all $1 < n < k$, $|\phi_{n,k}^{[n,k]}(x_k)| \leq \varepsilon'$. We set

$$s_{n,k} = \begin{cases} (\phi_{n,k}^1, \ldots, \phi_{n,k}^{[n,k]-1}), & \text{if } [\mathbb{N} \setminus \{1\}]^2 \subset A \\ (\phi_{n,k}^1, \ldots, \phi_{n,k}^{[n,k]}), & \text{if } [\mathbb{N} \setminus \{1\}]^2 \subset B \end{cases}.$$

Claim. There is an $M > 0$ such that for all $k \in \mathbb{N}$,

$$\#\{s_{n,k} : \ 2 \leq n \leq k - 1\} \leq M.$$

Let $(x_n)_n$ be bounded by some $c > 0$. Next fix any $k \in \mathbb{N}$ and consider the following two cases:

The first case is $[\mathbb{N} \setminus \{1\}]^2 \subset B$. In this case $\phi_{n,k}^{[n,k]} \in s_{n,k}$ and if $s_{n_1,k} \neq s_{n_2,k}$ then $\phi_{n_1,k}^{[n_1,k]}$ is incomparable to $\phi_{n_2,k}^{[n_2,k]}$ in the sense, that every two special functionals, extending $s_{n_1,k}$ and $s_{n_2,k}$ respectively, have disjoint sets of indices.

So let $s_{n_j,k}$, $1 \leq j \leq N$ all be different from each other and consider the $\sigma_{\mathcal{F}}$ special functionals $z_{n_j}^* = E_{n_j,k} y_{n_j,k}^*$, $1 \leq j \leq N$ where $E_{n_j,k} = (\max \operatorname{supp} \phi_{n_j,k}^{[n_j,k]}, \infty)$. According to the previous observation these functionals have pairwise disjoint indices. Moreover

$$(3) \qquad |z_{n_j}^*(x_k)| = |y_{n_j,k}^*(x_k) - \phi_{n_j,k}^{[n_j,k]}(x_k)| \geq \varepsilon - \varepsilon'$$

since $[\mathbb{N} \setminus \{1\}]^2 \subset B$ and $(x_n)_n$ is ε' separated.

Inequality (3) yields that

$$\left(\sum_{j=1}^{N} (z_{n_j}^*(x_k))^2 \right)^{1/2} \geq (\varepsilon - \varepsilon') N^{1/2}.$$

Therefore there are $(a_j)_{j=1}^{N}$ with $\sum_{j=1}^{n} a_j^2 \leq 1$ such that

$$\sum_{j=1}^{N} a_j z_{n_j}^*(x_k) \geq (\varepsilon - \varepsilon') N^{1/2}.$$

On the other hand, by the definition of the norm on $JT_{\mathcal{F}_2}$, $\sum_{j=1}^{N} a_j z_{n_j}^*(x_k) \leq \|x_k\| \leq c$. It follows that $N \leq (\frac{c}{\varepsilon - \varepsilon'})^2$ and this is the required upper estimate for N.

The second case is $[\mathbb{N}\setminus\{1\}]^2 \subset A$. As in the first case, if $1 < n_1 < n_2 < k$ and $s_{n_1,k} \neq s_{n_2,k}$ then $\phi_{n_1,k}^{[n_1,k]}$ and $\phi_{n_2,k}^{[n_2,k]}$ are incomparable and since $s_{n_1,k} \neq s_{n_2,k}$ they also have different indices. As in the first case let $s_{n_j,k}$, $1 \leq j \leq N$ all be different from each other and set $z_{n_j}^* E_{n_j,k} y_{n_j,k}^*$, $1 \leq j \leq N$ where in this case $E_{n,k} = [\min \operatorname{supp} \phi_{n_j,k}^{[n_j,k]}, \infty)$. By our previous observation it follows that these $\sigma_{\mathcal{F}}$ special functionals have pairwise disjoint indices. Notice also that $|z_{n_j}^*(x_k)| = |y_{n_j,k}^*(x_k)| \geq \varepsilon$. Therefore exactly as in the first case we obtain an upper estimate for N independent of k and this finishes the proof of the claim.

In the case where $[\mathbb{N}\setminus\{1\}]^2 \subset B$, $|s_{n,k}^*(x_n)| = |y_{n,k}^*(x_n)| \geq \varepsilon$.

In the case where $[\mathbb{N}\setminus\{1\}]^2 \subset A$, $|s_{n,k}^*(x_n)| = |y_{n,k}^*(x_n) - \phi_{n,k}^{[n,k]}(x_n)| \geq \varepsilon - \varepsilon'$.

In any case we have that $|s_{n,k}^*(x_n)| \geq \varepsilon - \varepsilon' > 0$.

Combining this with the previous claim, we get that for any $k \geq 3$ there are $z_{1,k}^*, \ldots, z_{M,k}^* \in \overline{\mathcal{S}}^{w^*}$ such that for any $1 < n < k$ there is $i \in [1, M]$ so that $|z_{i,k}^*(x_n)| \geq \varepsilon - \varepsilon'$.

Since $\overline{\mathcal{S}}^{w^*}$ is weak-* compact we can pass to an $L \in [\mathbb{N}]$ such that $(z_{i,k}^*)_{k \in L}$ is weak-* convergent to some $z_i^* \in \overline{\mathcal{S}}^{w^*}$. It is easy to see that in this case, for any $n \in \mathbb{N}$ there is an $i \in [1, M]$ so that $|z_i^*(x_n)| \geq \varepsilon - \varepsilon'$. Therefore there exists an infinite subset P of $\mathbb{N}$ and $1 \leq i_0 \leq M$ such that $|z_{i_0}^*(x_n)| \geq \varepsilon - \varepsilon'$ for every $n \in P$. It also follows that $z_{i_0}^*$ is an infinite $\sigma_{\mathcal{F}}$ special functional. These contradict our assumption that $\lim_n b^*(x_n) = 0$ for every infinite $\sigma_{\mathcal{F}}$ special functional b^*. $\qquad\square$

We now prove the following lemma about $JT_{\mathcal{F}_2}$:

Lemma 3.13. Let $x \in JT_{\mathcal{F}_2}$ with finite support and $\varepsilon > 0$. There exists $n \in \mathbb{N}$ such that if $y^* \sum_{k=1}^d a_k y_k^* \in \mathcal{F}_2$ with $\max\{|a_k| : 1 \leq k \leq d\} < \frac{1}{n}$, then $|y^*(x)| < \varepsilon$.

Proof. Let $\delta = \frac{\varepsilon}{2 \sum_{n \in \mathbb{N}} |x(n)|}$ and $\tau_j = \sup\{\|f\|_\infty : f \in F_j\}$. Since $\sum_{j=1}^\infty \tau_j \leq 1$, by the definition of a JTG family, there is a $j_0 \in \mathbb{N}$ such that $\sum_{j=j_0+1}^\infty \tau_j < \delta$. Let n be such that $\frac{1}{n} < \frac{\varepsilon}{2j_0 \|x\|}$.

Assume that $y^* = \sum_{k=1}^d a_k y_k^* \in \mathcal{F}_2$ with $\max\{|a_k| : 1 \leq k \leq d\} < \frac{1}{n}$. For every $k \in [1, d]$ let $y_k^* = y_{k,1}^* + y_{k,2}^*$ with $\operatorname{ind}(y_{k,1}^*) \subset \{1, \ldots, j_0\}$ and $\operatorname{ind}(y_{k,2}^*) \subset \{j_0 + 1, j_0 + 2, \ldots\}$. So we may write $y^* = \sum_{k=1}^d a_k y_{k,1}^* + \sum_{k=1}^d a_k y_{k,2}^*$. Notice now that for any $n \in \mathbb{N}$, $|\sum_{k=1}^d a_k y_{k,2}^*(n)| \leq \sum_{k=1}^d \|y_{k,2}^*\|_\infty$ and since $(\operatorname{ind}(y_{k,2}^*))_{k=1}^d$ are pairwise disjoint and all greater than j_0 we get that $\sum_{k=1}^d \|y_{k,2}^*\|_\infty < \delta$. Therefore $\|\sum_{k=1}^d a_k y_{k,2}^*\|_\infty < \delta$ and it follows that

$$(4) \qquad \left| \sum_{k=1}^d a_k y_{k,2}^*(x) \right| \leq \sum_{n \in \mathbb{N}} \delta |x(n)| = \frac{\varepsilon}{2}.$$

On the other hand since $(y_{k,1}^*)_{k=1}^d$ have pairwise disjoint indices, at most j_0 of them are non-zero and $|y_{k,1}^*(x)| \leq \|x\|$. Therefore $|\sum_{k=1}^d a_k y_{k,1}^*(x)| \leq j_0 \frac{1}{n}\|x\| < \frac{\varepsilon}{2}$. Combining this with (4) we get that $|y^*(x)| < \varepsilon$ as required. $\square$

We combine now Lemmas 3.12 and 3.13 to prove the following:

Proposition 3.14. Let $(x_n)_n$ be a weakly null separated sequence in $JT_{\mathcal{F}_2}$ with $\|x_n\|_{\mathcal{F}_2} \leq C$ for all n. Then for all $m \in \mathbb{N}$, there is $L \in [\mathbb{N}]$ such that for every $y^* \in \mathcal{F}_2$,

$$\#\{n \in L : |y^*(x_n)| \geq \frac{1}{m}\} \leq 66m^2C^2.$$

Proof. We may and do assume that $\{x_n : n \in \mathbb{N}\}$ is normalized. We set $\delta_1 = \frac{1}{4m}$ and we find $L_1 \in [\mathbb{N}]$ such that Lemma 3.12 is valid for $\varepsilon = \delta_1$. Then we set $n_1 = \min L_1$ and using Lemma 3.13 we find $n = r_1 \in \mathbb{N}$ such that the conclusion of Lemma 3.13 is valid for $\varepsilon = \delta_1$ and $x = x_{n_1}$. Then, after setting $\delta_2 = \min\{\frac{1}{8mr_1^2}, \delta_1\}$ we find $L_2 \in [\mathbb{N} \setminus \{n_1\}]$ such that Lemma 3.12 is valid for $\varepsilon = \delta_2$. We set $n_2 = \min L_2$ and we find $n = r_2 \in \mathbb{N}$ with $r_2 > r_1$ such that the conclusion of Lemma 3.13 is valid for $\varepsilon = \delta_2$ and $x \in \{x_{n_1}, x_{n_2}\}$.

Recursively, having defined $\delta_1 \geq \delta_2 \geq \cdots \geq \delta_{p-1}$, $L_1 \supset L_2 \supset L_3 \supset \cdots \supset L_{p-1}$, $n_1 < n_2 < \cdots < n_{p-1}$ and $r_1 < r_2 < \cdots < r_{p-1}$, we set $\delta_p = \min\{\frac{1}{4m2^{p-1}r_{p-1}^2}, \delta_{p-1}\}$ and we find $L_p \in [L_{p-1}\setminus\{n_{p-1}\}]$ such that Lemma 3.12 is valid for $\varepsilon = \delta_1$. We set $n_p = \min L_p$ and we find $n = r_p > r_{p-1}$ that the conclusion of Lemma 3.13 is satisfied for $\varepsilon = \delta_p$ and $x \in \{x_{n_1}, x_{n_2}, \ldots, x_{n_p}\}$. At the end we consider the set $L = \{n_1 < n_2 < \cdots < n_p < \cdots\}$.

The crucial properties of this construction are the following:

(1) If $\sum_{k=1}^\ell a_k y_k^* \in \mathcal{F}_2$ and $|a_k| < \frac{1}{r_p}$, for $k = 1, \ldots, \ell$ then we have that $|\sum_{k=1}^\ell a_k y_k^*(x_{n_i})| < \delta_p$ for all $i = 1, \ldots, p$.
(2) For every $x^* \in \mathbb{S}$ we have that $\#\{i \geq p : |x^*(x_{n_i})| \geq \delta_p\} \leq 2$.

We will make use of these two properties to prove the proposition.

Let $d = 66m^2$. It suffices to prove that if $n_{\ell_1} < n_{\ell_2} < \cdots < n_{\ell_d}$ and $y^* = \sum_{k=1}^\ell a_k y_k^* \in \mathcal{F}_2$, then there is an $1 \leq i \leq d$ such that $|y^*(x_{n_{\ell_i}})| < \frac{1}{m}$.

We set

$$A_1 = \{k \in [1, \ell] : |a_k| \geq \frac{1}{r_{\ell_1}}\},$$

$$A_p = \{k \in [1, \ell] : \frac{1}{r_{\ell_p}} \leq |a_k| < \frac{1}{r_{\ell_{p-1}}}\}, \quad \text{for } 1 < p < d$$

$$A_d\{k \in [1, \ell] : \frac{1}{r_{\ell_d}} > |a_k|\}.$$

Observe that for $p < d$, we have that $\#A_p \leq r_{\ell_p}^2$. By property (1) we have that for any $p \in [1, d)$,

$$(5) \qquad \left| \sum_{k \in \cup_{j>p} A_j} a_k y_k^*(x_{n_{\ell_p}}) \right| < \delta_{\ell_p} \leq \delta_1 \frac{1}{4m}.$$

Next, for $1 \leq j < p \leq d$ we set

$$B_{j,p} = \{k \in A_j : |y_k^*(x_{n_{\ell_p}})| \geq \delta_{\ell_j+1}\}$$

and then for $p \in (1, d]$ we define

$$B_p = \bigcup_{j<p} B_{j,p}.$$

Since for $p > j$ we have that $\ell_p \geq \ell_j + 1$, property (2) yields that for every $k \in A_j$ there exist at most two $B_{j,p}$'s containing k. Hence every $k \in \{1, \ldots, \ell\}$ belongs to at most two B_p's.

Next we shall estimate the term $\sum_{k \in \bigcup_{j<p} A_j \setminus B_p} |a_k y_k^*(x_{n_{\ell_p}})|$. Let $p \in (1, d]$. We have that

(6)
$$\sum_{k \in \bigcup_{j<p} A_j \setminus B_p} |a_k y_k^*(x_{n_{\ell_p}})| = \sum_{j=1}^{p-1} \sum_{k \in A_j \setminus B_{j,p}} |a_k y_k^*(x_{n_{\ell_p}})|$$

$$\leq \sum_{j=1}^{p-1} \#(A_j)\delta_{\ell_j+1} \leq \sum_{j=1}^{p-1} r_{\ell_j}^2 \frac{1}{4m2^{\ell_j} r_{\ell_j}^2} < \frac{1}{4m}.$$

We now argue that for at least $\frac{d}{2} + 1$ many of $\{x_{n_{\ell_1}}, \ldots, x_{n_{\ell_d}}\}$, we have that $|\sum_{k \in A_p} a_k y_k^*(x_{n_{\ell_p}})| < \frac{1}{4m}$. If this is not the case, then for at least $\frac{d}{2}$ many, $|\sum_{k \in A_p} a_k y_k^*(x_{n_{\ell_p}})| \geq \frac{1}{4m}$ and therefore $\sum_{k \in A_p} a_k^2 \geq \frac{1}{16m^2}$. Thus

$$\sum_{k=1}^{\ell} a_k^2 \sum_{p=1}^{d} \sum_{k \in A_p} a_k^2 \geq \frac{d}{2} \cdot \frac{1}{16m^2} = \frac{d}{32m^2}$$

which is a contradiction since $d = 66m^2$ and $\sum_{k=1}^{\ell} a_k^2 \leq 1$.

Now we shall prove that for at least $\frac{d}{2} + 1$ many of $\{x_{n_{\ell_1}}, \ldots, x_{n_{\ell_d}}\}$, $|\sum_{k \in B_p} a_k y_k^*(x_{n_{\ell_p}})| < \frac{1}{4m}$. Again if this is not the case, then for at least $\frac{d}{2}$ many $|\sum_{k \in B_p} a_k y_k^*(x_{n_{\ell_p}})| \geq \frac{1}{4m}$ and therefore $\sum_{k \in B_p} a_k^2 \geq \frac{1}{16m^2}$. Since every k appears in at most two B_p's, we have that

$$2 \geq 2\sum_{k=1}^{\ell} a_k^2 \geq \sum_{p=1}^{d} \sum_{k \in B_p} a_k^2 \geq \frac{d}{2} \cdot \frac{1}{16m^2} = \frac{d}{32m^2}$$

which is a contradiction.

These last two observations show that there exists at least one $p \in [1, d]$ such that both $|\sum_{k \in A_p} a_k y_k^*(x_{n_{\ell_p}})| < \frac{1}{4m}$ and $|\sum_{k \in B_p} a_k y_k^*(x_{n_{\ell_p}})| < \frac{1}{4m}$.

Combining this with (5) and (6), we get that for this particular p,

$$\left|\sum_{k=1}^{\ell} a_k y_k^*(x_{n_{\ell_p}})\right| \le \left|\sum_{k\in\cup_{j>p}A_j} a_k y_k^*(x_{n_{\ell_p}})\right| + \left|\sum_{k\in A_p} a_k y_k^*(x_{n_{\ell_p}})\right|$$

$$+ \left|\sum_{k\in B_p} a_k y_k^*(x_{n_{\ell_p}})\right| + \left|\sum_{k\in\cup_{j<p}A_j\setminus B_p} a_k y_k^*(x_{n_{\ell_p}})\right|$$

$$< \frac{1}{m}$$

as required. $\qquad\qquad\qquad\qquad\qquad\qquad\qquad\qquad\qquad\qquad\qquad\qquad\qquad\qquad\square$

4. THE SPACE $(\mathfrak{X}_{\mathcal{F}_2})_*$ AND THE SPACE OF THE OPERATORS $\mathcal{L}((\mathfrak{X}_{\mathcal{F}_2})_*)$

In this section we proceed to construct a HI space not containing a reflexive subspace. This space is $(\mathfrak{X}_{\mathcal{F}_2})_*$ where $\mathfrak{X}_{\mathcal{F}_2}$ is the strongly strictly singular HI extension (Sections 1 and 2) of the set $\mathcal{F}_2$. The set $\mathcal{F}_2$ is defined from a family $F = (F_j)_j$ as in Section 3. The proof that $(\mathfrak{X}_{\mathcal{F}_2})_*$ does not contain a reflexive subspace, uses the method of attractors and the key ingredient is the attractor functional and the attracting sequences introduced in Section 1. The structure of the quotients of $\mathfrak{X}_{\mathcal{F}_2}$ is also investigated.

The family $F = (F_j)_{j\in\mathbb{N}}$

We shall use the sequence of positive integers $(m_j)_j$, $(n_j)_j$ introduced in Definition 1.2 of strictly singular extensions which for convenience we recall:

- $m_1 = 2$ and $m_{j+1} = m_j^5$.
- $n_1 = 4$, and $n_{j+1} = (5n_j)^{s_j}$ where $s_j = \log_2 m_{j+1}^3$.

We set $F_0 = \{\pm e_n^* : n \in \mathbb{N}\}$ and for $j = 1, 2, \ldots$ we set

$$F_j = \left\{\frac{1}{m_{4j-3}^2}\sum_{i\in I}\pm e_i^* : \#(I) \le \frac{n_{4j-3}}{2}\right\} \cup \{0\}.$$

In the sequel we shall denote by $\mathfrak{X}_{\mathcal{F}_2}$ the HI extension of $JT_{\mathcal{F}_2}$ with ground set $\mathcal{F}_2$ defined by the aforementioned family $(F_j)_{j\in\mathbb{N}}$ as in Definition 3.5.

Proposition 4.1. The space $\mathfrak{X}_{\mathcal{F}_2}$ is a strongly strictly singular extension of $JT_{\mathcal{F}_2}(= Y_{\mathcal{F}_2})$.

Proof. Let $C > 0$. We select $j(C)$ such that $\frac{33}{2}m_{2j}^4 C^2 < n_{2j-1}$ for every $j \ge j(C)$ and we shall show that the integer $j(C)$ satisfies the conclusion of Definition 2.1.

Let $(x_n)_{n\in\mathbb{N}}$ be a block sequence in $\mathfrak{X}_{\mathcal{F}_2}$ such that $\|x_n\| \le C$ for all n, $\|x_n\|_\infty \to 0$ and $(x_n)_{n\in\mathbb{N}}$ is a weakly null sequence in $JT_{\mathcal{F}_2}$. It suffices to show that the sequence $(x_n)_{n\in\mathbb{N}}$ is separated (Definition 3.11). Indeed, then Proposition 3.14 and our choice of $j(C)$ yield that for every $j \ge j(C)$ there exists $L \in [\mathbb{N}]$ such that for every $y^* \in \mathcal{F}_2$ we have that $\#\{n \in L : |y^*(x_n)| > \frac{2}{m_{2j}^2}\} \le 66(\frac{m_{2j}^2}{2})^2 C^2 < n_{2j-1}$.

In order to show that the sequence $(x_n)_{n\in\mathbb{N}}$ is separated we start with the following easy observations:

(i) If $m_{4j_0-3}^2 > \frac{C}{\varepsilon} \# \operatorname{supp}(x)$ and $\|x\| \leq C$ then for every $\phi \in \bigcup_{j \geq j_0} F_j$ we

have that $|\phi(x)| \leq \varepsilon$.

(ii) If $\|x\|_\infty < \frac{2\varepsilon}{n_{4j_0-3}}$ and $\phi \in \bigcup_{j \leq j_0} F_j$ then $|\phi(x)| \leq \varepsilon$.

Let $L \in [\mathbb{N}]$ and $\varepsilon > 0$. Using (i) and (ii) we may inductively select $1 = j_0 < j_1 < j_2 < \cdots$ in $\mathbb{N}$ and $k_1 < k_2 < \cdots$ in L such that for each i and $\phi \in \bigcup_{j \notin [j_{i-1}, j_i)} F_j$ we have that $|\phi(x_{k_i})| < \varepsilon$. Setting $M = \{k_1, k_2, \ldots\}$ we have that the sequence $(x_n)_{n \in M}$ is ε-separated. Therefore the sequence $(x_n)_{n \in \mathbb{N}}$ is separated. $\square$

A consequence of the above proposition and the results of Sections 1 and 2 is the following:

Theorem 4.2.   (a) The space $\mathfrak{X}_{\mathcal{F}_2}$ is HI and reflexively saturated.
  (b) The predual $(\mathfrak{X}_{\mathcal{F}_2})_*$ is HI.
  (c) Every bounded linear operator $T : \mathfrak{X}_{\mathcal{F}_2} \to \mathfrak{X}_{\mathcal{F}_2}$ is of the form $T = \lambda I + S$, with S strictly singular and weakly compact.
  (d) Every bounded linear operator $T : (\mathfrak{X}_{\mathcal{F}_2})_* \to (\mathfrak{X}_{\mathcal{F}_2})_*$ is of the form $T = \lambda I + S$, with S strictly singular.

Proof. All the above properties are consequences of the fact that $\mathfrak{X}_{\mathcal{F}_2}$ is a strongly strictly singular extension of $JT_{\mathcal{F}_2}$. In particular (a) follows from Proposition 1.21 and Theorem 1.18, (b) follows from Theorem 2.12, (c) follows from Theorem 2.15 while (d) follows from Theorem 2.16. $\square$

Proposition 4.3. Let $(x_k, x_k^*)_{k=1}^{n_{4j-3}}$ be a $(18, 4j-3, 1)$ attracting sequence in $\mathfrak{X}_{\mathcal{F}_2}$ such that $\|x_{2k-1}\|_{\mathcal{F}_2} \leq \frac{2}{m_{4j-3}^2}$ for $k = 1, \ldots, n_{4j-3}/2$. Then

$$\left\| \frac{1}{n_{4j-3}} \sum_{k=1}^{n_{4j-3}} (-1)^{k+1} x_k \right\| \leq \frac{144}{m_{4j-3}^2}.$$

Proof. The conclusion follows by an application of Proposition 1.17 (ii) after checking that for every $g \in \mathcal{F}_2$ we have that $|g(x_{2k})| > \frac{2}{m_{4j-3}^2}$ for at most n_{4j-4} k's. From the fact that each x_{2k} is of the form e_l it suffices to show that for every $g \in \mathcal{F}_2$, the cardinality of the set $\{l : |g(e_l)| > \frac{2}{m_{4j-3}^2}\}$ is at most n_{4j-4}.

Let $g \in \mathcal{F}_2$, $g = \sum_{i=1}^{d} a_i g_i$ where $\sum_{i=1}^{d} a_i^2 \leq 1$ and $(g_i)_{i=1}^{d}$ are $\sigma_{\mathcal{F}}$ special functionals with disjoint indices. For each i we divide the functional g_i into two parts, $g_i = y_i^* + z_i^*$, with $\operatorname{ind}(y_i^*) \subset \{1, \ldots, j-1\}$ and $\operatorname{ind}(z_i^*) \subset \{j, j+1, \ldots\}$. For $l \notin \bigcup_{i=1}^{d} \operatorname{supp}(y_i^*)$ we have that $|g(e_l)| \leq \sum_{i=1}^{d} |z_i^*(e_l)| < \sum_{r=j}^{\infty} \frac{1}{m_{4r-3}^2} < \frac{2}{m_{4j-3}^2}$.

Since $\#\left(\bigcup_{i=1}^{d} \operatorname{supp}(y_i^*) \right) \leq \frac{n_1}{2} + \frac{n_5}{2} + \cdots \frac{n_{4j-7}}{2} < n_{4j-5}$ the conclusion follows.

The proof of the proposition is complete. $\square$

Definition 4.4. Let $\chi = (x_k, x_k^*)_{k=1}^{n_{4j-3}}$ be a $(18, 4j-3, 1)$ attracting sequence, with $\|x_{2k-1}\|_{\mathcal{F}_2} \leq \frac{2}{m_{4j-3}^2}$ for $1 \leq k \leq n_{4j-3}/2$. We set

$$g_\chi = \frac{1}{m_{4j-3}^2} \sum_{k=1}^{n_{4j-3}/2} x_{2k}^*$$

$$F_\chi = -\frac{1}{m_{4j-3}^2} \sum_{k=1}^{n_{4j-3}/2} x_{2k-1}^*$$

$$d_\chi = \frac{m_{4j-3}^2}{n_{4j-3}} \sum_{k=1}^{n_{4j-3}} (-1)^k x_k$$

Lemma 4.5. If χ is a $(18, 4j-3, 1)$ attracting sequence, $\chi = (x_k, x_k^*)_{k=1}^{n_{4j-3}}$, with $\|x_{2k-1}\|_{\mathcal{F}_2} \leq \frac{2}{m_{4j-3}^2}$ for $1 \leq k \leq n_{4j-3}/2$ then

(1) $\|g_\chi - F_\chi\| \leq \frac{1}{m_{4j-3}}$.

(2) $\frac{1}{2} = g_\chi(d_\chi) \leq \|d_\chi\| \leq 144$, and hence $\|g_\chi\| \geq \frac{1}{288}$.

Proof. (1) We have $g_\chi - F_\chi = \frac{1}{m_{4j-3}}\left(\frac{1}{m_{4j-3}} \sum_{k=1}^{n_{4j-3}} x_k^*\right)$. Since $(x_k^*)_{k=1}^{n_{4j-3}}$ is a

special sequence of length n_{4j-3}, the functional $\frac{1}{m_{4j-3}} \sum_{k=1}^{n_{4j-3}} x_k^*$ belongs to D_G

and hence to $B_{\mathfrak{X}_{\mathcal{F}_2}^*}$ The conclusion follows.

(2) It is straightforward from Definitions 1.15 and 4.4 that $g_\chi(d_\chi) = \frac{1}{2}$. Since $\|x_{2k-1}\|_{\mathcal{F}_2} \leq \frac{2}{m_{4j-3}^2}$ for $k = 1, \ldots, n_{4j-3}/2$, Proposition 4.3 yields that $\|d_\chi\| \leq 144$. Thus $\|g_\chi\| \geq \frac{g_\chi(d_\chi)}{\|d_\chi\|} \geq \frac{\frac{1}{2}}{144} = \frac{1}{288}$. $\square$

Lemma 4.6. Let Z be a block subspace of $(\mathfrak{X}_{\mathcal{F}_2})_*$. Also, let $\varepsilon > 0$ and $j > 1$. There exists a $(18, 4j-3, 1)$ attracting sequence $\chi = (x_k, x_k^*)_{k=1}^{n_{4j-3}}$ with $\sum_{k=1}^{n_{4j-3}/2} \|x_{2k-1}\|_{\mathcal{F}_2} < \frac{1}{n_{4j-3}^2}$ and $\text{dist}(F_\chi, Z) < \varepsilon$.

Proof. We select an integer j_1 such that $m_{2j_1}^{\frac{1}{2}} > n_{4j-3}$. From Lemma 2.10 we may select a $(18, 2j_1, 1)$ exact pair (x_1, x_1^*) with $\text{dist}(x_1^*, Z) < \frac{\varepsilon}{n_{4j-3}}$ and $\|x_1\|_{\mathcal{F}_2} \leq \frac{2}{m_{2j_1}}$. Let $2j_2 = \sigma(x_1^*)$. We select $l_2 \in \Lambda_{2j_2}$ and we set $x_2 = e_{l_2}$ and $x_2^* = e_{l_2}^*$.

We then set $2j_3 = \sigma(x_1^*, x_2^*)$ and we select, using Lemma 2.10, a $(18, 2j_3, 1)$ exact pair (x_3, x_3^*) with $x_2 < x_3$, $\text{dist}(x_3^*, Z) < \frac{\varepsilon}{n_{4j-3}}$ and $\|x_3\|_{\mathcal{F}_2} \leq \frac{2}{m_{2j_3}}$.

It is clear that we may inductively construct a $(18, 4j-3, 1)$ attracting sequence $\chi = (x_k, x_k^*)_{k=1}^{n_{4j-3}}$ such that $\sum_{k=1}^{n_{4j-3}/2} \|x_{2k-1}\|_{\mathcal{F}_2} \leq \sum_{k=1}^{n_{4j-3}/2} \frac{2}{m_{2j_{2k-1}}} < \frac{1}{n_{4j-3}^2}$ and $\text{dist}(x_{2k-1}^*, Z) < \frac{\varepsilon}{n_{4j-3}}$ for $1 \leq k \leq n_{4j-3}/2$. It follows that

$$\text{dist}(F_\chi, Z) \leq \frac{1}{m_{4j-3}^2} \sum_{k=1}^{n_{4j-3}/2} \text{dist}(x_{2k-1}^*, Z) < \varepsilon. \qquad \square$$

Theorem 4.7. The space $(\mathfrak{X}_{\mathcal{F}_2})_*$ is a Hereditarily James Tree (HJT) space. In particular it does not contain any reflexive subspace and every infinite dimensional subspace Z of $(\mathfrak{X}_{\mathcal{F}_2})_*$ has nonseparable second dual Z^{**}.

Proof. Since each subspace of $(\mathfrak{X}_{\mathcal{F}_2})_*$ has a further subspace isomorphic to a block subspace it is enough to consider a block subspace Z of $(\mathfrak{X}_{\mathcal{F}_2})_*$ and to show that Z has the James Tree property.

We select a $j_\emptyset \in \Xi_1$ with $j_\emptyset \geq 2$. We shall inductively construct a family $(\chi_a)_{a \in \mathcal{D}}$ of attracting sequences and a family $(j_a)_{a \in \mathcal{D}}$ of integers such that

- (i) If $a <_{lex} \beta$ then $d_{\chi_a} < d_{\chi_\beta}$.
- (ii) For every $\beta \in \mathcal{D}$, $\chi_\beta = \left(x_k^\beta, (x_k^\beta)^*\right)_{k=1}^{n_{4j_\beta-3}}$ is a $(18, 4j_\beta - 3, 1)$ attracting sequence with $\text{dist}(F_{\chi_\beta}, Z) < \frac{1}{m_{4j_\beta-3}}$ and $\|x_{2k-1}^\beta\|_{\mathcal{F}_2} \leq \frac{2}{m_{4j_\beta-3}^2}$ for $k = 1, \ldots, n_{4j_\beta-3}/2$.
- (iii) If $\beta \in \mathcal{D}$ with $\beta \neq \emptyset$ then $j_\beta = \sigma_{\mathcal{F}}((g_{\chi_a})_{a<\beta})$.

The induction runs on the lexicographical ordering of $\mathcal{D}$. In the first step, i.e. for $\beta = \emptyset$, we select a $(18, 2j_\emptyset - 1, 1)$ attracting sequence with $\text{dist}(F_{\chi_\emptyset}, Z) < \frac{1}{m_{4j_\emptyset-3}}$ and $\|x_{2k-1}^\emptyset\|_{\mathcal{F}_2} \leq \frac{2}{m_{4j_\emptyset-3}^2}$ for $k = 1, \ldots, n_{4j_\emptyset-3}/2$. In the general inductive step, we assume that $(j_a)_{a<_{lex}\beta}$ and $(\chi_a)_{a<_{lex}\beta}$ have been constructed for some $\beta \in \mathcal{D}$. Since $\{a \in \mathcal{D} : a < \beta\} \subset \{a \in \mathcal{D} : a <_{lex} \beta\}$, the attracting sequences $(\chi_a)_{a<\beta}$ have already been constructed so we may set $j_\beta = \sigma_{\mathcal{F}}((g_{\chi_a})_{a<\beta})$. Denoting by β^- the immediate predecessor of β in the lexicographical ordering, we select, using Lemma 4.6, a $(18, 2j_\beta - 1, 1)$ attracting sequence $\chi_\beta = \left(x_k^\beta, (x_k^\beta)^*\right)_{k=1}^{n_{4j_\beta-3}}$ with $d_{\chi_{\beta^-}} < d_{\chi_\beta}$ such that $\text{dist}(F_{\chi_\beta}, Z) < \frac{1}{m_{4j_\beta-3}}$ and $\|x_{2k-1}^\beta\|_{\mathcal{F}_2} \leq \frac{2}{m_{4j_\beta-3}^2}$ for $1 \leq k \leq n_{4j_\beta-3}/2$. The inductive construction is complete.

For each branch b of the dyadic tree the sequence $(g_{\chi_a})_{a \in b}$ is a $\sigma_{\mathcal{F}}$ special sequence. Thus the series $\sum_{a \in \mathcal{D}} g_{\chi_a}$ converges in the w^* topology to a $\sigma_{\mathcal{F}}$ special functional $g_b \in \overline{G}^{w^*} \subset \overline{D_G}^{w^*} = B_{\mathfrak{X}_{\mathcal{F}_2}^*}$.

For each $\beta \in \mathcal{D}$ we select a $z_\beta^* \in Z$ such that $\|z_\beta^* - F_{\chi_\beta}\| < \frac{1}{m_{4j_\beta-3}}$. Then Lemma 4.5 (1) yields that $\|z_\beta^* - g_{\chi_\beta}\| \leq \|z_\beta^* - F_{\chi_\beta}\| + \|F_{\chi_\beta} - g_{\chi_\beta}\| < \frac{2}{m_{4j_\beta-3}}$. Now let b be a branch of the dyadic tree. Since $\sum_{a \in b} \|z_a^* - g_{\chi_a}\| < \sum_{a \in b} \frac{2}{m_{4j_a-3}} < \frac{3}{m_{4j_\emptyset-3}} < \frac{1}{1152}$ it follows that the series $\sum_{a \in \mathcal{D}} z_a^*$ is also w^* convergent and its w^* limit $z_b^* \in Z^{**}$ satisfies $\|z_b^* - g_b\| < \frac{1}{1152}$. This actually yields that the block sequence $(z_a^*)_{a \in \mathcal{D}}$ defines a James Tree structure in the subspace Z.

The family $\{z_b^* : b$ a branch of $\mathcal{D}\}$ is a family in Z^{**} with the cardinality of the continuum. We complete the proof of the theorem by showing that for $b \neq b'$ we have that $\|z_b^* - z_{b'}^*\| \geq \frac{1}{576}$. Let $b \neq b'$ be two branches of the dyadic tree. We select $a \in \mathcal{D}$ with $a \in b \setminus b'$ (i.e. a is an initial part of b but not of

b'). Then our construction and Lemma 4.5 (2) yield that

$$\|z_b^* - z_{b'}^*\| \geq \|g_b - g_{b'}\| - \|z_b^* - g_b\| - \|z_{b'}^* - g_{b'}\|$$

$$> \frac{(g_b - g_{b'})(d_{\chi_a})}{\|d_{\chi_a}\|} - \frac{1}{1152} - \frac{1}{1152}$$

$$\geq \frac{g_{\chi_a}(d_{\chi_a})}{144} - \frac{1}{576} = \frac{\frac{1}{2}}{144} - \frac{1}{576} = \frac{1}{576}.$$

$\square$

Proposition 4.8. For every block subspace $Y = \overline{\mathrm{span}}\{y_n : n \in \mathbb{N}\}$ of $\mathfrak{X}_{\mathcal{F}_2}$ there exist a further block subspace $Y' = \overline{\mathrm{span}}\{y_n' : n \in \mathbb{N}\}$ and a block subspace $Z = \overline{\mathrm{span}}\{z_k : k \in \mathbb{N}\}$ of $\mathfrak{X}_{\mathcal{F}_2}$ such that the following are satisfied. The space Z is reflexive, the spaces Y' and Z are disjointly supported (i.e. $\mathrm{supp}\, z_k \cap \mathrm{supp}\, y_n' = \emptyset$ for all n, k) and the space $X = \overline{\mathrm{span}}(\{z_k : k \in \mathbb{N}\} \cup \{y_n' : n \in \mathbb{N}\})$ has nonseparable dual.

Proof. The proof is similar to that of Theorem 4.7. Let Y be a block subspace of $\mathfrak{X}_{\mathcal{F}_2}$. Using Proposition 1.14 we may inductively construct (the induction runs on the lexicographic order of the dyadic tree $\mathcal{D}$) a family $(\chi_a)_{a \in \mathcal{D}}$ of attracting sequences and a family $(j_a)_{a \in \mathcal{D}}$ of integers such that the following conditions are satisfied:

  (i) If $a <_{lex} \beta$ then $d_{\chi_a} < d_{\chi_\beta}$.
  (ii) For every $\beta \in \mathcal{D}$, $\chi_\beta = \big(x_k^\beta, (x_k^\beta)^*\big)_{k=1}^{n_{4j_\beta}-3}$ is a $(18, 2j_\beta - 1, 1)$ attracting sequence with $x_{2k-1}^\beta \in Y$ and $\|x_{2k-1}^\beta\|_{\mathcal{F}_2} \leq \frac{2}{m_{4j_\beta-3}}$ for $k = 1, \dots, n_{4j_\beta}-3/2$.
  (iii) $j_\emptyset \in \Xi_1$ with $j_\emptyset \geq 2$, while if $\beta \in \mathcal{D}$ with $\beta \neq \emptyset$, then $j_\beta = \sigma_{\mathcal{F}}((g_{\chi_\emptyset})_{a<\beta})$.

For each $a \in \mathcal{D}$ we set $z_a = \frac{2m_{4j_a-3}}{n_{4j_a-3}} \sum\limits_{k=1}^{n_{4j_a}-3/2} x_{2k}^a$ and we consider the space $Z = \overline{\mathrm{span}}\{z_a : a \in \mathcal{D}\}$.

We first observe that for each $a \in \mathcal{D}$ the functional $f_a = \frac{1}{m_{4j_a-3}} \sum\limits_{k=1}^{n_{4j_a}-3} (x_k^a)^*$ belongs to $D_G \subset B_{\mathfrak{X}_{\mathcal{F}_2}}$, hence $\|z_a\| \geq f_a(z_a) = 1$. On the other hand we have that $\|z_a\|_{\mathcal{F}_2} \leq \frac{2}{m_{4j_a-3}}$. Indeed, let $g = \sum\limits_{i=1}^{d} a_i g_i \in \mathcal{F}_2$ (i.e. $\sum\limits_{i=1}^{d} a_i^2 \leq 1$ while $(g_i)_{i=1}^{d}$ and $\sigma_{\mathcal{F}}$ special functionals with pairwise disjoint indices). For each $i = 1, \dots, d$ let $g_i = y_i^* + z_i^*$ with $\mathrm{ind}(y_i^*) \subset \{1, \dots, j_a - 1\}$ and $\mathrm{ind}(z_i^*) \subset$

$\{j_a, j_a + 1, \dots\}$. Then

$$
\begin{aligned}
|g(z_a)| &\leq \sum_{i=1}^{d} |y_i^*(z_a)| + \sum_{i=1}^{d} |z_i^*(z_a)| \\
&\leq \frac{2m_{4j_a-3}}{n_{4j_a-3}} \left(\frac{n_1}{2} + \dots + \frac{n_{4j_a-7}}{2} \right) + \frac{2m_{4j_a-3}}{n_{4j_a-3}} \sum_{r=j_a}^{\infty} \frac{n_{4j_a-3}}{2} \frac{1}{m_{4r-3}^2} \\
&\leq \frac{2}{m_{4j_a-3}}.
\end{aligned}
$$

It follows that $\sum_{a \in \mathcal{D}} \frac{\|z_a\|_{\mathcal{F}_2}}{\|z_a\|} \leq \sum_{a \in \mathcal{D}} \frac{2}{m_{4j_a-3}} < \frac{1}{2}$ which yields that the space $Z = \overline{\text{span}}\{z_a : a \in \mathcal{D}\}$ is reflexive (see Proposition 1.21).

For every branch b of the dyadic tree, the functional g_b which is defined to be the w^* sum of the series $\sum_{a \in \mathcal{D}} g_{\chi_a}$ belongs to $\overline{\mathcal{F}_2}^{w^*} \subset B_{\mathfrak{X}_{\mathcal{F}_2}}$. The family $\{g_b|_X :$ b a branch of $\mathcal{D}\}$ is a family of X^* with the cardinality of the continuum. For $b \neq b'$, selecting $a \in b \setminus b'$ the vector $d_{\chi_a} = \frac{m_{4j_a-3}^2}{n_{4j_a-3}} \sum_{k=1}^{n_{4j_a-3}} (-1)^k x_k^a$ belongs to $Y + Z$ while from Lemma 4.5 we have that $\|d_{\chi_a}\| \leq 144$. Thus $\|g_b|_X - g_{b'}|_X\|_{X^*} \geq \frac{g_b(d_{\chi_a}) - g_{b'}(d_{\chi_a})}{\|d_{\chi_a}\|} \geq \frac{\frac{1}{2} - 0}{144} = \frac{1}{288}$.

Therefore X^* is nonseparable. $\qquad\square$

Lemma 4.9. If $S : (\mathfrak{X}_{\mathcal{F}_2})_* \to (\mathfrak{X}_{\mathcal{F}_2})_*$ is a strictly singular operator then its conjugate operator $S^* : \mathfrak{X}_{\mathcal{F}_2} \to \mathfrak{X}_{\mathcal{F}_2}$ is also strictly singular.

Proof. From Theorem 2.15 the operator S^* takes the form $S^* = \lambda I_{\mathfrak{X}_{\mathcal{F}_2}} + W$ with $\lambda \in \mathbb{R}$ and $W : \mathfrak{X}_{\mathcal{F}_2} \to \mathfrak{X}_{\mathcal{F}_2}$ a strictly singular and weakly compact operator. We have to show that $\lambda = 0$.

The operator $W^* : \mathfrak{X}_{\mathcal{F}_2}^* \to \mathfrak{X}_{\mathcal{F}_2}^*$ is also weakly compact, while $W^* = S^{**} - \lambda I_{\mathfrak{X}_{\mathcal{F}_2}^*}$ which yields that $W^*((\mathfrak{X}_{\mathcal{F}_2})_*) \subset (\mathfrak{X}_{\mathcal{F}_2})_*$. These facts, in conjunction to the fact that $(\mathfrak{X}_{\mathcal{F}_2})_*$ contains no reflexive subspace (Theorem 4.7), imply that the restriction $W^*|_{(\mathfrak{X}_{\mathcal{F}_2})_*}$ is strictly singular. Thus, since $\lambda I_{(\mathfrak{X}_{\mathcal{F}_2})_*} = S - W^*|_{(\mathfrak{X}_{\mathcal{F}_2})_*}$ with both $S, W^*|_{(\mathfrak{X}_{\mathcal{F}_2})_*}$ being strictly singular, we get that $\lambda = 0$. $\qquad\square$

Corollary 4.10. Every bounded linear operator $T : (\mathfrak{X}_{\mathcal{F}_2})_* \to (\mathfrak{X}_{\mathcal{F}_2})_*$ takes the form $T = \lambda I + W$ with $\lambda \in \mathbb{R}$ and W a weakly compact operator.

Proof. We know from Theorem 2.16 that $T = \lambda I + W$ with W a strictly singular operator. Lemma 4.9 yields that W^* is also strictly singular. From Theorem 2.15 we get that W^* is weakly compact, hence W is weakly compact. $\qquad\square$

Theorem 4.11. Let Z be a w^* closed subspace of $\mathfrak{X}_{\mathcal{F}_2}$ of infinite codimension such that for every $i = 1, 2, \dots$ we have that

(7) $$\liminf_{k \in \Lambda_i} \text{dist}(e_k, Z) = 0$$

$((\Lambda_i)_{i\in\mathbb{N}}$ are the sets appearing in Definition 1.3). Then every infinite dimensional subspace of $\mathfrak{X}_{\mathcal{F}_2}/Z$ has nonseparable dual.

Proof. We denote by Q the quotient operator $Q : \mathfrak{X}_{\mathcal{F}_2} \to \mathfrak{X}_{\mathcal{F}_2}/Z$ and we recall that since Z is w^* closed, $Z_\perp$ 1-norms $\mathfrak{X}_{\mathcal{F}_2}/Z$. Let Y be a closed subspace of $\mathfrak{X}_{\mathcal{F}_2}$ with $Z \subset Y$ such that Y/Z is infinite dimensional; we shall show that $(Y/Z)^*$ is nonseparable.

For a given $j \in \mathbb{N}$ using Lemma 2.20 and our assumption (7) are able to construct a $(18, 4j - 3, 1)$ attracting sequence $\chi = (x_k, x_k^*)_{k=1}^{n_{4j-3}}$ such that each one of the sums $\sum \text{dist}(x_{2k-1}, Y)$, $\sum \text{dist}(x_{2k-1}^*, Z_\perp)$, $\sum \|x_{2k-1}\|_{\mathcal{F}_2}$, $\sum \text{dist}(x_{2k}, Z)$ is as small as we wish. Setting $d_\chi^1 = \dfrac{m_{4j-3}^2}{n_{4j-3}} \displaystyle\sum_{k=1}^{n_{4j-3}/2} x_{2k-1}$ we get that Qd_χ is almost equal to Qd_χ^1 which almost belongs to Y/Z. Also F_χ almost belongs to $Z_\perp$, while $F_\chi(d_\chi) = \frac{1}{2}$ and $\|F_\chi - g_\chi\| \le \frac{1}{m_{4j-3}}$.

Using these estimates we are able to construct a dyadic tree $(\chi_a)_{a\in\mathcal{D}}$ of attracting sequences and a family $(j_a)_{a\in\mathcal{D}}$ of integers satisfying

(i) If $a <_{lex} \beta$ then $d_{\chi_a} < d_{\chi_\beta}$.

(ii) For every $\beta \in \mathcal{D}$, $\chi_\beta = \left(x_k^\beta, (x_k^\beta)^*\right)_{k=1}^{n_{4j_\beta-3}}$ is a $(18, 4j_\beta - 3, 1)$ attracting sequence with $\text{dist}(F_{\chi_\beta}, Z_\perp) < \frac{1}{m_{4j_\beta-3}}$, $\text{dist}(Qd_{\chi_\beta}, Y/Z) < \frac{1}{m_{4j_\beta-3}}$ and $\|x_{2k-1}^\beta\|_{\mathcal{F}_2} \le \frac{2}{m_{4j_\beta-3}^2}$ for $k = 1, \ldots, n_{4j_\beta-3}/2$.

(iii) If $\beta \in \mathcal{D}$ with $\beta \ne \emptyset$ then $j_\beta = \sigma_{\mathcal{F}}((g_{\chi_a})_{a<\beta})$, while $j_\emptyset \in \Xi_1$ with $j_\emptyset \ge 3$.

For every $\beta \in \mathcal{D}$ we select $H_\beta \in Z_\perp$ with $\|H_\beta - F_{\chi_\beta}\| < \frac{1}{m_{4j_\beta-3}}$ and then for every branch b of $\mathcal{D}$ we denote by h_b the w^* limit of the series $\displaystyle\sum_{\beta\in b} H_\beta$.

Using the above estimates, and arguing similarly to the proof of Theorem 4.7, we obtain that $\{h_b|Y : b$ is a branch of $\mathcal{D}\}$ is a discrete family in $(Y/Z)^*$ and therefore $(Y/Z)^*$ is nonseparable. $\qquad\square$

Remark 4.12. Actually it can be shown that the space $\mathfrak{X}_{\mathcal{F}_2}/Z$ is HJT.

Corollary 4.13. There exists a partition of the basis $(e_n^*)_{n\in\mathbb{N}}$ of $(\mathfrak{X}_{\mathcal{F}_2})_*$ into two sets $(e_n^*)_{n\in L_1}$, $(e_n^*)_{n\in L_2}$ such that setting $X_{L_1} = \overline{\text{span}}\{e_n^* : n \in L_1\}$, $X_{L_2} = \overline{\text{span}}\{e_n^* : n \in L_2\}$ both $X_{L_1}^*$, $X_{L_2}^*$ are HI with no reflexive subspace.

Proof. We choose $L_1 \in [\mathbb{N}]$ such that the sets $\Lambda_i \cap L_1$ and $\Lambda_i \setminus L_1$ are infinite for each i and we set $L_2 = \mathbb{N} \setminus L_1$. The spaces $X_{L_i} = \overline{\text{span}}\{e_n^* : n \in L_i\}$, $i = 1, 2$ satisfy the desired properties. Indeed, since $X_{L_1}^*$ is isometric to $\mathfrak{X}_{\mathcal{F}_2}/\overline{\text{span}}\{e_n : n \in L_2\}$, Theorem 4.11 yields that $X_{L_1}^*$ has no reflexive subspace while from Theorem 2.21 we get that it is an HI space. For $X_{L_2}^*$ the proof is completely analogous. $\qquad\square$

5. THE STRUCTURE OF $\mathfrak{X}_{\mathcal{F}_2}^*$ AND A VARIANT OF $\mathfrak{X}_{\mathcal{F}_2}$

In the present section the structure of $\mathfrak{X}_{\mathcal{F}_2}^*$ is studied. This space is not HI since for every subspace Y of $(\mathfrak{X}_{\mathcal{F}_2})_*$ the space ℓ_2 embeds into Y^{**}. We also present a variant of $\mathfrak{X}_{\mathcal{F}_2}$, denoted $\mathfrak{X}_{\mathcal{F}_2'}$, such that $\mathfrak{X}_{\mathcal{F}_2'}^*/(\mathfrak{X}_{\mathcal{F}_2'})_*$ is isomorphic to $\ell_2(\Gamma)$ which yields some peculiar results on the structure of $\mathfrak{X}_{\mathcal{F}_2'}$ and $\mathfrak{X}_{\mathcal{F}_2'}^*$. Another variant of $\mathfrak{X}_{\mathcal{F}_2}$ yielding a HI dual not containing reflexive subspace is also discussed. It is well known that, in JT (James tree space) the quotient space JT^*/JT_* is isometric to $\ell_2(\Gamma)$. It seems unlikely to have the same property for $\mathfrak{X}_{\mathcal{F}_2}$. The main difficulty concerns the absence of biorthogonality between disjoint $\sigma_{\mathcal{F}}$ special functionals. However the next Proposition indicates that in some cases phenomena analogous to those in JT also occur.

Proposition 5.1. Let $(b_n^*)_n$ be a disjoint family of $\sigma_{\mathcal{F}}$ special functionals each one defined by an infinite special sequence $b_n = (f_1^n, \ldots, f_k^n, \ldots)$. Assume furthermore that for each (n,k) there exists a $(18, 4j_{(n,k)} - 3, 1)$ attracting sequence $\chi_{(n,k)} = (x_\ell^{(n,k)}, (x_\ell^{(n,k)})^*)_{\ell=1}^{n_{4j_{(n,k)}-3}}$, (Definition 1.15) with $\|x_{2\ell-1}^{(n,k)}\|_{\mathcal{F}_2} < \frac{1}{n_{4j_{(n,k)}-3}^2}$ and $f_n^k = g_{\chi_{(n,k)}}$, (Definition 4.4).

Then $(b_n^*)_n$ is equivalent to the standard ℓ_2-basis.

Let's provide a short description of the proof. We start with the following lemma:

Lemma 5.2. Let $\chi = (x_k, x_k^*)_{k=1}^{n_{4j-3}}$ be a $(18, 4j - 3, 1)$ attracting sequence such that $\|x_{2k-1}\|_{\mathcal{F}_2} < \frac{1}{n_{4j-3}^2}$ for all k. Then for every $\phi \in \mathcal{F}_2$ of the form

$$\phi = \sum_{i=1}^{d} a_i \phi_i \text{ with } j \notin \cup_{i=1}^{d} \text{ind}(\phi_i) \text{ we have that } |\phi(d_\chi)| < \frac{1}{m_{4j-3}}. \text{ (Recall that}$$

$d_\chi = \frac{m_{4j-3}^2}{n_{4j-3}} \sum_{k=1}^{n_{4j-3}} (-1)^k x_k$, see Definition 4.4).

Proof. We set $d_\chi^1 = \frac{m_{4j-3}^2}{n_{4j-3}} \sum_{k=1}^{n_{4j-3}/2} x_{2k-1}$ and $d_\chi^2 = \frac{m_{4j-3}^2}{n_{4j-3}} \sum_{k=1}^{n_{4j-3}/2} x_{2k}$. From our assumption that $\|x_{2k-1}\|_{\mathcal{F}_2} < \frac{1}{n_{4j-3}^2}$ for every k, we get that $|\phi(d_\chi^1)| \leq \frac{m_{4j-3}^2}{n_{4j-3}} \frac{n_{4j-3}}{2} \frac{1}{n_{4j-3}^2}$.

If $f \in F_i$ for some $i < j$ we have that $|f(d_\chi^2)| \leq \frac{1}{m_{4i-3}^2} \frac{m_{4j-3}^2}{n_{4j-3}} \frac{n_{4i-3}}{2}$, while for $f \in F_i$ with $i > j$ we have that $|f(d_\chi^2)| \leq \frac{1}{m_{4i-3}^2} \frac{m_{4j-3}^2}{n_{4j-3}} \frac{n_{4j-3}}{2}$.

Therefore

$$
\begin{aligned}
|\phi(d_\chi)| &\leq |\phi(d_\chi^1)| + |\sum_{i=1}^{d} a_i \phi_i(d_\chi^2)| \leq |\phi(d_\chi^1)| + \sum_{i=1}^{d} |\phi_i(d_\chi^2)| \\
&\leq |\phi(d_\chi^1)| + \sum_{i<j} \sup\{|f(d_\chi^2)| : f \in F_i\} + \sum_{i>j} \sup\{|f(d_\chi^2)| : f \in F_i\} \\
&\leq \frac{m_{4j-3}^2}{2n_{4j-3}^2} + \frac{m_{4j-3}^2}{n_{4j-3}} \sum_{i<j} \frac{n_{4i-3}}{2m_{4i-3}^2} + \frac{m_{4j-3}^2}{2} \sum_{i>j} \frac{1}{m_{4i-3}^2} \\
&< \frac{1}{m_{4j-3}}.
\end{aligned}
$$

$\square$

The content of the above lemma is that each b^*, defined by an infinite $\sigma_{\mathcal{F}}$ special sequence b as in the previous proposition, is almost biorthogonal to any other $(b')^*$ which is disjoint from b.

Next we describe the main steps in the proof of Proposition 5.1.

Proof of Proposition 5.1: The proof follows the main lines of the proof of Lemma 11.3 of [AT1]. Given $(a_i)_{i=1}^d$ with $a_i \in \mathbb{Q}$ such that $\sum_{i=1}^d a_i^2 = 1$, we have that $\sum_{i=1}^d a_i b_i^* \in \mathcal{F}_2$ hence $\| \sum_{i=1}^d a_i b_i^* \| \le 1$.

In order to complete the proof we shall show that

(8)
$$\frac{1}{1000} \le \| \sum_{i=1}^d a_i b_i^* \|$$

which yields the desired result.

To establish (8), we choose $k \in \mathbb{N}$ with $\frac{(5n_{2k}-1)^{\log_2(m_{2k})}}{n_{2k}} < \frac{\varepsilon}{4d}$ and then we choose $\{l_t^i : 1 \le i \le d, \ 1 \le t \le n_{2k}\}$ such that setting $x_{(t,i)} = d_{\chi_{(i,l_t^i)}}$ the following conditions are satisfied. First, the sequence $(x_{(t,i)})_{1 \le i \le d, \ 1 \le t \le n_{2k}}$ ordered lexicographically (i.e. $(t,i) <_{lex} (t',i')$ iff $t < t'$ or $t = t'$ and $i < i'$) is a $(144, \varepsilon)$ R.I.S. with associated sequence $4j'_{(t,i)} - 3 := 4j_{(i,l_t^i)} - 3$ while $m_{4j'_{(1,1)}-3} > \frac{2dn_{2k}}{\varepsilon}$.

We set $z_i = \frac{1}{n_{2k}} \sum_{t=1}^{n_{2k}} x_{(t,i)}$ for $i = 1, \ldots, d$. In order to prove (8) it is enough to show that

(i) $(\sum_{r=1}^d a_r b_r^*)(\sum_{i=1}^d a_i z_i) > \frac{1}{2} - \varepsilon$.

(ii) $\| \sum_{i=1}^d a_i z_i \| \le 288$.

(i) is an easy consequence of Lemma 5.2. Indeed

$$
\begin{aligned}
(\sum_{r=1}^d a_r b_r^*)(\sum_{i=1}^d a_i z_i) &= \sum_{r=1}^d a_r^2 b_r^*(z_r) + \sum_{i=1}^d \sum_{r \ne i} a_r b_r^*(z_i) \\
&\ge \frac{1}{2} - \frac{1}{n_{2k}} \sum_{i=1}^d |a_i| \cdot |(\sum_{r \ne i} a_r b_r^*)(\sum_{t=1}^{n_{2k}} x_{(t,i)})| \\
&\ge \frac{1}{2} - \frac{1}{n_{2k}} \sum_{i=1}^d |a_i|(\sum_{t=1}^{n_{2k}} \frac{1}{4m_{4j'_{(t,i)}-3}}) \\
&\ge \frac{1}{2} - \frac{1}{n_{2k}} \frac{2d}{m_{4j'_{(1,1)}-3}} > \frac{1}{2} - \varepsilon.
\end{aligned}
$$

For each (t,i) we set $k_{(t,i)} = \min \operatorname{supp} x_{(t,i)}$ we set and $s_i = \{k_{(t,i)} : t = 1, 2, \ldots, n_{2k}\}$. We consider the set

$$\mathcal{H}_2 = \{e_n^* : n \in \mathbb{N}\} \cup \{\sum_{i=1}^{d} \sum_{j} \lambda_{i,j} s_{i,j}^* : \lambda_{i,j} \in \mathbb{Q}, \sum_{i=1}^{d} \sum_{j} \lambda_{i,j}^2 \leq 1, \text{ where}$$

$$(s_{i,j})_j \text{ are disjoint subintervals of } s_i\}$$

and the norming set D' of space $T[\mathcal{H}_2, (\mathcal{A}_{5n_j}, \frac{1}{m_j})_{j \in \mathbb{N}}]$.

We also set $\tilde{z}_i = \frac{1}{n_{2k}} \sum_{t=1}^{n_{2k}} e_{k_{(t,i)}}^*$ for $i = 1, \ldots, d$.

Claim. For every $f \in D_{\mathcal{F}_2}$ (where $D_{\mathcal{F}_2}$ is the norming set of the space $\mathfrak{X}_{\mathcal{F}_2}$) there exist an $h \in D'$ with nonnegative coordinates such that $|f(\sum_{i=1}^{d} a_i z_i)| \leq 288 h(\sum_{i=1}^{d} |a_i| \tilde{z}_i) + \varepsilon$.

The proof of the above claim is obtained using similar methods to the proof of the basic inequality (Proposition A.5).

Arguing in a similar manner to the corresponding part of Lemma 11.3 of [AT1] we shall show that $h(\sum_{i=1}^{d} |a_i| \tilde{z}_i) \leq 1 + \varepsilon$. We may assume that the functional h admits a tree $T_h = (h_a)_{a \in \mathcal{A}}$ (see Definition A.1) such that each h_a is either of type 0 (then $h_a \in \mathcal{H}_2$) or of type I, and moreover that the coordinates of each h_a are nonnegative. Let $(g_{a_s})_{s=1}^{s_0}$ be the functionals corresponding to the maximal elements of the tree $\mathcal{A}$. We denote by $\preceq$ the ordering of the tree $\mathcal{A}$. Let

$$A = \left\{ s \in \{1, 2, \ldots, s_0\} : \prod_{\gamma \prec a_s} \frac{1}{w(h_\gamma)} \leq \frac{1}{m_{2k}} \right\}$$

$$B = \{1, 2, \ldots, s_0\} \setminus A$$

and set $h_A = h|_{\bigcup_{s \in A} \operatorname{supp} g_{a_s}}$, $h_B = h|_{\bigcup_{s \in B} \operatorname{supp} g_{a_s}}$.

We have that $h_A(\tilde{z}_i) \leq \frac{1}{m_j}$ for each i thus

$$(9) \qquad h_A(\sum_{i=1}^{d} |a_i| \tilde{z}_i) \leq \frac{1}{m_{2k}} \sum_{i=1}^{d} |a_i| \leq \frac{d}{m_j} < \frac{\varepsilon}{2}.$$

It remains to estimate the value $h_B(\sum_{i=1}^{d} |a_i| \tilde{z}_i)$. We observe that

$$\sum_{i=1}^{d} |a_i| \tilde{z}_i = \sum_{t=1}^{n_{2k}} \frac{1}{n_{2k}} (\sum_{i=1}^{d} |a_i| e_{k_{(t,i)}}).$$

We set

$$E_1 = \Big\{ t \in \{1, 2, \ldots, n_{2k}\} : \text{ the set } \{k_{(t,1)}, k_{(t,2)}, \ldots, k_{(t,d)}\} \text{ is contained}$$

$$\text{in } \operatorname{ran} g_{a_s} \text{ for some } s \in B \text{ or does not intersect any } \operatorname{ran} g_{a_s}, \ s \in B \Big\}$$

$$E_2 = \{1, 2, \ldots, n_{2k}\} \setminus E_1.$$

For each $s = 1, 2, \ldots, s_0$ set $\theta_s = \frac{1}{n_{2k}} \# \big\{ t : \{l_t^1, l_t^2, \ldots, l_t^d\} \subset \operatorname{ran} g_{a_s} \big\}$ and observe that $\sum_{s \in B} \theta_s \leq 1$.

We first estimate the quantity $g_{a_s}(\sum_{t \in E_1} \frac{1}{n_{2k}}(\sum_{i=1}^d |a_i| e_{k_{(t,i)}}))$ for $s \in B$. Each g_{a_s} being in $\mathcal{H}_2$ takes the form $g_{a_s} = \sum_i \sum_j \lambda_{i,j} s_{i,j}^*$. For $1 \leq i' \leq d$ we

get that $(\sum_j \lambda_{i',j} s_{i',j}^*)(\sum_{t \in E_1} \frac{1}{n_{2k}}(\sum_{i=1}^d |a_i| e_{k_{(t,i)}})) |\tilde{a}_{i'}| (\sum_j \lambda_{i',j} s_{i',j}^*)(\sum_{t \in E_1} \frac{1}{n_{2k}} e_{k_{(t,i)}})$

$\leq |\tilde{a}_{i'}| (\max_j \lambda_{i',j}) \theta_s$. Thus

$$g_{a_s}\Big(\sum_{t \in E_1} \frac{1}{n_{2k}}\Big(\sum_{i=1}^d |a_i| e_{k_{(t,i)}}\Big)\Big) \leq \theta_s \sum_{i=1}^d |\tilde{a}_i| \max_j \lambda_{i,j}$$

$$\leq \theta_s\Big(\sum_{i=1}^d \max_j \lambda_{i,j}^2\Big)^{\frac{1}{2}}\Big(\sum_{i=1}^d |\tilde{a}_i|^2\Big)^{\frac{1}{2}} \leq \theta_s.$$

Therefore
(10)

$$h_B\Big(\sum_{t \in E_1} \frac{1}{n_{2k}}\Big(\sum_{i=1}^d |a_i| e_{k_{(t,i)}}\Big)\Big) \leq \sum_{s \in B} g_{a_s}\Big(\sum_{t \in E_1} \frac{1}{n_{2k}}\Big(\sum_{i=1}^d |a_i| e_{k_{(t,i)}}\Big)\Big) \leq \sum_{s \in B} \theta_s \leq 1.$$

From the definition of the set E_2, the set $\{k_{(t,1)}, k_{(t,2)}, \ldots, k_{(t,d)}\}$, for each $t \in E_2$, intersects at least one but is not contained in any $\operatorname{ran} g_{a_s}$, $s \in B$. Also as in the proof of Lemma A.4 we get that $\#(B) \leq (5n_{2k-1})^{\log_2(m_{2k})}$. These yield that $\#(E_2) \leq 2(5n_{2k-1})^{\log_2(m_{2k})}$. Therefore from our choice of k we derive that
(11)

$$h_B\Big(\sum_{t \in E_2} \frac{1}{n_{2k}}\Big(\sum_{i=1}^d |a_i| e_{k_{(t,i)}}\Big)\Big) \leq \Big(\sum_{t \in E_2} \frac{1}{n_{2k}}\Big)\Big(\sum_{i=1}^d |a_i|\Big) < \frac{2(5n_{2k-1})^{\log_2(m_{2k})}}{n_{2k}} < \frac{\varepsilon}{2}.$$

From (9),(10) and (11), we conclude that

$$h\Big(\sum_{i=1}^d |a_i| e_{k_{(t,i)}}\Big) \leq h_A\Big(\sum_{i=1}^d |a_i| \tilde{z}_i\Big) + h_B\Big(\sum_{t \in E_1} \frac{1}{n_{2k}}\Big(\sum_{i=1}^d |a_i| e_{k_{(t,i)}}\Big)\Big)$$

$$+ h_B\Big(\sum_{t \in E_2} \frac{1}{n_{2k}}\Big(\sum_{i=1}^d |a_i| e_{k_{(t,i)}}\Big)\Big) \leq \frac{\varepsilon}{2} + 1 + \frac{\varepsilon}{2} = 1 + \varepsilon.$$

$\square$

As a consequence we obtain the following:

Theorem 5.3. For every infinite dimensional subspace Y of $(\mathfrak{X}_{\mathcal{F}_2})_*$, the space ℓ_2 is isomorphic to a subspace of Y^{**}.

A variant of $\mathfrak{X}_{\mathcal{F}_2}$

Next we shall indicate how we can obtain a space $\mathfrak{X}_{\mathcal{F}_2'}$ similar to $\mathfrak{X}_{\mathcal{F}_2}$ satisfying the additional property that $\mathfrak{X}_{\mathcal{F}_2'}^*/(\mathfrak{X}_{\mathcal{F}_2'})_*$ is isomorphic to $\ell^2(\Gamma)$. Notice that such a space has the following peculiar property:

Proposition 5.4. Granting that $\mathfrak{X}_{\mathcal{F}_2'}^*/(\mathfrak{X}_{\mathcal{F}_2'})_*$ is isomorphic to $\ell^2(\Gamma)$, every infinite dimensional w^*-closed subspace Z of $\mathfrak{X}_{\mathcal{F}_2'}^*$ is either nonseparable or isomorphic to ℓ_2.

Proof. Let $Q : \mathfrak{X}_{\mathcal{F}_2'}^* \to \mathfrak{X}_{\mathcal{F}_2'}^*/(\mathfrak{X}_{\mathcal{F}_2'})_*$ be the quotient map. There are two cases. If there exists a subspace $Z' \hookrightarrow Z$ of finite codimension such that $Q|_{Z'}$ is an isomorphism, then Z is isomorphic to ℓ_2. If not then there exists a normalized block sequence $(v_n)_{n\in\mathbb{N}}$ in $(\mathfrak{X}_{\mathcal{F}_2'})_*$ such that $\sum_{n=1}^{\infty} \text{dist}(v_n, Z) < \frac{1}{3456}$. Setting $V = \overline{\text{span}}\{v_n : n \in \mathbb{N}\}$ we observe that $\text{dist}(S_V, Z) \leq \frac{1}{3456}$ hence, since Z is w^*-closed,

$$(12) \qquad \text{dist}(S_{\overline{V}^{w^*}}, Z) \leq \frac{1}{3456}.$$

As in the proof of Theorem 4.7 we consider a James Tree structure $(w_a)_{a\in\mathcal{D}}$ in V such that the corresponding family $\{w_b : b \in [\mathcal{D}]\}$ satisfies the following properties:

(i) $\|w_b\| \leq 2$ for every $b \in [\mathcal{D}]$.
(ii) For $b \neq b'$ in $[\mathcal{D}]$ we have that $\|w_b - w_{b'}\| \geq \frac{1}{576}$.

The above (i) and (12) yield that for every $b \in [\mathcal{D}]$ there exists $z_b \in Z$ such that

$$(13) \qquad \|z_b - w_b\| \leq \frac{1}{1728}$$

From (13) and the above (ii) we conclude that for $b \neq b'$ in $[\mathcal{D}]$ we have that $\|z_b - z_{b'}\| \geq \frac{1}{1728}$ which yields that Z is nonseparable. $\square$

The following summarizes some of the properties of the space $\mathfrak{X}_{\mathcal{F}_2'}$.

Corollary 5.5. There exists a separable Banach space $\mathfrak{X}_{\mathcal{F}_2'}$ such that

(i) The space $\mathfrak{X}_{\mathcal{F}_2'}$ is HI and reflexively saturated.
(ii) Every quotient of $\mathfrak{X}_{\mathcal{F}_2'}$ has a further quotient isomorphic to ℓ_2.
(iii) Every quotient of $\mathfrak{X}_{\mathcal{F}_2'}$ either has nonseparable dual or it is isomorphic to ℓ_2.
(iv) There exists a quotient of $\mathfrak{X}_{\mathcal{F}_2'}$ not containing reflexive subspaces.

Before presenting the definition of the norming set $D_{\mathcal{F}_2'}$ let's explain our motivation. First we observe that Proposition 5.1 yields that for a sequence $(b_n^*)_n$ satisfying the assumptions, the sequence $([b_n^*])_n$ in the quotient space $W = \mathfrak{X}_{\mathcal{F}_2}^*/(\mathfrak{X}_{\mathcal{F}_2})_*$ is equivalent to the ℓ_2 basis. This in particular yields that

W contains copies of $\ell_2(\Gamma)$ with $\#\Gamma$ equal to the continuum. Our intention is to define $\mathcal{F}_2' \subset \mathcal{F}_2$ and $D_{\mathcal{F}_2'} \subset D_{\mathcal{F}_2}$ such that every infinite $\sigma_\mathcal{F}$-special sequence $b = (f_1, f_2, \ldots, f_n, \ldots)$ satisfies the requirements of Proposition 5.1 with respect to the norm induced by the set $D_{\mathcal{F}_2'}$. Clearly if this is accomplished, then granting Proposition 5.1, the quotient $\mathfrak{X}^*_{\mathcal{F}_2'}/(\mathfrak{X}_{\mathcal{F}_2'})_*$ will be equivalent to $\ell_2(\Gamma)$.

The norm in the space $\mathfrak{X}_{\mathcal{F}_2'}$ is induced by a set $D_{\mathcal{F}_2'}$ which in turn, is recursively defined as $\cup_{n=0}^\infty D_n$. The key ingredient is that the ground set $\mathcal{F}_2'$, which is a subset of $\mathcal{F}_2$, is also defined inductively following the definition of D_n. Thus in each step we define the set S_n of the $\sigma_\mathcal{F}$-special sequences related to $\mathcal{F}_2$ and from this set, the set $\mathcal{F}_2^n$.

For $n = 0$, we set $S_0 = \emptyset$, $D_0 = \{\pm e_n^* : n \in \mathbb{N}\}$.

For $n = 1$ we set $S_1 = \cup_{j=1}^\infty F_j$, $\mathcal{F}_2^1$ is defined from S_1 and D_1 results from $D_0 \cup \mathcal{F}_2^1$ after applying the operations of Definition 1.2 and taking rational convex combinations.

Assume that S_n, $\mathcal{F}_2^n$, D_n have been defined such that every $\sigma_\mathcal{F}$ special sequence $(f_1, \ldots, f_d)$ in S_n satisfies $d \leq n$. The $\sigma_\mathcal{F}$ special sequence $f_1, \ldots, f_d$ in S_n is called $n + 1$-**extendable** if for each $1 \leq i \leq d$ there exists a $(18, 4j_i - 3, 1, D_n, \mathcal{F}_2^n)$ attracting sequence $\chi_i = (x_k, x_k^*)_{k=1}^{4n_{j_i} - 3}$, with $f_i = g_{\chi_i}$ (Definition 4.4). Here a $(18, 4j_i - 3, 1, D_n, \mathcal{F}_2^n)$ attracting sequence is defined as in Definition 1.15 where the norm of the underlying space is induced by the set D_n and moreover $\|x_{2k-1}\|_{D_n} \leq 18$ and $\|x_{2k-1}\|_{\mathcal{F}_2^n} \leq \frac{1}{n_{4j_i-3}^2}$.

Then we set $S_{n+1} = S_n \cup \{(f_1, \ldots, f_d) : (f_1, \ldots, f_{d-1})$ is a $n+1$-extendable $\sigma_\mathcal{F}$ special sequence$\}$.

Next we define $\mathcal{F}_2^{n+1}$ from S_{n+1} in the usual manner and then D_{n+1} from $D_n \cup \mathcal{F}_2^{n+1}$ as before.

This completes the inductive definition. We set $\mathcal{F}_2' = \cup_n \mathcal{F}_2^n$ and $D_{\mathcal{F}_2'} = \cup_n D_n$.

It is easy to see that for every $b = (f_n)_n$ such that $b^* \in \overline{\mathcal{F}_2'}^{w^*}$ the sequence $(f_n)_n$ satisfies the properties of Proposition 5.1 and this yields that indeed $\mathfrak{X}^*_{\mathcal{F}_2'}/(\mathfrak{X}_{\mathcal{F}_2'})_*$ is isomorphic to $\ell^2(\Gamma)$.

6. A NONSEPARABLE HI SPACE WITH NO REFLEXIVE SUBSPACE

In this section we proceed to the construction of a nonseparable HI space containing no reflexive subspace. The general scheme we shall follow is similar to the one used for the definition of $\mathfrak{X}_{\mathcal{F}_2}$. However there are two major differences. The first concerns saturation methods. In the present construction we shall use the operations $(S_{n_j}, \frac{1}{m_j})_j$ for appropriate sequences $(m_j)_j$, $(n_j)_j$. The James Tree space which will play the role of the ground space is also different from $JT_{\mathcal{F}_2}$. Indeed the ground set $\mathcal{F}_s'$ is built on a family $(F_j)_j$ which is related to the Schreier families $(S_{n_{4j-3}})_j$. Furthermore in $\mathcal{F}_s'$ we connect the $\sigma_\mathcal{F}$ special functionals with the use of the Schreier operation instead of taking ℓ_2 sums as in $\mathcal{F}_2$. Finally, $\mathcal{F}_s'$ is defined recursively as we did in the previous variant $\mathfrak{X}_{\mathcal{F}_2'}$ of $\mathfrak{X}_{\mathcal{F}_2}$. The spaces $(\mathfrak{X}_{\mathcal{F}_s'})_*$, $\mathfrak{X}_{\mathcal{F}_s'}$ share the same properties with $(\mathfrak{X}_{\mathcal{F}'_2})_*$, $\mathfrak{X}_{\mathcal{F}'_2}$. The difference occurs between $\mathfrak{X}^*_{\mathcal{F}'_2}$ and $(\mathfrak{X}_{\mathcal{F}'_2})^*$. Indeed, as we have seen $(\mathfrak{X}_{\mathcal{F}'_2})^*/(\mathfrak{X}_{\mathcal{F}'_2})_*$ is isomorphic to $\ell_2(\Gamma)$, while as it will be shown $(\mathfrak{X}_{\mathcal{F}'_s})^*/(\mathfrak{X}_{\mathcal{F}'_s})_*$ is isomorphic to $c_0(\Gamma)$ with $\#\Gamma$ equal to the continuum. The later actually yields all the desired properties for $(\mathfrak{X}_{\mathcal{F}'_s})^*$. Namely it is HI and it does not contain any reflexive subspace.

We recall the definition of $(\mathcal{S}_n)_n$, the first infinite sequence of the Schreier families. The first Schreier family $\mathcal{S}_1$ is the following

$$\mathcal{S}_1 = \{F \subset \mathbb{N} : \#F \leq \min F\} \cup \{\emptyset\}.$$

For $n \geq 1$ the definition goes as follows

$$\mathcal{S}_{n+1} = \left\{ F = \bigcup_{i=1}^d F_i : F_i \in \mathcal{S}_n \ F_i < F_{i+1}, \text{ for all } i \text{ and } d \leq \min F_1 \right\}.$$

Each $\mathcal{S}_n$ is, as can be easily verified by induction, compact, hereditary and spreading.

A finite sequence $(E_1, E_2, \ldots, E_k)$ of successive subsets of $\mathbb{N}$ is said to be $\mathcal{S}_n$ admissible, $n \in \mathbb{N}$, if $\{\min E_i : i = 1, \ldots, k\} \in \mathcal{S}_n$. A finite sequence $(f_1, f_2, \ldots, f_k)$ of vectors in c_{00} is said to be $\mathcal{S}_n$ admissible if the sequence $(\text{supp } f_1, \text{supp } f_2, \ldots, \text{supp } f_k)$ is $\mathcal{S}_n$ admissible.

We fix two sequences of integers $(m_j)_{j\in\mathbb{N}}$ and $(n_j)_{j\in\mathbb{N}}$ defined as follows:

- $m_1 = 2$ and $m_{j+1} = m_j^{m_j}$.
- $n_1 = 1$, and $n_{j+1} = 2^{2m_{j+1}} n_j$.

Definition 6.1. (basic special convex combinations) Let $\varepsilon > 0$ and $j \in \mathbb{N}$, $j > 1$. A convex combination $\sum_{k \in F} a_k e_k$ of the basis $(e_k)_{k \in \mathbb{N}}$ is said to be an (ε, j) basic special convex combination $((\varepsilon, j)$ B.S.C.C.) if

(1) $F \in \mathcal{S}_{n_j}$
(2) For every $P \in \mathcal{S}_{2 \log_2(m_j)(n_{j-1}+1)}$ we have that $\sum_{k \in P} a_k < \varepsilon$.
(3) The sequence $(a_k)_{k \in F}$ is a non increasing sequence of positive reals.

Remark 6.2. The basic special convex combinations have been used implicitly in [AD], their exact definition was given in [AMT] while they have systematically studied in [AT1].

Definition 6.3. (special convex combinations) Let $\varepsilon > 0$, $j \in \mathbb{N}$ with $j > 1$ and let $(x_k)_{k \in \mathbb{N}}$ be a block sequence of the standard basis. A convex combination $\sum_{k \in F} a_k x_k$ of the sequence $(x_k)_{k \in \mathbb{N}}$ is said to be an (ε, j) special convex combination $((\varepsilon, j)$ S.C.C.) of $(x_k)_{k \in \mathbb{N}}$ if $\sum_{k \in F} a_k e_{t_k}$ (where $t_k = \min \text{supp } x_k$ for each k) is an (ε, j) basic special convex combination.

Moreover, if $\sum_{k \in F} a_k x_k$ is a S.C.C. in a Banach space $(X, \| \ \|)$ such that $\|x_k\| \leq 1$ for all k and $\| \sum_{k \in F} a_k x_k \| \geq \frac{1}{2}$ we say that $\sum_{k \in F} a_k x_k$ is a seminormalized (ε, j) special convex combination of $(x_k)_{k \in \mathbb{N}}$.

Definition 6.4. We set $F_0 = \{\pm e_n^* : n \in \mathbb{N}\}$ while for $j = 1, 2, \ldots$ we set $F_j = \{\frac{1}{m_{4j-3}^2} \sum_{i \in I} \pm e_i^* : I \in \mathcal{S}_{n_{4j-3}}\} \cup \{0\}$. We also set $F = \bigcup_{j=0}^{\infty} F_j$.

Let's observe that the sequence $\mathcal{F} = (F_j)_{j=0}^{\infty}$ is a JTG family. The $\sigma_{\mathcal{F}}$ special sequences corresponding to this family are defined exactly as in Definition 3.2.

Definition 6.5. (σ coding, special sequences and attractor sequences) Let $\mathbb{Q}_s$ denote the set of all finite sequences $(\phi_1, \phi_2, \ldots, \phi_d)$ such that $\phi_i \in c_{00}(\mathbb{N})$, $\phi_i \neq 0$ with $\phi_i(n) \in \mathbb{Q}$ for all i, n and $\phi_1 < \phi_2 < \cdots < \phi_d$. We fix a pair Ω_1, Ω_2 of disjoint infinite subsets of $\mathbb{N}$. From the fact that $\mathbb{Q}_s$ is countable we are able to define a Gowers-Maurey type injective coding function $\sigma : \mathbb{Q}_s \to \{2j : j \in \Omega_2\}$ such that $m_{\sigma(\phi_1, \phi_2, \ldots, \phi_d)} > \max\{\frac{1}{|\phi_i(e_l)|} :$ $l \in \operatorname{supp}\phi_i, \ i = 1, \ldots, d\} \cdot \max \operatorname{supp} \phi_d$. Also, let $(\Lambda_i)_{i \in \mathbb{N}}$ be a sequence of pairwise disjoint infinite subsets of $\mathbb{N}$ with $\min \Lambda_i > m_i$.

(A) A finite sequence $(f_i)_{i=1}^d$ is said to be a $\mathcal{S}_{n_{4j-1}}$ **special sequence** provided that

   (i) $(f_1, f_2, \ldots, f_d) \in \mathbb{Q}_s$ and $(f_1, f_2, \ldots, f_d)$ is a $\mathcal{S}_{n_{4j-1}}$ admissible sequence, $f_i \in D_G$ for $i = 1, 2, \ldots, n_{4j-1}$.

   (ii) $w(f_1) = m_{2k}$ with $k \in \Omega_1$, $m_{2k}^{1/2} > n_{4j-1}$ and for each $1 \leq i < d$, $w(f_{i+1}) = m_{\sigma(f_1, \ldots, f_i)}$.

(B) A finite sequence $(f_i)_{i=1}^d$ is said to be a $\mathcal{S}_{n_{4j-3}}$ **attractor sequence** provided that

   (i) $(f_1, f_2, \ldots, f_d) \in \mathbb{Q}_s$ and $(f_1, f_2, \ldots, f_d)$ is a $\mathcal{S}_{n_{4j-3}}$ admissible sequence.

   (ii) $w(f_1) = m_{2k}$ with $k \in \Omega_1$, $m_{2k}^{1/2} > n_{4j-3}$ and $w(f_{2i+1}) = m_{\sigma(f_1, \ldots, f_{2i})}$ for each $1 \leq i < \frac{d}{2}$.

   (iii) $f_{2i} = e_{l_{2i}}^*$ for some $l_{2i} \in \Lambda_{\sigma(f_1, \ldots, f_{2i-1})}$, for $i = 1, \ldots, \frac{d}{2}$.

Definition 6.6. (The space $\mathfrak{X}_{\mathcal{F}_s'}$) In order to define the norming set D of the space $\mathfrak{X}_{\mathcal{F}'}$ we shall inductively define four sequences of subsets of $c_{00}(\mathbb{N})$, denoted as $(K_n)_{n \in \mathbb{N}}, (\tau_n)_{n \in \mathbb{N}}, (G_n)_{n \in \mathbb{N}}, (D_n)_{n \in \mathbb{N}}$.

We set $K_0 = F$ ($K_0^0 = F$, $K_0^j = \emptyset$, $j = 1, 2, \ldots$), $G_0 = F$, $\tau_0 = \emptyset$ and $D_0 = \operatorname{conv}_{\mathbb{Q}}(F)$. Suppose that $K_{n-1}, \tau_{n-1}, G_{n-1}$ and D_{n-1} have been defined. The inductive properties of $(K_n)_{n \in \mathbb{N}}, (\tau_n)_{n \in \mathbb{N}}, (G_n)_{n \in \mathbb{N}}, (D_n)_{n \in \mathbb{N}}$ are included in the inductive definition. We set

$$K_n^{2j} = K_{n-1}^{2j} \cup \{\frac{1}{m_{2j}} \sum_{i=1}^d f_i : f_1 < \cdots < f_d \text{ is } \mathcal{S}_{n_{2j}} \text{ admissible, } f_i \in D_{n-1}\}$$

$$K_n^{4j-3} = K_{n-1}^{4j-3} \cup \{\pm E(\frac{1}{m_{4j-3}} \sum_{i=1}^d f_i) : (f_1, \ldots, f_d) \text{ is a } \mathcal{S}_{n_{4j-3}} \text{ attractor}$$

$$\text{sequence, } f_i \in K_{n-1} \text{ and } E \text{ is an interval of } \mathbb{N}\}$$

$$K_n^{4j-1} = K_{n-1}^{4j-1} \cup \{\pm E(\frac{1}{m_{4j-1}} \sum_{i=1}^{d} f_i) : (f_1, \dots, f_d) \text{ is a } S_{n_{4j-1}} \text{ special}$$

$$\text{sequence, } f_i \in K_{n-1} \text{ and } E \text{ is an interval of } \mathbb{N}\}$$

$$K_n^0 = F$$

We set $K_n = \bigcup_{j=0}^{\infty} K_n^j$.

In order to define τ_n we need the following definition.

Definition 6.7. ((D_{n-1}, j) **exact functionals**) A functional $f \in F$ is said to be (D_{n-1}, j) exact if $f \in F_j$ and there exists $x \in c_{00}(\mathbb{N})$ with $\|x\|_{D_{n-1}} \leq 1000$, $\text{ran}(x) \subset \text{ran}(f)$, $f(x) = 1$ such that for every $i \neq j$, we have that $\|x\|_{F_i} \leq \frac{1000}{m_{4i-3}^2}$ if $i < j$ while $\|x\|_{F_i} \leq 1000 \frac{m_{4j-3}^2}{m_{4i-3}^2}$ if $i > j$.

We set

$$\tau_n = \{\pm E(\sum_{i=1}^{d} \phi_i) : d \leq n, E \text{ is an interval, } (\phi_i)_{i=1}^{d} \text{ is } \sigma_{\mathcal{F}} \text{ special}$$

$$\text{and each } \phi_i \text{ is } (D_{n-1}, \text{ind}(\phi_i)) \text{ exact}\}.$$

We recall that for $\Phi = \pm E(\sum_{i=1}^{d} \phi_i) \in \tau_n$, $\text{ind}(\Phi) = \{\text{ind}(\phi_i) : E \cap \text{ran}\,\phi_i \neq \emptyset\}$.

We set

$$G_n = \{\sum_{i=1}^{d} \varepsilon_i \Phi_i : \Phi_i \in \tau_n, \ \varepsilon_i \in \{-1, 1\}, \ \min \text{supp}\,\Phi_i \geq d,$$

$$(\text{ind}(\Phi_i))_{i=1}^{d} \text{ are pairwise disjoint}\}$$

We set $D_n = \text{conv}_{\mathbb{Q}}(K_n \cup G_n \cup D_{n-1})$.

We finally set $D = \bigcup_{n=0}^{\infty} D_n$. We also set $\tau = \bigcup_{n=0}^{\infty} \tau_n$, $\mathcal{F}'_s = \bigcup_{n=0}^{\infty} G_n$, $K = \bigcup_{n=0}^{\infty} K_n$. We set $K^j = \bigcup_{n=1}^{\infty} K_n^j$ for $j = 1, 2, \dots$. For $f \in K^j$ we write $w(f) = m_j$. We notice that $w(f)$ is not necessarily uniquely determined.

We also need the following definition.

Definition 6.8. ((D, j) **exact functionals**) A functional $f \in F$ is said to be (D, j) exact if $f \in F_j$ and there exists $x \in c_{00}(\mathbb{N})$ with $(\|x\|_D =)\|x\| \leq 1000$, $\text{ran}(x) \subset \text{ran}(f)$, $f(x) = 1$ such that for every $i \neq j$, we have that $\|x\|_{F_i} \leq \frac{1000}{m_{4i-3}^2}$ if $i < j$ while $\|x\|_{F_i} \leq 1000 \frac{m_{4j-3}^2}{m_{4i-3}^2}$ if $i > j$.

Remarks 6.9. (i) If the functional ϕ is (D_n, j) exact then it is also (D_k, j) exact for all $k \leq n$.

(ii) Let $(\phi_i)_{i\in\mathbb{N}}$ be a $\sigma_{\mathcal{F}}$ special sequence such that each ϕ_i is $(D, \mathrm{ind}(\phi_i))$ exact. Then each ϕ_i is $(D_n, \mathrm{ind}(\phi_i))$ exact for all n and $\sum_{i=1}^{d} \phi_i \in \tau_n \subset \tau$ for all $n \geq d$. It follows that $\sum_{i=1}^{\infty} \phi_i \in \overline{\mathcal{T}}^{w^*} \subset \overline{\mathcal{F}'_s}^{w^*} \subset \overline{D}^{w^*} = B_{\mathfrak{X}^*_{\mathcal{F}'_s}}$.

(iii) Let $(\phi_i)_{i\in\mathbb{N}}$ be a $\sigma_{\mathcal{F}}$ special sequence such that $\sum_{i=1}^{d} \phi_i \in \tau$ for all d. In this case we call the $\sigma_{\mathcal{F}}$ special sequence $(\phi_i)_{i\in\mathbb{N}}$ survivor and the functional $\Phi = \sum_{i=1}^{\infty} \phi_i$ a survivor $\sigma_{\mathcal{F}}$ special functional. Then each ϕ_i is (D_n, j_i) exact (where $j_i = \mathrm{ind}(\phi_i)$) for all n, thus for each n there exists $x_{i,n}$ with $\|x_{i,n}\|_{D_n} \leq 1000$, $\mathrm{ran}(x_{i,n}) \subset \mathrm{ran}(\phi_i)$, $\phi_i(x_{i,n}) = 1$ and such that for every $k \neq j_i$, we have that $\|x_{i,n}\|_{F_k} \leq \frac{1000}{m^2_{4k-3}}$ if $k < j$ while $\|x_{i,n}\|_{F_k} \leq 1000 \frac{m^2_{4j-3}}{m^2_{4k-3}}$ if $k > j$. Taking a subsequence of $(x_{i,n})_{n\in\mathbb{N}}$ norm converging to some x_i it is easily checked that $\|x_i\| \leq 1000$, $\phi_i(x_i) = 1$ while $\|x_i\|_{F_k} \leq \frac{1000}{m^2_{4k-3}}$ for $k < j$ and $\|x_i\|_{F_k} \leq 1000 \frac{m^2_{4j-3}}{m^2_{4k-3}}$ for $k > j$.

A sequence $(x_i)_{i\in\mathbb{N}}$ satisfying the above property is called a sequence witnessing that the $\sigma_{\mathcal{F}}$ special sequence $(\phi_i)_{i\in\mathbb{N}}$ (or the special functional $\Phi = \sum_{i=1}^{\infty} \phi_i$) is survivor.

Lemma 6.10. The norming set D of the space $\mathfrak{X}_{\mathcal{F}'_s}$ is the minimal subset of $c_{00}(\mathbb{N})$ satisfying the following conditions:

(i) $\mathcal{F}'_s \subset D$.

(ii) D is closed in the $(\mathcal{S}_{n_{2j}}, \frac{1}{m_{2j}})$ operations.

(iii) D is closed in the $(\mathcal{S}_{n_{4j-1}}, \frac{1}{m_{4j-1}})$ operations on $\mathcal{S}_{n_{4j-1}}$ special sequences.

(iv) D is closed in the $(\mathcal{S}_{n_{4j-3}}, \frac{1}{m_{4j-3}})$ operations on $\mathcal{S}_{n_{4j-3}}$ special sequences.

(v) D is symmetric, closed in the restrictions of its elements on intervals of $\mathbb{N}$ and rationally convex.

It is easily proved that the Schauder basis $(e_n)_{n\in\mathbb{N}}$ of the space $\mathfrak{X}_{\mathcal{F}'_s}$ is boundedly complete and that $\mathfrak{X}_{\mathcal{F}'_s}$ is an asymptotic ℓ_1 space. Since the space $JT_{\mathcal{F}'_s}$ is c_0 saturated (see Remark B.16 where we use the notation $JT_{\mathcal{F}_{\tau,s}}$ for such a space) we get the following.

Proposition 6.11. The identity operator $I : \mathfrak{X}_{\mathcal{F}'_s} \to JT_{\mathcal{F}_s}$ is strictly singular.

Remark 6.12. Applying the methods of [AT1] and taking into account that the identity operator $I : \mathfrak{X}_{\mathcal{F}'_s} \to JT_{\mathcal{F}_s}$ is strictly singular we may prove the following. For every $\varepsilon > 0$ and $j > 1$ every block subspace of $\mathfrak{X}_{\mathcal{F}'_s}$ contains a vector x which is a seminormalized (ε, j) S.C.C. with $\|x\|_{\mathcal{F}'_s} < \varepsilon$.

Definition 6.13. (exact pairs in $\mathfrak{X}_{\mathcal{F}'_s}$) A pair (x, f) with $x \in c_{00}$ and $f \in K$ is said to be a $(12, j, \theta)$ exact pair, where $j \in \mathbb{N}$, if the following conditions are satisfied:

(i) $1 \leq \|x\| \leq 12$, $f(x) = \theta$ and $\mathrm{ran}(f) = \mathrm{ran}(x)$.

(ii) For every $g \in K$ with $w(g) = m_i$ and $i < j$, we have that $|g(x)| \leq 24/m_i$.

(iii) For every sequence $(\phi_i)_i$ in K with $m_j < w(\phi_1) < w(\phi_2) < \cdots$ we have that $\sum_i |\phi_i(x)| \leq 12/m_j$.

Proposition 6.14. For every $j \in \mathbb{N}$, $\varepsilon > 0$ and every block subspace Z of $\mathfrak{X}_{\mathcal{F}'_s}$, there exists a $(12, 2j, 1)$ exact pair (z, f) with $z \in Z$ and $\|z\|_{\mathcal{F}'_s} < \varepsilon$.

Proof. Since the identity operator $I : \mathfrak{X}_{\mathcal{F}'_s} \to JT_{\mathcal{F}_s}$ is strictly singular we may assume, passing to a block subspace of Z, that $\|z\|_{\mathcal{F}'_s} \leq \frac{\varepsilon}{12}\|z\|$ for every $z \in Z$.

Let $(x_k)_{k \in \mathbb{N}}$ be a block sequence in Z such that $(x_k)_{k \in \mathbb{N}}$ is a $(2, \frac{1}{m_{2j}})$ R.I.S. and each x_k is a seminormalized $(\frac{1}{m_{j_k}}, j_k)$ S.C.C. Passing to a subsequence we may assume that $(b^*(x_k))_{k \in \mathbb{N}}$ converges for every $\sigma_{\mathcal{F}}$ branch b. We set $z_k = x_{2k-1} - x_{2k}$. Then $(z_k)_{k \in \mathbb{N}}$ is a $(4, \frac{1}{m_{2j}})$ R.I.S. such that $b^*(z_k) \to 0$ for every branch b.

We recall that each $g \in \mathcal{F}'_s$ has the form $g = \sum_{i=1}^{d} \varepsilon_i \Phi_i^*$ with $\varepsilon_i \in \{-1, 1\}$, $(\Phi_i^*)_{i=1}^{d} \in \tau$ with $\min \mathrm{supp}\, x_i^* \geq d$ and $(\mathrm{ind}(x_i^*))_{i=1}^{d}$ pairwise disjoint. We may assume, replacing $(z_k)_{k \in \mathbb{N}}$ by an appropriate subsequence, that for every $g \in G$ we have that the set $\{\min \mathrm{supp}\, z_k : |g(z_k)| > \frac{1}{m_{2j}}\}$ belongs to $\mathcal{S}_2$, the second Schreier family.

It follows now from Proposition 6.2 of [AT1] that if $z = \sum_{k \in F} a_k z_k$ is a $(1/m_{2j}^2, 2j)$ special convex combination of $(z_k)_{k \in \mathbb{N}}$ and f is of the form $f = 1/m_{2j} \sum_{k \in F} f_k$ where $f_k \in K$ with $f_k(z_k) = 1$ and $\mathrm{ran}(f_k) = \mathrm{ran}(z_k)$ then (z, f) is the desired $(12, 2j, 1)$ exact pair. $\square$

Definition 6.15. (dependent sequences and attracting sequences in $\mathfrak{X}_{\mathcal{F}'_s}$)

(A) A double sequence $(x_k, x_k^*)_{k=1}^{d}$ is said to be a $(C, 4j - 1, \theta)$ dependent sequence (for $C > 1$, $j \in \mathbb{N}$, and $0 \leq \theta \leq 1$) if there exists a sequence $(2j_k)_{k=1}^{d}$ of even integers such that the following conditions are fulfilled:

(i) $(x_k^*)_{k=1}^{d}$ is a $\mathcal{S}_{n_{4j}-1}$ special sequence with $w(x_k^*) = m_{2j_k}$ for each k.

(ii) Each (x_k, x_k^*) is a $(C, 2j_k, \theta)$ exact pair.

(iii) Setting $t_k = \min \mathrm{supp}\, x_k$, we have that $t_1 > m_{2j}$ and $\{t_1, \ldots, t_d\}$ is a maximal element of $\mathcal{S}_{n_{4j}-1}$. (Observe, for later use, that Remark 3.18 of [AT1] yields that there exist $(a_k)_{k=1}^{d}$ such that

$$\sum_{k=1}^{d} a_k e_{t_k} \text{ is a } (\frac{1}{m_{4j-1}^2}, 4j - 1) \text{ basic special convex combination}).$$

(B) A double sequence $(x_k, x_k^*)_{k=1}^d$ is said to be a $(C, 4j - 3, \theta)$ attracting sequence (for $C > 1$, $j \in \mathbb{N}$, and $0 \le \theta \le 1$) if there exists a sequence $(2j_k)_{k=1}^d$ of even integers such that the following conditions are fulfilled:

(i) $(x_k^*)_{k=1}^d$ is a $\mathcal{S}_{n_{4j-3}}$ attractor sequence with $w(x_{2k-1}^*) = m_{2j_{2k-1}}$ and $x_{2k}^* = e_{l_{2k}}^*$ where $l_{2k} \in \Lambda_{2j_{2k}}$ for all $k \le d/2$.

(ii) $x_{2k} = e_{l_{2k}}$.

(iii) Setting $t_k = \min \operatorname{supp} x_k$, we have that $t_1 > m_{2j}$ and $\{t_1, \dots, t_d\}$ is a maximal element of $\mathcal{S}_{n_{4j-3}}$. (Observe that Remark 3.18 of [AT1] yields that there exist $(a_k)_{k=1}^d$ such that $\sum_{k=1}^d a_k e_{t_k}$ is a $(\frac{1}{m_{4j-3}^2}, 4j - 3)$ basic special convex combination).

(iv) Each (x_{2k-1}, x_{2k-1}^*) is a $(C, 2j_{2k-1}, \theta)$ exact pair.

Proposition 6.16. The space $\mathfrak{X}_{\mathcal{F}_s'}$ is reflexively saturated and Hereditarily Indecomposable.

Proof. The proof that $\mathfrak{X}_{\mathcal{F}_s'}$ is reflexively saturated is a consequence of the fact that the identity operator $I : \mathfrak{X}_{\mathcal{F}_s'} \to JT_{\mathcal{F}_s}$ is strictly singular and its proof is identical to that of Proposition 1.21.

In order to show that the space $\mathfrak{X}_{\mathcal{F}_s'}$ is Hereditarily Indecomposable we consider a pair of block subspaces Y and Z and $\delta > 0$. We choose j such that $m_{4j-1} > \frac{192}{\delta}$.

Using Proposition 6.14 we may choose a $(12, 4j - 1, 1)$ dependent sequence $(x_k, x_k^*)_{k=1}^d$ such that $\|x_{2k-1}\|_{\mathcal{F}_s} < \frac{2}{m_{4j-1}^2}$ for all k while $x_{2k-1} \in Y$ if k is odd and $x_{2k-1} \in Z$ if k is even. ¿From the observation in Definition 6.15(A)(iii) there exist $(a_k)_{k=1}^d$ such that $\sum_{k=1}^d a_k e_{t_k}$ is a $(\frac{1}{m_{4j-1}^2}, 4j - 1)$ basic special convex combination (where $t_k = \min \operatorname{supp} x_k$). A variant of Proposition 1.17 (i) in terms of the space $\mathfrak{X}_{\mathcal{F}_s'}$ (using Proposition 6.2 of [AT1]) yields that $\| \sum_{k=1}^d (-1)^{k+1} a_k x_k \| \le \frac{96}{m_{4j-1}^2}$. On the other hand the functional

$$f = \frac{1}{m_{4j-1}} \sum_{k=1}^d x_k^*$$

belongs to the norming set D of the space $\mathfrak{X}_{\mathcal{F}_s'}$ and estimating $f(\sum_k a_k x_k)$ we get that $\| \sum_{k=1}^d a_k x_k \| \ge \frac{1}{m_{2j-1}}$.

Setting $y = \sum_{k \text{ odd}} a_k x_k$ and $z = \sum_{k \text{ even}} a_k x_k$ we have that $y \in Y$ and $z \in Z$ while from the above inequalities we get that $\|y - z\| \le \delta \|y + z\|$. Therefore $\mathfrak{X}_{\mathcal{F}_s'}$ is a Hereditarily Indecomposable space. $\square$

Proposition 6.17. The dual space $\mathfrak{X}_{\mathcal{F}_s'}^*$ is the norm closed linear span of the w^* closure of $\mathcal{F}_s'$ i.e.

$$\mathfrak{X}_{\mathcal{F}_s'}^* = \overline{\operatorname{span}}(\overline{\mathcal{F}_s'}^{w^*}).$$

Proof. Assume the contrary. Then using arguments similar to those of the proof of Proposition 1.19 we may choose a $x^* \in \mathcal{X}^*_{\mathcal{F}'_s}$ with $\|x^*\| = 1$, and a block sequence $(x_k)_{k \in \mathbb{N}}$ in $\mathcal{X}_{\mathcal{F}'_s}$ with $x^*(x_k) > 1$ and $\|x_k\| \leq 2$ such that $x_k \xrightarrow{w} 0$ in $JT_{\mathcal{F}_s}$. Observe that the action of x^* ensures that every convex combination of $(x_k)_{k \in \mathbb{N}}$ has norm greater than 1.

We may choose a convex block sequence $(y_k)_{k \in \mathbb{N}}$ of $(x_k)_{k \in \mathbb{N}}$ with $\|y_k\|_{\mathcal{F}_s} < \frac{\varepsilon}{2}$ where $\varepsilon = \frac{1}{m_4}$. We select a block sequence $(z_k)_{k \in \mathbb{N}}$ of $(y_k)_{k \in \mathbb{N}}$ such that each z_k is a convex combination of $(y_k)_{k \in \mathbb{N}}$ and such that $(z_k)_{k \in \mathbb{N}}$ is $(4, \varepsilon)$ R.I.S. This is possible if we consider each z_k to be a $(\frac{1}{m_{i_k}}, i_k)$ S.C.C. and $m_{i_{k+1}} \varepsilon > \max \operatorname{supp} z_k$ for an appropriate increasing sequence of integers $(i_k)_{k \in \mathbb{N}}$. We then consider $x = \sum a_k z_k$, an $(\varepsilon, 4)$ S.C.C. of $(z_k)_{k \in \mathbb{N}}$. A variant of Proposition 6.2(1a) of [AT1] yields that $\|x\| \leq \frac{20}{m_4} < 1$. On the other hand, since x is a convex combination of $(x_k)_{k \in \mathbb{N}}$ we get that $\|x\| > 1$, a contradiction. $\qquad \square$

Definition 6.18. Let $(x_n)_{n \in \mathbb{N}}$ be a bounded block sequence in $JT_{\mathcal{F}_s}$ and $\varepsilon > 0$. We say that $(x_n)_{n \in \mathbb{N}}$ is ε-*separated* if for every $\phi \in \cup_{j \in \mathbb{N}} F_j$

$$\#\{n : |\phi(x_n)| \geq \varepsilon\} \leq 1.$$

In addition, we say that $(x_n)_{n \in \mathbb{N}}$ is *separated* if for every $L \in [\mathbb{N}]$ and $\varepsilon > 0$ there exists an $M \in [L]$ such that $(x_n)_{n \in M}$ is ε-separated.

Lemma 6.19. Let $(x_n)_{n \in \mathbb{N}}$ be a weakly null separated sequence in $JT_{\mathcal{F}_s}$. Then for every $\varepsilon > 0$, there exists an $L \in [\mathbb{N}]$ such that for all $y^* \in \overline{\mathcal{T}}^{w^*}$,

$$\#\{n \in L : |y^*(x_n)| \geq \varepsilon\} \leq 2.$$

The proof of the above lemma is similar to that of Lemma 3.12.

Lemma 6.20. Let $(z_k)_{k \in \mathbb{N}}$ be a block sequence in $\mathcal{X}_{\mathcal{F}'_s}$ such that each z_k is a $(\frac{1}{m_{2j_k}}, 2j_k)$ special convex combination of a normalized block sequence, where $(j_k)_{k \in \mathbb{N}}$ is strictly increasing. Then the sequence $(z_k)_{k \in \mathbb{N}}$ is separated.

Proof. Given $\varepsilon > 0$ and $L \in [\mathbb{N}]$ we have to find an $M \in [L]$ such that for every $\phi \in \bigcup_{j \in N} F_j$ we have that $|\phi(z_k)| > \varepsilon$ for at most one $k \in M$. For simplicity in our notation we may assume, passing to a subsequence, that $\frac{1}{m_{2j_1}} < \varepsilon$ and $\max \operatorname{supp} z_{k-1} < \varepsilon m_{2j_k}$ for each k.

Now let $\phi \in \bigcup_{j \in N} F_j$. Then ϕ takes the form $\phi = \frac{1}{m^2_{4j-3}} \sum_{i \in I} \pm e^*_i$ with $I \in S_{n_{4j-3}}$ for some j. Let k_0 such that $2j_{k_0} < 4j - 3 < 2j_{k_0+1}$.

We have that $|\phi(z_k)| \leq \frac{1}{m^2_{4j-3}} \sum_{i \in I} |e^*_i(z_k)| < \frac{1}{m_{4j-3}} \frac{1}{m_{2j_{k_0}}} \# \operatorname{supp}(z_k) < \varepsilon$ for every $k < k_0$. Also for $k > k_0$ we get that $|\phi(z_k)| \leq \frac{2}{m_{2j_k}} < \varepsilon$. Thus the subsequence we have selected is ε-separated and this finishes the proof of the lemma. $\qquad \square$

Remark 6.21. Let's observe, for later use, that easy modifications of the previous proof yield that for a sequence $(z_k)_{k \in \mathbb{N}}$ as above the sequence $(z_{2k-1} - z_{2k})_{k \in \mathbb{N}}$ is also separated.

Lemma 6.22. Let $(z_k)_{k\in\mathbb{N}}$ be a weakly null separated sequence in $\mathfrak{X}_{\mathcal{F}'_s}$. Then for every $\varepsilon > 0$ there exists $L \in [\mathbb{N}]$ such that for every $\phi \in \mathcal{F}'_s$, $\{\min \operatorname{supp} z_k : k \in L, |\phi(z_k)| > \varepsilon\} \in \mathcal{S}_2$.

Proof. The proof is almost identical to the proof of Lemma 10.9 of [AT1]. For the sake of completeness we include the proof here.

Using Lemma 6.19 we construct a sequence $(L_k)_{k\in\mathbb{N}}$ of infinite subsets of the natural numbers such that the following conditions hold

(i) $\min \operatorname{supp} x_{l_1} \geq 3$.
(ii) $l_k = \min L_k \notin L_{k+1}$ and $L_{k+1} \subset L_k$ for each $k \in \mathbb{N}$.
(iii) For each $k \in \mathbb{N}$ if $p_k = \max \operatorname{supp} x_{l_k}$ then for every segment $s \in \overline{\mathcal{T}}^{w^*}$ we have that

$$\#\{n \in L_{k+1} : |s^*(x_n)| > \frac{\varepsilon}{p_k}\} \leq 2.$$

We set $L = \{l_1, l_2, l_3, \ldots\}$ and we claim that the set L satisfies the required condition.

Indeed, let $\phi = \sum_{i=1}^{d} \varepsilon_i s_i^* \in \mathcal{F}_s$ where segments $s_1, s_2, \ldots, s_d$ are in $\overline{\mathcal{T}}^{w^*}$, have pairwise disjoint sets of indices, $d \leq \min s_i$ and $\varepsilon_i \in \{-1, 1\}$ for each $i = 1, 2, \ldots, d$. We set

$$l_{i_0} = \min\{n \in L : \operatorname{supp} x_n \cap \operatorname{supp} \phi \neq \emptyset\}.$$

Observe that $d \leq p_{i_0}$. We set

$$F = \{n \in L : |\phi(x_n)| > \varepsilon\}.$$

We have that $F \subset \{l_{i_0}, l_{i_0+1}, l_{i_0+2}, \ldots\}$ thus $F \setminus \{l_{i_0}\} \subset L_{i_0+1}$. Also (iii) yields that for each $i = 1, 2, \ldots, d$ the set $F_i = \{n \in L_{i_0+1} : |s_i^*(x_n)| > \frac{\varepsilon}{p_{i_0}}\}$ has at most two elements.

We observe that

$$F \setminus \{l_{i_0}\} \subset \bigcup_{i=1}^{d} F_i.$$

Indeed if $n \in L_{i_0+1}$ and $n \notin \bigcup_{i=1}^{d} F_i$ then by our inductive construction $|s_i^*(x_n)| \leq \frac{\varepsilon}{p_{i_0}}$ for each $i = 1, 2, \ldots, d$ thus $|\sum_{i=1}^{d} \varepsilon_i s_i^*(x_n)| \leq d\frac{\varepsilon}{p_{i_0}}$ and it follows that $|\phi(x_n)| \leq \varepsilon$ therefore $n \notin F$.

We conclude that $\#(F \setminus \{l_{i_0}\}) \leq 2d$. Also $\min \operatorname{supp} x_n > p_{i_0} \geq d$ for each $n \in F \setminus \{l_{i_0}\}$ hence the set $\{\min \operatorname{supp} x_n : n \in F \setminus \{l_{i_0}\}\}$ is the union of two sets belonging to the first Schreier family $\mathcal{S}_1$. Since $\min \operatorname{supp} x_{l_{i_0}} \geq 3$ the set $\{\min \operatorname{supp} x_n : n \in F\}$ is the union of three sets of $\mathcal{S}_1$ and its minimum is greater or equal to 3. It follows that

$$\{\min \operatorname{supp} x_n : n \in F\} \in \mathcal{S}_2$$

which completes the proof of the Lemma. $\qquad\square$

Lemma 6.23. For every block subspace Z of $(\mathfrak{X}_{\mathcal{F}'_s})_*$ and and $j > 1$, $\varepsilon > 0$ there exists a $(60, 2j, 1)$ exact pair (z, z^*) such that $\mathrm{dist}(z^*, Z) < \varepsilon$ and $\|z\|_{\mathcal{F}'_s} < \frac{2}{m_{2j}}$.

Proof. As in the proof of Theorem 8.3 of [AT1] we may select a block sequence $(z_k)_{k \in \mathbb{N}}$ and a sequence $(f_k)_{k \in \mathbb{N}}$ in D such that

    (i) Each z_k is a $(\frac{1}{m_{2j_k}}, 2j_k)$ S.C.C. of a normalized block sequence and the sequence $(j_k)_k$ is strictly increasing.

    (ii) $f_k(z_k) > \frac{1}{3}$ and $\mathrm{ran}\, f_k = \mathrm{ran}\, z_k$.

    (iii) $\mathrm{dist}(f_k, Z) < \frac{1}{2^k}$.

We assume that the sequence $(z_k)_{k \in \mathbb{N}}$ is weakly Cauchy. Then the sequence $(z_{2k-1} - z_{2k})_{k \in \mathbb{N}}$ is weakly null while from Remark 6.21 this sequence is separated. From Lemma 6.22, for every $\varepsilon > 0$ there exists $L \in [\mathbb{N}]$ such that for every $\phi \in \mathcal{F}'_s$, $\{\min \mathrm{supp}(z_{2k-1} - z_{2k}) : k \in L, |\phi(z_{2k-1} - z_{2k})| > \varepsilon\} \in \mathcal{S}_2$.

    The rest of the proof follows the argument of Theorem 8.3 of the Memoirs monograph [AT1]. $\qquad\square$

Proposition 6.24. Every infinite dimensional subspace of $(\mathfrak{X}_{\mathcal{F}'_s})_*$ has non-separable second dual. In particular the space $(\mathfrak{X}_{\mathcal{F}'_s})_*$ contains no reflexive subspace.

Proof. Using Lemma 6.23 for every block subspace and every j we may select, similarly to Lemma 4.6, a $(60, 4j - 3, 1)$ attracting sequence $\chi = (x_k, x_k^*)_{k=1}^d$ with $\sum_k \|x_{2k-1}\|_{\mathcal{F}'_s} < \frac{1}{m_{4j-3}^2}$ and $\sum_k \mathrm{dist}(x_{2k-1}^*, Z) < \frac{1}{m_{4j-3}}$. We recall at this point (see Definition 6.15(B)(iii)) that there exists a sequence $(a_k)_{k=1}^d$ such that $\sum_{k=1}^d a_k e_{t_k}$ is a $(\frac{1}{m_{4j-3}^2}, 4j - 3)$ basic special convex combination (where $t_k = \min \mathrm{supp}\, x_k$).

    We set $F_\chi = -\frac{1}{m_{4j-3}^2} \sum_k x_{2k-1}^*$ and $g_\chi = \frac{1}{m_{4j-3}^2} \sum_k x_{2k}^*$. From the fact that $(x_k^*)_{k=1}^d$ is a $\mathcal{S}_{n_{4j-3}}$ special sequence we have that $\|\frac{1}{m_{2j-1}}(x_1^* + x_2^* + \cdots + x_d^*)\| \leq 1$ and since $g_\chi - F_\chi = \frac{1}{m_{2j-1}}(\frac{1}{m_{2j-1}}(x_1^* + x_2^* + \cdots + x_d^*))$ we get that $\|g_\chi - F_\chi\| \leq \frac{1}{m_{2j-1}}$. We also have that $\mathrm{dist}(F_\chi, Z) < \frac{1}{m_{4j-3}}$.

    Similarly to Proposition 7.5 of [AT1] and to Proposition 1.17 of the present paper, we may prove that $\|\sum_{k=1}^d (-1)^k a_k x_k\| \leq \frac{300}{m_{4j-3}^2}$. Observe also that $g_\chi(\sum_{k=1}^d (-1)^k a_k x_k) = \frac{1}{m_{4j-3}^2} \sum_k a_{2k} \geq \frac{1}{3m_{4j-3}^2}$. ¿From these inequalities it follows that there exists $1 \leq \theta_\chi \leq 900$ such that $g_\chi(d_\chi) = 1$ and $\|d_\chi\| \leq 900$ where $d_\chi = 3\theta_\chi m_{4j-3}^2 \sum_{k=1}^d (-1)^k a_k x_k$. It is also easily checked that $\|d_\chi\|_{F_i} \leq \frac{1000}{m_{4i-3}^2}$ for $i < j$ while $\|d_\chi\|_{F_i} \leq 1000 \frac{m_{4j-3}^2}{m_{4i-3}^2}$ if $i > j$. Thus the vector d_χ witnesses that the functional g_χ is (D, j) exact.

Using arguments similar to those of Theorem 4.7, for a given block subspace Z of $(\mathfrak{X}_{\mathcal{F}'_s})_*$ we construct a family $(\chi_a)_{a \in \mathcal{D}}$ ($\mathcal{D}$ is the dyadic tree) of dependent sequences with properties analogous to (i),(ii), (iii) of Theorem 4.7 and such that for every $a \in \mathcal{D}$ the functional g_{χ_a} is $(\mathcal{D}, j_a)$ exact. It follows that for every branch b of the dyadic tree the sum $\sum_{a \in b} g_{\chi_a}$ converges in the w^* topology to a survivor $\sigma_{\mathcal{F}}$ special functional $g_b \in B_{\mathfrak{X}^*_{\mathcal{F}'_s}}$ and there exists $z_b \in Z^{**}$ with $\|z_b - g_b\| < \frac{1}{1152}$. Then, as in the proof of Theorem 4.7 we obtain that Z^{**} is nonseparable. $\qquad\square$

Proposition 6.25. The space $(\mathfrak{X}_{\mathcal{F}'_s})_*$ is Hereditarily Indecomposable.

Proof. Let Y, Z be a pair of block subspaces of $(\mathfrak{X}_{\mathcal{F}'_s})_*$. For every $j > 1$, using Lemma 6.23, we are able to construct a $(60, 4j - 1, 1)$ dependent sequence $(x_k, x_k^*)_{k=1}^{n_{4j-1}}$ such that $\|x_k\|_{\mathcal{F}'_s} < \frac{1}{m_{4j-1}^2}$ while $\sum \text{dist}(x_{2k-1}^*, Y) <$ and $\sum \text{dist}(x_{2k}^*, Z) <$ for each k. From the observation in Definition 6.15(A)(iii) there exist $(a_k)_{k=1}^d$ such that $\sum_{k=1}^d a_k e_{t_k}$ is a $(\frac{1}{m_{4j-1}^2}, 4j-1)$ basic special convex combination. As in Proposition 6.16 we get that $\|\sum_{k=1}^d (-1)^{k+1} a_k x_k\| \le \frac{480}{m_{4j-1}^2}$.

We set $h_Y = \frac{1}{m_{4j-1}} \sum_{k \text{ odd}} x_k^*$ and $h_Z = \frac{1}{m_{4j-1}} \sum_{k \text{ even}} x_k^*$. The functional $h_Y + h_Z = \frac{1}{m_{4j-1}} \sum_{k=1}^d x_k^*$ belongs to the norming set D hence $\|h_Y + h_Z\| \le 1$.

On the other hand the action of $h_Y - h_Z$ to the vector $\sum_{k=1}^d (-1)^{k+1} a_k x_k$ yields that $\|h_Y - h_Z\| \ge \frac{m_{4j-1}}{480}$.

From the above estimates and since $\text{dist}(h_Y, Y) < 1$ and $\text{dist}(h_Z, Z) < 1$ we may choose $f_Y \in Y$ and $f_Z \in Z$ with $\|f_Y - f_Z\| \ge (\frac{m_{4j-1}}{1440} - \frac{2}{3})\|f_Y + f_Z\|$. Since this can be done for arbitrary large j we obtain that $(\mathfrak{X}_{\mathcal{F}'_s})_*$ is Hereditarily Indecomposable. $\qquad\square$

Proposition 6.26. The quotient space $\mathfrak{X}^*_{\mathcal{F}'_s}/(\mathfrak{X}_{\mathcal{F}'_s})_*$ is isomorphic to $c_0(\Gamma)$ where the set Γ coincides with the set of all survivor $\sigma_{\mathcal{F}}$ special sequences.

Proof. As follows from 6.17 the quotient space $\mathfrak{X}^*_{\mathcal{F}'_s}/(\mathfrak{X}_{\mathcal{F}'_s})_*$ is generated in norm by the classes of the elements of the set $\overline{\mathcal{F}'_s}^{w^*}$. Since clearly

$$\overline{\mathcal{F}'_s}^{w^*} = F \cup \{\sum_{i=1}^d \varepsilon_i \Phi_i : \Phi_i \in \tau, \ \varepsilon_i \in \{-1, 1\}, \ \min \text{supp} \, \Phi_i \ge d,$$

$$(\text{ind}(\Phi_i))_{i=1}^d \text{ are pairwise disjoint}\}$$

we get that

$$\mathfrak{X}^*_{\mathcal{F}'_s} = \overline{\text{span}}(\{e_n^* : n \in \mathbb{N}\} \cup \{\Phi : \Phi \text{ is a survivor } \sigma_{\mathcal{F}} \text{ special functional}\})$$

Thus $\mathfrak{X}^*_{\mathcal{F}'_s}/(\mathfrak{X}_{\mathcal{F}'_s})_* = \overline{\text{span}}\{\Phi + (\mathfrak{X}_{\mathcal{F}'_s})_* : \Phi \text{ is a survivor } \sigma_{\mathcal{F}} \text{ special functional}\}$
To prove that this space is isomorphic to $c_0(\Gamma)$ we shall show that for every

choice $(\Phi)_{i=1}^d$ of pairwise different survivor $\sigma_{\mathcal{F}}$ special functionals and every choice of signs $(\varepsilon_i)_{i=1}^d$ we have that

$$(14) \qquad \frac{1}{2000} \leq \| \sum_{i=1}^d \varepsilon_i(\Phi_i + (\mathfrak{X}_{\mathcal{F}'_s})_*) \| \leq 1.$$

We have that $\| \sum_{i=1}^d \varepsilon_i(\Phi_i + (\mathfrak{X}_{\mathcal{F}'_s})_*) \| = \lim_k \| \sum_{i=1}^d \varepsilon_i(E_k \Phi_i) \|$ where for each k, $E_k = \{k, k+1, \ldots\}$. The right part of inequality (14) follows directly, since for all but finite k the functional $E_k(\sum_{i=1}^d \varepsilon_i \Phi_i)$ belongs to $\overline{\mathcal{F}'_s}^{w^*} \subset B_{\mathfrak{X}^*_{\mathcal{F}'_s}}$.

For each $i = 1, \ldots, d$ let $\Phi_i = \sum_{l=1}^\infty \phi_l^i$ with $(\phi_l^i)_{l \in \mathbb{N}}$ a survivor $\sigma_{\mathcal{F}}$ special sequence and and let (x_l^i) be a sequence witnessing this fact (see Remark 6.9 (iii)). We choose k_0 such that $(\text{ind}(E_{k_0} \Phi_i))_{i=1}^d$ are pairwise disjoint and $\min(\bigcup_{i=1}^d \text{ind}(E_{k_0} \Phi_i)) = r_0$ with $m_{2r_0-1} > 10^{10}$.

Let $k \geq k_0$. We choose t such that $\text{ran}(\phi_t^1) \subset E_k$ and let $\text{ind}(\phi_t^1) = l_0$. We get that $\sum_{i=2}^d |\Phi_i(x_t^1)| \leq \sum_{r=r_0}^{l_0-1} \frac{1000}{m_{4r-3}^2} + \sum_{r=l_0+1}^\infty \frac{1000 m_{2l_0-1}^2}{m_{4r-3}^2} < \frac{1}{2}$. We thus get that

$$\| E_k(\sum_{i=1}^d \varepsilon_i \Phi_i) \| \geq \frac{1}{1000}(\Phi_1(x_t^1) - \sum_{i=2}^d |\Phi_i(x_t^1)|) > \frac{1}{1000}(1 - \frac{1}{2}) = \frac{1}{2000}.$$

The proof of the proposition is complete. $\qquad \square$

Theorem 6.27. There exists a Banach space $\mathfrak{X}_{\mathcal{F}'_s}$ satisfying the following properties:

(i) The space $\mathfrak{X}_{\mathcal{F}'_s}$ is an asymptotic ℓ_1 space with a boundedly complete Schauder basis $(e_n)_{n \in \mathbb{N}}$ and is Hereditarily Indecomposable and reflexively saturated.

(ii) The predual space $(\mathfrak{X}_{\mathcal{F}'_s})_* = \overline{\text{span}}\{e_n^* : n \in \mathbb{N}\}$ is Hereditarily Indecomposable and each infinite dimensional subspace of $(\mathfrak{X}_{\mathcal{F}'_s})_*$ has nonseparable second dual. In particular the space $(\mathfrak{X}_{\mathcal{F}'_s})_*$ contains no reflexive subspace.

(iii) The dual space $\mathfrak{X}_{\mathcal{F}'_s}^*$ is nonseparable, Hereditarily Indecomposable and contains no reflexive subspace.

(iv) Every bounded linear operator $T : X \to X$ where $X = (\mathfrak{X}_{\mathcal{F}'_s})_*$ or $X = \mathfrak{X}_{\mathcal{F}'_s}$ or $X = \mathfrak{X}_{\mathcal{F}'_s}^*$ takes the form $T = \lambda I + W$ with W a weakly compact operator. In particular each $T : \mathfrak{X}_{\mathcal{F}'_s}^* \to \mathfrak{X}_{\mathcal{F}'_s}^*$ is of the form $T = Q^* + K$ with $Q : \mathfrak{X}_{\mathcal{F}'_s} \to \mathfrak{X}_{\mathcal{F}'_s}$ and K a compact operator, hence $T = \lambda I + R$ with R an operator with separable range.

Proof. As we have observed the Schauder basis $(e_n)_{n \in \mathbb{N}}$ of $\mathfrak{X}_{\mathcal{F}'_s}$ is boundedly complete and $\mathfrak{X}_{\mathcal{F}'_s}$ is asymptotic ℓ_1. In Proposition 6.16 we have shown that $\mathfrak{X}_{\mathcal{F}'_s}$ is reflexively saturated and Hereditarily Indecomposable. The facts that

$(\mathfrak{X}_{\mathcal{F}'_s})_*$ is Hereditarily Indecomposable and that every subspace of it has non-separable second dual have been shown in Proposition 6.24 and Proposition 6.25.

From the facts that the quotient space $\mathfrak{X}^*_{\mathcal{F}'_s}/(\mathfrak{X}_{\mathcal{F}'_s})_*$ is isomorphic to $c_0(\Gamma)$ and $(\mathfrak{X}_{\mathcal{F}'_s})_*$ is Hereditarily Indecomposable and taking into account that $\mathfrak{X}_{\mathcal{F}'_s}$, being a Hereditarily Indecomposable space, contains no isomorphic copy of ℓ_1 we get that the dual space $\mathfrak{X}^*_{\mathcal{F}'_s}$ is also Hereditarily Indecomposable (Corollary 1.5 of [AT1]). Since $\mathfrak{X}^*_{\mathcal{F}'_s}$ is Hereditarily Indecomposable and contains a subspace (which is $(\mathfrak{X}_{\mathcal{F}'_s})_*$ with no reflexive subspace we conclude that $\mathfrak{X}^*_{\mathcal{F}'_s}$ also does not have any reflexive subspace.

Using similar arguments to those of the proof of Theorems 2.15, 2.16 and Corollary 4.10 we may prove that every bounded linear operator $T : (\mathfrak{X}_{\mathcal{F}'_s})_* \to (\mathfrak{X}_{\mathcal{F}'_s})_*$ and every bounded linear operator $T : \mathfrak{X}_{\mathcal{F}'_s} \to \mathfrak{X}_{\mathcal{F}'_s}$ takes the form $T = \lambda I + W$ with W a strictly singular and weakly compact operator. Since $\mathfrak{X}_{\mathcal{F}'_s}$ contains no isomorphic copy of ℓ_1 and $\mathfrak{X}^{**}_{\mathcal{F}'_s}$ is isomorphic to $\mathfrak{X}_{\mathcal{F}'_s} \oplus \ell_1(\Gamma)$ Proposition 1.7 of [AT1] yields that every bounded linear operator $T : \mathfrak{X}^*_{\mathcal{F}'_s} \to \mathfrak{X}^*_{\mathcal{F}'_s}$ is of the form $T = Q^* + K$ with $Q : \mathfrak{X}_{\mathcal{F}'_s} \to \mathfrak{X}_{\mathcal{F}'_s}$ and K a compact operator. From the form of the operators of $\mathfrak{X}_{\mathcal{F}'_s}$ we have mentioned before we conclude that T takes the form $T = \lambda I + R$ with R a weakly compact operator and hence of separable range.   □

Remark 6.28. It is worth mentioning that the key ingredient to obtain $\mathfrak{X}^*_{\mathcal{F}'_s}/(\mathfrak{X}_{\mathcal{F}'_s})_*$ isomorphic to $c_0(\Gamma)$ which actually yields the HI property of $\mathfrak{X}^*_{\mathcal{F}'_s}$ is that in the ground set $\mathcal{F}'_s$ we connect the $\sigma_{\mathcal{F}}$ special functionals using the Schreier operation. This forces us to work with the saturation families $(\mathcal{S}_{n_j}, \frac{1}{m_j})_j$ instead of $(\mathcal{A}_{n_j}, \frac{1}{m_j})_j$. The reason for this is that working with $\mathcal{F}'_s$ built on $(F_j)_j$ with $F_j = \{\frac{1}{m^2_{4j-3}} \sum_{i \in I} \pm e^*_i : \#(I) \leq \frac{n_{4j-3}}{2}\}$ the extension with attractors of this ground set $\mathcal{F}'_s$ based on $(\mathcal{A}_{n_j}, \frac{1}{m_j})_j$ is not strongly strictly singular.

However there exists an alternative way of connecting the $\sigma_{\mathcal{F}}$ special functionals lying between the Schreier operation and the ℓ_2 sums. This yields the James Tree space $JT_{\mathcal{F}_{2,s}}$ defined and studied in Appendix D. It is easy to check that the corresponding HI extension with attractors $\mathfrak{X}_{\mathcal{F}'_{2,s}}$ of $JT_{\mathcal{F}_{2,s}}$ is a strictly singular one either we work on in the frame of $(\mathcal{A}_{n_j}, \frac{1}{m_j})_j$ or of $(\mathcal{S}_{n_j}, \frac{1}{m_j})_j$. It is open whether the corresponding space $\mathfrak{X}^*_{\mathcal{F}'_{2,s}}$ contains ℓ_2 or not. If it does not contain ℓ_2 then $\mathfrak{X}^*_{\mathcal{F}'_{2,s}}$ will be also a nonseparable HI space not containing any reflexive subspace with the additional property that $\mathfrak{X}^*_{\mathcal{F}'_{2,s}}/(\mathfrak{X}_{\mathcal{F}'_{2,s}})_*$ is isomorphic to $\ell_2(\Gamma)$.

7. A HJT SPACE WITH UNCONDITIONALLY AND REFLEXIVELY SATURATED DUAL

This section concerns the definition of the space $\mathfrak{X}^{us}_{\mathcal{F}_2}$ namely a separable space with a boundedly complete basis which is reflexive and unconditionally saturated and its predual $(\mathfrak{X}^{us}_{\mathcal{F}_2})_*$ is HJT space hence it does not contain any reflexive subspace. This construction starts with the ground set $\mathcal{F}_2$ used in Section 4. In the extensions we use only attractors for which we eliminate a sufficient part of their conditional structure. The proof of the property that $\mathfrak{X}^{us}_{\mathcal{F}_2}$ is unconditionally saturated follows the arguments of [AM],[AT2] while the HJT property of $(\mathfrak{X}^{us}_{\mathcal{F}_2})_*$ results from the remaining part of the conditional structure of the attractors. We additionally show that $(\mathfrak{X}^{us}_{\mathcal{F}_2})_*$ is HI.

Let $\mathbf{Q}$ be the set of all finitely supported scalar sequences with rational coordinates, of maximum modulus 1 and nonempty support. We set

$$\mathbf{Q_s} = \{(x_1, f_1, \ldots, x_n, f_n) : x_i, f_i \in \mathbb{Q}, \ i = 1, \ldots, n$$
$$\operatorname{ran}(x_i) \cup \operatorname{ran}(f_i) < \operatorname{ran}(x_{i+1}) \cup \operatorname{ran}(f_{i+1}), \ i = 1, \ldots, n-1\}.$$

For $\phi = (x_1, f_1, \ldots, x_n, f_n) \in \mathbf{Q_s}$ and $l \leq n$ we denote by ϕ_l the sequence $(x_1, f_1, \ldots, x_l, f_l)$. We consider an injective coding function $\sigma : \mathbf{Q_s} \to \{2j : j \in \mathbb{N}\}$ such that for every $\phi = (x_1, f_1, \ldots, x_n, f_n) \in \mathbf{Q_s}$

$$\sigma(x_1, f_1, \ldots, x_{n-1}, f_{n-1}) < \sigma(x_1, f_1, \ldots, x_n, f_n)$$
$$\text{and } \max\{\operatorname{ran}(x_n) \cup \operatorname{ran}(f_n)\} \leq m_{\sigma(\phi)}^{\frac{1}{2}}.$$

The norming set D^{us} of the space $\mathfrak{X}^{us}_{\mathcal{F}_2}$ will be defined as $D^{us} = \bigcup_{n=0}^{\infty} D_n$ after defining inductively two sequences $(K_n)_{n=0}^{\infty}$, $(D_n)_{n=0}^{\infty}$ of subsets of $c_{00}(\mathbb{N})$ with $D_n = \operatorname{conv}_{\mathbb{Q}}(K_n)$.

Let $\mathcal{F}_2$ be the set defined in the beginning of the third section. We set

$$K_0 = \mathcal{F}_2 \quad \text{and} \quad D_0 = \operatorname{conv}_{\mathbb{Q}}(K_0).$$

Assume that K_{n-1} and D_{n-1} have been defined. Then for each $j \in \mathbb{N}$ we set

$$K_n^{2j} = K_{n-1}^{2j} \cup \{\frac{1}{m_{2j}} \sum_{i=1}^{d} f_i : f_1 < \cdots < f_d, \ f_i \in D_{n-1}, \ d \leq n_{2j}\}.$$

For fixed $j \in \mathbb{N}$ we consider the collection of all sequences $\phi = (x_i, f_i)_{i=1}^{n_{4j-3}}$ satisfying the following conditions:

(i) $x_1 = e_{l_1}$ and $f_1 = e_{l_1}^*$ for some $l_1 \in \Lambda_{2j_1}$ where j_1 is an integer with $m_{2j_1}^{1/2} > n_{4j-3}$.

(ii) For $1 \leq i \leq n_{4j-3}/2$, $f_{2i} \in K_{n-1}^{\sigma(\phi_{2i-1})}$ and $\|x_{2i}\|_{K_{n-1}} \leq \frac{18}{m_{\sigma(\phi_{2i-1})}}$.

(iii) For $1 \leq i < n_{4j-3}/2$, $x_{2i+1} = e_{l_{2i+1}}$ and $f_{2i+1} = e_{l_{2i+1}}^*$ for some $l_{2i+1} \in \Lambda_{\sigma(\phi_{2i})}$.

For every ϕ satisfying (i),(ii) and (iii) we define the set

$$K_{n,\phi}^{4j-3} = \{\frac{\pm 1}{m_{4j-3}} E\Big(\sum_{i=1}^{n_{4j-3}/2} (\lambda_{f'_{2i}} f_{2i-1} + f'_{2i})\Big) : E \text{ is an interval of } \mathbb{N},$$
$$f'_{2i} \in K_{n-1}^{\sigma(\phi_{2i-1})}, \ \lambda_{f'_{2i}} = f'_{2i}(m_{\sigma(\phi_{2i-1})} x_{2i}),$$
$$(x_{2i-1}, f_{2i-1}, x_{2i}, f'_{2i})_{i=1}^{n_{4j-3}/2} \in \mathbf{Q_s}\}.$$

We define
$$K_n^{4j-3} = \cup\{K_{n,\phi}^{4j-3} : \phi \text{ satisfies conditions (i), (ii), (iii)}\} \cup K_{n-1}^{4j-3}$$
and we set
$$K_n = \bigcup_j (K_n^{2j} \cup K_n^{4j-3}) \quad \text{and} \quad D_n = \text{conv}_{\mathbb{Q}}(K_n).$$

We finally set
$$K^{us} = \bigcup_{n=0}^{\infty} K_n \qquad \text{and} \qquad D^{us} = \bigcup_{n=0}^{\infty} D_n$$

The space $\mathfrak{X}_{\mathcal{F}_2}^{us}$ is the completion of the space $(c_{00}, \| \;\|_{D^{us}})$ where
$$\|x\|_{D^{us}} = \sup\{f(x) : f \in D^{us}\}.$$

Using the same arguments as those in Proposition 4.1 we get the following.

Lemma 7.1. The identity operator $I : \mathfrak{X}_{\mathcal{F}_2}^{us} \to JT_{\mathcal{F}_2}$ is strongly strictly singular (Definition 2.1).

Definition 7.2. The sequence
$$\phi = (x_1, f_1, x_2, f_2, x_3, f_3, x_4, f_4, \ldots, x_{n_{4j-3}}, f_{n_{4j-3}}) \in \mathbf{Q}_s$$
is said to be a n_{4j-3} attracting sequence provided that

(i) $x_1 = e_{l_1}$ and $f_1 = e_{l_1}^*$ for some $l_1 \in \Lambda_{2j_1}$ where j_1 is an integer with $m_{2j_1}^{1/2} > n_{4j-3}$.

(ii) For $1 \le i \le n_{4j-3}/2$, $(m_{\sigma(\phi_{2i-1})}x_{2i}, f_{2i})$ is a $(18, \sigma(\phi_{2i-1}), 1)$ exact pair (Definition 1.9) while $\displaystyle\sum_{i=1}^{n_{4j-3}/2} \|x_{2i}\|_{\mathcal{F}_2} < \frac{1}{n_{4j-3}}$.

(iii) For $1 \le i < n_{4j-3}/2$, $x_{2i+1} = e_{l_{2i+1}}$ and $f_{2i+1} = e_{l_{2i+1}}^*$ for some $l_{2i+1} \in \Lambda_{\sigma(\phi_{2i})}$.

We consider the vectors d_ϕ in $\mathfrak{X}_{\mathcal{F}_2}^{us}$, and g_ϕ, F_ϕ in $(\mathfrak{X}_{\mathcal{F}_2}^{us})_*$ as they are defined in Definition 4.4. Let also notice, for later use, that the analogue of Lemma 4.5 remains valid.

Lemma 7.3. For every block subspace Z of the predual space $(\mathfrak{X}_{\mathcal{F}_2}^{us})_*$ and every $j \in N$ there exists a n_{4j-3} attracting sequence
$$\phi = (x_1, f_1, x_2, f_2, \ldots, x_{n_{4j-3}}, f_{n_{4j-3}}) \quad \text{with} \quad \sum_{i=1}^{n_{4j-3}/2} \text{dist}(f_{2i}, Z) < \frac{1}{m_{4j-3}^2}.$$

Proof. Since the identity $I : \mathfrak{X}_{\mathcal{F}_2}^{us} \to JT_{\mathcal{F}_2}$ is strongly strictly singular (7.1), we may construct, using the analogue of Lemma 2.10 in terms of $\mathfrak{X}_{\mathcal{F}_2}^{us}$, the desired attracting sequence. □

Lemma 7.4. Let $\chi = (x_{2k}, x_{2k}^*)_{k=1}^{n_{4j-3}/2}$ be a $(18, 4j-3, 1)$ attracting sequence such that $\displaystyle\sum_{k=1}^{n_{4j-3}/2} \|x_{2k-1}\|_{\mathcal{F}_2} < \frac{1}{m_{4j-3}^2}$. Then for every branch b such that $j \notin \text{ind}(b)$ we have that $|b^*(d_\chi)| < \frac{3}{m_{4j-3}}$.

Proof. Let $b = (f_1, f_2, f_3, \ldots)$ be a branch (i.e. $(f_i)_{i \in \mathbb{N}}$ is a $\sigma_{\mathcal{F}}$ special sequence). We recall that $d_\chi = \frac{m_{4j-3}^2}{n_{4j-3}} \sum_{k=1}^{n_{4j-3}} (-1)^k x_k$ and we set

$$d_1 = -\frac{m_{4j-3}^2}{n_{4j-3}} \sum_{k=1}^{n_{4j-3}/2} x_{2k-1} \text{ and } d_2 = \frac{m_{4j-3}^2}{n_{4j-3}} \sum_{k=1}^{n_{4j-3}/2} x_{2k}.$$

Our assumption $\sum_{k=1}^{n_{4j-3}/2} \|x_{2k-1}\|_{\mathcal{F}_2} < \frac{1}{m_{4j-3}^2}$ yields that

(15)

$$|b^*(d_1)| \leq \frac{m_{4j-3}^2}{n_{4j-3}} \sum_{k=1}^{n_{4j-3}/2} |b^*(x_{2k-1})| \leq \frac{m_{4j-3}^2}{n_{4j-3}} \sum_{k=1}^{n_{4j-3}/2} \|x_{2k-1}\|_{\mathcal{F}_2} < \frac{1}{n_{4j-3}}.$$

We decompose b^* as $b^* = x^* + y^*$ with $\operatorname{ind}(x^*) \subset \{1, \ldots, j-1\}$ and $\operatorname{ind}(y^*) \subset \{j+1, j+2, \ldots\}$. We recall that an $f \in F$ with $\inf(f) = l$ is of the form $f = \frac{1}{m_{4l-3}^2} \sum_{i \in \operatorname{supp}(f)} \pm e_i^*$ with $\operatorname{supp}(f) \leq n_{4l-3}/2$. Thus $\operatorname{supp}(x^*) \leq \frac{n_1}{2} + \frac{n_5}{2} + \cdots + \frac{n_{4j-7}}{2} < n_{4j-4}$. Hence

(16)
$$|x^*(d_2)| \leq \frac{m_{4j-3}^2}{n_{4j-3}} n_{4j-4} < \frac{1}{m_{4j-3}}.$$

On the other hand we have that $\|y^*\|_\infty \leq \frac{1}{m_{4j+1}}$, therefore

(17)
$$|y^*(d_2)| \leq \frac{1}{m_{4j+1}} \cdot \frac{m_{4j-3}^2}{n_{4j-3}} \cdot \frac{n_{4j-3}}{2} < \frac{1}{m_{4j-3}}.$$

From (15),(16) and (17) we obtain that $|b^*(d_\chi)| < \frac{3}{m_{4j-3}}$. $\square$

Proposition 7.5. The space $(\mathfrak{X}_{\mathcal{F}_2}^{us})_*$ is Hereditarily Indecomposable.

Proof. Let Z_1, Z_2 be a pair of block subspaces in $(\mathfrak{X}_{\mathcal{F}_2}^{us})_*$ and let $0 < \delta < 1$.

We may inductively construct, using Lemma 7.3, a sequence $(\chi_r)_{r \in \mathbb{N}}$ such that the following conditions are satisfied.

(i) Each $\chi_r = (x_k^r, (x_k^r)^*)_{k=1}^{n_{4j_r - 3}}$ is a $(18, 4j_r - 3, 1)$ attracting sequence with $\sum_{k=1}^{n_{2j_r - 1}} \|x_{2k-1}^r\|_{\mathcal{F}_2} < \frac{1}{m_{4j_r - 3}^2}$ and additionally $\operatorname{dist}(F_{\chi_r}, Z_1) < \frac{1}{m_{2j_r - 1}}$ if r is odd, while $\operatorname{dist}(F_{\chi_r}, Z_2) < \frac{1}{m_{2j_r - 1}}$ if r is even.

(ii) $(d_{\chi_r})_{r \in \mathbb{N}}$ is a block sequence.

(iii) For $r > 1$, $j_r = \sigma_{\mathcal{F}}(g_{\chi_1}, \ldots, g_{\chi_{r-1}})$.

Claim. The sequence $(d_{\chi_{2r-1}} - d_{\chi_{2r}})_{r \in \mathbb{N}}$ is a weakly null sequence in $\mathfrak{X}_{\mathcal{F}_2}$.

Proof of the claim. From the analogue of Proposition 1.19 the space $\mathfrak{X}_{\mathcal{F}_2}^*$ is the closed linear span of the pointwise closure $\overline{\mathcal{F}_2}^{w^*}$ of the set $\mathcal{F}_2$. From the observation after Theorem 3.7 we have that

$$\overline{\mathcal{F}_2}^{w^*} = F_0 \cup \Big\{ \sum_{i=1}^{\infty} a_i x_i^* : \sum_{i=1}^{\infty} a_i^2 \leq 1, \ (x_i^*)_{i=1}^d \text{ are } \sigma_{\mathcal{F}} \text{ special functionals}$$

$$\text{with } (\text{ind}(x_i^*))_{i=1}^d \text{ pairwise disjoint } \min \operatorname{supp} x_i^* \geq d \Big\}.$$

Thus it is enough to show that $b^*(d_{\chi_{2r-1}} - d_{\chi_{2r}}) \xrightarrow{r \to \infty} 0$ for every branch b.

Let b be an arbitrary branch. If $b = (g_{\chi_1}, g_{\chi_2}, g_{\chi_3}, g_{\chi_4}, \dots)$ from we obtain that $g(d_{\chi_{2r-1}} - d_{\chi_{2r}}) = g_{\chi_{2r-1}}(d_{\chi_{2r-1}}) - g_{\chi_{2r}}(d_{\chi_{2r}}) = \frac{1}{2} - \frac{1}{2} = 0$ for every r. If $b \neq (g_{\chi_1}, g_{\chi_2}, g_{\chi_3}, g_{\chi_4}, \dots)$ the injectivity of the coding function $\sigma_{\mathcal{F}}$ yields that there exist $r_0 \in \mathbb{N}$ such that $j_r \notin \text{ind}(b^*)$ for all $r > 2r_0$. Hence, for $r > r_0$, Lemma 7.4 yields that $|b(d_{\chi_{2r-1}} - d_{\chi_{2r}})| \leq |b(d_{\chi_{2r-1}})| + |b(d_{\chi_{2r}})| < \frac{3}{m_{4j_{2r-1}-3}} + \frac{3}{m_{4j_{2r}-3}} < \frac{4}{m_{4j_{2r-1}-3}}$ and therefore $b^*(d_{\chi_{2r-1}} - d_{\chi_{2r}}) \xrightarrow{r \to \infty} 0$.

The proof of the claim is complete. □

It follows from the claim that there exists a convex combination of the sequence $(d_{\chi_{2r-1}} - d_{\chi_{2r}})_{r \in \mathbb{N}}$ with norm less than $\frac{\delta}{3}$; let $(a_r)_{r=1}^d$ be nonnegative reals with $\sum_{r=1}^{d} a_r = 1$ such that $\| \sum_{r=1}^{d} a_r (d_{\chi_{2r-1}} - d_{\chi_{2r}}) \| < \frac{\delta}{3}$.

We set $g = \sum_{r=1}^{2d} g_{\chi_r}$ and $g' = \sum_{r=1}^{d} (g_{\chi_{2r-1}} - g_{\chi_{2r}})$. Since $g \in \mathcal{F}_2$ we have that $\|g\| \leq 1$. On the other hand

$$\|g'\| \geq \frac{g'\Big(\sum_{r=1}^{d} a_r (d_{\chi_{2r-1}} - d_{\chi_{2r}}) \Big)}{\| \sum_{r=1}^{d} a_r (d_{\chi_{2r-1}} - d_{\chi_{2r}}) \|}$$

$$> \frac{\sum_{r=1}^{d} a_r (g_{\chi_{2r-1}} - g_{\chi_{2r}})(d_{\chi_{2r-1}} - d_{\chi_{2r}})}{\frac{\delta}{3}} = \frac{\sum_{r=1}^{d} a_r (\frac{1}{2} + \frac{1}{2})}{\frac{\delta}{3}} = \frac{3}{\delta}.$$

For each $r \leq 2d$ with r odd we select $z_r^* \in Z_1$ such that $\|z_r^* - F_{\chi_r}\| < \frac{1}{m_{4j_r-3}}$, while for r even we select $z_r^* \in Z_2$ such that $\|z_r^* - F_{\chi_r}\| < \frac{1}{m_{4j_r-3}}$. We set

$$F_1 = \sum_{r=1}^{d} z_{2r-1}^* (\in Z_1) \quad \text{and} \quad F_2 = \sum_{r=1}^{d} z_{2r}^* (\in Z_2).$$

From our choice of z_r^* and the analogue of Lemma 4.5 we get that

(18)
$$\sum_{r=1}^{2d} \|z_r - g_{\chi_r}\| \leq \sum_{r=1}^{2d} (\|z_r - F_{\chi_r}\| + \|F_{\chi_r} - g_{\chi_r}\|) \leq \sum_{r=1}^{2d} \Big(\frac{1}{m_{4j_r-3}} + \frac{1}{m_{4j_r-3}} \Big) < 1.$$

From (18) we obtain that $\|(F_1 + F_2) - g\| < 1$ and $\|(F_1 - F_2) - g'\| < 1$. Thus, the facts that $\|g\| \leq 1$ and $\|g'\| > \frac{3}{\delta}$ yield that $\|F_1 + F_2\| < 2$ and $\|F_1 - F_2\| > \frac{3}{\delta} - 1 > \frac{2}{\delta}$, therefore $\|F_1 + F_2\| < \delta \|F_1 + F_2\|$. The proof of the proposition is complete. □

Proposition 7.6. The space $(\mathfrak{X}^{us}_{\mathcal{F}_2})_*$ is HJT. In particular the space $(\mathfrak{X}^{us}_{\mathcal{F}_2})_*$ contains no reflexive subspace and every infinite dimensional subspace Z of $(\mathfrak{X}^{us}_{\mathcal{F}_2})_*$ has nonseparable second dual Z^{**}.

Proof. The proof is identical to that of Theorem 4.7. $\qquad\square$

Theorem 7.7. The space $\mathfrak{X}^{us}_{\mathcal{F}_2}$ has the following properties

(i) Every subspace Y of $\mathfrak{X}^{us}_{\mathcal{F}_2}$ contains a further subspace Z which is reflexive and has an unconditional basis.

(ii) The predual $(\mathfrak{X}^{us}_{\mathcal{F}_2})_*$ of the space $\mathfrak{X}^{us}_{\mathcal{F}_2}$ is Hereditarily Indecomposable and has no reflexive subspace.

Proof. First, the identity operator $I : \mathfrak{X}^{us}_{\mathcal{F}_2} \to JT_{\mathcal{F}_2}$, being strongly strictly singular, is strictly singular (Proposition 2.3). Therefore the space $\mathfrak{X}^{us}_{\mathcal{F}_2}$ is reflexively saturated. Let Z be an arbitrary block subspace of the space $\mathfrak{X}^{us}_{\mathcal{F}_2}$. From the fact that $I : \mathfrak{X}^{us}_{\mathcal{F}_2} \to JT_{\mathcal{F}_2}$ is strictly singular we may choose a block sequence $(z_k)_{k\in\mathbb{N}}$ in Z with $\|z_k\| = 1$ and $\sum_{k=1}^{\infty} \|z_k\|_{\mathcal{F}_2} < \frac{1}{16}$. We may prove that $(z_k)_{k\in\mathbb{N}}$ is an unconditional basic sequence following the procedure used in the proof of Proposition 3.6 of [AT2].

The facts that the space $(\mathfrak{X}^{us}_{\mathcal{F}_2})_*$ is Hereditarily Indecomposable and has no reflexive subspace have been proved in Propositions 7.5 and 7.6. $\qquad\square$

Defining the norming set of the present section using $\mathcal{F}_s$ (instead of $\mathcal{F}_2$) in the first inductive step (namely in the definition of K_0) and using the saturation methods $(\mathcal{S}_{n_j}, \frac{1}{m_j})_j$ (instead of $(\mathcal{A}_{n_j}, \frac{1}{m_j})_j$) we produce a Banach space $\mathfrak{X}^{us}_{\mathcal{F}_s}$ which is unconditionally saturated while its predual and its dual share similar properties with the space $\mathfrak{X}_{\mathcal{F}_s}$ of Section 6. Namely we have the following.

Theorem 7.8. There exists a Banach space $\mathfrak{X}^{us}_{\mathcal{F}_s}$ with the properties:

(i) The predual $(\mathfrak{X}^{us}_{\mathcal{F}_s})_*$ of $\mathfrak{X}^{us}_{\mathcal{F}_s}$ is HI and every infinite dimensional subspace of $(\mathfrak{X}^{us}_{\mathcal{F}_s})_*$ has nonseparable second dual. In particular $(\mathfrak{X}^{us}_{\mathcal{F}_s})_*$ contains no reflexive subspace.

(ii) The space $\mathfrak{X}^{us}_{\mathcal{F}_s}$ is unconditionally and reflexively saturated.

(iii) The dual space $(\mathfrak{X}^{us}_{\mathcal{F}_s})^*$ is nonseparable HI and contains no reflexive subspace.

APPENDIX A. THE AUXILIARY SPACE AND THE BASIC INEQUALITY

The basic inequality is the main tool in providing upper bounds for the action of functionals on certain vectors of X_G. It has appeared in several variants in previous works like [AT1], [ALT], [ArTo]. In this section we present another variant which mainly concerns the case of strongly strictly singular extensions and in particular we provide the proof of Proposition 1.7 stated in Section 1. The proof of the present variant follows the same lines as the previous ones.

Definition A.1. The tree T_f of a functional $f \in W$. Let $f \in D$. By a tree of f (or tree corresponding to the analysis of f) we mean a finite family $T_f = (f_a)_{a \in \mathcal{A}}$ indexed by a finite tree $\mathcal{A}$ with a unique root $0 \in \mathcal{A}$ such that the following conditions are satisfied:

1. $f_0 = f$ and $f_a \in D$ for all $a \in \mathcal{A}$.
2. An $a \in \mathcal{A}$ is maximal if and only if $f_a \in G$.
3. For every $a \in \mathcal{A}$ which is not maximal, denoting by S_a the set of the immediate successors of a, exactly one of the following holds:

   (a) $S_a = \{\beta_1, \dots, \beta_d\}$ with $f_{\beta_1} < \cdots < f_{\beta_d}$ and there exists $j \in \mathbb{N}$ such that $d \le n_j$ and $f_a = \frac{1}{m_j} \sum_{i=1}^{d} f_{\beta_i}$ (recall that in this case we say that f_a is of type I).

   (b) $S_a = \{\beta_1, \dots, \beta_d\}$ and there exists a family of positive rationals $\{r_{\beta_i} : i = 1, \dots, d\}$ with $\sum_{i=1}^{d} r_{\beta_i} = 1$ such that $f_a = \sum_{i=1}^{d} r_{\beta_i} f_{\beta_i}$. Moreover for all $i = 1, \dots, d$, $\operatorname{ran} f_{\beta_i} \subset \operatorname{ran} f_a$. (recall that in this case we say that f_a is of type II).

It is obvious that every $f \in D$ has a tree which is not necessarily unique.

Definition A.2. (The auxiliary space T_{j_0}) Let $j_0 > 1$ be fixed. We set $C_{j_0} = \{\sum_{i \in F} \pm e_i^* : \#(F) \le n_{j_0-1}\}$.

The auxiliary space T_{j_0} is the completion of $(c_{00}(\mathbb{N}), \| \|_{D_{j_0}})$ where the norming set D_{j_0} is defined to be the minimal subset of $c_{00}(\mathbb{N})$ which (i) Contains C_{j_0}. (ii) It is closed under $(\mathcal{A}_{5n_j}, \frac{1}{m_j})$ operations for all $j \in \mathbb{N}$. (iii) It is rationally convex.

Observe that the Schauder basis $(e_n)_{n \in \mathbb{N}}$ of T_{j_0} is 1-unconditional.

Remark A.3. Let D'_{j_0} be the minimal subset of $c_{00}(\mathbb{N})$ which (i) Contains C_{j_0}. (ii) Is closed under $(\mathcal{A}_{5n_j}, \frac{1}{m_j})$ operations for all $j \in \mathbb{N}$. We notice that each $f \in D'_{j_0}$ has a tree $(f_a)_{a \in \mathcal{A}}$ in which for $a \in \mathcal{A}$ which is not maximal, f is the result of an $(\mathcal{A}_{5n_j}, \frac{1}{m_j})$ operation (for some j) of the functionals $(f_\beta)_{\beta \in S_a}$.

It can be shown that D'_{j_0} is also a norming set for the space T_{j_0} and that for every $j \in \mathbb{N}$ we have that $\operatorname{conv}_{\mathbb{Q}}\{f \in D_{j_0} : w(f) = m_j\} = \operatorname{conv}_{\mathbb{Q}}\{f \in D'_{j_0} : w(f) = m_j\}$. For proofs of similar results in a different context we refer to [AT1] (Lemma 3.5).

Lemma A.4. Let $j_0 \in \mathbb{N}$ and $f \in D'_{j_0}$. Then for every family $k_1 < k_2 < \ldots < k_{n_{j_0}}$ we have that

$$(19) \qquad |f(\frac{1}{n_{j_0}} \sum_{l=1}^{n_{j_0}} e_{k_l})| \le \begin{cases} \frac{2}{m_i \cdot m_{j_0}}, & \text{if } w(f) = m_i, \ i < j_0 \\ \frac{1}{m_i}, & \text{if } w(f) = m_i, \ i \ge j_0 \end{cases}$$

In particular $\|\frac{1}{n_{j_0}} \sum_{l=1}^{n_{j_0}} e_{k_l}\|_{D_{j_0}} \le \frac{1}{m_{j_0}}$.

If we additionally assume that the functional f admits a tree $(f_\alpha)_{\alpha \in \mathcal{A}}$ such that $w(f_\alpha) \neq m_{j_0}$ for every $\alpha \in \mathcal{A}$, then we have that

$$(20) \quad |f(\frac{1}{n_{j_0}} \sum_{l=1}^{n_{j_0}} e_{k_l})| \leq \begin{cases} \frac{2}{m_i \cdot m_{j_0}^2}, & \text{if } w(f) = m_i, \ i < j_0 \\ \frac{1}{m_i}, & \text{if } w(f) = m_i, \ i > j_0 \end{cases} \leq \frac{1}{m_{j_0}^2}.$$

Proof. We first prove the following claim.

Claim. Let $h \in D'_{j_0}$. Then

(i) $\#\{k : |h(e_k)| > \frac{1}{m_{j_0}}\} < (5n_{j_0-1})^{\log_2(m_{j_0})}$.

(ii) If the functional h has a tree $(h_a)_{a \in \mathcal{A}}$ with $w(h_a) \neq m_{j_0}$ for each $a \in \mathcal{A}$ then
$$\#\{k : |h(e_k)| > \frac{1}{m_{j_0}^2}\} < (5n_{j_0-1})^{2\log_2(m_{j_0})}.$$

Proof of the claim. We shall prove only part (i) of the claim, as the proof of (ii) is similar. Let $(h_a)_{a \in \mathcal{A}}$ be a tree of h and let n be its height (i.e. the length of its maximal branch). We may assume that $|h(e_k)| > \frac{1}{m_{j_0}}$ for all $k \in \text{supp } h$. Let $h = h_0, h_1, \ldots, h_n$ be a maximal branch (then $h_n \in C_{j_0}$) and let $k \in \text{supp } h_n$. Then $\frac{1}{m_{j_0}} < |h(e_k)| = \prod_{l=0}^{n-1} \frac{1}{w(h_l)} \leq \frac{1}{2^n}$, hence $n \leq \log_2(m_{j_0}) - 1$.

On the other hand, since $|h(e_k)| > \frac{1}{m_{j_0}}$ for all $k \in \text{supp } h$, each h_a with a non maximal is a result of an $(\mathcal{A}_{5n_j}, \frac{1}{m_j})$ operation for $j \leq j_0-1$. An inductive argument yields that for $i \leq n$ the cardinality of the set $\{h_a : |a| = i\}$ is less or equal to $(5n_{j_0-1})^i$. The facts that $n \leq \log_2(m_{j_0}) - 1$ and that each element of $g \in C_{j_0}$ has $\#(\text{supp}(g)) \leq n_{j_0-1}$ yield that $\#(\text{supp}(h)) \leq n_{j_0-1}(5n_{j_0-1})^{\log_2(m_{j_0})-1} < (5n_{j_0-1})^{\log_2(m_{j_0})}$.

The proof of the claim is complete. □

We pass to the proof of the lemma. The case $w(f) = m_i, \ i \geq j_0$ is straightforward. Let $f \in D'_{j_0}$ with $w(f) = m_i, \ i < j$. Then $f = \frac{1}{m_i} \sum_{t=1}^{d} f_t$ where $f_1 < \cdots < f_d$ belong to D'_{j_0} and $d \leq n_i$.

For $t = 1, \ldots, d$ we set $H_t = \{k : |f_t(e_k)| > \frac{1}{m_{j_0}}\}$. Part (i) of the claim yields that $\#(H_t) < (5n_{j_0-1})^{\log_2(m_{j_0})}$. Thus, setting $H = \bigcup_{t=1}^{d} H_t$, we get that $\#(H) < d(5n_{j_0-1})^{\log_2(m_{j_0})} \leq (5n_{j_0-1})^{\log_2(m_{j_0})+1}$. Therefore

$$|f(\frac{1}{n_{j_0}} \sum_{l=1}^{n_{j_0}} e_{k_l})| \leq \frac{1}{m_i}\left(|(\sum_{t=1}^{d} f_t)_{|H}(\frac{1}{n_{j_0}} \sum_{l=1}^{n_{j_0}} e_{k_l})|\right)$$

$$+ \frac{1}{m_i}\left(|(\sum_{t=1}^{d} f_t)_{|(\mathbb{N}\setminus H)}(\frac{1}{n_{j_0}} \sum_{l=1}^{n_{j_0}} e_{k_l})|\right)$$

$$\leq \frac{1}{m_i}\#(H)\frac{1}{n_{j_0}} + \frac{1}{m_i}\frac{1}{m_{j_0}} < \frac{2}{m_i m_{j_0}}.$$

The second part is proved similarly by using part (ii) of the claim. □

Proposition A.5. (The basic inequality) Let $(x_k)_{k \in \mathbb{N}}$ be a (C, ε) R.I.S. in X_G and $j_0 > 1$ such that for every $g \in G$ the set $\{k : |g(x_k)| > \varepsilon\}$ has cardinality at most $n_{j_0 - 1}$. Let $(\lambda_k)_{k \in \mathbb{N}} \in c_{00}$ be a sequence of scalars. Then for every $f \in D$ of type I we can find g_1, such that either $g_1 = h_1$ or $g_1 = e_t^* + h_1$ with $t \notin \operatorname{supp} h_1$ where $h_1 \in \operatorname{conv}_{\mathbb{Q}}\{h \in D'_{j_0} : w(h) = w(f)\}$ and $g_2 \in c_{00}(\mathbb{N})$ with $\|g_2\|_\infty \leq \varepsilon$ with g_1, g_2 having nonnegative coordinates and such that

$$(21) \qquad |f(\sum \lambda_k x_k)| \leq C(g_1 + g_2)(\sum |\lambda_k| e_k).$$

If we additionally assume that for every $h \in D$ with $w(h) = m_{j_0}$ and every interval E of the natural numbers we have that

$$(22) \qquad |h(\sum_{k \in E} \lambda_k x_k)| \leq C(\max_{k \in E} |\lambda_k| + \varepsilon \sum_{k \in E} |\lambda_k|)$$

then, if $w(f) \neq m_{j_0}$, h_1 may be selected satisfying additionally the following property: $h_1 = \sum r_l \tilde{h}_l$ with $r_l \in \mathbb{Q}^+$, $\sum r_l = 1$ and for each l the functional $\tilde{h}_l$ belongs to D'_{j_0} with $w(\tilde{h}_l) = w(f)$ and admits a tree $T_{\tilde{h}_l} = (f_a^l)_{a \in \mathcal{C}_l}$ with $w(f_a^l) \neq m_{j_0}$ for all $a \in \mathcal{C}_l$.

Proof. The proof in the general case (where (22) is not assumed) and in the special case (where we assume (22)) is actually the same. We shall give the proof only in the special case. The proof in the general case arises by omitting any reference to distinguishing cases whether a functional has weight m_{j_0} or not and treating the functionals with $w(f) = m_{j_0}$ as for any other j.

We fix a tree $T_f = (f_a)_{a \in \mathcal{A}}$ of f. Before passing to the proof we adopt some useful notation and state two lemmas. Their proofs can be found in [AT1] (Lemmas 4.4 and 4.5).

Definition A.6. For each $k \in \mathbb{N}$ we define the set $\mathcal{A}_k$ as follows:

$$\mathcal{A}_k = \Big\{ a \in \mathcal{A} \text{ such that } f_a \text{ is not of type } II \text{ and}$$

(i) $\operatorname{ran} f_a \cap \operatorname{ran} x_k \neq \emptyset$

(ii) $\forall\, \gamma < a$ if f_γ is of type I then $w(f_\gamma) \neq m_{j_0}$

(iii) $\forall\, \beta \leq a$ if $\beta \in S_\gamma$ and f_γ is of type I
then $\operatorname{ran} f_\beta \cap \operatorname{ran} x_k = \operatorname{ran} f_\gamma \cap \operatorname{ran} x_k$

(iv) if $w(f_a) \neq m_{j_0}$ then for all $\beta \in S_a$

$$\operatorname{ran} f_\beta \cap \operatorname{ran} x_k \subsetneqq \operatorname{ran} f_a \cap \operatorname{ran} x_k \Big\}$$

The next lemma describes the properties of the set $\mathcal{A}_k$.

Lemma A.7. For every $k \in \mathbb{N}$ we have the following:

(i) If $a \in \mathcal{A}$ and f_a is of type II then $a \notin \mathcal{A}_k$.
(Hence $\mathcal{A}_k \subset \{a \in \mathcal{A} : f_a \text{ is of type } I \text{ or } f_a \in G\}$.)
(ii) If $a \in \mathcal{A}_k$, then for every $\beta < a$ if f_β is of type I then $w(f_\beta) \neq m_{j_0}$.
(iii) If $\mathcal{A}_k$ is not a singleton then its members are incomparable members of the tree $\mathcal{A}$. Moreover if a_1, a_2 are two different elements of $\mathcal{A}_k$

and β is the (necessarily uniquely determined) maximal element of $\mathcal{A}$ satisfying $\beta < a_1$ and $\beta < a_2$ then f_β is of type II.

(iv) If $a \in \mathcal{A}$ is such that supp $f_a \cap$ ran $x_k \neq \emptyset$ and $\gamma \notin \mathcal{A}_k$ for all $\gamma < a$ then there exists $\beta \in \mathcal{A}_k$ with $a \leq \beta$. In particular if supp $f \cap$ ran $x_k \neq \emptyset$ then $\mathcal{A}_k \neq \emptyset$.

Definition A.8. For every $a \in \mathcal{A}$ we define $D_a = \bigcup_{\beta \geq a}\{k : \beta \in \mathcal{A}_k\}$.

Lemma A.9. According to the notation above we have the following:

(i) If supp $f \cap$ ran $x_k \neq \emptyset$ then $k \in D_0$ (recall that 0 denotes the unique root of $\mathcal{A}$ and $f = f_0$). Hence $f(\sum \lambda_k x_k) = f(\sum_{k \in D_0} \lambda_k x_k)$.

(ii) If f_a is of type I with $w(f_a) = m_{j_0}$ then D_a is an interval of $\mathbb{N}$.

(iii) If f_a is of type I with $w(f_a) \neq m_{j_0}$ then

$$\left\{\{k\} : k \in D_a \setminus \bigcup_{\beta \in S_a} D_\beta\right\} \cup \{D_\beta : \beta \in S_a\}$$

is a family of successive subsets of $\mathbb{N}$. Moreover for every $k \in D_a \setminus \bigcup_{\beta \in S_a} D_\beta$ (i.e. for k such that $a \in \mathcal{A}_k$) such that supp $f_a \cap$ ran $x_k \neq \emptyset$ there exists a $\beta \in S_a$ such that either min supp $x_k \leq$ max supp $f_\beta <$ max supp x_k or min supp $x_k <$ min supp $f_\beta \leq$ max supp x_k.

(iv) If f_a is of type II, $\beta \in S_a$ and $k \in D_a \setminus D_\beta$ then supp $f_\beta \cap$ ran $x_k = \emptyset$ and hence $f_\beta(x_k) = 0$.

Recall that we have fixed a tree $(f_a)_{a \in \mathcal{A}}$ for the given f. We construct two families $(g_a^1)_{a \in \mathcal{A}}$ and $(g_a^2)_{a \in \mathcal{A}}$ such that the following conditions are fulfilled.

(i) For every $a \in \mathcal{A}$ such that f_a is not of type II, $g_a^1 = h_a$ or $g_a^1 = e_{k_a}^* + h_a$ with $t_a \notin$ supp h_a, where $h_a \in \text{conv}_{\mathbb{Q}}(D'_{j_0})$ and $g_a^2 \in c_{00}(\mathbb{N})$ with $\|g_a^2\|_\infty \leq \varepsilon$.

(ii) For every $a \in \mathcal{A}$, supp $g_a^1 \subset D_a$ and supp $g_a^2 \subset D_a$ and the functionals g_a^1, g_a^2 have nonnegative coordinates.

(iii) For $a \in \mathcal{A}$ with $f_a \in G$ and $D_a \neq \emptyset$ we have that $g_a^1 \in C_{j_0}$.

(iv) For f_a of type II with $f = \sum_{\beta \in S_a} r_\beta f_\beta$ (where $r_\beta \in \mathbb{Q}^+$ for every $\beta \in S_a$ and $\sum_{\beta \in S_a} r_\beta = 1$) we have $g_a^1 = \sum_{\beta \in S_a} r_\beta g_\beta^1$ and $g_a^2 = \sum_{\beta \in S_a} r_\beta g_\beta^2$.

(v) For f_a of type I with $w(f) = m_{j_0}$ we have $g_a^1 = e_{k_a}^*$ where $k_a \in D_a$ is such that $|\lambda_{k_a}| = \max_{k \in D_a} |\lambda_k|$ and $g_a^2 = \sum_{k \in D_a} \varepsilon e_k^*$.

(vi) For f_a of type I with $w(f) = m_j$ for $j \neq j_0$ we have $g_a^1 = h_a$ or $g_a^1 = e_{k_a}^* + h_a$ with $h_a \in \text{conv}_{\mathbb{Q}}\{h \in D'_{j_0} : w(h) = m_j\}$ and $k_a \notin$ supp h_a.

(vii) For every $a \in \mathcal{A}$ the following inequality holds:

$$|f_a(\sum_{k \in D_a} \lambda_k x_k)| \leq C(g_a^1 + g_a^2)(\sum_{k \in D_a} |\lambda_k| e_k).$$

When the construction of $(g_a^1)_{a \in \mathcal{A}}$ and $(g_a^2)_{a \in \mathcal{A}}$ has been accomplished, we set $g_1 = g_0^1$ and $g_2 = g_0^2$ (where 0 is the root of $\mathcal{A}$ and $f = f_0$) and we observe that these are the desired functionals. To show that such $(g_a^1)_{a \in \mathcal{A}}$ and $(g_a^2)_{a \in \mathcal{A}}$

exist we use finite induction starting with $a \in \mathcal{A}$ which are maximal and in the general inductive step we assume that g_β^1, g_β^2 have been defined for all $\beta > a$ satisfying the inductive assumptions and we define g_a^1 and g_a^2.

1 $\overset{st}{=}$ **inductive step**

Let $a \in \mathcal{A}$ which is maximal. Then $f_a \in G$. If $D_a = \emptyset$ we define $g_a^1 = 0$ and $g_a^2 = 0$. If $D_a \neq \emptyset$ we set

$$E_a = \{k \in D_a : |f_a(x_k)| > \varepsilon\} \quad \text{and} \quad F_a = D_a \setminus E_a.$$

From our assumption we have that $\#(E_a) \leq n_{j_0-1}$ and we define

$$g_a^1 = \sum_{k \in E_a} e_k^* \quad \text{and} \quad g_a^2 = \sum_{k \in F_a} \varepsilon e_k^*.$$

We observe that $g_a^1 \in C_{j_0}$ and $\|g_a^2\|_\infty \leq \varepsilon$. Inequality (vii) is easily checked (see Proposition 4.3 of [AT1]).

General inductive step

Let $a \in \mathcal{A}$ and suppose that g_γ^1 and g_γ^2 have been defined for every $\gamma > a$ satisfying the inductive assumptions. If $D_a = \emptyset$ we set $g_a^1 = 0$ and $g_a^2 = 0$. In the remainder of the proof we assume that $D_a \neq \emptyset$. We consider the following three cases:

1 $\overset{st}{=}$ **case** The functional f_a is of type II.

Let $f_a = \sum_{\beta \in S_a} r_\beta f_\beta$ where $r_\beta \in \mathbb{Q}^+$ are such that $\sum_{\beta \in S_a} r_\beta = 1$. In this case, we have that $D_a = \bigcup_{\beta \in S_a} D_\beta$. We define

$$g_a^1 = \sum_{\beta \in S_a} r_\beta g_\beta^1 \quad \text{and} \quad g_a^2 = \sum_{\beta \in S_a} r_\beta g_\beta^2.$$

For the proof of inequality (vii) see Proposition 4.3 of [AT1].

2 $\overset{nd}{=}$ **case** The functional f_a is of type I with $w(f) = m_{j_0}$.

In this case D_a is an interval of the natural numbers (Lemma A.9(ii)). Let $k_a \in D_a$ be such that $|\lambda_{k_a}| = \max_{k \in D_a} |\lambda_k|$. We define

$$g_a^1 = e_{k_a}^* \quad \text{and} \quad g_a^2 = \sum_{k \in D_a} \varepsilon e_k^*.$$

Inequality (vii) is easily established.

3 $\overset{rd}{=}$ **case** The functional f_a is of type I with $w(f) = m_j$ for $j \neq j_0$.

Then $f_a = \frac{1}{m_j} \sum_{\beta \in S_a} f_\beta$ and the family $\{f_\beta : \beta \in S_a\}$ is a family of successive functionals with $\#(S_a) \leq n_j$. We set

$$
\begin{aligned}
E_a \;&=\; \{k : a \in \mathcal{A}_k \text{ and } \operatorname{supp} f_a \cap \operatorname{ran} x_k \neq \emptyset\} \\
&(= \{k \in D_a \setminus \bigcup_{\beta \in S_a} D_\beta : \operatorname{supp} f_a \cap \operatorname{ran} x_k \neq \emptyset\}).
\end{aligned}
$$

We consider the following partition of E_a.

$$E_a^2 = \{k \in E_a : m_{j_{k+1}} \leq m_j\} \quad \text{and} \quad E_a^1 = E_a \setminus E_a^2.$$

We define

$$g_a^2 = \sum_{k \in E_a^2} \varepsilon e_k^* + \sum_{\beta \in S_a} g_\beta^2.$$

Observe that $\|g_a^2\|_\infty \leq \varepsilon$. Let $E_a^1 = \{k_1 < k_2 < \cdots < k_l\}$. From the definition of E_a^1 we get that $m_j < m_{j_{k_2}} < \cdots < m_{j_{k_l}}$. We set

$$k_a = k_1 \quad \text{and} \quad g_a^1 = e_{k_a}^* + h_a \quad \text{where} \quad h_a = \frac{1}{m_j}\left(\sum_{i=2}^{l} e_{k_i}^* + \sum_{\beta \in S_a} g_\beta^1\right)$$

(The term $e_{k_a}^*$ does not appear if $E_a^1 = \emptyset$.)

For the verification of inequality (vii) see Proposition 4.3 of [AT1].

It remains to show that $h_a \in \text{conv}_{\mathbb{Q}}\{h \in D_{j_0}' : w(h) = m_j\}$ By the second part of Lemma A.9(iii), for every $k \in E_a$ there exists an element of the set $N = \{\min \text{supp} f_\beta, \max \text{supp} f_\beta : \beta \in S_a\}$ belonging to $\text{ran } x_k$. Hence $\#(E_a^1) \leq \#(E_a) \leq 2n_j$.

We next show that $h_a \in \text{conv}_{\mathbb{Q}}\{g \in D_{j_0}' : w(g) = m_j\}$. We first examine the case that for every $\beta \in S_a$ the functional f_β is not of type II. Then for every $\beta \in S_a$ one of the following holds:

(i) $f_\beta \in G$. In this case $g_\beta^1 \in C_{j_0}$ (by the first inductive step).

(ii) f_β is of type I with $w(f_\beta) = m_{j_0}$. In this case $g_\beta^1 = e_{k_\beta}^* \in D_{j_0}'$.

(iii) f_β is of type I with $w(f_\beta) = m_j$ for $j \neq j_0$. In this case $g_\beta^1 = e_{k_\beta}^* + h_\beta$ (or $g_\beta^1 = h_\beta$) where $h_\beta \in \text{conv}_{\mathbb{Q}}(D_{j_0}')$ and $k_\beta \notin \text{supp } h_\beta$. We set $E_\beta^1 = \{n \in \mathbb{N} : n < k_\beta\}$, $E_\beta^2 = \{n \in \mathbb{N} : n > k_\beta\}$ and $h_\beta^1 = E_\beta^1 h_\beta$, $h_\beta^2 = E_\beta^2 h_\beta$. The functionals h_β^1, $e_{t_\beta}^*$, h_β^2 are successive and belong to $D_{j_0} = \text{conv}_{\mathbb{Q}}(D_{j_0}')$.

We set

$$
\begin{aligned}
T_a^1 &= \{\beta \in S_a : f_\beta \in G\} \\
T_a^2 &= \{\beta \in S_a : f_\beta \text{ of type } I \text{ and } w(f_\beta) = m_{j_0}\} \\
T_a^3 &= \{\beta \in S_a : f_\beta \text{ of type } I \text{ and } w(f_\beta) \neq m_{j_0}\}.
\end{aligned}
$$

The family of successive (see Lemma A.9(iii)) functionals of D_{j_0},

$$\{e_{k_i}^* : i = 2, \ldots, l\} \cup \{g_\beta^1 : \beta \in T_a^1\} \cup \{g_\beta^1 : \beta \in T_a^2\} \cup$$
$$\cup \{h_\beta^1 : \beta \in T_a^3\} \cup \{e_{k_\beta}^* : \beta \in T_a^3\} \cup \{h_\beta^2 : \beta \in T_a^3\}$$

has cardinality $\leq 5n_j$, thus we get that $h_a \in D_{j_0}$ with $w(h_a) = m_j$. Therefore from Remark A.3 we get that

$$h_a \in \text{conv}_{\mathbb{Q}}\{h \in D_{j_0}' : w(h) = m_j\}.$$

For the case that for some $\beta \in S_a$ the functional f_β is of type II see [AT1] Proposition 4.3. $\qquad\square$

Proof of Proposition 1.7. The proof is an application of the basic inequality (Proposition A.5) and Lemma A.4. Indeed, let $f \in D$ with $w(f) = m_i$.

Proposition A.5 yields the existence of a functional h_1 with $h_1 \in \mathrm{conv}_{\mathbb{Q}}\{h \in D'_{j_0} : w(h) = m_i\}$, a $t \in \mathbb{N}$ and a $h_2 \in c_{00}(\mathbb{N})$ with $\|h_2\|_\infty \leq \varepsilon$, such that

$$|f(\frac{1}{n_{j_0}} \sum_{k=1}^{n_{j_0}} x_k)| \leq C(e_t^* + h_1 + h_2)(\frac{1}{n_{j_0}} \sum_{k=1}^{n_{j_0}} e_k).$$

If $i \geq j_0$ we get that $|f(\frac{1}{n_{j_0}} \sum_{k=1}^{n_{j_0}} x_k)| \leq C(\frac{1}{n_{j_0}} + \frac{1}{m_i} + \varepsilon) < \frac{C}{n_{j_0}} + \frac{C}{m_i} + C\varepsilon$. If $i < j_0$, using Lemma A.4 we get that $|f(\frac{1}{n_{j_0}} \sum_{k=1}^{n_{j_0}} x_k)| \leq C(\frac{1}{n_{j_0}} + \frac{2}{m_i \cdot m_{j_0}} + \varepsilon) < \frac{3C}{m_i \cdot m_{j_0}}$.

In order to prove 2) let $(b_k)_{k=1}^{n_{j_0}}$ be scalars with $|b_k| \leq 1$ such that (2) is satisfied. Then condition (22) of the basic inequality is satisfied for the linear combination $\frac{1}{n_{j_0}} \sum_{k=1}^{n_{j_0}} b_k x_k$ and thus for every $f \in D$ with $w(f) = m_i$, $i \neq j_0$, there exist a $t \in \mathbb{N}$ and $h_1, h_2 \in c_{00}(\mathbb{N})$ with h_1, h_2 having nonnegative coordinates and $\|h_2\|_\infty \leq \varepsilon$ such that

$$
\begin{aligned}
|f(\frac{1}{n_{j_0}} \sum_{k=1}^{n_{j_0}} b_k x_k)| &\leq C(e_t^* + h_1 + h_2)(\frac{1}{n_{j_0}} \sum_{k=1}^{n_{j_0}} |b_k| e_k) \\
&\leq C(e_t^* + h_1 + h_2)(\frac{1}{n_{j_0}} \sum_{k=1}^{n_{j_0}} e_k)
\end{aligned}
$$

with h_1 being a rational convex combination $h_1 = \sum r_l \tilde{h}_l$ and for each l the functional $\tilde{h}_l$ belongs to D'_{j_0} with $w(\tilde{h}_l) = m_i$ and has a tree $T_{\tilde{h}_l} = (f_a^l)_{a \in \mathcal{C}_l}$ with $w(f_a^l) \neq m_{j_0}$ for all $a \in \mathcal{C}_l$. Using the second part of Lemma A.4 we deduce that

$$|f(\frac{1}{n_{j_0}} \sum_{k=1}^{n_{j_0}} b_k x_k)| \leq C(\frac{1}{n_{j_0}} + \frac{1}{m_{j_0}^2} + \varepsilon) < \frac{4C}{m_{j_0}^2}.$$

For $f \in D$ with $w(f) = m_{j_0}$ it follows from condition (2) that $|f(\frac{1}{n_{j_0}} \sum_{k=1}^{n_{j_0}} b_k x_k)| \leq \frac{C}{n_{j_0}}(1 + \frac{2}{m_{j_0}^2} n_{j_0}) < \frac{4C}{m_{j_0}^2}$. $\qquad \square$

APPENDIX B. THE JAMES TREE SPACES $JT_{\mathcal{F}_{2,s}}$, $JT_{\mathcal{F}_2}$ AND $JT_{\mathcal{F}_s}$

In this part we continue the study of the James Tree spaces initialized in Section 3. We give a slightly different definition of JTG sequences and then we define the space $JT_{\mathcal{F}_2}$ exactly as in section 3. We also define the spaces $JT_{\mathcal{F}_s}$, $JT_{\mathcal{F}_{2,s}}$. We prove that $JT_{\mathcal{F}_2}$ is ℓ_2 saturated while $JT_{\mathcal{F}_s}$, $JT_{\mathcal{F}_{2,s}}$ are c_0 saturated. We also give an example of $JT_{\mathcal{F}_2}$, defined for a precise family $(F_j)_j$ such that the basis $(e_n)_{n \in \mathbb{N}}$ of the space is normalized weakly null and for every subsequence $(e_n)_{n \in M}$, $M \in [\mathbb{N}]$ the space $X_M = \overline{\mathrm{span}}\{e_n : n \in M\}$ has nonseparable dual. As we have mentioned before the study of the James Tree spaces does not require techniques related to HI constructions.

Definition B.1. (JTG families) A sequence $(F_j)_{j=0}^\infty$ of subsets of $c_{00}(\mathbb{N})$ is said to be a *James Tree Generating family* (JTG family) provided that it satisfies the following conditions:

(A) $F_0 = \{\pm e_n^* : n \in \mathbb{N}\}$ and each F_j is nonempty, countable, symmetric, compact in the topology of pointwise convergence and closed under restrictions to intervals of $\mathbb{N}$.

(B) Setting $\tau_j = \sup\{\|f\|_\infty : f \in F_j\}$, the sequence $(\tau_j)_{j \in \mathbb{N}}$ is strictly decreasing and $\sum_{j=1}^{\infty} \tau_j \le 1$.

(C) For every block sequence $(x_k)_{k \in \mathbb{N}}$ of $c_{00}(\mathbb{N})$, every $j = 0, 1, 2, \ldots$ and every $\delta > 0$ there exists a vector $x \in \text{span}\{x_k : k \in \mathbb{N}\}$ such that

$$\delta \cdot \sup\{f(x) : f \in \bigcup_{i=0}^{\infty} F_i\} > \sup\{f(x) : f \in F_j\}.$$

We set $F = \bigcup_{j=0}^{\infty} F_j$. The set F defines a norm $\| \ \|_F$ on $c_{00}(\mathbb{N})$ by the rule

$$\|x\|_F = \sup\{f(x) : f \in F\}.$$

The space Y_F is the completion of the space $(c_{00}(\mathbb{N}), \| \ \|_F)$.

Examples B.2. We provide some examples of JTG families.

(i) The first example is what we call the Maurey-Rosenthal JTG family, related to the first construction of a normalized weakly null sequence with no unconditional subsequence ([MR]). In particular the norming set for their example is the set $F = \bigcup_{j=0}^{\infty} F_j$ together with the $\sigma_{\mathcal{F}}$ special functionals resulting from the family F. We proceed defining the sets $(F_j)_{j=0}^{\infty}$.

Let $(k_j)_{j \in \mathbb{N}}$ be a strictly increasing sequence of integers such that

$$\sum_{j=1}^{\infty} \sum_{n \ne j} \min\{\frac{\sqrt{k_n}}{\sqrt{k_j}}, \frac{\sqrt{k_j}}{\sqrt{k_n}}\} \le 1.$$

We set $F_0 = \{\pm e_n^* : n \in \mathbb{N}\}$ while for $j = 1, 2, \ldots$ we set

$$F_j = \{\frac{1}{\sqrt{k_j}}(\sum_{i \in F} \pm e_i^*) : \emptyset \ne F \subset \mathbb{N}, \ \#(F) \le k_j\} \cup \{0\}.$$

The above conditions (1) and (2) for the sequence $(k_j)_{j \in \mathbb{N}}$ easily yield that $(F_j)_{j=0}^{\infty}$ is a JTG family.

(ii) The second example is the family introduced in Section 4. For completeness we recall its definition. Let $(m_j)_{j \in \mathbb{N}}$ and $(n_j)_{j \in \mathbb{N}}$ defined as follows:
  - $m_1 = 2$ and $m_{j+1} = m_j^5$.
  - $n_1 = 4$, and $n_{j+1} = (5n_j)^{s_j}$ where $s_j = \log_2 m_{j+1}^3$.

We set $F_0 = \{\pm e_n^* : n \in \mathbb{N}\}$ and for $j = 1, 2, \ldots$ we set

$$F_j = \{\frac{1}{m_{2j-1}^2}\sum_{i \in I} \pm e_i^* : \#(I) \le \frac{n_{2j-1}}{2}\} \cup \{0\}.$$

We shall show that the sequence $(F_j)_{j=0}^{\infty}$ is a JTG family. Conditions (A), (B) of Definition B.1 are obviously satisfied. Suppose that condition (C) fails. Then for some $j \in \mathbb{N}$, there exists a block sequence $(x_k)_{k \in \mathbb{N}}$ in $c_{00}(\mathbb{N})$ with $\|x_k\|_F = 1$ and a $\delta > 0$ such that $\delta \| \sum a_k x_k \|_F \leq \| \sum a_k x_k \|_{F_j}$ for every sequence of scalars $(a_k)_{k \in \mathbb{N}} \in c_{00}(\mathbb{N})$. We observe that $\|x_k\|_{\infty} \geq \frac{2\delta m_{2j-1}^2}{n_{2j-1}}$ for all k. Indeed, if $\|x_k\|_{\infty} < \frac{2\delta m_{2j-1}^2}{n_{2j-1}}$ then for every $f \in F_j$, $f = \frac{1}{m_{2j-1}^2} \sum_{i \in I} \pm e_i^*$, with $\#(I) \leq \frac{n_{2j-1}}{2}$ we would have that $|f(x_k)| \leq \frac{1}{m_{2j-1}^2} \sum_{i \in I} |e_i^*(x_k)| <$

$\frac{1}{m_{2j-1}^2} \frac{n_{2j-1}}{2} \frac{2\delta m_{2j-1}^2}{n_{2j-1}} = \delta$ which yields that $\|x_k\|_{F_j} < \delta$, a contradiction. Hence for each k we may select a $t_k \in \operatorname{supp} x_k$ such that $|e_{t_k}^*(x_k)| \geq \frac{2\delta m_{2j-1}^2}{n_{2j-1}}$. Since the sequence $(\frac{n_{2i-1}}{m_{2i-1}^2})_{i \in \mathbb{N}}$ increases to infinity we may choose a $j' \in \mathbb{N}$ such that $\delta^2 \frac{n_{2j'-1}}{m_{2j'-1}^2} > (\frac{n_{2j-1}}{m_{2j-1}^2})^2$. We consider the vector $y = \sum_{k=1}^{n_{2j'-1}/2} x_k$. We have that $\delta \|y\|_F \geq \delta \frac{1}{m_{2j'-1}^2} \sum_{k=1}^{n_{2j'-1}/2} |e_{t_k}^*(x_k)| \geq$

$\delta \frac{1}{m_{2j'-1}^2} \frac{n_{2j'-1}}{2} \frac{2\delta m_{2j-1}^2}{n_{2j-1}} > \frac{1}{m_{2j-1}^2} n_{2j-1}$. On the other hand $\|y\|_{F_j} \leq \frac{1}{m_{2j-1}^2} \frac{n_{2j-1}}{2}$, a contradiction.

(iii) Another example of a JTG family has been given in Definition 6.4 and we have used it to define the ground set for the space $\mathfrak{X}_{\mathcal{F}_s}$.

Remarks B.3.   (i) The standard basis $(e_n)_{n \in \mathbb{N}}$ of $c_{00}(\mathbb{N})$ is a normalized bimonotone Schauder basis of the space Y_F.

(ii) The set F is compact in the topology of pointwise convergence. Indeed, let $(f_n)_{n \in \mathbb{N}}$ be a sequence in F. There are two cases. Either there exists $j_0 \in \mathbb{N}$ such that the set F_{j_0} contains a subsequence of $(f_n)_{n \in \mathbb{N}}$ in which case the compactness of F_{j_0} yields the existence of a further subsequence converging pointwise to some $f \in F_{j_0}$, otherwise if no such j_0 exists, then we may find a subsequence $(f_{k_n})_{n \in \mathbb{N}}$ and a strictly increasing sequence $(i_n)_{n \in \mathbb{N}}$ of integers with $f_{k_n} \in F_{i_n}$ and thus $\|f_{k_n}\|_{\infty} \leq \tau_{i_n}$ for all n. Since condition (B) of Definition B.1 yields that $\tau_n \to 0$ we get that $f_{k_n} \xrightarrow{p} 0 \in F$.

(iii) The fact that F is countable and compact yields that the space $(C(F), \| \ \|_{\infty})$ is c_0 saturated [BP]. It follows that the space Y_F is also c_0 saturated, since Y_F is isometric to a subspace of $(C(F), \| \ \|_{\infty})$.

(iv) For each j we consider the seminorm $\| \ \|_{F_j} : c_{00}(\mathbb{N}) \to \mathbb{R}$ defined by $\|x\|_{F_j} = \sup\{|f(x)| : f \in F_j\}$. In general $\| \ \|_{F_j}$ is not a norm. Defining Y_{F_j} to be the completion of the space $(c_{00}(\mathbb{N}) , \| \ \|_{F_j})$, condition (C) of Definition B.1 is equivalent to saying that the identity operator $I : Y_F \to Y_{F_j}$ is strictly singular.

Furthermore, observe that setting $H_j = \cup_{i=0}^{j} F_i$ the identity operator $I : Y_F \to Y_{H_j}$ (Y_{H_j} is similarly defined) is also strictly singular.

Indeed, let $(x_k)_{k\in\mathbb{N}}$ be a block sequence of $c_{00}(\mathbb{N})$ and let $\delta > 0$. We choose a block sequence $(x_k^0)_{k\in\mathbb{N}}$ of $(x_k)_{k\in\mathbb{N}}$ with $\|x_k^0\|_F = 1$ and $\sum_{k=1}^{\infty} \|x_k^0\|_{F_1} < \delta$. Then for every $x \in \text{span}\{x_k^0 : k \in \mathbb{N}\}$ we have that $\delta\|x\|_F \geq \|x\|_{F_0}$. We then select a block sequence $(x_k^1)_{k\in\mathbb{N}}$ of $(x_k^0)_{k\in\mathbb{N}}$ such that $\delta\|x\|_F \geq \|x\|_{F_1}$ for every $x \in \text{span}\{x_k^1 : k \in \mathbb{N}\}$. Following this procedure, after $j + 1$ steps we may select a block sequence $(x_k^j)_{k\in\mathbb{N}}$ of $(x_k)_{k\in\mathbb{N}}$ such that $\delta\|x\|_F \geq \|x\|_{F_i}$ for $i = 1, \ldots, j$ and thus $\delta\|x\|_F \geq \|x\|_{H_j}$ for every $x \in \text{span}\{x_k^j : k \in \mathbb{N}\}$.

Next using the $\sigma_{\mathcal{F}}$ coding defined in Definition 3.2 we introduce the $\sigma_{\mathcal{F}}$ special sequences and functionals in the same manner as in Definition 3.3. For a $\sigma_{\mathcal{F}}$ special functional x^* the index $\text{ind}(x^*)$ has the analogous meaning. Finally we denote by $\mathcal{S}$ the set of all finitely supported $\sigma_{\mathcal{F}}$ special functionals.

The next proposition is an immediate consequence of the above definition and describes the tree-like interference of two $\sigma_{\mathcal{F}}$ special sequences.

Proposition B.4. Let $(f_i)_i$, $(h_i)_i$ be two distinct $\sigma_{\mathcal{F}}$ special sequences. Then $\text{ind}(f_i) \neq \text{ind}(h_j)$ for $i \neq j$ while there exists i_0 such that $f_i = h_i$ for all $i < i_0$ and $\text{ind}(f_i) \neq \text{ind}(h_i)$ for $i > i_0$.

Definition B.5. (The norming sets $\mathcal{F}_{2,s}$, $\mathcal{F}_2$, $\mathcal{F}_s$) Let $(F_j)_{j=0}^{\infty}$ be a JTG family. We set

$$\mathcal{F}_2 = F_0 \cup \left\{ \sum_{k=1}^{d} a_k x_k^* : a_k \in \mathbb{Q}, \sum_{k=1}^{d} a_k^2 \leq 1, x_k^* \in \mathcal{S} \cup \bigcup_{i=1}^{\infty} F_i, k = 1, \ldots, d \right.$$
$$\text{with } (\text{ind}(x_k^*))_{k=1}^{d} \text{ pairwise disjoint} \Big\}$$

$$\mathcal{F}_{2,s} = F_0 \cup \left\{ \sum_{k=1}^{d} a_k x_k^* : a_k \in \mathbb{Q}, \sum_{k=1}^{d} a_k^2 \leq 1, x_k^* \in \mathcal{S} \cup \bigcup_{i=1}^{\infty} F_i, k = 1, \ldots, d \right.$$
$$\text{with } (\text{ind}(x_k^*))_{k=1}^{d} \text{ pairwise disjoint and } \min \text{supp} \, x_k^* \geq d \Big\},$$

and

$$\mathcal{F}_s = F_0 \cup \left\{ \sum_{k=1}^{d} \varepsilon_k x_k^* : \varepsilon_1, \ldots, \varepsilon_d \in \{-1, 1\}, x_k^* \in \mathcal{S} \cup \bigcup_{i=1}^{\infty} F_i, k = 1, \ldots, d \right.$$
$$\text{with } (\text{ind}(x_i^*))_{i=1}^{d} \text{ pairwise disjoint and } \min \text{supp} \, x_i^* \geq d, d \in \mathbb{N} \}.$$

The space $JT_{\mathcal{F}_{2,s}}$ is defined as the completion of the space $(c_{00}(\mathbb{N}), \| \cdot \|_{\mathcal{F}_{2,s}})$, the space $JT_{\mathcal{F}_2}$ is defined to be the completion of the space $(c_{00}(\mathbb{N}), \| \cdot \|_{\mathcal{F}_2})$ while $JT_{\mathcal{F}_s}$ the completion of $(c_{00}(\mathbb{N}), \| \cdot \|_{\mathcal{F}_s})$ (where $\|x\|_{\mathcal{F}_*} = \sup\{f(x) : f \in \mathcal{F}\}$ for $x \in c_{00}(\mathbb{N})$, for either $\mathcal{F}_* = \mathcal{F}_2$ or $\mathcal{F}_* = \mathcal{F}_{2,s}$ or $\mathcal{F}_* = \mathcal{F}_s$). For a functional $f \in \mathcal{F}_* \setminus F_0$ of the form $f = \sum_{k=1}^{l} a_k x_k^*$ the set of its indices $\text{ind}(f)$ is defined to be the set $\text{ind}(f) = \bigcup_{k=1}^{l} \text{ind}(x_k^*)$.

Remark B.6. The standard basis $(e_n)_{n\in\mathbb{N}}$ of $c_{00}(\mathbb{N})$ is a normalized bimonotone Schauder basis for the space $JT_{\mathcal{F}_*}$.

Let's observe that the only difference between the definition of $\mathcal{F}_{2,s}$ and that of $\mathcal{F}_2$ is the way we connect the $\sigma_{\mathcal{F}}$ special functionals. In the case of $\mathcal{F}_2$ the $\sigma_{\mathcal{F}}$ special functionals are connected more freely than in $\mathcal{F}_{2,s}$ and obviously $\mathcal{F}_{2,s} \subset \mathcal{F}_2$. This difference leads the spaces $JT_{\mathcal{F}_{2,s}}$ and $JT_{\mathcal{F}_2}$ to have extremely different structures. We study the structure of these two spaces as well as the structure of $JT_{\mathcal{F}_s}$. Namely we have the following theorem.

Theorem B.7.   (i) The space $JT_{\mathcal{F}_{2,s}}$ is c_0 saturated.
   (ii) The space $JT_{\mathcal{F}_2}$ is ℓ_2 saturated.
   (iii) The space $JT_{\mathcal{F}_s}$ is c_0 saturated.

Proposition B.4 yields that the set of all finite $\sigma_{\mathcal{F}}$ special sequences is naturally endowed with a tree structure. The set of infinite branches of this tree structure is identified with the set of all infinite $\sigma_{\mathcal{F}}$ special sequences. For such a branch $b = (f_1, f_2, \ldots)$ the functional $b^* = \lim\limits_{d} \sum\limits_{i=1}^{d} f_i$ (where the limit is taken in the pointwise topology) is a cluster point of the sets $\mathcal{F}_2$, $\mathcal{F}_{2,s}$ and $\mathcal{F}_s$ and hence belongs to the unit balls of the dual spaces $JT_{\mathcal{F}_2}^*$, $JT_{\mathcal{F}_{2,s}}^*$ and $JT_{\mathcal{F}_s}^*$. Let also point out that a $\sigma_{\mathcal{F}}$ special functional x^* is either finite or takes the form Eb^* for some branch b and some interval E. Furthermore, it is easy to check that the set $\{Ex^* : E \text{ interval}, x^* \; \sigma_{\mathcal{F}} \text{ special functional}\}$ is closed in the pointwise topology.

Our main goal in this section is to prove Theorem B.7. Many of the Lemmas used in proving this theorem are common for $JT_{\mathcal{F}_{2,s}}$, $JT_{\mathcal{F}_2}$ and $JT_{\mathcal{F}_s}$. For this reason it is convenient to use the symbol $\mathcal{F}_*$ when stating or proving a property which is valid for $\mathcal{F}_* = \mathcal{F}_{2,s}$, $\mathcal{F}_* = \mathcal{F}_2$ and $\mathcal{F}_* = \mathcal{F}_s$.

Lemma B.8. The identity operator $I : JT_{\mathcal{F}_*} \to Y_F$ is strictly singular.

Proof. Assume the contrary. Then there exists a block subspace Y of $JT_{\mathcal{F}_*}$ such that the identity operator $I : (Y, \| \; \|_{\mathcal{F}_*}) \to (Y, \| \; \|_F)$ is an isomorphism. Since Y_F is c_0 saturated (Remark B.3 (iii)) we may assume that $(Y, \| \; \|_{\mathcal{F}_*})$ is is spanned by a block basis which is equivalent to the standard basis of c_0. Using property (C) of Definition B.1 and Remark B.3 (iv) we inductively choose a normalized block sequence $(x_n)_{n\in\mathbb{N}}$ in $(Y, \| \; \|_{\mathcal{F}_*})$ and a strictly increasing sequence $(j_n)_{n\in\mathbb{N}}$ of integers such that for some δ determined by the isomorphism, the following hold:

   (i) $\|x_n\|_{F_{j_n}} > \delta$.
   (ii) $\|x_{n+1}\|_{\bigcup_{k=1}^{j_n} F_k} < \delta$.

From (i) and (ii) and the definition of each $\mathcal{F}_*$ we easily get that $\|x_1 + \cdots + x_n\| \xrightarrow{n} \infty$. This is a contradiction since $(x_n)_{n\in\mathbb{N}}$, being a normalized block basis of a sequence equivalent to the standard basis of c_0, is also equivalent to the standard basis of c_0. $\quad\square$

The following lemma, although it refers exclusively to the functional b^*, its proof is crucially depended on the fact that in $\mathcal{F}_*$ we connect the special functionals under certain norms. A similar result is also obtained in [AT1] (Lemma 10.6).

Lemma B.9. Let $(x_n)_{n \in \mathbb{N}}$ be a bounded block sequence in $JT_{\mathcal{F}_*}$. Then there exists an $L \in [\mathbb{N}]$ such that for every branch b the limit $\lim\limits_{n \in L} b^*(x_n)$ exists. In particular, if the sequence $(x_n)_{n \in \mathbb{N}}$ is seminormalized (i.e. $\inf \|x_n\|_{\mathcal{F}_*} > 0$) and $L = \{l_1 < l_2 < l_3 < \cdots\}$ then the sequence $y_n = \frac{x_{l_{2n-1}} - x_{l_{2n}}}{\|x_{l_{2n-1}} - x_{l_{2n}}\|}$ satisfies $\|y_n\|_{\mathcal{F}_*} = 1$ and $\lim\limits_{n} b^*(y_n) = 0$ for every branch b.

Proof. We first prove the following claim.

Claim. For every $\varepsilon > 0$ and $M \in [\mathbb{N}]$ there exists $L \in [M]$ and a finite collection of branches $\{b_1, \ldots, b_l\}$ such that for every branch b with $b \notin \{b_1, \ldots, b_l\}$ we have that $\limsup\limits_{n \in L} |b^*(x_n)| \le \varepsilon$.

Proof of the claim. Assume the contrary. Then we may inductively construct a sequence $M_1 \supset M_2 \supset M_3 \cdots$ of infinite subsets of $\mathbb{N}$ and a sequence $b_1, b_2, b_3, \ldots$ of pairwise different branches satisfying $|b_i^*(x_n)| > \varepsilon$ for all $n \in M_i$.

We set $C = \sup\limits_{n} \|x_n\|_{\mathcal{F}_*}$ and we consider $k > \frac{C}{\varepsilon}$. Since the branches $b_1, b_2, \ldots, b_{k^2}$ are pairwise different we may choose an infinite interval E with $\min E \ge k^2$ such that the functionals $(Eb_i^*)_{i=1}^{k^2}$ have disjoint indices. We also consider any $n \in M_{k^2}$ with $\operatorname{supp} x_n \subset E$ and we set $\varepsilon_i = \operatorname{sgn} b_i^*(x_n)$ for $i = 1, 2, \ldots, k^2$. Then the functional $f = \sum\limits_{i=1}^{k^2} \frac{\varepsilon_i}{k} b_i^*$ belongs to $B_{JT_{\mathcal{F}_*}^*}$. Therefore

$$\|x_n\|_{\mathcal{F}_*} \ge f(x_n) = \sum_{i=1}^{k^2} \frac{1}{k} |b_i^*(x_n)| \ge \sum_{i=1}^{k^2} \frac{1}{k} \cdot \varepsilon = k \cdot \varepsilon > C,$$

a contradiction completing the proof of the claim. $\qquad\square$

Using the claim we inductively select a sequence $L_1 \supset L_2 \supset L_3 \supset \cdots$ of infinite subsets of $\mathbb{N}$ and a sequence $B_1, B_2, B_3, \ldots$ of finite collections of branches such that for every branch $b \notin B_i$ we have that $|b^*(x_n)| < \frac{1}{i}$ for all $n \in L_i$. We then choose a diagonal set L_0 of the nested sequence $(L_i)_{i \in \mathbb{N}}$. Then for every branch b not belonging to $B = \bigcup\limits_{i=1}^{\infty} B_i$ we have that $\lim\limits_{n \in L_0} b^*(x_n) = 0$. Since the set B is countable, we may choose, using a diagonalization argument, an $L \in [L_0]$ such that the sequence $(b^*(x_n))_{n \in L}$ converges for every $b \in F$. The set L clearly satisfies the conclusion of the lemma. $\qquad\square$

Combining Lemma B.8 and Lemma B.9 we get the following.

Corollary B.10. Every block subspace of $JT_{\mathcal{F}_*}$ contains a block sequence $(y_n)_{n \in \mathbb{N}}$ such that $\|y_n\|_{\mathcal{F}_*} = 1$, $\|y_n\|_F \xrightarrow{n \to \infty} 0$ and $b^*(y_n) \xrightarrow{n \to \infty} 0$ for every branch b.

Lemma B.11. Let Y be a block subspace of $JT_{\mathcal{F}_*}$ and let $\varepsilon > 0$. Then there exists a finitely supported vector $y \in Y$ such that $\|y\|_{\mathcal{F}_*} = 1$ and $|x^*(y)| < \varepsilon$ for every $\sigma_{\mathcal{F}}$ special functional x^*.

Proof. Assume the contrary. Then there exists a block subspace Y of $JT_{\mathcal{F}_*}$ and an $\varepsilon > 0$ such that

$$(23) \qquad \varepsilon \cdot \|y\|_{\mathcal{F}_*} \leq \sup\{|x^*(y)| : \ x^* \text{ is a } \sigma_{\mathcal{F}} \text{ special functional}\}$$

for every $y \in Y$. Let $q > \frac{8}{\varepsilon^2}$. From Corollary B.10 we may select a block sequence $(y_n)_{n\in\mathbb{N}}$ in Y such that $\|y_n\|_{\mathcal{F}_*} = 1$, $\|y_n\|_F \xrightarrow{n\to\infty} 0$ and $b^*(y_n) \xrightarrow{n\to\infty} 0$ for every branch b. Observe also that $(y_n)_{n\in\mathbb{N}}$ is a separated sequence (Definition 3.11) hence from Lemma 3.12 we may assume passing to a subsequence that for every $\sigma_{\mathcal{F}}$ special functional x^* we have that $|x^*(y_n)| \geq \frac{1}{q^2}$ for at most two y_n. (Although Lemma 3.12 is stated for $JT_{\mathcal{F}_2}$ with small modifications in the proof remains valid for either $\mathcal{F}_s$ or $\mathcal{F}_{2,s}$.)

We set $t_1 = 1$. From (23) there exists a $\sigma_{\mathcal{F}}$ special functional y_1^* with $\operatorname{ran} y_1^* \subset \operatorname{ran} y_{t_1}$ such that $|y_1^*(y_{t_1})| > \frac{\varepsilon}{2}$. Setting $d_1 = \max \operatorname{ind} y_1^*$ we select t_2 such that $\|y_{t_2}\|_F < \frac{\varepsilon}{4d_1}$. Let z_2^* be a $\sigma_{\mathcal{F}}$ special functional with $\operatorname{ran} z_2^* \subset \operatorname{ran} y_{t_2}$ such that $|z_2^*(y_{t_2})| > \frac{3\varepsilon}{4}$. We write $z_2^* = x_2^* + y_2^*$ with $\operatorname{ind} x_2^* \subset \{1, \ldots, d_1\}$ and $\operatorname{ind} y_2^* \subset \{d_1 + 1, \ldots\}$. We have that $|x_2^*(y_{t_2})| \leq d_1 \|y_{t_2}\|_F < \frac{\varepsilon}{4}$ and thus $|y_2^*(y_{t_2})| > \frac{\varepsilon}{2}$. Following this procedure we select a finite collection $(t_n)_{n=1}^{q^2}$ of integers and a finite sequence of $\sigma_{\mathcal{F}}$ special functionals $(y_n^*)_{n=1}^{q^2}$ with $\operatorname{ran} y_n^* \subset \operatorname{ran} y_{t_n}$ and $|y_n^*(y_{t_n})| > \frac{\varepsilon}{2}$ such that the sets of indices $(\operatorname{ind}(y_n^*))_{n=1}^{q^2}$ are pairwise disjoint. We consider the vector $y = y_{t_1} + y_{t_2} + \ldots + y_{t_{q^2}}$.

We set $\varepsilon_i = \operatorname{sgn} y_n^*(y_{t_n})$ for $i = 1, \ldots, q^2$. The functional $f = \sum_{n=1}^{q^2} \frac{\varepsilon_n}{q} y_n^*$ belongs to $\mathcal{F}_{2,s}$ ($\subset \mathcal{F}_2$), while $qf \in \mathcal{F}_s$. Therefore

$$(24) \qquad \|y\|_{\mathcal{F}_*} \geq f(y_1 + y_2 + \ldots + y_{q^2}) \geq \frac{1}{q} \sum_{n=1}^{q^2} |y_n^*(y_{t_n})| > q\frac{\varepsilon}{2} > \frac{4}{\varepsilon}.$$

It is enough to show that $\sup\{|x^*(y)| : \ x^* \text{ is a } \sigma_{\mathcal{F}} \text{ special functional}\} \leq 3$ so as to derive a contradiction with (23) and (24). Let x^* be a $\sigma_{\mathcal{F}}$ special functional. Then from our assumptions that $|x^*(y_n)| > \frac{1}{q^2}$ for at most two y_n,

$$|x^*(y)| \leq 2 + (q^2 - 2)\frac{1}{q^2} < 3.$$

The proof of the lemma is complete. $\qquad \square$

Theorem B.12. Let Y be a subspace of either $JT_{\mathcal{F}_{2,s}}$ or of $JT_{\mathcal{F}_s}$. Then for every $\varepsilon > 0$, there exists a subspace of Y which $1 + \varepsilon$ isomorphic to c_0.

Proof. Let Y be a block subspace of $JT_{\mathcal{F}_{2,s}}$ or of $JT_{\mathcal{F}_s}$ and let $\varepsilon > 0$. Using Lemma B.11 we may inductively select a normalized block sequence $(y_n)_{n\in\mathbb{N}}$ in Y such that, setting $d_n = \max \operatorname{supp} y_n$ for each n and $d_0 = 1$, $|x^*(y_n)| < \frac{\varepsilon}{2^n d_{n-1}}$ for every $\sigma_{\mathcal{F}}$ special functional x^*.

We claim that $(y_n)_{n \in \mathbb{N}}$ is $1 + \varepsilon$ isomorphic to the standard basis of c_0. Indeed, let $(\beta_n)_{n=1}^N$ be a sequence of scalars. We shall show that $\max_{1 \le n \le N} |\beta_n| \le$

$\| \sum_{n=1}^N \beta_n y_n \|_{\mathcal{F}_*} \le (1 + \varepsilon) \max_{1 \le n \le N} |\beta_n|$ for either $\mathcal{F}_* = \mathcal{F}_{2,s}$ or $\mathcal{F}_* = \mathcal{F}_s$. We may assume that $\max_{1 \le n \le N} |\beta_n| = 1$. The left inequality follows directly from the bimonotonicity of the Schauder basis $(e_n)_{n \in \mathbb{N}}$ of $JT_{\mathcal{F}_*}$.

To see the right inequality we consider an arbitrary $g \in \mathcal{F}_*$. Then there exist $d \in \mathbb{N}$, $(x_i^*)_{i=1}^d$ in $\mathcal{S} \cup (\bigcup_{i=1}^\infty F_i)$ with $(\mathrm{ind}(x_i^*))_{i=1}^d$ pairwise disjoint and $\min \mathrm{supp}\, x_i^* \ge d$, such that $g = \sum_{i=1}^d a_i x_i^*$ with $\sum_{i=1}^d a_i^2 \le 1$ in the case $\mathcal{F}_* = \mathcal{F}_{2,s}$, while $a_i \in \{-1, 1\}$ in the case $\mathcal{F}_* = \mathcal{F}_s$. Let n_0 be the minimum integer n such that $d \le d_n$. Since $\min \mathrm{supp}\, g \ge d > d_{n_0-1}$ we get that $g(y_n) = 0$ for $n < n_0$. In either case we get that

$$g(\sum_{n=1}^N \beta_n y_n) \le |g(y_{n_0})| + \sum_{n=n_0+1}^N |g(y_n)| \le 1 + \sum_{n=n_0+1}^N \sum_{i=1}^d |x_i^*(y_n)|$$

$$< 1 + \sum_{n=n_0+1}^N d \frac{\varepsilon}{2^n d_{n-1}} < 1 + \sum_{n=n_0+1}^N \frac{\varepsilon}{2^n} < 1 + \varepsilon.$$

The proof of the theorem is complete. $\qquad \square$

Lemma B.13. For every $x \in c_{00}(\mathbb{N})$ and every $\varepsilon > 0$ there exists $d \in \mathbb{N}$ (denoted $d = d(x, \varepsilon)$) such that for every $g \in \mathcal{F}_2 \setminus F_0$ with $\mathrm{ind}(g) \cap \{1, \ldots, d\} = \emptyset$ we have that $|g(x)| < \varepsilon$.

Proof. Let $C = \|x\|_{\ell_1}$ be the ℓ_1 norm of the vector x. We choose $d \in \mathbb{N}$ such that $\sum_{l=d+1}^\infty \tau_l^2 < (\frac{\varepsilon}{C})^2$. Now let $g = \sum_{i=1}^k a_i x_i^* \in \mathcal{F}_2$, such that $\mathrm{ind}(g) \cap \{1, \ldots, d\} = \emptyset$. Each x_i^* takes the form $x_i^* = \sum_{j=1}^{r_i} x_{i,j}^*$, where for each i either $r_i = 1$ and $x_{i,1} \in \bigcup_{i=1}^\infty F_i$ or $(x_{i,j})_{j=1}^{r_i}$ is a $\sigma_{\mathcal{F}}$ special sequence, and the indices $(\mathrm{ind}(x_{i,j}^*))_{i,j}$ are pairwise different elements of $\{d+1, d+2, \ldots\}$. We get that

$$|g(x)| \le \sum_{i=1}^k |a_i| \cdot |x_i^*(x)| \le (\sum_{i=1}^k |a_i|^2)^{1/2} (\sum_{i=1}^k |x_i^*(x)|^2)^{1/2}$$

$$\le 1 \cdot (\sum_{i=1}^k \|x\|_{\ell_1}^2 \|x_i^*\|_\infty^2)^{1/2} \le C (\sum_{l=d+1}^\infty \tau_l^2)^{1/2} < \varepsilon.$$

$\qquad \square$

Theorem B.14. For every subspace Y of $JT_{\mathcal{F}_2}$ and every $\varepsilon > 0$ there exists a subspace of Y which is $1 + \varepsilon$ isomorphic to ℓ_2.

Proof. Let Y be a block subspace of $JT_{\mathcal{F}_2}$ and let $\varepsilon > 0$. We choose a sequence $(\varepsilon_n)_{n \in \mathbb{N}}$ of positive reals satisfying $\sum_{n=1}^{\infty} \varepsilon_n < \frac{\varepsilon}{2}$. We shall produce a block sequence $(x_n)_{n \in \mathbb{N}}$ in Y and a strictly increasing sequence of integers $(d_n)_{n \in \mathbb{N}}$ such that

(i) $\|x_n\|_{\mathcal{F}_2} = 1$.

(ii) For every $\sigma_{\mathcal{F}}$ special functional x^* we have that $|x^*(x_n)| < \frac{\varepsilon_n}{3d_{n-1}}$.

(iii) If $g = \sum_{i=1}^{k} a_i y_i^* \in \mathcal{F}_2$ is such that $\mathrm{ind}(g) \cap \{1, 2, \ldots, d_n\} = \emptyset$, then $|g(x_n)| < \varepsilon_n$.

The construction is inductive. We choose an arbitrary finitely supported vector $x_1 \in Y$ with $\|x_1\|_{\mathcal{F}_2} = 1$ and we set $d_1 = d(x_1, \varepsilon_1)$ (see the notation in the statement of Lemma B.13). Then, using Lemma B.11 we select a vector $x_2 \in Y \cap c_{00}(\mathbb{N})$ with $x_1 < x_2$ such that $\|x_2\|_{\mathcal{F}_2} = 1$ and $|x^*(x_2)| < \frac{\varepsilon_2}{3d_1}$ for every $\sigma_{\mathcal{F}}$ special functional x^*. We set $d_2 = d(x_2, \varepsilon_2)$. It is clear how the inductive construction proceeds. We shall show that for every sequence of scalars $(\beta_n)_{n=1}^{N}$ we have that

$$(25) \qquad (1 - \varepsilon)\Big(\sum_{n=1}^{N} \beta_n^2 \Big)^{1/2} \leq \Big\| \sum_{n=1}^{N} \beta_n x_n \Big\|_{\mathcal{F}_2} \leq (1 + \varepsilon)\Big(\sum_{n=1}^{N} \beta_n^2 \Big)^{1/2}.$$

We may assume that $\sum_{n=1}^{N} \beta_n^2 = 1$.

We first show the left hand inequality of (25). For each n we choose $g_n \in \mathcal{F}_2$, $g_n = \sum_{i=1}^{l_n} a_{n,i} g_{n,i}$, with $\mathrm{ran}\, g_n \subset \mathrm{ran}\, x_n$ and

$$(26) \qquad g_n(x_n) > 1 - \frac{\varepsilon}{3}.$$

For each (n, i) we write the functional $g_{n,i}$ as the sum of three successive functionals, $g_{n,i} = x_{n,i}^* + y_{n,i}^* + z_{n,i}^*$ such that $\mathrm{ind}(x_{n,i}^*) \subset \{1, \ldots, d_{n-1}\}$, $\mathrm{ind}(y_{n,i}^*) \subset \{d_{n-1} + 1, \ldots, d_n\}$ and $\mathrm{ind}(z_{n,i}^*) \subset \{d_n + 1, \ldots\}$. From the choice of the vector x_n and the definition $x_{n,i}^*$ we get that

$$(27) \qquad \Big| \sum_{i=1}^{l_n} a_{n,i} x_{n,i}^*(x_n) \Big| \leq \sum_{i=1}^{l_n} |x_{n,i}^*(x_n)| \leq d_{n-1} \cdot \|x_n\|_F < d_{n-1} \frac{\varepsilon_n}{3d_{n-1}} < \frac{\varepsilon}{3}.$$

The definition of the number d_n yields also that

$$(28) \qquad \Big| \sum_{i=1}^{l_n} a_{n,i} z_{n,i}^*(x_n) \Big| < \frac{\varepsilon}{3}.$$

From (26), (27) and (28) we get that $g_n'(x_n) > 1 - \varepsilon$ where the functional $g_n' = \sum_{i=1}^{l_n} a_{n,i} y_{n,i}^*$, belongs to $\mathcal{F}_2$, satisfies $\mathrm{ran}(g_n') \subset \mathrm{ran}(g_n) \subset \mathrm{ran}(x_n)$ and $\mathrm{ind}(g_n') \subset \{d_{n-1} + 1, \ldots, d_n\}$. We consider the functional $g = \sum_{n=1}^{N} \beta_n g_n' =$

$\sum_{n=1}^{N} \sum_{i=1}^{l_n} \beta_n a_{n,i} y_{n,i}^*$. Since $\sum_n \sum_i (\beta_n a_{n,i})^2 \leq 1$ and the sets $(\mathrm{ind}(y_{n,i}^*))_{n,i}$ are pairwise disjoint we get that $g \in \overline{\mathcal{F}_2}^p \subset B_{JT_{\mathcal{F}_2}^*}$. Therefore

$$\| \sum_{n=1}^{N} \beta_n x_n \|_{\mathcal{F}_2} \geq g(\sum_{n=1}^{N} \beta_n x_n) \sum_{n=1}^{N} \beta_n^2 g_n'(x_n) > 1 - \varepsilon.$$

We next show the right hand inequality of (25). Let $(\beta_n)_{n=1}^{N}$ be any sequence of scalars such that $\sum_{n=1}^{N} \beta_n^2 \leq 1$. We consider an arbitrary $f \in \mathcal{F}_2$, and we shall show that $f(\sum_{n=1}^{N} \beta_n x_n) \leq 1 + \varepsilon$. Let $f = \sum_{i=1}^{k} a_i x_i^*$, where $(x_i^*)_{i=1}^{k}$ belong to $\mathcal{S} \cup (\bigcup_{i=1}^{\infty} F_i)$ with pairwise disjoint sets of indices and $\sum_{i=1}^{k} a_i^2 \leq 1$.

We partition the set $\{1, 2, \ldots, k\}$ in the following manner. We set

$$A_1 = \{i \in \{1, 2, \ldots, k\} : \mathrm{ind}(x_i^*) \cap \{1, 2, \ldots, d_1\} \neq \emptyset\}.$$

If $A_1, \ldots, A_{n-1}$ have been defined we set

$$A_n = \{i \in \{1, 2, \ldots, k\} : \mathrm{ind}(x_i^*) \cap \{1, 2, \ldots, d_n\} \neq \emptyset\} \setminus \bigcup_{i=1}^{n-1} A_i.$$

Finally we set $A_{N+1} = \{1, 2, \ldots, k\} \setminus \bigcup_{i=1}^{N} A_i$.

The sets $(A_n)_{n=1}^{N+1}$ are pairwise disjoint and $\#(\bigcup_{i=1}^{n} A_i) \leq d_n$ for $n = 1, 2, \ldots, N$. We set

$$f_{A_n} = \sum_{i \in A_n} a_i x_i^*, \qquad i = 1, 2, \ldots, N+1.$$

It is clear that $\|f_{A_n}\|_{JT_{\mathcal{F}_2}^*} \leq \left(\sum_{i \in A_n} a_i^2 \right)^{1/2}$ for each n.

Let $n \in \{1, 2, \ldots, N\}$ be fixed. We have that

$$(29) \qquad |f(\beta_n x_n)| \leq \sum_{l=1}^{n-1} |f_{A_l}(x_n)| + |f_{A_n}(\beta_n x_n)| + |\sum_{l=n+1}^{N+1} f_{A_l}(x_n)|.$$

From condition (ii) we get that

$$(30) \qquad \sum_{l=1}^{n-1} |f_{A_l}(x_n)| \leq \sum_{i \in \cup_{l=1}^{n-1} A_l} |x_i^*(x_n)| < d_{n-1} \cdot \frac{\varepsilon_n}{d_{n-1}} = \varepsilon_n.$$

On the other hand
$$\mathrm{ind}(\sum_{l=n+1}^{N+1} f_{A_l}) \cap \{1, \ldots, d_n\} = \mathrm{ind}(\sum_{l=n+1}^{N+1} \sum_{i \in A_l} a_i x_i^*) \cap \{1, \ldots, d_n\} = \emptyset \text{ and thus}$$

condition (iii) yields that

$$(31) \qquad |\sum_{l=n+1}^{N+1} f_{A_l}(x_n)| < \varepsilon_n.$$

Inequalities (29),(30) and (31) yield that

$$|f(\beta_n x_n)| < |f_{A_n}(\beta_n x_n)| + 2\varepsilon_n.$$

Therefore

$$
\begin{aligned}
|f(\sum_{n=1}^{N} \beta_n x_n)| \;\; &\leq \;\; \sum_{n=1}^{N} |f(\beta_n x_n)| \leq \sum_{n=1}^{N} \left(|f_{A_n}(\beta_n x_n)| + 2\varepsilon_n \right) \\
&\leq \;\; \sum_{n=1}^{N} |\beta_n| |f_{A_n}(x_n)| + 2 \sum_{n=1}^{N} \varepsilon_n \\
&< \;\; \left(\sum_{n=1}^{N} |\beta_n|^2 \right)^{1/2} \left(\sum_{n=1}^{N} |f_{A_n}(x_n)|^2 \right)^{1/2} + \varepsilon \\
&\leq \;\; 1 \cdot \left(\sum_{n=1}^{N} \sum_{i \in A_n} a_i^2 \right)^{1/2} + \varepsilon \leq 1 + \varepsilon.
\end{aligned}
$$

The proof of the theorem is complete. $\qquad\qquad\qquad\qquad\qquad\qquad$ □

Proposition B.15. The dual space $JT^*_{\mathcal{F}_*}$ is equal to the closed linear span of the set containing $(e^*_n)_{n \in \mathbb{N}}$ and b^* for every branch b,

$$JT^*_{\mathcal{F}_*} = \overline{\operatorname{span}}(\{e^*_n : n \in \mathbb{N}\} \cup \{b^* : b \text{ is a } \sigma_{\mathcal{F}} \text{ branch}\}).$$

Moreover the Schauder basis $(e_n)_{n \in \mathbb{N}}$ of the space $JT_{\mathcal{F}_*}$ is weakly null.

Proof. Since the space $JT_{\mathcal{F}_*}$ is c_0 saturated for $\mathcal{F}_* = \mathcal{F}_{2,s}$ or $\mathcal{F}_* = \mathcal{F}_s$ (Theorem B.12) or ℓ_2 saturated (for $\mathcal{F}_* = \mathcal{F}_2$) it contains no isomorphic copy of ℓ_1. Haydon's theorem yields that the unit ball of $JT^*_{\mathcal{F}_*}$ is the norm closed convex hull of its extreme points. Since the set $\mathcal{F}_*$ is the norming set of the space $JT_{\mathcal{F}_*}$ we have that $B_{JT^*_{\mathcal{F}_*}} = \overline{\operatorname{conv}(\mathcal{F}_*)}^{w^*}$ hence $\operatorname{Ext}(B_{JT^*_{\mathcal{F}_*}}) \subset \overline{\mathcal{F}_*}^{w^*}$. We thus get that $JT^*_{\mathcal{F}_*} = \overline{\operatorname{span}}(\overline{\mathcal{F}_*}^{w^*})$.

We observe that

$$
\overline{\mathcal{F}_{2,s}}^{w^*} = F_0 \cup \{ \sum_{i=1}^{d} a_i x^*_i : \sum_{i=1}^{d} a_i^2 \leq 1,\ (x^*_i)_{i=1}^{d} \text{ are } \sigma_{\mathcal{F}} \text{ special functionals}
$$

$$\text{with } (\operatorname{ind}(x^*_i))_{i=1}^{d} \text{ pairwise disjoint and } \min \operatorname{supp} x^*_i \geq d \},$$

$$
\overline{\mathcal{F}_2}^{w^*} = F_0 \cup \{ \sum_{i=1}^{\infty} a_i x^*_i : \sum_{i=1}^{\infty} a_i^2 \leq 1,\ (x^*_i)_{i=1}^{\infty} \text{ are } \sigma_{\mathcal{F}} \text{ special functionals}
$$

$$\text{with } (\operatorname{ind}(x^*_i))_{i=1}^{\infty} \text{ pairwise disjoint} \}.$$

and

$$\overline{\mathcal{F}_s}^{w^*} = F_0 \cup \Big\{ \sum_{i=1}^{d} \varepsilon_i x_i^* : \varepsilon_i \in \{-1, 1\}, (x_i^*)_{i=1}^{d} \text{ are } \sigma_{\mathcal{F}} \text{ special functionals}$$

with $(\mathrm{ind}(x_i^*))_{i=1}^{d}$ pairwise disjoint and $\min \mathrm{supp}\, x_i^* \geq d\}$,

The first and third equality follow easily. For the second the arguments are similar to Lemma 8.4.5 of [Fa].

The first part of the proposition for the cases $\mathcal{F}_* = \mathcal{F}_{2,s}$ or $\mathcal{F}_* = \mathcal{F}_s$ follows directly while for the case $\mathcal{F}_* = \mathcal{F}_2$ it is enough to observe that $\| \sum_{i=1}^{\infty} a_i x_i^* \|_{JT_{\mathcal{F}_2}^*} \leq \big(\sum_{i=1}^{\infty} a_i^2 \big)^{1/2}$ for every $g = \sum_{i=1}^{\infty} a_i x_i^* \in \mathcal{F}_2$. Therefore

$$JT_{\mathcal{F}_*}^* = \overline{\mathrm{span}}(\{e_n^* : n \in \mathbb{N}\} \cup \{b^* : b \text{ branch}\}).$$

From the first part of the proposition, to show that the basis $(e_n)_{n \in \mathbb{N}}$ is weakly null, it is enough to show that $b^*(e_n) \xrightarrow{n \to \infty} 0$ for every branch b. But if $b = (f_1, f_2, f_3, \ldots)$ is an arbitrary branch then the sequence $k_n = \mathrm{ind}(f_n)$ is strictly increasing and hence, since $\|f_n\|_{\infty} \leq \tau_{k_n}$, the conclusion follows. $\square$

Remark B.16. Let $(F_j)_{j=0}^{\infty}$ be a JTG family (Definition B.1). If τ is a subfamily of the family of finite $\sigma_{\mathcal{F}}$ special functionals such that $F \subset \tau$ and $Ex^* \in \tau$ for every $x^* \in \tau$ and interval E of $\mathbb{N}$, then, setting

$$\mathcal{F}_{\tau,s} = \Big\{ \sum_{i=1}^{d} \varepsilon_i x_i^* : \varepsilon_1, \ldots, \varepsilon_d \in \{-1, 1\}, x_1^*, \ldots, x_d^* \in \tau$$

with $\mathrm{ind}(x_i^*)_{i=1}^{d}$ pairwise disjoint and $\min \mathrm{supp}\, x_i^* \geq d, d \in \mathbb{N}\}$.

the space $JT_{\mathcal{F}_{\tau,s}}$, which is defined to be the completion of $(c_{00}(\mathbb{N}), \| \ \|_{\mathcal{F}_{\tau,s}})$, is also c_0 saturated.

Theorem B.17. There exists a Banach space X with a weakly null Schauder basis $(e_n)_{n \in \mathbb{N}}$ such that X is ℓ_2 saturated (c_0 saturated) and for every $M \in [\mathbb{N}]$ the space $X_M = \overline{\mathrm{span}}\{e_n : n \in M\}$ has nonseparable dual.

A similar result has been also obtained by E. Odell in [O] using a different approach.

Proof. Let $(F_j)_{j=0}^{\infty}$ be the Maurey-Rosenthal JTG family (Example B.2 (i)). As we have seen the space $X = JT_{\mathcal{F}_2}$ (Definition B.5) has a normalized weakly null Schauder basis $(e_n)_{n \in \mathbb{N}}$ (Proposition B.15) and it is ℓ_2 saturated (Theorem B.14).

Let now $M \in [\mathbb{N}]$. We inductively construct $(x_a, f_a, j_a)_{a \in \mathcal{D}}$, where $\mathcal{D}$ is the dyadic tree and the induction runs on the lexicographical order of $\mathcal{D}$, such that the following conditions are satisfied:

(i) For every $a \in \mathcal{D}$ there exists $F_a \subset M$ with $\#(F_a) = k_{j_a}$ such that
$$x_a = \frac{1}{\sqrt{k_{j_a}}} \sum_{i \in F_a} e_i \quad \text{and} \quad f_a = \frac{1}{\sqrt{k_{j_a}}} \sum_{i \in F_a} e_i^*.$$

(ii) $j_\emptyset \in \Xi_1$ with $j_\emptyset \geq 2$ while for $a \in \mathcal{D}$, $a \neq \emptyset$, $j_a = \sigma_{\mathcal{F}}((f_\beta)_{\beta < a})$.

(iii) If $a <_{lex} \beta$ then $F_a < F_\beta$.

Our construction yields that for every branch b of the dyadic tree the sequence $(f_a)_{a \in b}$ is a $\sigma_{\mathcal{F}}$ special sequence. Hence the w^* sum $g_b = \sum\limits_{a \in b} f_a$ is a member of $\overline{S}^{w^*}$ and thus it belongs to the unit ball of $JT_{\mathcal{F}_2}$. We shall show that $\|g_b|_{X_M} - g_{b'}|_{X_M}\|_{X_M^*} \geq \frac{1}{2}$ for infinite branches $b \neq b'$ of the dyadic tree.

We first observe that for every $a \in \mathcal{D}$ and $f \in F_j$ we have that $|f(x_a)| \leq \min\{\frac{\sqrt{k_j}}{\sqrt{k_{j_a}}}, \frac{\sqrt{k_{j_a}}}{\sqrt{k_j}}\}$. Thus, if $g = \sum\limits_{i=1}^{d} a_i x_i^* \in \mathcal{F}_2$ (Definition B.5), then

$$|g(x_a)| \leq \sum_{i=1}^{d} |x_i^*(x_a)| \leq \sum_{j < j_a} \frac{\sqrt{k_j}}{\sqrt{k_{j_a}}} + 1 + \sum_{j > j_a} \frac{\sqrt{k_{j_a}}}{\sqrt{k_j}} \leq 1 + 1 = 2.$$

We conclude that $x_a \in X_M$ with $\|x_a\|_{\mathcal{F}_2} \leq 2$.

Therefore, if $b \neq b'$ are infinite branches of $\mathcal{D}$ and $a \in b \setminus b'$ then

$$\|g_b|_{X_M} - g_{b'}|_{X_M}\|_{X_M^*} \geq \frac{(g_b - g_{b'})(x_a)}{\|x_a\|_{\mathcal{F}_2}} \geq \frac{f_a(x_a)}{2} = \frac{1}{2}.$$

The c_0 saturated space of the statement is the space $X = JT_{\mathcal{F}_{2,s}}$ and the proof is the same. $\square$

REFERENCES

[AD] S.A. Argyros and I. Deliyanni, *Examples of asymptotic ℓ_1 Banach spaces*, Trans. Amer. Math. Soc. **349**, (1997), 973–995.

[AF] S.A. Argyros and V. Felouzis, *Interpolating hereditarily indecomposable Banach spaces*, J. Amer. Math. Soc. **13**, (2000), no. 2, 243–294.

[AGT] S.A. Argyros, I. Gasparis, A. Tolias (in preparation).

[ALT] S.A. Argyros, J. Lopez-Abad and S. Todorcevic, *A class of Banach spaces with few non-strictly singular operators*, J. Funct. Anal. **222**, (2005), no. 2, 306–384.

[AM] S.A. Argyros and A. Manoussakis, *An Indecomposable and Unconditionally Saturated Banach space*, Studia Math., **159** (2003), 1–32.

[AMT] S.A. Argyros, S. Mercourakis and A. Tsarpalias, *Convex unconditionality and summability of weakly null sequences*, Israel J. of Math., **107**, (1998), 157–193.

[ArTo] S.A. Argyros and S. Todorcevic, *Ramsey Methods in Analysis*, Advance Cources in Mathematics CRM Barcelona, Birkhäuser, (2004).

[AT1] S.A. Argyros, A. Tolias, *Methods in the Theory of Hereditarily Indecomposable Banach Spaces*, Memoirs of the AMS, **170**, (2004), no 806, 114pp.

[AT2] S.A.Argyros, A.Tolias, *Indecomposability and Unconditionality in duality*, Geom. and Funct. Anal., **14**, (2004), 247–282.

[BP] C. Bessaga, A. Pełczyński, *Spaces of continuous functions IV*, Studia Math., **19**, (1960), 53–62.

[Fa] M. J. Fabian, *Gâteaux differentiability of convex functions and topology*, Can. Math. Soc. Series of monographs and advanced texts.

[Fe] V. Ferenczi, *Quotient Hereditarily Indecomposable Banach spaces*, Canad. J. Math., **51**, (1999), 566–584.

[G1] W.T. Gowers, *A Banach space not containing c_0, ℓ_1 or a reflexive subspace*, Trans. Amer. Math. Soc., **344**, (1994), 407–420.

[G2] W.T. Gowers, *An Infinite Ramsey Theorem and Some Banach-Space Dichotomies*, Ann. of Math., **156**, (2002), 797–833.

[GM1] W.T. Gowers, B. Maurey, *The Unconditional basic Sequence Problem*, Journal of A.M.S., **6**, (1993), 851–874.

[GM2] W.T. Gowers, B. Maurey, *Banach spaces with small spaces of operators*, Math. Ann., **307**, (1997), 543–568.
[Ha]   R.G. Haydon, *Some more characterizations of Banach spaces not containing ℓ_1*, Math. Proc. Cambridge Phil. Soc., **80**, (1976), 269–276.
[J]   R.C. James, *A separable somewhat reflexive space with nonseparable dual*, Bull. Amer. Math. Soc., **80**, (1974), 738–743.
[JR]   W.B. Johnson, H. P. Rosenthal, *On w^* basic sequences and their applications to the study of Banach spaces*, Studia Math., **43**, (1972), 77–92.
[MR]   B. Maurey, H. P. Rosenthal, *Normalized weakly null sequence with no unconditional subsequence*, Studia Math., **61**, (1977), no. 1, 77–98.
[O]   E. Odell, *A normalized weakly null sequence with no shrinking subsequence in a Banach space not containing ℓ_1*, Compositio Math., **41**, (1980), 287–295.
[S]   Th. Schlumprecht, *An arbitrarily distortable Banach space*, Israel J. Math. **76**, (1991), 81–95.

(S.A. Argyros) DEPARTMENT OF MATHEMATICS, NATIONAL TECHNICAL UNIVERSITY OF ATHENS
E-mail address: sargyros@math.ntua.gr

(A. D. Arvanitakis) DEPARTMENT OF MATHEMATICS, NATIONAL TECHNICAL UNIVERSITY OF ATHENS
E-mail address: aarva@math.ntua.gr

(A. Tolias) DEPARTMENT OF APPLIED MATHEMATICS, UNIVERSITY OF CRETE
E-mail address: atolias@tem.uoc.gr

THE DAUGAVET PROPERTY FOR LINDENSTRAUSS SPACES

ABSTRACT. A Banach space X is said to have the Daugavet property if every rank-one operator $T : X \longrightarrow X$ satisfies $\|\mathrm{Id} + T\| = 1 + \|T\|$. We give geometric characterizations of this property for Lindenstrauss spaces.

Julio Becerra Guerrero [1] and Miguel Martín [2]

1. INTRODUCTION

The study of the Daugavet property was inaugurated in 1961 when I. Daugavet [7] proved that every compact linear operator T on $C[0,1]$ satisfies the norm equality

(DE) $$\|\mathrm{Id} + T\| = 1 + \|T\|,$$

now known as the *Daugavet equation*. Over the years, the validity of this equation was proved for compact linear operators on various spaces, including $C(K)$ and $L_1(\mu)$ provided that K is perfect and μ does not have any atoms (see [18] for an elementary approach), and certain function algebras such as the disk algebra $A(\mathbb{D})$ or the algebra of bounded analytic functions H^∞ [19, 21]. In the nineties, new ideas were infused into the field, and the geometry of Banach spaces having the so-called Daugavet property was initiated. Let us recall that a Banach space X is said to have the *Daugavet property* [15] if every rank-one operator $T : X \longrightarrow X$ satisfies (DE), in which case, all weakly compact operators on X also satisfy (DE) (see [15, Theorem 2.3]). Therefore, this definition of Daugavet property coincides with those that gave a briefly appearance in [6, 1]. A good introduction to the the Daugavet equation is given in the books [2, 3] and the state-of-the-art on the subject can be found in the papers [15, 20]. For very recent results we refer the reader to [4, 5, 14, 16] and references therein.

Let us mention here several facts concerning the Daugavet property which are relevant to our discussion. It is clear that X has the Daugavet property whenever its topological dual X^* does, but the converse result is false (for instance, $X = C[0,1]$). It is known that a space with the Daugavet property cannot have the Radon-Nikodým property [21]; even more, every weakly open

2000 *Mathematics Subject Classification.* 46B04, 46B20.

Key words and phrases. Daugavet equation; Daugavet property; rough norm; Fréchet-smoothness; Lindenstrauss spaces.

[1]Partially supported by Junta de Andalucía grant FQM-199

[2]Partially supported by Spanish MCYT project number BFM2003-01681 and Junta de Andalucía grant FQM-185

subset of its unit ball has diameter 2 [17]. A space with the Daugavet property contains a copy of ℓ_1 [15], it does not have an unconditional basis [13] and it does not even embed into a space with an unconditional basis [15].

The Daugavet property is not always inherited by ultraproducts. Actually, given a Banach space X, every ultrapower $X_{\mathcal{U}}$, $\mathcal{U}$ a free ultrafilter on $\mathbb{N}$, has the Daugavet property if and only if X has the so-called *uniform Daugavet property*, a quantitative version of the Daugavet property introduced in [5], and which is strictly stronger than the usual Daugavet property [16, Theorem 3.3]. Even though, the basic examples of spaces with the Daugavet property ($C(K)$ with K perfect and $L_1[0,1]$) are in fact spaces with the uniform Daugavet property [5, Lemmas 6.6 and 6.7]. We refer to [11, 12] for definitions and basic results about ultraproducts of Banach spaces.

The aim of this note is to give geometric characterizations of the Daugavet property valid for the so-called *Lindenstrauss spaces* (i.e. Banach spaces whose dual is isometric to an $L_1(\mu)$ space) which remind those given in [4, Corollaries 4.1 and 4.4] for C^*-algebras. We apply them to prove that the Daugavet property passes to ultraproducts of Lindenstrauss spaces.

Let us fix notation and recall some common definitions.

Let X be a Banach space. The symbols B_X and S_X denote, respectively, the closed unit ball and the unit sphere of X, and we write $\text{ext}(C)$ to denote the set of extreme points of the convex set C. Let us fix u in S_X. We define the set $D(X, u)$ of all *states* of X relative to u by

$$D(X, u) := \{f \in B_{X^*} \ : \ f(u) = 1\},$$

which is a non-empty w^*-closed face of B_{X^*}. The norm of X is *smooth* at u if $D(X, u)$ reduces to a singleton, and it is *Fréchet-smooth* at $u \in S_X$ whenever there exists $\lim_{\alpha \to 0} \frac{\|u + \alpha x\| - 1}{\alpha}$ uniformly for $x \in B_X$. We define the *roughness of X at u* by the equality

$$\eta(X, u) \ := \ \limsup_{\|h\| \to 0} \frac{\|u + h\| + \|u - h\| - 2}{\|h\|}.$$

We remark that the absence of roughness of X at u (i.e., $\eta(X, u) = 0$) is nothing other than the Fréchet-smoothness of the norm of X at u [8, Lemma I.1.3]. Given $\delta > 0$, the Banach space X is said to be *δ-rough* if, for every u in S_X, we have $\eta(X, u) \geqslant \delta$. We say that X is *extremely rough* whenever it is 2-rough. A *slice* of B_X is a subset of the form

$$S(B_X, f, \alpha) = \{x \in B_X \ : \ \text{Re} \, f(x) > 1 - \alpha\},$$

where $f \in S_{X^*}$ and $0 < \alpha < 1$. If X is a dual space and f is actually taken from the predual, we say that $S(B_X, f, \alpha)$ is a *w^*-slice*. By [8, Proposition I.1.11], the norm of X is δ-rough if and only if every nonempty w^*-slice of B_{X^*} has diameter greater or equal than δ. A point $x \in S_X$ is said to be an *strongly exposed point* if there exists $f \in D(X, x)$ such that $\lim \|x_n - x\| = 0$ for every sequence (x_n) of elements of B_X such that $\lim \text{Re} \, f(x_n) = 1$ or, equivalently, if there is a point of Fréchet-smoothness in $D(X, x)$ (see [8, Corollary I.1.5]).

2. THE RESULTS

Our aim is to characterize those real or complex Lindenstrauss spaces with the Daugavet property. Since the Daugavet property passes from the dual of a Banach space to the space itself, one may wonder if it is enough to characterizes $L_1(\mu)$ spaces with the Daugavet property and then the mentioned result applies. Characterizations of the Daugavet property for $L_1(\mu)$ spaces can be obtained as a particular case of [4, Corollaries 4.2 and 4.4], where the work was done for von Neumann preduals. Let us state here this result for $L_1(\mu)$ spaces.

Let μ be a positive measure. Then, the following are equivalent:

   (i) $L_1(\mu)$ has the Daugavet property.
   (ii) Every weak-open subset of $B_{L_1(\mu)}$ has diameter 2.
   (iii) $B_{L_1(\mu)}$ has no strongly exposed points.
   (iv) $B_{L_1(\mu)}$ has no extreme points.
   (v) The measure μ does not have any atom.

Let us observe that, as an immediate consequence of (iv) above and the Krein-Milman Theorem, we have that an $L_1(\mu)$ space which is a dual space never has the Daugavet property. Therefore, the Daugavet property for a Lindenstrauss space never comes from its dual space, and we need to look for other kind of characterizations which depends not only on the measure μ but also on the particular form in which a given Lindenstrauss space is the predual of $L_1(\mu)$. For $C(K)$ spaces (which are very particular examples of Lindenstrauss spaces) this was done in [4, Corollaries 4.1 and 4.4], where the Daugavet property was characterized for C^*-algebras. Let us write here the result for $C(K)$ spaces.

Let K be a Hausdorff compact topological space. Then, the following are equivalent:

   (i) $C(K)$ has the Daugavet property.
   (ii) The norm of $C(K)$ is extremely rough.
   (iii) The norm of $C(K)$ is not Fréchet-smooth at any point.
   (iv) The space K does not have any isolated point.

Just remembering that the space K is isometric to a quotient of the topological space $\mathrm{ext}\left(B_{C(K)^*}\right)$ endowed with the weak* topology, one realizes that the above result characterizes the Daugavet property of $C(K)$ either in terms of the geometry of the space or in terms of the way in which $C(K)$ is a predual of $C(K)^*$.

The aim of this paper is to give characterizations of the Daugavet property for Lindenstrauss spaces analogous to the ones given above for $C(K)$ spaces. Of course, we have to translate the meaning of (iv) to an arbitrary Lindenstrauss space, and the above paragraph gives us the idea to do so for an arbitrary Banach space.

Definition 2.1. Given a Banach space X, we define the equivalence relation $f \sim g$ if and only if f and g are linearly dependent elements of $\mathrm{ext}\left(B_{X^*}\right)$, and we endowed the quotient space $\mathrm{ext}\left(B_{X^*}\right)/\sim$ with the quotient topology of the weak* topology.

In [19, Theorem 3.5], D. Werner proved that a Lindenstrauss space X for which $\text{ext}\,(B_{X^*})\,/\sim$ does not have any isolated point has the Daugavet property. We will show that this condition is actually a characterization. To do so, we need the following geometrical result, which may be of independent interest.

Proposition 2.2. *Let X be a Banach space and let f be an extreme point of B_{X^*} such that its equivalence class is an isolated point in $\text{ext}\,(B_{X^*})\,/\sim$. Then, f is a w^*-strongly exposed point of B_{X^*}.*

Proof. We may find a w^*-neighborhood U of f in B_{X^*} such that whenever $g \in \text{ext}\,(B_{X^*})$ belongs to U, then $f \sim g$. By Choquet's Lemma (see [10, Lemma 3.40], for instance), we may certainly suppose that U is a w^*-open slice of B_{X^*}; i.e., there are $x \in S_X$ and $0 < \alpha_0 < 1$ such that

$$(1) \qquad g \in \text{ext}\,(B_{X^*}), \ g \in S(B_{X^*}, x, \alpha_0) \quad \Longrightarrow \quad f \sim g.$$

We claim that, for $0 < \alpha \leqslant \alpha_0$ and $y \in S_X$ satisfying $\|y - x\| < \alpha$, there exists a modulus-one scalar ω_y such that $D(X, y)$ reduces to the singleton $\{\omega_y f\}$ and

$$\|\omega_y f - \omega_x f\| < \sqrt{2\,\alpha}\,.$$

Let us observe that this claim finish the proof, since it implies that every selector of the duality mapping is norm to norm continuous at x, which gives that the norm of X is Fréchet-smooth at x (see [9, Theorem II.2.1]) and then, $\omega_x f$ (and hence f) is w^*-strongly exposed (see [8, Corollary I.1.5]).

Let us prove the claim. If $\|y - x\| < \alpha$, every $g \in D(X, y)$ satisfies

$$(2) \qquad \text{Re}\,g(x) = \text{Re}\,g(y) - \big(\text{Re}\,g(y) - \text{Re}\,g(x)\big) \geqslant 1 - \|x - y\| > 1 - \alpha$$

and so, $D(X, y)$ is contained in $S(B_{X^*}, x, \alpha) \subset S(B_{X^*}, x, \alpha_0)$. Then, every extreme point of the w^*-closed face $D(X, y)$ (remaining extreme in B_{X^*}) is a multiple of f by Eq. (1). Since only one multiple of f can be in the face $D(X, y)$ and, being w^*-compact, $D(X, y)$ is the w^*-closed convex hull of its extreme points, we get $D(X, y) = \{\omega_y f\}$ for a suitable modulus-one scalar ω_y. Finally, on one hand, since $|f(x)| = 1$, we have that

$$\|\omega_x f - \omega_y f\| = |\omega_x - \omega_y| = |\omega_x f(x) - \omega_y f(x)| = |1 - \omega_y f(x)|.$$

On the other hand, Eq. (2) says that $\text{Re}\,\omega_y f(x) > 1 - \alpha$ and so, an straight-forward computation gives that

$$|1 - \omega_y f(x)| < \sqrt{2\,\alpha}\,. \qquad \qquad \square$$

Remarks 2.3.

(a) In the real case, the proof of Proposition 2.2 actually gives a stronger result. Namely, *let X be a real Banach space and let f be a w^*-isolated point of $\text{ext}\,(B_{X^*})$. Then, the face of the unit ball*

$$\{x \in B_X \ : \ f(x) = 1\}$$

has non-empty interior (relative to S_X), which implies that f is w^*-strongly exposed.

(b) Let us comment that no proper face of the unit ball of a complex Banach space has interior points relative to S_X, so the above result does not hold for complex spaces. Anyhow, a sight to the proof of Proposition 2.2 allows us to state the following improvement. *Let X be a complex Banach space and let f be an extreme point of B_{X^*} such that its equivalent class is isolated in* ext $(B_{X^*}) / \sim$. *Then, there exists an open subset U of S_X such that the norm of X is Fréchet-smooth at any point of U and each derivative is a multiple of f.*

We are now ready to state the main result of the paper.

Theorem 2.4. *Let X be a Lindenstrauss space. Then, the following are equivalent:*

(i) *X has the Daugavet property.*

(ii) *The norm of X is extremely rough.*

(iii) *The norm of X is not Fréchet-smooth at any point.*

(iv) ext $(B_{X^*}) / \sim$ *does not have any isolated point.*

Proof. The implications $(i) \Rightarrow (ii) \Rightarrow (iii)$ are clear and valid for general Banach spaces, and Proposition 2.2 gives $(iii) \Rightarrow (iv)$. Finally, $(iv) \Rightarrow (i)$ follows from [19, Theorem 3.5]. $\qquad\square$

Remark 2.5. It is worth mentioning that the above geometric characterizations are not valid for arbitrary Banach spaces. On one hand, *the norm of ℓ_1 is extremely rough (and so ℓ_1 has no points of Fréchet-smoothness), but ℓ_1 does not have the Daugavet property.* On the other hand, *for $X = \ell_2$ the set* ext $(B_{X^*}) / \sim$ *does not have any w^*-isolated point, but ℓ_2 is reflexive and, therefore, it does not have the Daugavet property.*

The above theorem and the fact that the class of Lindenstrauss spaces is closed under ultraproducts, gives us the following result.

Corollary 2.6. *The ultraproduct of every family of Lindenstrauss spaces with the Daugavet property also has the Daugavet property. In particular, the Daugavet and the uniform Daugavet properties are equivalent for Lindenstrauss spaces.*

Proof. Since the class of Lindenstrauss spaces is closed under arbitrary ultraproducts [12, Proposition 2.1], the result follows from Theorem 2.4 and the fact that the roughness of the norm is inherited under arbitrary ultraproducts [4, Lemma 5.1]. $\qquad\square$

Acknowledgments: The authors would like to express their gratitude to the organizers of the *V Conference on Banach spaces* (Cáceres, Spain, September 2004), for offering the opportunity to write this note.

REFERENCES

[1] Y. A. ABRAMOVICH, C. D. ALIPRANTIS, AND O. BURKINSHAW, The Daugavet equation in uniformly convex Banach spaces, *J. Funct. Analysis* **97** (1991), 215–230.

[2] Y. ABRAMOVICH AND C. ALIPRANTIS, *An invitation to Operator Theory*, Graduate Texts in Math. **50**, AMS, Providence, RI, 2002.

[3] Y. ABRAMOVICH AND C. ALIPRANTIS, *Problems in Operator Theory*, Graduate Texts in Math. **51**, AMS, Providence, RI, 2002.

[4] J. BECERRA GUERRERO AND M. MARTÍN, The Daugavet property of C^*-algebras, JB^*-triples, and of their isometric preduals, *J. Funct. Anal.* **224** (2005), 316–337.

[5] D. BILIK, V. KADETS, R. V. SHVIDKOY, AND D. WERNER, Narrow operators and the Daugavet property for ultraproduts, *Positivity* **9** (2005), 45–62.

[6] P. CHAUVEHEID, On a property of compact operators in Banach spaces, *Bull. Soc. Roy. Sci. Liège* **51** (1982), 371–378.

[7] I. K. DAUGAVET, On a property of completely continuous operators in the space C, *Uspekhi Mat. Nauk* **18** (1963), 157–158 (Russian).

[8] R. DEVILLE, G. GODEFROY, AND V. ZIZLER, *Smoothness and Renormings in Banach spaces*, Pitman Monographs **64**, New York 1993.

[9] J. DIESTEL, *Geometry of Banach Spaces: Selected Topics*, Springer-Verlag, New York, 1975.

[10] M. FABIAN, P. HABALA, P. HÁJEK, V. MONTESINOS, J. PELANT, AND V. ZIZLER, *Functional Analysis and infinite-dimensional geometry*, CMS Books in Mathematics **8**, Springer-Verlag, New York, 2001.

[11] S. HEINRICH, Ultraproducts in Banach space theory, *J. Reine Angew. Math.* **313** (1980), 72–104.

[12] S. HEINRICH, Ultraproducts of L_1-predual spaces, *Fund. Math.* **113** (1981), 221–234.

[13] V. M. KADETS, Some remarks concerning the Daugavet equation, *Quaestiones Math.* **19** (1996), 225–235.

[14] V. KADETS, N. KALTON, AND D. WERNER, Remarks on rich subspaces of Banach spaces, *Studia Math.* **159** (2003), 195–206.

[15] V. M. KADETS, R. V. SHVIDKOY, G. G. SIROTKIN, AND D. WERNER, Banach spaces with the Daugavet property, *Trans. Amer. Math. Soc.* **352** (2000), 855–873.

[16] V. KADETS AND D. WERNER, A Banach space with the Schur and the Daugavet property, *Proc. Amer. Math. Soc.* **132** (2004), 1765–1773.

[17] R. V. SHVIDKOY, Geometric aspects of the Daugavet property, *J. Funct. Anal.* **176** (2000), 198–212.

[18] D. WERNER, An elementary approach to the Daugavet equation, *Interaction between functional analysis, harmonic analysis, and probability (Columbia, MO, 1994)*, 449–454, Lecture Notes in Pure and Appl. Math. **175**, Dekker, New York, 1996.

[19] D. WERNER, The Daugavet equation for operators on function spaces, *J. Funct. Anal.* **143** (1997), 117–128.

[20] D. WERNER, Recent progress on the Daugavet property, *Irish Math. Soc. Bull.* **46** (2001), 77-97.

[21] P. WOJTASZCZYK, Some remarks on the Daugavet equation, *Proc. Amer. Math. Soc.* **115** (1992), 1047–1052.

(**Julio Becerra Guerrero**) DEPARTAMENTO DE MATEMÁTICA APLICADA, ESCUELA UNIVERSITARIA DE ARQUITECTURA TÉCNICA, UNIVERSIDAD DE GRANADA, 18071 GRANADA, SPAIN, juliobg@ugr.es

(**Miguel Martín**) DEPARTAMENTO DE ANÁLISIS MATEMÁTICO, FACULTAD DE CIENCIAS, UNIVERSIDAD DE GRANADA, 18071 GRANADA, SPAIN, mmartins@ugr.es

WEAKLY NULL SEQUENCES IN THE BANACH SPACE $C(K)$

I. GASPARIS, E. ODELL, AND B. WAHL

ABSTRACT. The hierarchy of the block bases of transfinite normalized averages of a normalized Schauder basic sequence is introduced and a criterion is given for a normalized weakly null sequence in $C(K)$, the Banach space of scalar valued functions continuous on the compact metric space K, to admit a block basis of normalized averages equivalent to the unit vector basis of c_0, the Banach space of null scalar sequences. As an application of this criterion, it is shown that every normalized weakly null sequence in $C(K)$, for countable K, admits a block basis of normalized averages equivalent to the unit vector basis of c_0.

1. INTRODUCTION

We study normalized weakly null sequences in the spaces $C(K)$ where K is a compact metric space. When K is uncountable, $C(K)$ is isomorphic to $C([0,1])$ ([33], [37], [13]), while for every countable compact metric space K there exist unique countable ordinals α and β with $C(K)$ (linearly) isometric to $C(\alpha)$ [32] and isomorphic (i.e., linearly homeomorphic) to $C(\omega^{\omega^\beta})$ [16] (in the sequel, for an ordinal α we let $C(\alpha)$ denote $C([1, \alpha])$, the Banach space of scalar valued functions, continuous on the ordinal interval $[1, \alpha]$ endowed with the order topology).

Every normalized weakly null sequence (f_n) in $C(K)$ for countable K, admits a basic shrinking subsequence ([14], [18]) that is, a subsequence (f_{k_n}) which is a Schauder basis for its closed linear span and whose corresponding sequence of biorthogonal functionals is a Schauder basis for the dual of the closed linear subspace generated by (f_{k_n}).

It is shown in [31] that while (f_n) must admit an unconditional subsequence in $C(\omega^\omega)$, it need not admit an unconditional subsequence in $C(\omega^{\omega^2})$.

We remark here that if a normalized basic sequence in $C(K)$ for countable K has no weakly null subsequence, then it admits no unconditional subsequence since such a subsequence would have a further subsequence *equivalent* (this term is explained below) to the unit vector basis of ℓ_1 and $C(K)$ has dual isometric to ℓ_1 which is separable.

1991 *Mathematics Subject Classification.* (2000) Primary: 46B03. Secondary: 06A07, 03E02.

Key words and phrases. $C(K)$ space, weakly null sequence, unconditional sequence, Schreier sets.

Since $C(\alpha)$ is c_0-saturated for all ordinals α [38] (a Banach space is c_0-saturated provided all of its infinite-dimensional subspaces contain an isomorph of c_0), some *block basis* of (f_n) is equivalent to the unit vector basis of c_0.

We recall here that if (e_n) is a Schauder basic sequence in a Banach space then a non-zero sequence (u_n) is called a block basis of (e_n), if there exist finite sets (F_n), with $\max F_n < \min F_{n+1}$ for all n, and scalars (a_n) with $a_i \neq 0$ for all $i \in F_n$ and $n \in \mathbb{N}$ such that $u_n = \sum_{i \in F_n} a_i e_i$, for all $n \in \mathbb{N}$. We then call F_n the *support* of u_n. We shall adopt the notation $u_1 < u_2 < \ldots$ to indicate that (u_n) is a block basis of (e_n) such that $\max \operatorname{supp} u_n < \min \operatorname{supp} u_{n+1}$, for all $n \in \mathbb{N}$. We also recall that two basic sequences (x_n), (y_n) are equivalent provided the map T sending x_n to y_n for all $n \in \mathbb{N}$, extends to an isomorphism between the closed linear spans X and Y of (x_n) and (y_n), respectively. In the case T only extends to a bounded linear operator from X into Y, we say (x_n) *dominates* (y_n).

Our main results are presented mostly in Sections 3 and 6. We show in Corollary 6.8 that if (f_n) is normalized weakly null in $C(\omega^{\omega^\xi})$, one can always find c_0 as a block basis of *normalized α-averages* of (f_n) for some $\alpha \leq \xi$, and a quantified description of α is given. Note that the proof given in [38] of the fact that $C(\omega^{\omega^\xi})$ is c_0-saturated is an existential one that is, it only provides the existence of a block basis of (f_n) equivalent to the unit vector basis of c_0 without giving any information about the support of the blocks or the scalar coefficients involved. A normalized 1-average of $(f_m)_{m \in M}$ (where $M = (m_i)$ is an infinite subsequence of $\mathbb{N}$) is a vector $x = (\sum_{i=1}^{m_1} f_{m_i})/\|\sum_{i=1}^{m_1} f_{m_i}\|$. Thus we have that the support of x is a maximal S_1-set in M where S_1 is the first Schreier class (see the definition of Schreier classes in the next section). A 2-average is similarly defined by averaging a block basis of 1-averages so that the support is a maximal S_2-set. This is carried out for all $\alpha < \omega_1$, as in the construction of the Schreier classes S_α, yielding the hierarchy of normalized α-averages of (f_n). The details are presented in Section 5. We should mention here that an alternative proof of Corollary 6.8 is obtained in the recent paper [9].

Section 3 includes the following results. We show in Theorem 3.8 and Corollary 3.9 that if a normalized weakly null sequence (f_n) in $C(\omega^{\omega^\xi})$ is S_ξ-unconditional (see Definition 2.1 and the comments after it) then it admits an unconditional subsequence. This result, combined with that of [31] and [35] on Schreier unconditional sequences, yields an easier proof of the aforementioned fact about weakly null sequences in $C(\omega^\omega)$ [31]. Indeed, as is observed in [31] (see [35] for a proof), every normalized weakly null sequence in a Banach space admits, for every $\epsilon > 0$, a subsequence that is S_1-unconditional with constant $2 + \epsilon$. It follows from this and Theorem 3.8 that every normalized weakly null sequence in $C(\omega^\omega)$ admits an unconditional subsequence. Another consequence of Theorem 3.8 is that the example of a normalized weakly null sequence in $C(\omega^{\omega^2})$ without unconditional subsequence [31], fails to admit an S_2-unconditional subsequence although of course it admits S_1-unconditional subsequences. This shows the optimality of the result in [31], [35] on Schreier unconditional sequences.

We show in Theorem 3.11 that if (χ_{G_n}) is a weakly null sequence of indicator functions in some space $C(K)$ then there exist $\xi < \omega_1$ and a subsequence of (χ_{G_n}) which is equivalent to a subsequence of the unit vector basis of the generalized Schreier space X^ξ ([1], [2]) (see Notation 3.3). We thus obtain a quantitative version of Rosenthal's unpublished result, that a weakly null sequence of indicator functions in some space $C(K)$ admits an unconditional subsequence (cf. also [10] and [8] for another proof of this result). Theorem 3.11 is extended in [11], to the setting of weakly null sequences in $C(K, X)$ with X finite dimensional.

In Section 6 we give a sufficient condition for a normalized weakly null sequence in some $C(K)$ space to admit a block basis of normalized averages equivalent to the unit vector basis of c_0. We show in Theorem 6.1 that if (f_n) is normalized weakly null in $C(K)$ and there exist a summable sequence of positive scalars (ϵ_n) and a subsequence (f_{m_n}) of (f_n) satisfying $\{n \in \mathbb{N} : |f_{m_n}(t)| \geq \epsilon_{m_n}\}$ is finite for all $t \in K$, then there exist $\xi < \omega_1$ and a block basis of normalized ξ-averages of (f_n) which is equivalent to the unit vector basis of c_0. There are two consequences of Theorem 6.1. The first, Corollary 6.8, has been already discussed. The second one is Corollary 6.3, which gives a quantitative version of a special case of Elton's famous result on extremely weakly unconditionally convergent sequences [22] (cf. also [23], [25], [4] for related results). It was shown in [22] that if (x_n) is a normalized basic sequence in some Banach space and the series $\sum_n |x^*(x_n)|$ converges for every extreme point x^* in the ball of X^*, then some block basis of (x_n) is equivalent to the unit vector basis of c_0. We show in Corollary 6.3 that if (f_n) is a normalized basic sequence in some $C(K)$ space satisfying $\sum_n |f_n(t)|$ converges for all $t \in K$, then there exist $\xi < \omega_1$ and a block basis of normalized ξ-averages of (f_n) which is equivalent to the unit vector basis of c_0.

Finally, Sections 4 and 5 contain a number of technical results on α-averages which are used in Section 6.

Some of the results contained in this paper were obtained in B. Wahl's thesis [41] written under the supervision of E. Odell.

2. PRELIMINARIES

We shall make use of standard Banach space facts and terminology as may be found in [30]. c_{00} is the vector space of the ultimately vanishing scalar sequences. If X is any set, we let $[X]^{<\infty}$ denote the set of its finite subsets, while $[X]$ stands for the set of all infinite subsets of X. If $M \in [\mathbb{N}]$, we shall adopt the convenient notation $M = (m_i)$ to denote the increasing enumeration of the elements of M.

A family $\mathcal{F} \subset [\mathbb{N}]^{<\infty}$ is *hereditary* if $G \in \mathcal{F}$ whenever $G \subset F$ and $F \in \mathcal{F}$. $\mathcal{F}$ is *spreading* if for every $\{m_1 < \cdots < m_k\} \in \mathcal{F}$ and all choices $n_1 < \cdots < n_k$ in $\mathbb{N}$ with $m_i \leq n_i$ ($i \leq k$), we have that $\{n_1, \ldots, n_k\} \in \mathcal{F}$. $\mathcal{F}$ is *compact*, if it is compact with respect to the topology of pointwise convergence in $[\mathbb{N}]^{<\infty}$.

$\mathcal{F}$ is *regular* if it possesses all three aforementioned properties and contains all singletons. A regular family $\mathcal{F}$ is said to be *stable*, provided that $F \in \mathcal{F}$ is a maximal, under inclusion, member of $\mathcal{F}$ if there exists $n > \max F$ with $F \cup \{n\} \notin \mathcal{F}$.

If E and F are finite subsets of $\mathbb{N}$, we write $E < F$ when $\max E < \min F$. Given families $\mathcal{F}_1$ and $\mathcal{F}_2$ whose elements are finite subsets of $\mathbb{N}$, we define ([1], [6], [36]) their *convolution* to be the family

$$\mathcal{F}_2[\mathcal{F}_1]\{\cup_{i=1}^n G_i : n \in \mathbb{N},\, G_1 < \cdots < G_n,\, G_i \in \mathcal{F}_1 \,\forall\, i \le n,$$
$$(\min G_i)_{i=1}^n \in \mathcal{F}_2\} \cup \{\emptyset\}.$$

It is not hard to see that $\mathcal{F}_2[\mathcal{F}_1]$ is regular (resp. stable), whenever each $\mathcal{F}_i$ is.

It turns out that for a regular family $\mathcal{F}$ there exists a countable ordinal ξ such that the ξ-th Cantor-Bendixson derivative $\mathcal{F}^{(\xi)}$ of $\mathcal{F}$ is equal to $\{\emptyset\}$. Hence $\mathcal{F}$ is homeomorphic to $[1, \omega^\xi]$, by the Mazurkiewicz-Sierpinski theorem [32]. We then say that $\mathcal{F}$ is of *order* ξ. If we define $\mathcal{F}^+\{F \in [\mathbb{N}]^{<\infty} : F \backslash \{\min F\} \in \mathcal{F}\}$, then it is not hard to see, using the Mazurkiewicz-Sierpinski theorem [32], that $\mathcal{F}^+$ is regular (and stable if $\mathcal{F}$ is) of order $\xi + 1$. It can be shown that if $\mathcal{F}_i$ is regular of order ξ_i, $i = 1, 2$, then $\mathcal{F}_2[\mathcal{F}_1]$ is of order $\xi_1 \xi_2$.

Notation. Given $\mathcal{F} \subset [\mathbb{N}]^{<\infty}$ and $M \in [\mathbb{N}]$, we set $\mathcal{F}[M] = \{F \cap M : F \in \mathcal{F}\}$. Clearly, $\mathcal{F}[M]$ is hereditary (resp. compact), if $\mathcal{F}$ is.

We shall now recall the transfinite definition of the Schreier families S_ξ, $\xi < \omega_1$ ([1]). First, given a countable ordinal α we associate to it a sequence of successor ordinals, $(\alpha_n + 1)$, in the following manner: If α is a successor ordinal we let $\alpha_n = \alpha - 1$ for all n. In case α is a limit ordinal, we choose $(\alpha_n + 1)$ to be a strictly increasing sequence of ordinals tending to α.

Now set $S_0 = \{\{n\} : n \in \mathbb{N}\} \cup \{\emptyset\}$ and $S_1 = \{F \subset \mathbb{N} : |F| \le \min F\} \cup \{\emptyset\}$. Note that $S_1 = S_1[S_0]$. Let $\xi < \omega_1$ and assume S_α has been defined for all $\alpha < \xi$. If ξ is a successor ordinal, say $\xi = \zeta + 1$, define

$$S_\xi = S_1[S_\zeta].$$

In the case ξ is a limit ordinal, let $(\xi_n + 1)$ be the sequence of successor ordinals associated to ξ. Set

$$S_\xi = \cup_n \{F \in S_{\xi_n + 1} : n \le \min F\} \cup \{\emptyset\}.$$

It is shown in [1] that the Schreier family S_ξ is regular of order ω^ξ for all $\xi < \omega_1$. It is shown in [24] that the Schreier families are stable.

Definition 2.1 ([31], [35]). *A normalized basic sequence (x_n) in a Banach space is said to be Schreier unconditional, if there exists a constant $C > 0$ such that $\|\sum_{n \in F} a_n x_n\| \le C \|\sum_n a_n x_n\|$, for every $F \subset \mathbb{N}$ with $|F| \le \min F$, and all choices of finitely supported scalar sequences (a_n).*

It has been already mentioned in the introductory section that every normalized weakly null sequence admits, for every $\epsilon > 0$, a subsequence that is Schreier unconditional with constant $2 + \epsilon$.

The concept of Schreier unconditionality can be generalized in the following manner: Consider a hereditary family $\mathcal{F}$ of finite subsets of $\mathbb{N}$ containing the singletons. A normalized basic sequence (x_n) is now called $\mathcal{F}$-*unconditional*, if there exists a constant $C > 0$ such that $\| \sum_{n \in F} a_n x_n \| \le C \| \sum_n a_n x_n \|$, for every $F \in \mathcal{F}$ and all choices of finitely supported scalar sequences (a_n).

3. UPPER SCHREIER ESTIMATES

In this section we show that every normalized weakly null sequence in $C(K)$, K a countable compact metric space, admits a subsequence dominated by a subsequence of the unit vector basis of a certain Schreier space (see the relevant definition after the statement of Theorem 3.10).

Recall, [16], that for every countable compact metric space K, there exists a unique countable ordinal α with $C(K)$ isomorphic to $C(\omega^{\omega^\alpha})$. Since most of the properties of weakly null sequences in $C(K)$ that we shall be interested in, are isomorphic invariants, there will be no loss of generality in assuming that $K = [1, \omega^\xi]$, for some $\xi < \omega_1$. As is has been already mentioned in the previous section, every regular family $\mathcal{F}$ of order ξ (this means $\mathcal{F}^{(\xi)} = \{\emptyset\}$) is homeomorphic to the ordinal interval $[1, \omega^\xi]$. Moreover, it is easy to construct by transfinite induction, a regular family of order ξ, for all $\xi < \omega_1$. We can thus identify $C(\omega^\xi)$ with $C(\mathcal{F})$, for every regular family of order ξ.

The advantage of such a representation is that one can easily construct a monotone, shrinking Schauder basis of $C(\mathcal{F})$, the so-called *node basis* [3]. Indeed, let $(\alpha_n)_{n=1}^\infty$ be an enumeration of the elements of $\mathcal{F}$, compatible with the natural partial ordering of $\mathcal{F}$ given by initial segment inclusion. This means that whenever α_m is a proper initial segment of α_n, then $m < n$. In particular, $\alpha_1 = \emptyset$. Such an enumeration is for instance, the anti-lexicographic enumeration of the elements of $\mathcal{F}$, i.e., $F \prec G$ if and only if either $\max F < \max G$, or $F \setminus \{\max F\} \prec G \setminus \{\max G\}$, for all F, G in $\mathcal{F}$.

Given $\alpha \in \mathcal{F}$, set $G_\alpha = \{\beta \in \mathcal{F} : \alpha \le \beta\}$, where $\alpha \le \beta$ means that α is an initial segment of β. Clearly, G_α is a clopen subset of $\mathcal{F}$ for every $\alpha \in \mathcal{F}$. The sequence $(\chi_{G_{\alpha_n}})_{n=1}^\infty$ is called the node basis of $C(\mathcal{F})$. It is not hard to check that $(\chi_{G_{\alpha_n}})_{n=1}^\infty$ is a normalized, monotone, shrinking Schauder basis for $C(\mathcal{F})$ [3].

Proposition 3.1. *Let $\mathcal{F}$ be a regular family and $u_1 < u_2 < \dots$ be a block basis of the node basis $(\chi_{G_{\alpha_n}})_{n=1}^\infty$ of $C(\mathcal{F})$. Then there exist positive integers $n_1 < n_2 < \dots$ with the following property: For every $\gamma \in \mathcal{F}$, $\{n_i : i \in \mathbb{N}, u_{n_i}(\gamma) \ne 0\} \in \mathcal{F}^+$.*

Proof. Define $F_n = \{\alpha_i : i \in \operatorname{supp} u_n\}$, for all $n \in \mathbb{N}$. Clearly, the F_n's are pairwise disjoint, finite subsets of $\mathcal{F}$. We observe that whenever $\alpha_i \in F_n$ and $\alpha_j \in F_m$ satisfy $\alpha_i \le \alpha_j$, then $n \le m$. This is so since $\alpha_i \le \alpha_j$ implies that $i \le j$ and, subsequently, that $u_n \le u_m$. Hence, $n \le m$.

We next choose inductively, integers $2 = n_1 < n_2 < \dots$ such that $\max \beta < n_{i+1}$ for every $\beta \in F_{n_i}$ and all $i \in \mathbb{N}$ (where, $\max \beta$ denotes the largest element of the finite subset β of $\mathbb{N}$). We claim (n_i) is as desired. Indeed, let $\gamma \in \mathcal{F}$.

Then

$$\{n_i : i \in \mathbb{N}, \, u_{n_i}(\gamma) \neq 0\} \subset \{n_i : i \in \mathbb{N}, \, \exists \beta \in F_{n_i}, \, \beta \leqslant \gamma\},$$

for writing $u_{n_i} = \sum_{\beta \in F_{n_i}} \lambda_\beta \chi_{G_\beta}$ for some suitable choice of scalars $(\lambda_\beta)_{\beta \in F_{n_i}}$, we see that $u_{n_i}(\gamma) \neq 0$ implies $\chi_{G_\beta}(\gamma)1$, for some $\beta \in F_{n_i}$ with $\beta \leqslant \gamma$. In particular, $\{n_i : i \in \mathbb{N}, \, u_{n_i}(\gamma) \neq 0\}$ is finite. Let now $\{n_{i_1} < \cdots < n_{i_k}\}$ be an enumeration of $\{n_i : i \in \mathbb{N}, \, u_{n_i}(\gamma) \neq 0\}$, and choose $\beta_j \in F_{n_{i_j}}$ with $\beta_j \leqslant \gamma$, for all $j \leq k$. Since $\{\beta_1, \ldots, \beta_k\}$ is well-ordered with respect to the partial ordering $\leqslant$ of $\mathcal{F}$ (all the β_j's are initial segments of γ), our preliminary observation yields $\beta_1 < \cdots < \beta_k$. Note that $\beta_1 \neq \emptyset$. By the choices made, $\max \beta_j < n_{i_j+1} \leq n_{i_{j+1}}$ for all $j \leq k$. Because $\mathcal{F}$ is hereditary and spreading, we infer that $\{n_{i_2}, \ldots, n_{i_k}\} \in \mathcal{F}$ whence $\{n_i : i \in \mathbb{N}, \, u_{n_i}(\gamma) \neq 0\} \in \mathcal{F}^+$, as required. $\square$

Corollary 3.2. *Suppose K is homeomorphic to $[1, \omega^\xi]$, $\xi < \omega_1$, and that (f_i) is a normalized weakly null sequence in $C(K)$. Let $\mathcal{F}$ be a regular family of order ξ. Then for every $N \in [\mathbb{N}]$ and every non-increasing sequence of positive scalars (ϵ_i), there exists $M \in [N]$, $M = (m_i)$, such that for every $t \in K$ the set $\{m_i : i \in \mathbb{N}, \, |f_{m_i}(t)| \geq \epsilon_i\}$ belongs to $\mathcal{F}^+$.*

Proof. We identify $C(\mathcal{F})$ with $C(K)$ and apply Proposition 3.1 to find a normalized, shrinking, monotone Schauder basis (e_i) for $C(K)$ with the following property: For every block basis $u_1 < u_2 < \ldots$ of (e_i) there exist positive integers $n_1 < n_2 < \ldots$ such that for all $t \in K$, $\{n_i : i \in \mathbb{N}, \, u_{n_i}(t) \neq 0\} \in \mathcal{F}^+$.

Now let (f_i) be normalized weakly null in $C(K)$. A classical perturbation result [14] yields a subsequence (f_{l_i}) of (f_i) and a block basis (u_i) of (e_i), $u_1 < u_2 < \ldots$, such that $l_i \in N$ and $\|f_{l_i} - u_i\| < \epsilon_i/2$, for all $i \in \mathbb{N}$. We next choose positive integers $n_1 < n_2 < \ldots$ such that $\{n_i : i \in \mathbb{N}, \, u_{n_i}(t) \neq 0\} \in \mathcal{F}^+$, for all $t \in K$. Set $m_i = l_{n_i}$, for all $i \in \mathbb{N}$. It is not hard to check using the spreading property of $\mathcal{F}$, that $M = (m_i)$ satisfies the desired conclusion. $\square$

Notation 3.3. *Let $\mathcal{F}$ be a regular family and let (e_i) denote the unit vector basis of c_{00}. We define a norm $\| \cdot \|_\mathcal{F}$ on c_{00} by the rule*

$$\Big\| \sum_i a_i e_i \Big\|_\mathcal{F} = \sup \Big\{ \sum_{i \in F} |a_i| : F \in \mathcal{F} \Big\}, \text{ for all } (a_i) \in c_{00}.$$

The completion of $(c_{00}, \| \cdot \|_\mathcal{F})$ is a Banach space having (e_i) as a normalized, unconditional, shrinking, monotone Schauder basis (see [1], [2]). When $\mathcal{F} = S_\xi$, the ξ-th Schreier class, we obtain the generalized Schreier space X^ξ introduced in [1], [2].

Our next result yields that every normalized weakly null sequence in $C(\omega^{\omega^\xi})$ admits a subsequence dominated by a subsequence of the unit vector basis of the generalized Schreier space X^ξ.

Proposition 3.4. *Suppose K is homeomorphic to $[1, \omega^\xi]$, $\xi < \omega_1$, and that (f_i) is a normalized weakly null sequence in $C(K)$. Let $\mathcal{F}$ be a regular family* *of order ξ. Given $0 < \epsilon < 1$, there exists $M \in [\mathbb{N}]$, $M = (m_i)$, such that*

$$\Big\| \sum_i a_i f_{m_i} \Big\| \leq \frac{2}{1 - \epsilon} \sup\Big\{ \Big\| \sum_{i \in F} a_i f_{m_i} \Big\| : F \subset \mathbb{N}, \ (m_i)_{i \in F} \in \mathcal{F} \Big\}$$

$$\leq \frac{2}{1 - \epsilon} \Big\| \sum_i a_i e_{m_i} \Big\|_{\mathcal{F}}, \ \text{for all } (a_i) \in c_{00}.$$

Proof. We may assume that (f_i) is 2-basic. Choose a decreasing sequence of positive scalars (ϵ_i) such that $\sum_i \epsilon_i < \epsilon/3$. We next choose $M \in [\mathbb{N}]$, $M = (m_i)$, satisfying the conclusion of Corollary 3.2 applied to (f_i) and the scalar sequence (ϵ_i).

Let $(a_i) \in c_{00}$ be such that $\|\sum_i a_i f_{m_i}\| = 1$, and let $t \in K$ satisfy $|\sum_i a_i f_{m_i}(t)| = 1$. Since $\{m_i : i \in \mathbb{N}, |f_{m_i}(t)| \geq \epsilon_i\}$ belongs to $\mathcal{F}^+$, we obtain

$$1 \leq 2 \sup\Big\{ \Big\| \sum_{i \in F} a_i f_{m_i} \Big\| : F \subset \mathbb{N}, \ (m_i)_{i \in F} \in \mathcal{F} \Big\} + \epsilon$$

from which the assertion of the proposition follows. $\qquad\square$

An immediate consequence of Proposition 3.4 is the next

Corollary 3.5. *Suppose K is homeomorphic to $[1, \omega^{\omega^\xi}]$, $\xi < \omega_1$, and that (f_i) is a normalized weakly null sequence in $C(K)$. If (f_i) admits a subsequence which is a c_0^ξ-spreading model (see Definition 4.4), then it also admits a subsequence equivalent to the c_0-basis.*

Remark 3.6. *S. Argyros has discovered an alternate proof of Corollary 3.2. He shows that given a weakly null sequence (f_i) in $C(\omega^\xi)$ and a summable sequence of positive scalars (ϵ_i) then, by identifying $C(\omega^\xi)$ with $C(\mathcal{F})$, one can select positive integers $1 = m_1 < m_2 < \ldots$ such that if $|f_{m_i}(F)| \geq \epsilon_i$ for some $i \in \mathbb{N}$ and $F \in \mathcal{F}$, then $F \cap (m_{i-1}, m_{i+1}) \neq \emptyset$ ($m_0 = 0$). Therefore, $\{m_{2i} : i \in \mathbb{N}, |f_{m_{2i}}(F)| \geq \epsilon_{2i}\} \in \mathcal{F}^+$, for every $F \in \mathcal{F}$ which clearly implies Corollary 3.2.*

Remark 3.7. *Proposition 9 and Lemma 13 in [29] yield that for a normalized weakly null sequence (f_i) in $C(\omega^\xi)$ there exist a subsequence (f_{m_i}), a compact hereditary family $\mathcal{D}$ with $\mathcal{D}^{(\xi+1)} = \emptyset$ and a constant $d > 0$ such that $\|\sum_i a_i f_{m_i}\| \leq d \sup\{\|\sum_{i \in A} a_i f_{m_i}\| : A \in \mathcal{D}\}$ for every $(a_i) \in c_{00}$.*

Theorem 3.8. *Suppose K is homeomorphic to $[1, \omega^\xi]$, $\xi < \omega_1$, and that (f_i) is a normalized weakly null sequence in $C(K)$. Let $\mathcal{F}$ be a regular family of order ξ. Assume (f_i) is $\mathcal{F}$-unconditional. Then (f_i) has an unconditional subsequence.*

Proof. Suppose (f_i) is $\mathcal{F}$-unconditional with constant $C > 0$. This means that $\|\sum_{i \in F} a_i f_i\| \leq C \|\sum_i a_i f_i\|$, for all $F \in \mathcal{F}$ and every $(a_i) \in c_{00}$. Let $M = (m_i)$ satisfy the conclusion of Proposition 3.4, for (f_i) and $\mathcal{F}$ with $\epsilon = 1/2$. We claim that (f_{m_i}) is unconditional. Indeed, let $(a_i) \in c_{00}$ and

$I \in [\mathbb{N}]$. Proposition 3.4 yields

$$\left\|\sum_{i \in I} a_i f_{m_i}\right\| \le 4 \sup\{\|\sum_{i \in F \cap I} a_i f_{m_i}\| : F \subset \mathbb{N}, (m_i)_{i \in F} \in \mathcal{F}\}.$$

Since $\mathcal{F}$ is hereditary and (f_i) is $\mathcal{F}$-unconditional, we have that

$$\|\sum_{i \in F \cap I} a_i f_{m_i}\| \le C\|\sum_i a_i f_{m_i}\|, \text{ whenever } (m_i)_{i \in F} \in \mathcal{F}.$$

Therefore, $\|\sum_{i \in I} a_i f_{m_i}\| \le 4C\|\sum_i a_i f_{m_i}\|$ which proves the claim. This completes the proof. $\square$

¿From Theorem 3.8 we easily obtain the next

Corollary 3.9. *A normalized weakly null sequence in $C(\omega^{\omega^\xi})$, $\xi < \omega_1$, admits an unconditional subsequence if, and only if, it admits a subsequence which is S_ξ-unconditional.*

Theorem 3.10. *Let (f_i) be a normalized weakly null sequence in $C(\omega^{\omega^\xi})$, $\xi < \omega_1$. Assume that (f_i) is an ℓ_1^ξ-spreading model. Then (f_i) admits a subsequence equivalent to a subsequence of the unit vector basis of X^ξ, the generalized Schreier space of order ξ (see Notation 3.3).*

We recall that $X^0 = c_0$ while X^1 was implicitly considered by Schreier [40]. The generalized Schreier spaces X^ξ, $\xi < \omega_1$, were introduced in [1], [2]. They can be thought as the the higher ordinal unconditional analogs of c_0.

We also recall ([10]), that a normalized basic sequence (x_n) is said to be an ℓ_1^ξ-*spreading model*, $\xi < \omega_1$, if there is a constant $\delta > 0$ such that $\|\sum_{n \in F} a_n x_n\| \ge \delta \sum_{n \in F} |a_n|$, for every $F \in S_\xi$ and all choices of scalars $(a_n)_{n \in F}$. Saying (x_n) is an ℓ_1^ξ-spreading model means that ℓ_1 is a spreading model for the space generated by some subsequence of (x_n), in the sense of [17], [12], [34]. ℓ_1^ξ-spreading models are instrumental in the study of *asymptotic* ℓ_1-spaces [36]. It is shown in [7] that a weakly null sequence which is an ℓ_1^ξ-spreading model, admits a subsequence which is S_ξ-unconditional. The unit vector basis of X^ξ is an ℓ_1^ξ-spreading model with constant $\delta = 1$.

Proof of Theorem 3.10. We first apply Proposition 3.4 with $\epsilon = 1/2$, to obtain an infinite subset $M = (m_i)$ of $\mathbb{N}$ with $\|\sum_i a_i f_{m_i}\| \le 4\|\sum_i a_i e_{m_i}\|_{S_\xi}$ for all $(a_i) \in c_{00}$, where (e_i) denotes the unit vector basis of X^ξ. On the other hand, as (f_i) is an ℓ_1^ξ-spreading model, there exists a constant $\delta > 0$ such that

$$\|\sum_i a_i f_{m_i}\| \ge \delta\|\sum_i a_i e_{m_i}\|_\xi, \text{ for all } (a_i) \in c_{00}.$$

We infer from the preceding inequalities that (f_{m_i}) and (e_{m_i}) are equivalent.
                                                                                                    $\square$

Our final result in this section yields a quantitative version of Rosenthal's result, that a weakly null (in $C(K)$) sequence of indicator functions of clopen subsets of a compact Hausdorff space K, admits an unconditional subsequence (cf. also [10] and [8] for another proof of this result).

Theorem 3.11. *Let K be a compact Hausdorff space. Suppose that (f_n) is a normalized weakly null sequence in $C(K)$ such that there exists $\epsilon > 0$ with the property $f_n(t) = 0$ or $|f_n(t)| \geq \epsilon$ for all $t \in K$ and $n \in \mathbb{N}$. Then there exist $\xi < \omega_1$ and a subsequence of (f_n) equivalent to a subsequence of the natural Schauder basis of X^ξ.*

Proof. We first employ the results of [1] in order to find the smallest countable ordinal η for which there is a subsequence (f_{m_n}) of (f_n), such that no subsequence of (f_{m_n}) is an ℓ_1^η-spreading model. Such an ordinal exists because (f_n) is weakly null. We claim that η is a successor ordinal. To see this we shall need a result from [7] (Corollary 3.6) which states that a weakly null sequence (f_n) in a $C(K)$ space admits a subsequence which is an ℓ_1^α-spreading model, for some $\alpha < \omega_1$ if, and only if, there exist a constant $\delta > 0$ and $L \in [\mathbb{N}]$, $L = (l_n)$, so that for every $F \in S_\alpha$ there exists $t \in K$ satisfying $|f_{l_n}(t)| \geq \delta$, for all $n \in F$.

Define $G_n = \{t \in K : f_n(t) \neq 0\}$. Our assumptions yield $G_n = \{t \in K : |f_n(t)| \geq \epsilon\}$, for all $n \in \mathbb{N}$. Observe that for every $\alpha < \eta$ and $P \in [\mathbb{N}]$, there exists $Q \in [P]$, $Q = (q_n)$, so that (f_{q_n}) is an ℓ_1^α-spreading model. It follows now, from the previously cited result of [7], that for every $\alpha < \eta$ and $P \in [\mathbb{N}]$, there exists $Q \in [P]$, $Q = (q_n)$, so that for every $F \in S_\alpha$, $\cap_{n \in F} G_{q_n} \neq \emptyset$. This in turn yields that every subsequence of (f_{m_n}) admits, for every $\alpha < \eta$, a further subsequence which is an ℓ_1^α-spreading model with constant independent of α and the particular subsequence. Were η a limit ordinal, we would have that some subsequence of (f_{m_n}) is an ℓ_1^η-spreading model, contrary to our assumption.

Hence, $\eta = \xi + 1$, for some $\xi < \omega_1$. Let (e_n) be the natural basis of X^ξ. We show that some subsequence of (f_{m_n}) is equivalent to a subsequence of (e_n). Because $\xi < \eta$, we can assume without loss of generality, after passing to a subsequence if necessary, that (f_{m_n}) is an ℓ_1^ξ-spreading model and thus there exists a constant $\rho > 0$ such that $\| \sum_n a_n f_{m_n} \| \geq \rho \| \sum_n a_n e_n \|_{S_\xi}$ for all $(a_n) \in c_{00}$. Define

$$\mathcal{F} = \{F \in [\mathbb{N}]^{<\infty} : \cap_{i \in F} G_{m_i} \neq \emptyset\}.$$

Clearly, $\mathcal{F}$ is hereditary. It is shown in [7], based on the fact that no subsequence of (f_{m_n}) is an $\ell_1^{\xi+1}$-spreading model, that there exist $L \in [\mathbb{N}]$, $L = (l_n)$, and $d \in \mathbb{N}$ so that every member of $\mathcal{F}[L]$ is contained in the union of d members of $S_\xi[L]$. Let $k_n = m_{l_n}$, for all $n \in \mathbb{N}$. We deduce from our preceding work that $\| \sum_n a_n f_{k_n} \| \leq d \| \sum_n a_n e_{l_n} \|_{S_\xi}$, for every $(a_n) \in c_{00}$. Therefore, (f_{k_n}) and (e_{l_n}) are equivalent. $\qquad\square$

4. NORMALIZED AVERAGES OF A BASIC SEQUENCE

Let $\vec{s} = (e_n)$ be a normalized basic sequence in a Banach space, and let $\mathcal{F}$ be a regular and stable family. We shall introduce a hierarchy

$$\{(\alpha_n^{\mathcal{F}, \vec{s}, M})_{n=1}^\infty, M \in [\mathbb{N}], \alpha < \omega_1\}$$

of normalized block bases of $\vec{s}$, similar to that of the repeated averages introduced in [10]. The latter however consists of convex block bases of $\vec{s}$, not necessarily normalized.

We fix a normalized basic sequence $\vec{s} = (e_n)$ and a regular and stable family $\mathcal{F}$. To simplify our notation, we shall write α_n^M instead of $\alpha_n^{\mathcal{F},\vec{s},M}$. We shall next define, by transfinite induction on $\alpha < \omega_1$, a family of normalized block bases $(\alpha_n^M)_{n=1}^{\infty}$ of $\vec{s}$, where $M \in [\mathbb{N}]$, so that the following properties are fulfilled for every $\alpha < \omega_1$ and $M \in [\mathbb{N}]$:

(1) $\alpha_n^M < \alpha_{n+1}^M$, for all $n \in \mathbb{N}$.
(2) $M = \cup_n \operatorname{supp} \alpha_n^M$, for all $M \in [\mathbb{N}]$.

If $\alpha = 0$ and $M = (m_n)$ set $\alpha_n^M = e_{m_n}$, for all $n \in \mathbb{N}$.

Suppose $(\beta_n^N)_{n=1}^{\infty}$ has been defined so that (1) and (2), above, are satisfied for all $\beta < \alpha$ and $N \in [\mathbb{N}]$. Let $M \in [\mathbb{N}]$. In order to define $(\alpha_n^M)_{n=1}^{\infty}$, assume first that α is successor, say $\alpha = \beta + 1$. Let k_1 be the unique integer such that the set $\{\min \operatorname{supp} \beta_i^M : i \le k_1\}$ is a maximal member of $\mathcal{F}$. We define

$$\alpha_1^M = \Big(\sum_{i=1}^{k_1} \beta_i^M\Big) / \Big\|\sum_{i=1}^{k_1} \beta_i^M\Big\|.$$

Suppose that $\alpha_1^M < \cdots < \alpha_n^M$ have been defined and that the union of their supports forms an initial segment of M. Set

$$M_{n+1} = \{m \in M : \max \operatorname{supp} \alpha_n^M < m\}.$$

Let k_{n+1} be the unique integer such that the set $\{\min \operatorname{supp} \beta_i^{M_{n+1}} : i \le k_{n+1}\}$ is a maximal member of $\mathcal{F}$. We define

$$\alpha_{n+1}^M = \Big(\sum_{i=1}^{k_{n+1}} \beta_i^{M_{n+1}}\Big) / \Big\|\sum_{i=1}^{k_{n+1}} \beta_i^{M_{n+1}}\Big\|.$$

This completes the definition of $(\alpha_n^M)_{n=1}^{\infty}$ when α is a successor ordinal. Note that the construction described above can be carried out because $\mathcal{F}$ is stable. (1) and (2) are now satisfied by $(\alpha_n^M)_{n=1}^{\infty}$.

Now suppose α is a limit ordinal. Let $(\alpha_n + 1)$ be the sequence of successor ordinals associated to α. Let $M \in [\mathbb{N}]$ and set $m_1 = \min M$. In case $m_1 = 1$, set $\alpha_1^M = e_1$. If $m_1 > 1$, define

$$\alpha_1^M = u^M / \|u^M\|, \text{ where } u^M = (1/m_1)e_{m_1} + [\alpha_{m_1}]_1^{M \setminus \{m_1\}}.$$

Suppose that $\alpha_1^M < \cdots < \alpha_n^M$ have been defined and that the union of their supports forms an initial segment of M. Set

$$M_{n+1} = \{m \in M : \max \operatorname{supp} \alpha_n^M < m\}$$

and $m_{n+1} = \min M_{n+1}$. Define

$$\alpha_{n+1}^M = u^{M_{n+1}} / \|u^{M_{n+1}}\|, \text{ where}$$

$$u^{M_{n+1}} = (1/m_{n+1})e_{m_{n+1}} + [\alpha_{m_{n+1}}]_1^{M_{n+1} \setminus \{m_{n+1}\}}.$$

Note that $\alpha_{n+1}^M = \alpha_1^{M_{n+1}}$. This completes the definition $(\alpha_n^M)_{n=1}^\infty$ when α is a limit ordinal. It is clear that (1) and (2) are satisfied.

Remark 4.1. *In case* $\mathcal{F} = S_1$, *the first Schreier family, it is not hard to see that* $\operatorname{supp} \alpha_n^M \in S_\alpha$, *for all* $\alpha < \omega_1$, *all* $M \in [\mathbb{N}]$ *and all* $n \in \mathbb{N}$.

The next lemma is an immediate consequence of the preceding definition.

Lemma 4.2. *Let* $\alpha < \omega_1$, $M \in [\mathbb{N}]$ *and* $n \in \mathbb{N}$. *Then there exists* $N \in [\mathbb{N}]$ *such that* $\alpha_n^M = \alpha_1^N$.

Our next result will be applied later, in conjunction with the infinite Ramsey theorem, in order to determine if there exists a block basis of the form (α_n^M), equivalent to the c_0-basis.

Lemma 4.3. *Let* $\alpha < \omega_1$, $M \in [\mathbb{N}]$ *and* $n \in \mathbb{N}$. *Let* $L_i \in [\mathbb{N}]$ *and* $k_i \in \mathbb{N}$, *for* $i \leq n$, *be so that* $\alpha_{k_1}^{L_1} < \cdots < \alpha_{k_n}^{L_n}$ *and* $\cup_{i=1}^n \operatorname{supp} \alpha_{k_i}^{L_i}$ *is an initial segment of* M. *Then* $\alpha_i^M = \alpha_{k_i}^{L_i}$, *for all* $i \leq n$.

Proof. By Lemma 4.2 we may assume that $k_i = 1$ for all $i \leq n$. We prove the assertion of the lemma by transfinite induction on α. The case $\alpha = 0$ is trivial. Suppose the assertion holds for all ordinals smaller than α, and all $M \in [\mathbb{N}]$ and $n \in \mathbb{N}$. Let $M \in [\mathbb{N}]$. We prove the assertion for α by induction on n. If $n = 1$, we first consider the case of α being a successor ordinal, say $\alpha = \beta + 1$. We know from the definitions that

$$\operatorname{supp} \alpha_1^{L_1} = \cup_{j=1}^{p_1} \operatorname{supp} \beta_j^{L_1},$$

where $\{\min \operatorname{supp} \beta_j^{L_1} : j \leq p_1\}$ is a maximal member of $\mathcal{F}$. In particular, the set $\cup_{j=1}^{p_1} \operatorname{supp} \beta_j^{L_1}$ is an initial segment of M. The induction hypothesis on β now implies that $\beta_j^M = \beta_j^{L_1}$, for all $j \leq p_1$. It follows now that $\alpha_1^M = \alpha_1^{L_1}$.

To complete the case $n = 1$, we consider the possibility that α is a limit ordinal. Let $(\alpha_n + 1)$ be the sequence of ordinals associated to α and suppose that $m = \min M$. Then $m = \min \operatorname{supp} \alpha_1^{L_1}$ and so $m = \min L_1$. In case $m = 1$ we have, trivially, $\alpha_1^M = \alpha_1^{L_1} = e_m$. When $m > 1$, $u^M = (1/m)e_m + [\alpha_m]_1^{M \setminus \{m\}}$, $u^{L_1} = (1/m)e_m + [\alpha_m]_1^{L_1 \setminus \{m\}}$ and $\alpha_1^M = u^M / \|u^M\|$, $\alpha_1^{L_1} = u^{L_1} / \|u^{L_1}\|$.

It follows that $\operatorname{supp} [\alpha_m]_1^{L_1 \setminus \{m\}}$ is an initial segment of $M \setminus \{m\}$, and so we infer from the induction hypothesis applied to α_m, that $[\alpha_m]_1^{L_1 \setminus \{m\}} = [\alpha_m]_1^{M \setminus \{m\}}$. Thus $\alpha_1^M = \alpha_1^{L_1}$ which completes the case $n = 1$.

Assume now the assertion holds for $n - 1$ and write $M = \cup_{i=1}^n \operatorname{supp} \alpha_1^{L_i} \cup N$, where $\cup_{i=1}^n \operatorname{supp} \alpha_1^{L_i}$ is an initial segment of M, which is disjoint from N. The induction hypothesis for $n - 1$ yields $\alpha_i^M = \alpha_1^{L_i}$ for all $i < n$. Hence $M = \cup_{i=1}^{n-1} \operatorname{supp} \alpha_i^M \cup P$, where $P = \operatorname{supp} \alpha_1^{L_n} \cup N$. It follows from the definition that $\alpha_n^M = \alpha_1^P$. But now the case $n = 1$ guarantees that $\alpha_1^P = \alpha_1^{L_n}$ and the assertion of the lemma is settled. $\qquad\square$

Terminology. Let (e_n) be a normalized Schauder basic sequence in a Banach space and let $\mathcal{F}$ be a regular family. A finite block basis $u_1 < \cdots <$

u_m of (e_n) is said to be $\mathcal{F}$-*admissible* if $\{\min \operatorname{supp} u_i : i \leq m\} \in \mathcal{F}$. It is called *maximally* $\mathcal{F}$-admissible, if $\mathcal{F}$ is additionally assumed to be stable and $\{\min \operatorname{supp} u_i : i \leq m\}$ is a maximal member of $\mathcal{F}$.

Definition 4.4. *A normalized block basis (u_n) of (e_n) with $u_1 < u_2 < ...$ is a c_0^ξ-spreading model, if there exists a constant $C > 0$ such that $\| \sum_{i \in F} a_i u_i \| \leq C \max_{i \in F} |a_i|$, for every $F \in [\mathbb{N}]^{<\infty}$ with $(u_i)_{i \in F}$ S_ξ-admissible, and every choice of scalars $(a_i)_{i \in F}$.*

In what follows we fix a normalized basic sequence $\vec{s} = (e_n)$ and a regular and stable family $\mathcal{F}$. We abbreviate $\alpha_n^{\mathcal{F}, \vec{s}, M}$ to α_n^M.

Terminology. Suppose that $\alpha < \omega_1$ and $M \in [\mathbb{N}]$. An α-*average* of (e_n) supported by M, is any vector of the form α_1^L for some $L \in [M]$.

In the sequel we shall make use of the infinite Ramsey theorem [20], [34] and so we recall its statement. $[\mathbb{N}]$ is endowed with the topology of pointwise convergence.

Theorem 4.5. *Let $\mathcal{A}$ be an analytic subset of $[\mathbb{N}]$. Then there exists $N \in [\mathbb{N}]$ so that either $[N] \subset \mathcal{A}$, or $[N] \cap \mathcal{A} = \emptyset$.*

Our next result is inspired by an unpublished result of W.B. Johnson (see [34]).

Lemma 4.6. *Let α and γ be countable ordinals and suppose there exists $N \in [\mathbb{N}]$ such that for every $M \in [N]$ there exists a block basis of α-averages of (e_n), supported by M, which is a c_0^γ-spreading model. Then there exist $M \in [N]$ and a constant $C > 0$ so that $\| \sum_{i=1}^{n_L} \alpha_i^L \| \leq C$, for every $L \in [M]$, where n_L stands for the unique integer satisfying $\{\min \operatorname{supp} \alpha_i^L : i \leq n_L\}$ is maximal in S_γ.*

Proof. Define $\mathcal{D}_k = \{L \in [N] : \| \sum_{i=1}^{n_L} \alpha_i^L \| \leq k\}$, for all $k \in \mathbb{N}$. $\mathcal{D}_k$ is closed in the topology of pointwise convergence, thanks to Lemma 4.3. We claim that there exist $k \in \mathbb{N}$ and $M \in [N]$ so that $[M] \subset \mathcal{D}_k$. The assertion of the lemma clearly follows once this claim is established. Were the claim false, then Theorem 4.5 would yield a nested sequence $M_1 \supset M_2 \supset ...$ of infinite subsets of N such that $[M_k] \cap \mathcal{D}_k = \emptyset$, for all $k \in \mathbb{N}$. Choose an infinite sequence of integers $m_1 < m_2 < ...$ with $m_i \in M_i$ for all $i \in \mathbb{N}$. Set $M = (m_i)$. Since $M \in [N]$ our assumptions yield a block basis (u_i) of α-averages of (e_i), supported by M, which is a c_0^γ-spreading model. Therefore there exists a constant $C > 0$ such that $\| \sum_{i \in F} u_i \| \leq C$, whenever $(u_i)_{i \in F}$ is S_γ-admissible. Choose $k \in \mathbb{N}$ with $k > C$. Then choose $i_0 \in \mathbb{N}$ so that $\operatorname{supp} u_i \subset M_k$, for all $i > i_0$. If we set $L = \cup_{i=i_0+1}^\infty \operatorname{supp} u_i$, then $L \in [M_k]$, and $\alpha_i^L = u_{i+i_0}$, for all $i \in \mathbb{N}$, by Lemma 4.3. Hence, $L \notin \mathcal{D}_k$. However,

$$\| \sum_{i=1}^{n_L} \alpha_i^L \| \| \sum_{i=1}^{n_L} u_{i_0+i} \| \leq C < k$$

which is a contradiction. $\square$

5. CONVOLUTION OF TRANSFINITE AVERAGES

We fix a normalized 2-basic, shrinking sequence $\vec{s} = (e_i)$ in some Banach space. We shall often make use of the following result established in [35]: Given $\epsilon > 0$ there exists $M \in [\mathbb{N}]$ such that for every finitely supported scalar sequence $(a_i)_{i \in M}$ with $\|\sum_{i \in M} a_i e_i\| = 1$, we have $\max_{i \in M} |a_i| \leq 1 + \epsilon$. For the rest of this section, we let $\mathcal{F} = S_1$. *All transfinite averages of $\vec{s}$ will be taken with respect to $\mathcal{F}$.* As in the previous section, α_n^M abbreviates $\alpha_n^{\mathcal{F}, \vec{s}, M}$.

The purpose of the present section is to deal with the following problem: Let α and β be countable ordinals and suppose that (u_i) is a block basis of $(\alpha + \beta)$-averages of $\vec{s}$. Does there exist a block basis (v_i) of α-averages of $\vec{s}$ such that (u_i) is a block basis of β-averages of (v_i) ?

It follows directly from the definitions that this is indeed the case when $\beta < \omega$. However, if β is an infinite ordinal, the preceding question has, in general, a negative answer.

In Proposition 5.9, we give a partially affirmative answer to this question which, roughly speaking, states that every $(\alpha + \beta)$ average of $\vec{s}$ can be represented as a finite sum $\sum_{i=1}^{n} \lambda_i w_i$, where $w_1 < \cdots < w_n$ is an S_β-admissible block basis of α-averages of $\vec{s}$ and $(\lambda_i)_{i=1}^{n}$ is a sequence of positive scalars which are almost equal each other. We employ this result in order to prove the following theorem about transfinite c_0-spreading models of $\vec{s}$, which will in turn be applied in subsequent sections. In the sequel, when we refer to a block basis we shall always mean a block basis of $\vec{s}$. Also all transfinite averages will be taken with respect to $\vec{s}$.

Theorem 5.1. *Let α and β be countable ordinals and $N \in [\mathbb{N}]$. Suppose that for every $P \in [N]$ there exists $M \in [P]$ such that no block basis of α-averages supported by M is a c_0^β-spreading model. Then for every $P \in [N]$ and $\epsilon > 0$ there exists $Q \in [P]$ with the following property: Every $(\alpha + \beta)$-average u supported by Q admits a decomposition $u = \sum_{i=1}^{n} \lambda_i u_i$, where $u_1 < \cdots < u_n$ is a normalized block basis and $(\lambda_i)_{i=1}^{n}$ is a sequence of positive scalars such that*

(1) *There exists $I \subset \{1, \ldots, n\}$ with $(u_i)_{i \in I}$ S_β-admissible, and such that u_i is an α-average for all $i \in I$, while $\|\sum_{i \in \{1,\ldots,n\} \setminus I} \lambda_i u_i\|_{\ell_1} < \epsilon$.*
(2) *$\max_{i \in I} \lambda_i < \epsilon$.*

Recall that if $\sum_{i=1}^{n} a_i e_i$ is a finite linear combination of $\vec{s}$ then we denote by $\|\sum_{i=1}^{n} a_i e_i\|_{\ell_1}$ the quantity $\sum_{i=1}^{n} |a_i|$. To prove this theorem we shall need to introduce some terminology.

Definition 5.2. *Let α and β be countable ordinals and $\epsilon > 0$. A normalized block u is said to admit an $(\epsilon, \alpha, \beta)$-decomposition, if there exist normalized blocks $u_1 < \cdots < u_n$ and positive scalars $(\lambda_i)_{i=1}^{n}$ with $u = \sum_{i=1}^{n} \lambda_i u_i$ and so that the following conditions are satisfied:*

(1) *There exists $I \subset \{1, \ldots, n\}$ with $(u_i)_{i \in I}$ S_β-admissible, and such that u_i is an α-average for all $i \in I$, while $\|\sum_{i \in \{1,\ldots,n\} \setminus I} \lambda_i u_i\|_{\ell_1} < \epsilon$.*
(2) *$|\lambda_i - \lambda_j| < \epsilon$ for all i and j in I.*

Terminology. The quantity $\max_{i \in I} \lambda_i$ is called the *weight* of the decomposition. If u is an $(\alpha + \beta)$-average admitting an $(\epsilon, \alpha, \beta)$-decomposition, $u = \sum_{i=1}^{n} \lambda_i u_i$, satisfying (1), (2), above, and $I \subset \{1, \ldots, n\}$ is as in (1), then every subset of $\{\min \operatorname{supp} u_i : i \in I\}$ will be called an $(\epsilon, \alpha, \beta)$-*admissible subset of* $\mathbb{N}$ *resulting from* u. It is clear that the collection of all $(\epsilon, \alpha, \beta)$-admissible subsets of $\mathbb{N}$ resulting from some (not necessarily the same) $(\alpha + \beta)$-average (for some fixed choices of ϵ, α, β), forms a hereditary family.

Lemma 5.3. *Let $P \in [\mathbb{N}]$ and $\epsilon > 0$. Assume that for every $L \in [P]$ there exists an $(\alpha + \beta)$-average supported by L which admits an $(\epsilon, \alpha, \beta)$-decomposition. Then there exists $Q \in [P]$ such that every $(\alpha + \beta)$-average supported by Q admits an $(\epsilon, \alpha, \beta)$-decomposition.*

Proof. Let

$$\mathcal{D} = \{L \in [P] : [\alpha + \beta]_1^L \text{ admits an } (\epsilon, \alpha, \beta) - \text{decomposition}\}.$$

Lemma 4.3 yields that $\mathcal{D}$ is closed in the topology of pointwise convergence. Theorem 4.5 now implies the existence of some $Q \in [P]$ such that either $[Q] \subset \mathcal{D}$, or $[Q] \cap \mathcal{D} = \emptyset$. Our assumptions rule out the second alternative for Q. Hence $[Q] \subset \mathcal{D}$ which proves the lemma. $\square$

In the next series of lemmas (Lemma 5.4 and Lemma 5.5), we describe some criteria for embedding a Schreier family into an appropriate hereditary family of finite subsets of $\mathbb{N}$. These criteria, as well as their proofs, are variants of similar results contained in [10], [7]. We shall therefore omit the proofs and refer the reader to the aforementioned papers (see for instance Propositions 2.3.2 and 2.3.6 in [10], or Theorems 2.11 and 2.13 in [7]). These lemmas will be applied in the proof of Proposition 5.9, which constitutes the main step towards the proof of Theorem 5.1.

Notation. Let $\mathcal{F}$ be a family of finite subsets of $\mathbb{N}$ and $M \in [\mathbb{N}]$. Let $M = (m_i)$ be the increasing enumeration of M. We set $\mathcal{F}(M) = \{\{m_i : i \in F\} : F \in \mathcal{F}\}$. Clearly, $\mathcal{F}(M) \subset \mathcal{F}$ if $\mathcal{F}$ is spreading. We also recall that $\mathcal{F}[M] = \{F \cap M : F \in \mathcal{F}\}$. Finally, for every $L \in [\mathbb{N}]$ and $\alpha < \omega_1$, we let $(F_i^{\alpha}(L))_{i=1}^{\infty}$ denote the unique decomposition of L into successive, maximal members of S_{α}.

Lemma 5.4. *Suppose that $1 \leq \xi < \omega_1$, $\mathcal{D}$ is a hereditary family of finite subsets of $\mathbb{N}$ and $N \in [\mathbb{N}]$. Assume that for every $n \in \mathbb{N}$ and $P \in [N]$ there exists $L \in [P]$ such that $\cup_{i=1}^{n}(F_i^{\xi}(L) \setminus \{\min F_i^{\xi}(L)\}) \in \mathcal{D}$. Then there exists $M \in [N]$ such that $S_{\xi+1}(M) \subset \mathcal{D}$.*

Lemma 5.5. *Suppose that $\mathcal{D}$ is a hereditary family of finite subsets of $\mathbb{N}$ and $N \in [\mathbb{N}]$. Let $\xi < \omega_1$ be a limit ordinal and let (α_n) be an increasing sequence of ordinals tending to ξ. Assume there exists a sequence $M_1 \supset M_2 \supset \ldots$ of infinite subsets of N such that $S_{\alpha_n}(M_n) \subset \mathcal{D}$, for all $n \in \mathbb{N}$. Then there exists $M \in [N]$ such that $S_{\xi}(M) \subset \mathcal{D}$.*

In the sequel we shall make use of the following permanence property of Schreier families established in [36]:

Lemma 5.6. *Suppose that $\alpha < \beta < \omega_1$. Then there exists $n \in \mathbb{N}$ such that for every $F \in S_\alpha$ with $n \leq \min F$ we have $F \in S_\beta$.*

We shall also make repeated use of the following result from [5]:

Lemma 5.7. *For every $N \in [\mathbb{N}]$ there exists $M \in [N]$ such that for every $\alpha < \omega_1$ and $F \in S_\alpha[M]$ we have $F \setminus \{\min F\} \in S_\alpha(N)$.*

Lemma 5.7 combined with Proposition 3.2 in [36] yields the next

Lemma 5.8. *Let α and β be countable ordinals and $N \in [\mathbb{N}]$. Then there exists $M \in [N]$ such that*

(1) *For every $F \in S_\beta[S_\alpha][M]$ we have $F \setminus \{\min F\} \in S_{\alpha+\beta}$.*
(2) *For every $F \in S_{\alpha+\beta}[M]$ we have $F \setminus \{\min F\} \in S_\beta[S_\alpha]$.*

Proposition 5.9. *Let α and β be countable ordinals and $N \in [\mathbb{N}]$. Then given $\epsilon > 0$ and $P \in [N]$ there exist $Q \in [P]$ and $R \in [Q]$ such that*

(1) *Every $(\alpha + \beta)$-average supported by Q admits an $(\epsilon, \alpha, \beta)$ decomposition.*
(2) *For every $F \in S_\beta[R]$, $F \setminus \{\min F\}$ is an $(\epsilon, \alpha, \beta)$-admissible set resulting from some $(\alpha + \beta)$-average supported by Q.*

Proof. Fix $\alpha < \omega_1$. We prove the assertion of the proposition by transfinite induction on β. The case $\beta = 1$ follows directly from the definitions since every $(\alpha + 1)$-average admits an $(\epsilon, \alpha, 1)$-decomposition. In fact, in this case, we may take $Q = P$ and $R = \{\min \operatorname{supp} \alpha_i^P : i \in \mathbb{N}\}$ and check that (1) and (2) hold.

Now let $\beta > 1$ and suppose the assertion holds for all ordinals smaller than β. Assume first β is a successor ordinal and let $\beta - 1$ be its predecessor. Let $\epsilon > 0$ and $P \in [N]$ be given and choose a sequence of positive scalars (δ_i) such that $\sum_i \delta_i < \epsilon/4$. Let $M \in [P]$. The induction hypothesis for $\beta - 1$ yields infinite subsets $R_1 \subset Q_1$ of M satisfying (1) and (2) for $(\delta_1, \alpha, \beta - 1)$. Choose a maximal member F_1 of $S_{\beta-1}$ with $F_1 \subset R_1$. We may choose an $(\alpha + \beta - 1)$-average u_1, supported by Q_1 and such that $F_1 \setminus \{\min F_1\}$ is $(\delta_1, \alpha, \beta - 1)$-admissible resulting from u_1.

Choose $M_2 \in [M]$ with $\min M_2 > \max \operatorname{supp} u_1$. Arguing similarly, we choose a maximal member F_2 of $S_{\beta-1}$ with $F_2 \subset M_2$, and an $(\alpha + \beta - 1)$-average u_2 supported by M_2, which admits a $(\delta_2, \alpha, \beta-1)$-decomposition from which $F_2 \setminus \{\min F_2\}$ is resulting. We continue in this fashion and obtain a sequence $F_1 < F_2 < \ldots$, of successive maximal members of $S_{\beta-1}[M]$, and a block basis $u_1 < u_2 < \ldots$, of $(\alpha + \beta - 1)$-averages supported by M such that for all $i \in \mathbb{N}$,

$$(5.1) \qquad u_i \text{ admits a } (\delta_i, \alpha, \beta - 1) - \text{decomposition.}$$

$$(5.2) \qquad F_i \setminus \{\min F_i\} \text{ is } (\delta_i, \alpha, \beta - 1) - \text{admissible, resulting from } u_i.$$

We next let, for all $i \in \mathbb{N}$, d_i denote the weight of the $(\delta_i, \alpha, \beta-1)$-decomposition of u_i, from which $F_i \setminus \{\min F_i\}$ is resulting. Clearly, $d_i \in (0, 3]$. Therefore, without loss of generality, by passing to a subsequence if necessary, we may assume that

$$(5.3) \qquad |d_i - d_j| < \epsilon/4, \text{ for all } i, j \text{ in } \mathbb{N}.$$

Now let $n \in \mathbb{N}$ and choose $n < i_1 < \cdots < i_m$ such that $(u_{i_k})_{k=1}^m$ is maximally S_1-admissible. Set $u = (\sum_{k=1}^m u_{i_k}) / \| \sum_{k=1}^m u_{i_k} \|$. It is clear that u is an $(\alpha + \beta)$-average supported by M. It is easy to check, using (5.1) and (5.3), that u admits an $(\epsilon, \alpha, \beta)$-decomposition. On the other hand, (5.2) implies that $\cup_{k=1}^n (F_{i_k} \setminus \{\min F_{i_k}\})$ is $(\epsilon, \alpha, \beta)$-admissible, resulting from u.

Taking into account the stability of $S_{\beta-1}$, we conclude the following: Given $n \in \mathbb{N}$ and $M \in [P]$

$$(5.4) \qquad \text{There exists an } (\alpha + \beta) - \text{average } u \text{ supported by } M$$
$$\text{which admits an } (\epsilon, \alpha, \beta) - \text{decomposition.}$$

$$(5.5) \qquad \text{There exists } L \in [M] \text{ such that } \cup_{i=1}^n (F_i^{\beta-1}(L) \setminus \{\min F_i^{\beta-1}(L)\})$$
$$\text{is } (\epsilon, \alpha, \beta) - \text{admissible, resulting from } u.$$

(Recall that for $\gamma < \omega_1$, $(F_i^\gamma(L))_{i=1}^\infty$ denotes the unique decomposition of L into consecutive, maximal members of S_γ.)

Lemma 5.3 and (5.4) now yield some $Q \in [P]$ satisfying (1) for $(\epsilon, \alpha, \beta)$. Let $\mathcal{D}$ denote the hereditary family of the $(\epsilon, \alpha, \beta)$-admissible subsets of Q resulting from some $(\alpha + \beta)$-average supported by Q. We infer from (5.5) that for every $n \in \mathbb{N}$ and $M \in [Q]$ there exists $L \in [M]$ such that $\cup_{i=1}^n (F_i^{\beta-1}(L) \setminus \{\min F_i^{\beta-1}(L)\}) \in \mathcal{D}$. We deduce from Lemma 5.4 that $S_\beta(R_0) \subset \mathcal{D}$ for some $R_0 \in [Q]$. Employing Lemma 5.7, we find $R \in [R_0]$ such that $F \setminus \{\min F\} \in \mathcal{D}$, for all $F \in S_\beta[R]$. Thus Q and R satisfy (1) and (2) for $(\epsilon, \alpha, \beta)$, when β is a successor ordinal.

We now consider the case of β being a limit ordinal. We may choose an increasing sequence of ordinals (β_n) having β as its limit, and such that $(\alpha + \beta_n + 1)$ is the sequence of successor ordinals associated to the limit ordinal $\alpha + \beta$. Let $\epsilon > 0$ and $P \in [N]$ be given. Let $M \in [P]$ and choose $m \in M$ with $1/m < \epsilon/4$. Next choose $M_1 \in [M]$ with $m < \min M_1$ and such that $S_{\beta_m}[M_1] \subset S_\beta$ (see Lemma 5.6). We now apply the induction hypothesis for β_m to obtain an $(\alpha + \beta_m)$-average v supported by M_1 and admitting an $(\epsilon/4, \alpha, \beta_m)$-decomposition. It is clear that $u = ((1/m)e_m + v) / \|(1/m)e_m + v\|$, is an $(\alpha + \beta)$-average supported by M and admitting an $(\epsilon, \alpha, \beta)$-decomposition. Note also that if F is $(\epsilon/4, \alpha, \beta_m)$-admissible resulting from v, then it is also $(\epsilon, \alpha, \beta)$-admissible resulting from u.

It follows now, by lemma 5.3, that there exists $Q \in [P]$ such that (1) holds for $(\epsilon, \alpha, \beta)$. Next choose positive integers $k_1 < k_2 < \ldots$ such that $S_{\beta_n}[k_n, \infty) \subset S_\beta$ (see Lemma 5.6), for all $n \in \mathbb{N}$. Successive applications of the inductive hypothesis applied to each β_n and Lemma 5.7, yield infinite subsets $Q_1 \supset R_1 \supset Q_2 \supset R_2 \supset \ldots$ of Q with $k_n < \min Q_n$ and such that each member of $S_{\beta_n}(R_n)$ is an $(\epsilon/4, \alpha, \beta_n)$-admissible set resulting from some $(\alpha + \beta_n)$-average supported by Q_n, for all $n \in \mathbb{N}$. Let $\mathcal{D}$ denote the hereditary family of the $(\epsilon, \alpha, \beta)$-admissible subsets of Q resulting from some $(\alpha + \beta)$-average supported by Q. Our preceding argument shows that $S_{\beta_n}(R_n) \subset \mathcal{D}$, as long as $n \in Q$ and $1/n < \epsilon/4$. We deduce now from Lemma 5.5, that there exists $R_0 \in [Q]$ such that $S_\beta(R_0) \subset \mathcal{D}$. Once again, Lemma 5.7 yields some $R \in [R_0]$ with the property $F \setminus \{\min F\} \in \mathcal{D}$, for all $F \in S_\beta[R]$. Hence, $Q \supset R$ satisfy (1) and (2) for $(\epsilon, \alpha, \beta)$, when β is a limit ordinal. This completes the inductive step and the proof of the proposition. $\qquad\square$

We shall also need Elton's nearly unconditional theorem ([21], [34]).

Theorem 5.10. *Let (f_i) be a normalized weakly null sequence in some Banach space. There exists a subsequence (f_{m_i}) of (f_i) with the following property: For every $0 < \delta \leq 1$ there exists a constant $C(\delta) > 0$ such that $\|\sum_{i \in F} a_i f_{m_i}\| \leq C(\delta)\|\sum_i a_i f_{m_i}\|$, for every finitely supported scalar sequence (a_i) in $[-1, 1]$ and every $F \subset \{i \in \mathbb{N} : |a_i| \geq \delta\}$.*

Proof of Theorem 5.1. Let $P \in [N]$ and $\epsilon > 0$. Set

$$\mathcal{D} = \{L \in [P] : [\alpha + \beta]_1^L \text{ admits an}$$
$$(\epsilon, \alpha, \beta) - \text{decomposition of weight smaller than } \epsilon\}.$$

Lemma 4.3 yields that $\mathcal{D}$ is closed in the topology of pointwise convergence. The theorem asserts that $[Q] \subset \mathcal{D}$, for some $Q \in [P]$. Suppose this is not the case and choose, according to Theorem 4.5, $Q_0 \in [P]$ such that $[Q_0] \cap \mathcal{D} = \emptyset$. Next choose $Q_1 \in [Q_0]$ such that no block basis of α-averages supported by Q_1 is a c_0^β-spreading model. Let $M \in [Q_1]$. We infer from Proposition 5.9 that there exist infinite subsets $R \subset Q$ of M such that

(5.6)   Every $(\alpha + \beta) - \text{average supported by } Q$ admits an
          $(\epsilon/2, \alpha, \beta) - \text{decomposition.}$

(5.7)   If $F \in S_\beta[R]$, then $F \setminus \{\min F\}$ is $(\epsilon/2, \alpha, \beta) - \text{admissible}$
          resulting from some $(\alpha + \beta) - \text{average supported by } Q$.

Choose a maximal member F of $S_\beta[R]$. (5.6) and (5.7) allow us to find normalized blocks $u_1 < \cdots < u_n$, positive scalars $(\lambda_i)_{i=1}^n$ and $I \subset \{1, \ldots, n\}$ such that

(5.8) $\displaystyle\sum_{i=1}^n \lambda_i u_i$ is an $(\alpha + \beta) - \text{average supported by } Q$,

$(u_i)_{i \in I}$ is $S_\beta - \text{admissible}$ and $F \setminus \{\min F\} \subset \{\min \operatorname{supp} u_i : i \in I\}$,

u_i is an $\alpha - \text{average for all } i \in I$ and $\|\sum_{i \in \{1,\ldots,n\} \setminus I} \lambda_i u_i\|_{\ell_1} < \epsilon/2$,

$|\lambda_i - \lambda_j| < \epsilon/2$, for all i, j in I.

Since $\sum_{i=1}^n \lambda_i u_i$ is supported by $Q \subset Q_0$, and $[Q_0] \cap \mathcal{D} = \emptyset$, we must have that $\max_{i \in I} \lambda_i \geq \epsilon$. We deduce from (5.8) that

$$\lambda_i \geq \epsilon/2 \text{ for all } i \in I.$$

Set $J_0 = \{i \in I : \min F < \min \operatorname{supp} u_i\}$ and note that (5.8) implies that $F \setminus \{\min F\} \subset \{\min \operatorname{supp} u_i : i \in J_0\}$. It follows now, since F is maximal in S_β, that $\{\min F\} \cup \{\min \operatorname{supp} u_i : i \in J_0\}$ contains a maximal member of S_β as a subset and therefore, as S_β is stable, there exists an initial segment J of J_0 such that $\{\min F\} \cup \{\min \operatorname{supp} u_i : i \in J\}$ is a maximal member of S_β. Note also that $\|\sum_{i \in J} \lambda_i u_i\| \leq 3$.

Summarizing, given $M \in [Q_1]$ we found a block basis of α-averages $v_1 < \cdots < v_k$, supported by M, $m \in M$ with $m < \min \operatorname{supp} v_1$, and scalars $(\mu_i)_{i=1}^k$ in $[\epsilon/2, 2]$ so that

(5.9) $\{m\} \cup \{\min \operatorname{supp} v_i : i \leq k\}$ is maximal in S_β and $\|\sum_{i=1}^k \mu_i v_i\| \leq 3.$

Define

$$\mathcal{D}_1 = \Big\{ L \in [Q_1] : \exists (\mu_i)_{i=1}^k \subset [\epsilon/2, 2], \ \|\sum_{i=1}^k \mu_i \alpha_i^{L \setminus \{\min L\}}\| \leq 3, \text{ and}$$

$$\{\min L\} \cup \{\min \operatorname{supp} \alpha_i^{L \setminus \{\min L\}} : i \leq k\} \text{ is maximal in } S_\beta \Big\}.$$

Lemma 4.3 and the stability of S_β yield that $\mathcal{D}_1$ is closed in the topology of pointwise convergence. We now infer from (5.9) that every $M \in [Q_1]$ contains some $L \in \mathcal{D}_1$ as a subset. Thus, we deduce from Theorem 4.5 that there exists $M_0 \in [Q_1]$ with $[M_0] \subset \mathcal{D}_1$.

Now let $L \in [M_0]$ and denote by n_L the unique integer such that $(\alpha_i^L)_{i=1}^{n_L}$ is maximally S_β-admissible. Because $L \in \mathcal{D}_1$, we must have that

(5.10) $$\|\sum_{i=1}^{n_L} \mu_i \alpha_i^L\| \leq 4, \text{ for some choice of scalars}$$

$(\mu_i)_{i=1}^{n_L}$ in the interval $[\epsilon/2, 2]$.

Set $g_i = \alpha_i^{M_0}$, for all $i \in \mathbb{N}$. Then (g_i) is a normalized weakly null sequence, as $\vec{s}$ is assumed to be shrinking. Theorem 5.10 now yields a constant $C > 0$ and a subsequence of (g_i) (which, for clarity, is still denoted by (g_i)), such that

$$\|\sum_{i \in G} a_i g_i\| \leq C \|\sum_{i=1}^\infty a_i g_i\|,$$

for every finitely supported scalar sequence (a_i) in $[-2, 2]$ and every $G \subset \{i \in \mathbb{N} : |a_i| \geq \epsilon/2\}$. It follows from this, Lemma 4.3 and (5.10) that, whenever $F \in [\mathbb{N}]^{<\infty}$ is so that $(g_i)_{i \in F}$ is maximally S_β-admissible, then we have some choice of scalars $(\mu_i)_{i \in F}$ in $[\epsilon/2, 2]$ such that

$$\|\sum_{i \in F} \sigma_i \mu_i g_i\| \leq 8C,$$

for every choice of signs $(\sigma_i)_{i \in F}$. We conclude from the above, that some subsequence of (g_i) is a c_0^β-spreading model. Lemma 4.3 finally implies that there is some $L \in [M_0]$ (and thus $L \in [Q_1]$) such that (α_i^L) is a c_0^β-spreading model, contradicting the choice of Q_1. Therefore, we must have that $[Q] \subset \mathcal{D}$, for some $Q \in [P]$, and the proof of the theorem is now complete. $\qquad\square$

6. TRANSFINITE AVERAGES OF WEAKLY NULL SEQUENCES IN $C(K)$ EQUIVALENT TO THE UNIT VECTOR BASIS OF c_0

In this section we present the following

Theorem 6.1. *Let K be a compact metric space and let (f_n) be a normalized, basic sequence in $C(K)$. Suppose that there exist $M \in [\mathbb{N}]$ and a summable sequence of positive scalars (ϵ_n) such that for all $t \in K$, the set $\{n \in M : |f_n(t)| \geq \epsilon_n\}$ is finite. Then there exist $\xi < \omega_1$ and a block basis of ξ-averages of (f_n) equivalent to the unit vector basis of c_0.*

(Note that all transfinite averages of (f_n) are considered with respect to $\mathcal{F} = S_1$.)

Remark 6.2. *The hypotheses in Theorem 6.1 imply that $\sum_{n \in M} |f_n(t)|$ is a convergent series, for all $t \in K$. It follows then from Rainwater's theorem [39], that every normalized block basis of $(f_n)_{n \in M}$ is weakly null and therefore, the subsequence $(f_n)_{n \in M}$ of (f_n) is shrinking. Moreover, the convergence of the series $\sum_{n \in M} |f_n(t)|$ for all $t \in K$, implies that some block basis of $(f_n)_{n \in M}$ is equivalent to the unit vector basis of c_0. This is a special case of a famous result, due to J. Elton [22], which states that if (x_n) is a normalized basic sequence in some Banach space and the series $\sum_n |x^*(x_n)|$ converges for every extreme point x^* in the ball of X^*, then some block basis of (x_n) is equivalent to the unit vector basis of c_0. An alternate proof of this special case of Elton's theorem is given in [25]. See also [23], [4] for related results. We wish to indicate however, as our next corollary shows, that this special case of Elton's theorem is also a consequence of Theorem 6.1. Hence, our result may be viewed as a quantitative version of this special case of Elton's theorem.*

Corollary 6.3. *Let (f_n) be a normalized basic sequence in $C(K)$ such that $\sum_n |f_n(t)|$ is a convergent series, for all $t \in K$. Then there exist $\xi < \omega_1$ and a block basis of ξ-averages of (f_n) equivalent to the unit vector basis of c_0.*

The proof is given at the end of this section.

The ordinal ξ that appears in the conclusion of Theorem 6.1, is related to the complexity of the compact family $\{F \in [M]^{<\infty} : \exists t \in K \text{ with } |f_n(t)| \geq \epsilon_n, \forall n \in F\}$. It follows from Corollary 3.2, that every normalized weakly null sequence in $C(K)$, for K a countable compact metric space, admits a subsequence satisfying the hypotheses of Theorem 6.1. Moreover, if K is homeomorphic to $[1, \omega^{\omega^\alpha}]$, for some $\alpha < \omega_1$, then as is shown in Corollary 6.8, the ordinal ξ in the conclusion of Theorem 6.1 can be taken not to exceed α.

We shall next describe how to obtain the "optimal" ξ satisfying the conclusion of Theorem 6.1.

The following conventions hold throughout this section. K is a compact metric space and $\vec{s} = (f_n)$ is a normalized shrinking basic sequence in $C(K)$. We shall assume, without loss of generality, by passing to a subsequence if necessary, that $\vec{s}$ is 2-basic. We let $\mathcal{F} = S_1$. *All transfinite averages of $\vec{s}$ will be taken with respect to $\mathcal{F}$.* As in the previous section, α_n^M abbreviates $\alpha_n^{\mathcal{F}, \vec{s}, M}$. In the sequel, when we refer to a block basis we shall always mean a block basis of $\vec{s} = (f_n)$. Also all transfinite averages will be taken with respect to $\vec{s}$.

Definition 6.4.   (1) *Given $N \in [\mathbb{N}]$ and $1 \leq \alpha < \omega_1$, we say that N is α-large, if for every $\beta < \alpha$ and $M \in [N]$ there exists $L \in [M]$ such that no block basis of β-averages supported by L is a c_0^γ-spreading model, where $\beta + \gamma = \alpha$.*

(2) *Given $N \in [\mathbb{N}]$ set $\xi^N = \sup\{\alpha < \omega_1 : \exists$ an $\alpha - large\, M \in [N]\}$. Put $\xi^N = 0$, if this set is empty. Finally put $\xi^0 = \min\{\xi^N : N \in [\mathbb{N}]\}$.*

Note that if $\xi^0 = \xi^{N_0}$ for some $N_0 \in [\mathbb{N}]$, then $\xi^L = \xi^0$, for all $L \in [N_0]$. In fact, if $1 \leq \xi^0 < \omega_1$, then every infinite subset of N_0 is ξ^0-large.

Proposition 6.5. *Suppose that $\xi^N < \omega_1$, for some $N \in [\mathbb{N}]$. Then there exists a block basis of ξ^N-averages, supported by N, which is equivalent to the unit vector basis of c_0.*

We postpone the proof and observe that if $\xi^0 < \omega_1$ and $\xi^0 = \xi^{N_0}$, then Proposition 6.5 yields that every infinite subset of N_0 supports a block basis of ξ^0-averages, equivalent to the unit vector basis of c_0 and, moreover, it follows by our preceding comments, that ξ^0 is the smallest ordinal with this property. Therefore, the optimality of ξ^0 is considered in this sense. In order to prove Theorem 6.1, we need to introduce some more notation and terminology.

Definition 6.6.   (1) *Let $\beta < \alpha < \omega_1$, $p \in \mathbb{N}$ and $\epsilon > 0$. An α-average $u = \sum_i a_i f_i$, is said to be (β, p, ϵ)-large, if for every choice $I_1 < \cdots < I_k$ of k consecutive members of S_β, $k \leq p$, and all $t \in K$, we have $|\sum_{i \in I} a_i f_i(t)| \leq \epsilon + \sum_{i \notin I} a_i |f_i(t)|$, where $I = \cup_{j=1}^k I_j$.*

(2) *Let $N \in [\mathbb{N}]$, $1 \leq \alpha < \omega_1$ We say that N is α-nice if for every $\beta < \alpha$, every $M \in [N]$, every $p \in \mathbb{N}$ and all $\epsilon > 0$, there exists an α-average supported by M which is (β, p, ϵ)-large.*

The main step for proving Theorem 6.1 is

Theorem 6.7. *Suppose that $N \in [\mathbb{N}]$ is α-large, for some $1 \leq \alpha < \omega_1$. Then N is α-nice.*

We postpone the proof in order to give the

Proof of Theorem 6.1. Let

$$\mathcal{G} = \{F \in [\mathbb{N}]^{<\infty} : \exists t \in K \text{ with } |f_n(t)| \geq \epsilon_n, \, \forall n \in F\}.$$

Clearly, $\mathcal{G}$ is hereditary. The compactness of K and our assumptions, imply that $\mathcal{G}[M]$ is compact in the topology of pointwise convergence. It follows that there is a countable ordinal ζ such that $\mathcal{G}[M]^{(\zeta)}$ is finite. Write $\zeta = \omega^\gamma k + \eta$, for some $k \in \mathbb{N}$ and $\eta < \omega^\gamma$. We infer now by the result of [24], that there exists $N \in [M]$ with the property $\mathcal{G}[N] \subset S_{\gamma+1}$.

We claim that $\xi^N \leq \gamma + 1$ (see Definition 6.4). Indeed, were this claim false, we would choose $P \in [N]$ and a countable ordinal $\beta > \gamma + 1$ such that P is β-large. Theorem 6.7 then yields P is β-nice (see Definition 6.6). Next, let $\epsilon > 0$ and choose $Q \in [P]$ such that $\sum_{n \in Q} \epsilon_n < \epsilon/12$. Since $\gamma + 1 < \beta$ and P is β-nice, there exists a β-average $u = \sum_i a_i f_i$, supported by Q which is $(\gamma + 1, 1, \epsilon/2)$-large. This means

$$\left| \sum_{i \in I} a_i f_i(t) \right| \leq \epsilon/2 + \sum_{i \notin I} a_i |f_i(t)|,$$

for all $t \in K$ and every $I \in S_{\gamma+1}$. Given $t \in K$, put $\Lambda_t = \{n \in \mathbb{N} : |f_n(t)| \geq \epsilon_n\}$. Note that u is supported by N and so $\Lambda_t \cap \operatorname{supp} u \in S_{\gamma+1}$, for all $t \in K$, as $\Lambda_t \cap \operatorname{supp} u \in \mathcal{G}[N]$. Taking in account that $\|u\| = 1$, we have $0 \leq a_i \leq 3$, for all $i \in \mathbb{N}$. Hence,

$$|u(t)| \leq \left| \sum_{i \in \Lambda_t \cap \operatorname{supp} u} a_i f_i(t) \right| + \left| \sum_{i \notin \Lambda_t} a_i f_i(t) \right|$$

$$\leq \epsilon/2 + 2 \sum_{i \notin \Lambda_t} a_i |f_i(t)|$$

$$< \epsilon/2 + 6\epsilon/12 = \epsilon,$$

for all $t \in K$. Since ϵ was arbitrary, we have reached a contradiction. Therefore, our claim holds. In particular, $\xi^N < \omega_1$ and the assertion of the theorem is a consequence of Proposition 6.5. $\qquad \square$

Corollary 6.8. *Let (f_n) be a normalized weakly null sequence in $C(\omega^{\omega^\xi})$, $\xi < \omega_1$. Then there exist $\alpha \leq \xi$ and a block basis of α-averages of (f_n) equivalent to the unit vector basis of c_0.*

Proof. Set $K = [1, \omega^{\omega^\xi}]$. Corollary 3.2 yields $M \in [\mathbb{N}]$ and a summable sequence of positive scalars (ϵ_n) such that for all $t \in K$ the set $\{n \in M : |f_n(t)| \geq \epsilon_n\}$ belongs to S_ξ^+. In particular, $\Lambda_t \cap M$ is the union of two consecutive members of S_ξ. The argument in the proof of Theorem 6.1 shows that $\xi^M \leq \xi$. The assertion of the corollary now follows from Proposition 6.5. $\qquad \square$

We shall now give the proof of Proposition 6.5. This requires two lemmas.

Lemma 6.9. *Suppose that $1 \leq \alpha < \omega_1$. Let $m < n$ in $\mathbb{N}$ and $F \in [\mathbb{N}]^{<\infty}$ with $n < \min F$ be such that $\{n\} \cup F$ is a maximal member of S_α. Then $\{m\} \cup F \notin S_\alpha$.*

Proof. We use transfinite induction on α. When $\alpha = 1$, we must have that $|F| = n - 1$, in order for $\{n\} \cup F$ be maximal in S_1. Hence, $|\{m\} \cup F| = n > m = \min(\{m\} \cup F)$. Thus the assertion of the lemma holds in this case.

Next assume the assertion holds for all ordinals smaller than α ($\alpha > 1$). Suppose first α is a limit ordinal and let (α_n) be the sequence of successor ordinals associated to α. Since $\{n\} \cup F$ is maximal in S_α, we have that $\{n\} \cup F$ is maximal in S_{α_k}, for all $k \leq n$ such that $\{n\} \cup F \in S_{\alpha_k}$. Suppose we had

$\{m\} \cup F \in S_\alpha$. Then there is some $k \leq m$ such that $\{m\} \cup F \in S_{\alpha_k}$. We infer from the spreading property of S_{α_k}, as $m < n$, that $\{n\} \cup F \in S_{\alpha_k}$. Therefore, $\{n\} \cup F$ is maximal in S_{α_k}. The induction hypothesis applied on α_k now yields $\{m\} \cup F \notin S_{\alpha_k}$, a contradiction which proves the assertion when α is a limit ordinal.

We now assume α is a successor ordinal, say $\alpha = \beta + 1$. Since $\{n\} \cup F$ is maximal in S_α, there exist $F_1 < \cdots < F_n$, successive maximal members of S_β such that $\{n\} \cup F = \cup_{i=1}^n F_i$ (see [24]). We shall assume $m > 1$ or else the assertion holds since $\{1\}$ is maximal in every Schreier family and $F \neq \emptyset$. Note that the induction hypothesis on β implies that $G_1 = \{m\} \cup (F_1 \setminus \{n\}) \notin S_\beta$. It follows, as S_β is stable, that G_1 contains a maximal member H_1 of S_β as an initial segment, and so we may write $G_1 = H_1 \cup H_2$ with $H_2 \neq \emptyset$. Of course, $m = \min H_1$. Set $H = H_1 \cup \cup_{i=2}^m F_i$. Then H is maximal in S_α. This completes the proof of the lemma since H is a proper subset of $\{m\} \cup F$. $\square$

Lemma 6.10. *Let $P \in [\mathbb{N}]$, $\beta \leq \alpha < \omega_1$ and $\tau < \omega_1$. Assume that every block basis of β-averages supported by P is a c_0^γ-spreading model, where $\beta + \gamma = \alpha$, while every block basis of α-averages supported by P is a c_0^τ-spreading model. Then there exists $Q \in [P]$ such that every block basis of β-averages supported by Q is a $c_0^{\gamma+\tau}$-spreading model.*

Proof. We assume that both γ and τ are greater than or equal to 1, or else the assertion of the lemma is trivial. We also assume, without loss of generality thanks to Lemma 4.6, that there exists a constant $C > 0$ such that every block basis of β-averages (resp. α-averages) supported by P is a c_0^γ (resp. c_0^τ)-spreading model with constant C. We shall further assume, without loss of generality thanks to Lemma 5.8, that for every $F \in S_{\gamma+\tau}[P]$ we have $F \setminus \{\min F\} \in S_\tau[S_\gamma]$.

Let $M \in [P]$. Choose a sequence of positive scalars (δ_i) with $\sum_i \delta_i < 1/(4C)$. We apply Proposition 5.9, successively, to obtain the following objects:

   (1) A maximally S_τ-admissible block basis $v_1 < \cdots < v_n$ of α-averages, supported by M, with $\min M < \min \operatorname{supp} v_1$.
   (2) Successive, maximal members $F_1 < \cdots < F_n$ of $S_\gamma[M]$ such that $\max \operatorname{supp} v_i < \min F_{i+1}$, for all $i < n$.
   (3) Successive finite subsets of $\mathbb{N}$ $J_1 < \cdots < J_n$ such that for each $i \leq n$, there exist a normalized block basis $(u_j)_{j \in J_i}$, a subset I_i of J_i and positive scalars $(\lambda_j)_{j \in J_i}$ which satisfy the following properties:

$$(6.1) \qquad v_i = \sum_{j \in J_i} \lambda_j u_j, \text{ and } \left\| \sum_{j \in J_i \setminus I_i} \lambda_j u_j \right\|_{\ell_1} < \delta_i.$$

$$(6.2) \qquad (u_j)_{j \in I_i} \text{ is an } S_\gamma - \text{admissible block basis of } \beta - \text{averages}$$
$$\text{and } |\lambda_r - \lambda_s| < \delta_i, \text{ for all } r, s \text{ in } I_i.$$

$$(6.3) \qquad F_i \setminus \{\min F_i\} \subset \{\min \operatorname{supp} u_j : j \in I_i\}.$$

Our assumptions yield that $\| \sum_{i=1}^{n} v_i \| \leq C$ and that

$$1 - \delta_i \leq \left\| \sum_{j \in I_i} \lambda_j u_j \right\| \leq C \max_{j \in I_i} \lambda_j, \text{ for all } i \leq n.$$

(6.2) now implies

(6.4) $1/(2C) \leq \lambda_j \leq 3$, for all $j \in I_i$ and $i \leq n$.

We also obtain from (6.1) that

(6.5) $$\left\| \sum_{i=1}^{n} \sum_{j \in I_i} \lambda_j u_j \right\| \leq C + \sum_{i=1}^{n} \delta_i < 2C.$$

We next observe that for all $i < n$ and $j_0 \in I_i$, $\{\min \operatorname{supp} u_{j_0}\} \cup \{\min \operatorname{supp} u_j : j \in I_{i+1}\} \notin S_\gamma$. This is so since $F_{i+1} \setminus \{\min F_{i+1}\} \subset \{\min \operatorname{supp} u_j : j \in I_{i+1}\}$, (by (6.3)), $\max \operatorname{supp} v_i < \min F_{i+1}$, and thus, as a consequence of Lemma 6.9, we have that $\{\min \operatorname{supp} u_{j_0}\} \cup (F_{i+1} \setminus \{\min F_{i+1}\}) \notin S_\gamma$.

It follows from this that for all $i \leq n$ there exists an initial segment I_i^* of I_i (possibly, $I_i^* = \emptyset$) with $\max I_i^* < \max I_i$, such that $\{\min \operatorname{supp} u_j : j \in (I_i \setminus I_i^*) \cup I_{i+1}^*\}$ is a maximal member of S_γ, for all $i < n$. Note that $I_1^* = \emptyset$.

Set $T_i = (I_i \setminus I_i^*) \cup I_{i+1}^*$, for all $i < n$. Then $(u_j)_{j \in T_i}$ is maximally S_γ-admissible for all $i < n$. We also infer from (6.4) and (6.5) that

$$\left\| \sum_{j \in \cup_{i<n} T_i} \lambda_j u_j \right\| \leq 4C, \ \lambda_j \in [1/(2C), 3], \text{ for all } j \in \cup_{i<n} T_i.$$

Note also that $\min \operatorname{supp} u_{\min T_i} < \min \operatorname{supp} v_{i+1}$, for all $i < n$. Since $\min M < \min \operatorname{supp} v_1$ and $(v_i)_{i=1}^{n}$ is maximally S_τ-admissible, Lemma 6.9 and the spreading property of S_τ, yield that $\{\min M\} \cup \{\min \operatorname{supp} u_{\min T_i} : i < n\}$ is not a member of S_τ. Hence, by the stability of S_τ, there exists $m < n$ such that $\{\min M\} \cup \{\min \operatorname{supp} u_{\min T_i} : i \leq m\}$ is a maximal member of S_τ. Note also that $\| \sum_{j \in \cup_{i \leq m} T_i} \lambda_j u_j \| \leq 4C$ and $\lambda_j \in [1/(2C), 3]$, for all $j \in \cup_{i \leq m} T_i$.

Summarizing, given $M \in [P]$ there exists a maximally $S_\tau[S_\gamma]$-admissible block basis $(u_i)_{i=1}^{k}$ of β-averages, supported by M, and scalars $(\lambda_i)_{i=1}^{k}$ in $[1/(2C), 3]$ such that $\| \sum_{i=1}^{k} \lambda_i u_i \| \leq 5C$. Given $L \in [P]$ let n_L denote the unique integer such that $(\beta_i^L)_{i=1}^{n_L}$ is maximally $S_\tau[S_\gamma]$-admissible. Define

$$\mathcal{D} = \left\{ L \in [P] : \exists \, (\lambda_i)_{i=1}^{n_L} \subset [1/(2C), 3], \left\| \sum_{i=1}^{n_L} \lambda_i \beta_i^L \right\| \leq 5C \right\}.$$

Lemma 4.3 and the stability of $S_\tau[S_\gamma]$ yield that $\mathcal{D}$ is closed in the topology of pointwise convergence. We infer from our preceding discussion, that every $M \in [P]$ contains some $L \in \mathcal{D}$ as a subset. Thus, we deduce from Theorem 4.5 that there exists $M_0 \in [P]$ with $[M_0] \subset \mathcal{D}$. Arguing as in the last part of the proof of Theorem 5.1, using Theorem 5.10 and our assumptions on P, we obtain a block basis of β-averages which is a $c_0^{\gamma+\tau}$-spreading model. The assertion of the lemma now follows from Lemma 4.6. $\square$

Proof of Proposition 6.5. To simplify our notation, let us write ξ instead of ξ^N. We assert that for every $M \in [N]$ and all $\beta < \omega_1$ there exists a block basis of ξ-averages supported by M which is a c_0^β-spreading model. Once this is accomplished, the proposition will follow from the Kunen-Martin boundedness principle (see [19], [28]). To see this, let $N \in [\mathbb{N}]$. Given $n \in \mathbb{N}$, let $\mathcal{T}_n^N$ denote the family of those finite subsets of N that are initial segments of sets of the form $\cup_{i=1}^k \mathrm{supp}\, \xi_i^L$, for some $k \in \mathbb{N}$ and $L \in [N]$ such that $\|\sum_{i=1}^k \xi_i^L\| \leq n$. We claim there is some $n \in \mathbb{N}$ so that $\mathcal{T}_n^N$ is not compact in the topology of pointwise convergence. Otherwise, the Mazurkiewicz-Sierpinski theorem [32], yields $\zeta < \omega_1$ so that $\mathcal{T}_n^N$ is homeomorphic to a subset of $[1, \omega^{\omega^\zeta}]$, for all $n \in \mathbb{N}$. We may now choose, according to our assertion combined with Lemma 4.6, some $L_0 \in [N]$ and $n \in \mathbb{N}$ such that $(\xi_i^L)_{i=1}^\infty$ is a $c_0^{\zeta+1}$-spreading model with constant n, for all $L \in [L_0]$. It follows from this that for all $L \in [L_0]$, $\cup_{i=1}^{n_L} \mathrm{supp}\, \xi_i^L \in \mathcal{T}_n^N$, where n_L stands for the unique integer such that $(\xi_i^L)_{i=1}^{n_L}$ is maximally $S_{\zeta+1}$-admissible. Since S_α is homeomorphic to $[1, \omega^{\omega^\alpha}]$ for all $\alpha < \omega_1$ (see [1]), this implies that $S_{\zeta+1}$ is homeomorphic to a subset of $[1, \omega^{\omega^\zeta}]$ which is absurd. Hence, indeed, there is some $n \in \mathbb{N}$ with $\mathcal{T}_n^N$ non-compact. Subsequently, there exists $M \in [N]$, $M = (m_i)$, such that $\{m_1, \ldots, m_k\} \in \mathcal{T}_n^N$, for all $k \in \mathbb{N}$. We now infer from Lemma 4.3, that $\|\sum_{i=1}^k \xi_i^M\| \leq n$, for all $k \in \mathbb{N}$. Using an argument based on Theorem 4.5, similar to that in the proof of Lemma 4.6, we conclude that some block basis of ξ-averages is equivalent to the unit vector basis of c_0.

We shall next prove our initial assertion by transfinite induction on β. The assertion is trivial for $\beta = 0$. Assume $\beta \geq 1$ and that the assertion holds for all $M \in [N]$ and all ordinals smaller than β yet, for some $P \in [N]$ there is no block basis of ξ-averages, supported by P, which is a c_0^β-spreading model. We now show that P is $(\xi + \beta)$-large which, of course, is absurd.

To see this, first consider an ordinal $\gamma < \xi$ and let $M \in [P]$. Write $\xi = \gamma + \delta$. We claim that there exists $L \in [M]$ such that no block basis of γ-averages supported by L is a $c_0^{\delta+\beta}$-spreading model (note that $\gamma + (\delta + \beta) = \xi + \beta$). Were this claim false, then Lemma 4.6 would yield a constant $C > 0$ and $L_0 \in [M]$ such that, every block basis of γ-averages supported by L_0 is a $c_0^{\delta+\beta}$-spreading model with constant C. By employing Lemma 5.8 we may assume, without loss of generality, that for all $F \in S_\beta[S_\delta]$, $F \subset L_0$, we have $F \setminus \{\min F\} \in S_{\delta+\beta}$. But now, we shall exhibit a block basis of ξ-averages supported by L_0 (and thus also by P), which is a c_0^β-spreading model. Indeed, as $\xi = \gamma + \delta$, we may apply Proposition 5.9, successively, to obtain block bases $u_1 < u_2 < \ldots$ and $v_1 < v_2 < \ldots$ consisting of ξ and γ-averages, respectively, both supported by L_0; a sequence of positive scalars (λ_i) and a sequence $F_1 < F_2 < \ldots$ of successive finite subsets of $\mathbb{N}$ so that the following requirements are satisfied:

(1) $\|u_i - \sum_{j \in F_i} \lambda_j v_j\| < \epsilon_i$, for all $i \in \mathbb{N}$.
(2) $(v_j)_{j \in F_i}$ is S_δ-admissible and $\mathrm{supp}\, v_j \subset \mathrm{supp}\, u_i$, for all $j \in F_i$ and $i \in \mathbb{N}$.

In the above, (ϵ_i) is a summable sequence of positive scalars. Since $(\lambda_j)_{j \in \cup_i F_i}$ is bounded and (v_i) is a $c_0^{\delta+\beta}$-spreading model, our assumptions on L_0 readily imply that (u_i) is a block basis of ξ-averages supported by P which is a c_0^β-spreading model. This contradicts the choice of P. Therefore our claim holds.

Next, let $M \in [P]$, $\gamma < \beta$ and write $\beta = \gamma + \delta$. Note that $\xi + \beta = (\xi + \gamma) + \delta$. We now claim that there exists $L \in [M]$ such that no block basis of $(\xi + \gamma)$-averages supported by L is a c_0^δ-spreading model. If that were not the case then, thanks to Lemma 4.6, there would exist $L_0 \in [M]$ such that every block basis of $(\xi + \gamma)$-averages supported by L_0 is a c_0^δ-spreading model.

Since $\gamma < \beta$, the induction hypothesis combined with Lemma 4.6 implies the existence of some $L_1 \in [L_0]$ such that every block basis of ξ-averages supported by L_1 is a c_0^γ-spreading model. We deduce from Lemma 6.10 that some block basis of ξ-averages supported by L_0 (and thus also by P) is a $c_0^{\gamma+\delta}$-spreading model. Since $\beta = \gamma + \delta$, we contradict the choice of P. Therefore, this claim holds as well.

Summarizing, we showed that for every $\gamma < \xi + \beta$ and all $M \in [P]$ there exists $L \in [M]$ such that no block basis of γ-averages supported by L is a c_0^δ-spreading model, where $\gamma + \delta = \xi + \beta$. But this means $P \in [N]$ is $(\xi + \beta)$-large, contradicting the definition of ξ. The proof of the proposition is now complete. $\qquad\square$

In the next part of this section we give the proof of Theorem 6.7. We shall need a few technical lemmas.

Lemma 6.11. *Suppose that $N \in [\mathbb{N}]$ is α-nice (see Definition 6.6). Then for every $P \in [N]$, every $\beta < \alpha$, every $p \in \mathbb{N}$ and all $\epsilon > 0$, there exists $M \in [P]$ such that every α-average supported by M is (β, p, ϵ)-large.*

Proof. Define $\mathcal{D} = \{L \in [P] : \alpha_1^L$ is $(\beta, p, \epsilon) - \text{large}\}$. Lemma 4.3 yields $\mathcal{D}$ is closed in the topology of pointwise convergence. Because N is α-nice, we deduce that $[L] \cap \mathcal{D} \neq \emptyset$, for all $L \in [P]$. We infer now, from Theorem 4.5, that $[M] \subset \mathcal{D}$, for some $M \in [P]$. Clearly, M is as desired. $\qquad\square$

Lemma 6.12. *Suppose that $N_1 \supset N_2 \supset \ldots$ are infinite subsets of $\mathbb{N}$ and $\alpha_1 < \alpha_2 < \ldots$ are countable ordinals such that N_i is α_i-nice for all $i \in \mathbb{N}$. Let $N \in [\mathbb{N}]$ be such that $N \setminus N_i$ is finite, for all $i \in \mathbb{N}$. Then, N is α-nice, where $\alpha = \lim_i \alpha_i$.*

Proof. Let $M \in [N]$, $\beta < \alpha$, $p \in \mathbb{N}$ and $\epsilon > 0$. It suffices to find an α-average u supported by M which is (β, p, ϵ)-large. Choose a sequence of positive scalars (δ_i) with $\sum_i \delta_i < \epsilon/6$.

Let $k \in \mathbb{N}$ be such that $\beta < \alpha_k$. Since N_k is α_k-nice, we may apply Lemma 6.11, successively, to obtain infinite subsets $P_1 \supset P_2 \supset \ldots$ of $M \cap N_k$ such that, for all $i \in \mathbb{N}$, every α_k-average supported by P_i is (β, p, δ_i)-large. Next choose integers $p_1 < p_2 < \ldots$ such that $p_i \in P_i$, for all $i \in \mathbb{N}$, and set $P = (p_i)$.

We now employ Proposition 5.9 to find $Q \in [P]$ with the property that every α-average supported by Q admits an $(\epsilon/2, \alpha_k, \beta_k)$-decomposition (see

Definition 5.2), where $\alpha_k + \beta_k = \alpha$. Let u be an α-average supported by Q. Write $u = \sum_{i=1}^n \lambda_i u_i$, where $u_1 < \cdots < u_n$ are normalized blocks, $(\lambda_i)_{i=1}^n$ are positive scalars for which there exists $I \subset \{1, \ldots, n\}$ satisfying

$$u_i \text{ is an } \alpha_k - \text{average for all } i \in I, \text{ while } \left\| \sum_{i \in \{1,\ldots,n\}\backslash I} \lambda_i u_i \right\|_{\ell_1} < \epsilon/2.$$

If $u_i = \sum_s a_s^i f_s$, for $i \leq n$, then, clearly, $\sum_{i \in \{1,\ldots,n\}\backslash I} \lambda_i \sum_s a_s^i < \epsilon/2$.

We are going to show that u is (β, p, ϵ)-large. To this end, let J be the union of less than, or equal to, p consecutive members of S_β and let $t \in K$. Write $I = \{i_1 < \ldots < i_m\}$. Observe that u_{i_j} is an α_k-average supported by P_j and thus by the choice of P_j,

$$\left| \sum_{s \in J} a_s^{i_j} f_s(t) \right| \leq \delta_j + \sum_{s \notin J} a_s^{i_j} |f_s(t)|, \text{ for all } j \leq m.$$

Therefore, letting $I^c = \{1, \ldots, n\} \backslash I$,

$$\left| \sum_{i=1}^n \lambda_i \sum_{s \in J} a_s^i f_s(t) \right| \leq \left| \sum_{i \in I^c} \lambda_i \sum_{s \in J} a_s^i f_s(t) \right| + \left| \sum_{i \in I} \lambda_i \sum_{s \in J} a_s^i f_s(t) \right|$$

$$\leq \sum_{i \in I^c} \lambda_i \sum_s a_s^i + \sum_{i \in I} \lambda_i \left| \sum_{s \in J} a_s^i f_s(t) \right|$$

$$\leq \epsilon/2 + \sum_{j=1}^m \lambda_{i_j} \left(\delta_j + \sum_{s \notin J} a_s^{i_j} |f_s(t)| \right)$$

$$\leq \epsilon/2 + 3 \sum_{j=1}^{|I|} \delta_j + \sum_{i=1}^n \lambda_i \sum_{s \notin J} a_s^i |f_s(t)|$$

$$\leq \epsilon + \sum_{i=1}^n \lambda_i \sum_{s \notin J} a_s^i |f_s(t)|.$$

The proof of the lemma is now complete. $\square$

Lemma 6.13. *Let $u_1 < \cdots < u_n$ be a normalized finite block basis of (f_i). Write $u_i = \sum_s a_s^i f_s$, and set $k_i = \max \operatorname{supp} u_i$ for all $i \leq n$. Let $\alpha < \omega_1$ and denote by $(\alpha_j + 1)_{j=1}^\infty$ the sequence of ordinals associated to α. Let $\mathcal{G}$ be a hereditary and spreading family, and $(\delta_i)_{i=1}^n$ be a sequence of non-negative scalars. Suppose that $J \in \mathcal{G}[S_\alpha]$ satisfies the following property: If $2 \leq i \leq n$ is so that $J \cap \operatorname{supp} u_i$ is contained in the union of less than, or equal to, k_{i-1} consecutive members of S_{α_j}, for some $j \leq k_{i-1}$ then,*

$$\left| \sum_{s \in J} a_s^i f_s(t) \right| \leq \delta_i + \sum_{s \notin J} |a_s^i| |f_s(t)|, \text{ for all } t \in K.$$

Then for every scalar sequence $(b_i)_{i=1}^n$ and all $t \in K$, we have the estimate

$$(6.6) \quad \Big| \sum_{i=1}^{n} b_i \sum_{s \in J} a_s^i f_s(t) \Big| \leq \max \Big\{ \Big| \sum_{i \in I} b_i u_i(t) \Big| : (u_i)_{i \in I} \text{ is } \mathcal{G}^+ - admissible \Big\}$$

$$+ \Big(\sum_{i=1}^{n} \delta_i \Big) \max_{i \leq n} |b_i| + \sum_{i=1}^{n} |b_i| \sum_{s \notin J} |a_s^i| |f_s(t)|.$$

Proof. We may assume that $J \cap \cup_{i=1}^n \operatorname{supp} u_i \neq \emptyset$, or else the assertion of the lemma is trivial. We may thus write $J \cap \cup_{i=1}^n \operatorname{supp} u_i \cup_{l=1}^p J_l$, where $J_1 < \cdots < J_p$ are non-empty members of S_α with $\{\min J_l : l \leq p\} \in \mathcal{G}$.

Define $I_l = \{i \leq n : r(u_i) \cap J_l \neq \emptyset\}$ (where $r(u_i)$ denotes the range of u_i) and $i_l = \min I_l$, for all $l \leq p$. Put $I = \{i_l : l \leq p\}$ and let I^c be the complement of I in $\{1, \ldots, n\}$. Then $(u_i)_{i \in I}$ is $\mathcal{G}^+$-admissible.

Indeed, set $L_i = \{l \leq p : i_l = i\}$, for all $i \in I$. Observe that L_i is an interval and that $L_i < L_{i'}$ for all $i < i'$ in I. Hence, $\min J_{\min L_i} \leq \max \operatorname{supp} u_i$, for all $i \in I$. Since $\mathcal{G}$ is hereditary and spreading, we infer that $(k_i)_{i \in I} \in \mathcal{G}$. It follows now, by the spreading property of $\mathcal{G}$, that $(u_i)_{i \in I \setminus \{\min I\}}$ is $\mathcal{G}$-admissible.

Next assume that $i \in I^c \cap \cup_{l \leq p} I_l$. Then there is a unique $l \leq p$ with $i \in I_l$. Otherwise, $r(u_i) \cap J_l \neq \emptyset$ for at least two distinct l's, and so $i \in I$.

It follows now that $J \cap \operatorname{supp} u_i = J_l \cap \operatorname{supp} u_i$, for some $l \leq p$. Note that $i_l < i$ and that $J_l \cap r(u_{i_l}) \neq \emptyset$. Therefore $\min J_l \leq k_{i_l}$. We deduce from this that $J_l \in S_{\alpha_j+1}$ for some $j \leq k_{i_l}$ and, subsequently, that J_l is contained in the union of less than or equal to k_{i_l} consecutive members of S_{α_j}, for some $j \leq k_{i_l}$. The same holds for $J \cap \operatorname{supp} u_i$ and as $i_l < i$, we infer from our hypothesis, that

$$\Big| \sum_{s \in J} a_s^i f_s(t) \Big| \leq \delta_i + \sum_{s \notin J} |a_s^i| |f_s(t)|, \text{ for all } i \in I^c \text{ and } t \in K.$$

Now let $(b_i)_{i=1}^n$ be any scalar sequence and let $t \in K$. Then

$$\sum_{i=1}^{n} b_i \sum_{s \in J} a_s^i f_s(t) \sum_{i \in I} b_i \sum_{s \in J} a_s^i f_s(t) + \sum_{i \in I^c} b_i \sum_{s \in J} a_s^i f_s(t).$$

Our preceding discussions yield

$$(6.7) \quad \Big| \sum_{i \in I^c} b_i \sum_{s \in J} a_s^i f_s(t) \Big| \leq \sum_{i \in I^c} |b_i| \Big| \sum_{s \in J} a_s^i f_s(t) \Big|$$

$$\leq \sum_{i \in I^c} |b_i| \Big(\delta_i + \sum_{s \notin J} |a_s^i| |f_s(t)| \Big)$$

$$\leq \Big(\max_{i \leq n} |b_i| \Big) \sum_{i \in I^c} \delta_i + \sum_{i \in I^c} |b_i| \sum_{s \notin J} |a_s^i| |f_s(t)|$$

and

(6.8) $|\sum_{i \in I} b_i \sum_{s \in J} a_s^i f_s(t)||\sum_{i \in I} b_i \big(u_i(t) - \sum_{s \notin J} a_s^i f_s(t)\big)|$

$$\leq |\sum_{i \in I} b_i u_i(t)| + \sum_{i \in I} |b_i| \sum_{s \notin J} |a_s^i||f_s(t)|.$$

Combining (6.7) with (6.8) we obtain (6.6), since $(u_i)_{i \in I}$ is $\mathcal{G}^+$-admissible.   □

Lemma 6.14. *Suppose that $N \in [\mathbb{N}]$ is α-nice and that there exist $\Gamma \in [N]$ and $\gamma < \omega_1$ such that no block basis of α-averages supported by Γ is a c_0^γ-spreading model. Then there exist $M \in [N]$ and $1 \leq \beta \leq \gamma$ such that M is $(\alpha + \beta)$-nice.*

Proof. Define

$\beta = \min\{\psi < \omega_1 :\ \exists\, \Psi \in [N]$ such that no block basis of

α − averages supported by Ψ is a c_0^ψ − spreading model$\}$.

Our assumptions yield $1 \leq \beta \leq \gamma$. Choose $M \in [N]$ such that no block basis of α-averages supported by M is a c_0^β-spreading model. We are going to show that M is $(\alpha + \beta)$-nice. Let $M_0 \in [M]$ and $\tau < \alpha + \beta$. Let $p \in \mathbb{N}$ and $\epsilon > 0$. We shall exhibit an $(\alpha + \beta)$-average supported by M_0 which is (τ, p, ϵ)-large. Choose a decreasing sequence of positive scalars (δ_i) such that $\sum_i \delta_i < \epsilon/6$.

We first consider the case $\tau < \alpha$. Because N is α-nice, we may apply Lemma 6.11, successively, to obtain infinite subsets $P_1 \supset P_2 \supset \ldots$ of M_0 such that, for all $i \in \mathbb{N}$, every α-average supported by P_i is (τ, p, δ_i)-large. Choose integers $p_1 < p_2 < \ldots$ such that $p_i \in P_i$, for all $i \in \mathbb{N}$, and set $P_0 = (p_i)$. Proposition 5.9 now yields an $(\alpha + \beta)$-average u supported by P_0 and admitting an $(\epsilon/2, \alpha, \beta)$-decomposition (see Definition 5.2). In particular, there exist normalized blocks $u_1 < \cdots < u_n$, positive scalars $(\lambda_i)_{i=1}^n$ and $I \subset \{1, \ldots, n\}$ such that $u = \sum_{i=1}^n \lambda_i u_i$, u_i is an α-average for all $i \in I$ and $\|\sum_{i \in \{1, \ldots, n\} \setminus I} \lambda_i u_i\|_{\ell_1} < \epsilon/2$. Let J be the union of less than, or equal to, p consecutive members of S_τ, and let $t \in K$. By repeating the argument in the last part of the proof of Lemma 6.12 we conclude that u is (τ, p, ϵ)-large. This proves the assertion when $\tau < \alpha$.

Next suppose $\alpha \leq \tau < \alpha + \beta$ and choose $\zeta < \beta$ with $\tau = \alpha + \zeta$. Recall that the definition of β implies that every infinite subset of M_0 supports a block basis of α-averages which is a c_0^ζ-spreading model. Hence, thanks to Lemma 4.6, there will be no loss of generality in assuming that for some positive constant C, every block basis of α-averages supported by M_0 is a c_0^ζ-spreading model with constant C. We shall further assume, because of Lemma 5.8, that for every $F \in S_\tau[M_0]$ we have $F \setminus \{\min F\} \in S_\zeta[S_\alpha]$.

Let $(\alpha_j + 1)$ be the sequence of ordinals associated to α. We shall construct $m_1 < m_2 < \ldots$ in M_0 with the following property: If $n \in \mathbb{N}$ and $j \leq m_n$, then every α-average supported by $\{m_i :\ i > n\}$ is $(\alpha_j, m_n, \delta_n)$-large. This construction is done inductively as follows: Choose $m_1 \in M_0$. Apply Lemma 5.6 to find $L_1 \in [M_1]$ with $m_1 < \min L_1$ and such that $S_{\alpha_j}[L_1] \subset S_{\alpha_{m_1}}$ for all

$j \leq m_1$. We then employ Lemma 6.11, as N is α-nice, to obtain $M_1 \in [L_1]$ such that every α-average supported by M_1 is $(\alpha_{m_1}, m_1, \delta_1)$-large. It follows that every α-average supported by M_1 is $(\alpha_j, m_1, \delta_1)$-large, for all $j \leq m_1$. Set $m_2 = \min M_1$.

Suppose $n \geq 2$ and that we have selected integers $m_1 < \cdots < m_n$ in M_0, and infinite subsets $M_1 \supset \cdots \supset M_{n-1}$ of M_0 with $m_{i+1} = \min M_i$ and such that every α-average supported by M_i is $(\alpha_j, m_i, \delta_i)$-large for all $j \leq m_i$ and $i < n$.

We next choose, by Lemma 5.6, $L_n \in [M_{n-1}]$ with $m_n < \min L_n$ and such that $S_{\alpha_j}[L_n] \subset S_{\alpha_{m_n}}$, for all $j \leq m_n$. Because N is α-nice, Lemma 6.11 allows us to select $M_n \in [L_n]$ such that every α-average supported by M_n is $(\alpha_j, m_n, \delta_n)$-large for all $j \leq m_n$. Set $m_{n+1} = \min M_n$. This completes the inductive step. Evidently, $m_1 < m_2 < \ldots$ satisfy the required property.

We set $P = (m_n)$. The preceding construction yields the following fact that will be used later in the course of the proof: Suppose v is an α-average supported by P and $\min \operatorname{supp} v = m_n$, for some $n \geq 2$, then v is $(\alpha_j, m_{n-1}, \delta_{n-1})$-large, for all $j \leq m_{n-1}$.

Recall that no block basis of α-averages supported by P is a c_0^β-spreading model. Let $0 < \delta < \epsilon/(p(C+1)+3)$ and apply Theorem 5.1 to find an $(\alpha + \beta)$-average u supported by P, normalized blocks $u_1 < \cdots < u_n$, positive scalars $(\lambda_i)_{i=1}^n$ and $I \subset \{1, \ldots, n\}$ such that $u = \sum_{i=1}^n \lambda_i u_i$, u_i is an α-average for all $i \in I$, $\|\sum_{i \in \{1,\ldots,n\}\setminus I} \lambda_i u_i\|_{\ell_1} < \delta$ and $\max_{i \in I} \lambda_i < \delta$. We show u is (τ, p, ϵ)-large which will finish the proof of the lemma. Set

$$\mathcal{G} = \{F \in [\mathbb{N}]^{<\infty} : \exists\, F_1 < \cdots < F_p \text{ in } S_\zeta^+,\ F \subset \cup_{i=1}^p F_i\}.$$

Then $\mathcal{G}$ is a hereditary and spreading family.

Let $J \subset M_0$ be the union of less than, or equal to, p consecutive members of S_τ, and let $t \in K$. Our assumptions on M_0 yield $J \in \mathcal{G}[S_\alpha]$. Let $\{i_1 < \cdots < i_m\}$ be an enumeration of I and put $m_{d_k} = \max \operatorname{supp} u_{i_k}$, for all $k \leq m$. It has been already remarked that u_{i_k} is $(\alpha_j, m_{d_k-1}, \delta_{d_k-1})$-large, for all $2 \leq k \leq m$ and $j \leq m_{d_k-1}$. It follows that the hypotheses of Lemma 6.13 are fulfilled for the block basis $u_{i_1} < \cdots < u_{i_m}$ and the given $J \subset M_0$, with "δ_1"$= 0$ and "δ_k"$= \delta_{d_k-1}$ for $2 \leq k \leq m$. Writing $u_i = \sum_s a_s^i f_s$, for all $i < n$, we infer from (6.6) that

$$\left| \sum_{i \in I} \lambda_i \sum_{s \in J} a_s^i f_s(t) \right| \leq \max\left\{ \left| \sum_{i \in E} \lambda_i u_i(t) \right| : E \subset I,\ (u_i)_{i \in E} \text{ is }\right.$$
$$\left.\mathcal{G}^+ - \text{admissible}\right\}$$
$$+ \left(\sum_{i=1}^\infty \delta_i\right) \max_{i \in I} \lambda_i + \sum_{i \in I} \lambda_i \sum_{s \notin J} a_s^i |f_s(t)|.$$

Note that when $(u_i)_{i \in E}$ is $\mathcal{G}^+$-admissible, we have

$$\left\| \sum_{i \in E} \lambda_i u_i \right\| \leq (p(C+1)+1) \max_{i \in E} \lambda_i < (p(C+1)+1)\delta.$$

Hence,

$$\left|\sum_{i \in I} \lambda_i \sum_{s \in J} a_s^i f_s(t)\right| < \left(p(C+1)+2\right)\delta + \sum_{i \in I} \lambda_i \sum_{s \notin J} a_s^i |f_s(t)|.$$

Next, put $I^c = \{1, \ldots, n\} \setminus I$. Then,

$$\sum_{i \in I^c} \lambda_i \sum_s a_s^i < \delta, \text{ as } \left\|\sum_{i \in I^c} \lambda_i u_i\right\|_{\ell_1} < \delta.$$

Combining the preceding estimates we conclude

$$\left|\sum_{i=1}^n \lambda_i \sum_{s \in J} a_s^i f_s(t)\right| \leq \sum_{i \in I^c} \lambda_i \sum_s a_s^i + \left|\sum_{i \in I} \lambda_i \sum_{s \in J} a_s^i f_s(t)\right|$$

$$< \delta + \left(p(C+1)+2\right)\delta + \sum_{i \in I} \lambda_i \sum_{s \notin J} a_s^i |f_s(t)|$$

$$< \epsilon + \sum_{i=1}^n \lambda_i \sum_{s \notin J} a_s^i |f_s(t)|.$$

Therefore, u is (τ, p, ϵ)-large. This completes the proof. $\qquad \square$

We are now ready for the

Proof of Theorem 6.7. We claim that every infinite subset of N contains a further infinite subset which is α-nice. If this claim holds, then evidently, N is itself α-nice. So suppose on the contrary, that the claim is false and choose $N_0 \in [N]$ having no infinite subset which is α-nice. We now claim that there exist $1 \leq \beta_1 < \alpha$ and $N_1 \in [N_0]$ which is β_1-nice. Indeed, define

$$\beta_1 = \min\{\zeta < \omega_1 : \exists M \in [N_0] \text{ such that no block basis of}$$

$$0 - \text{averages supported by } M \text{ is a } c_0^\zeta - \text{spreading model}\}.$$

Since N is α-large, α belongs to the set and so $1 \leq \beta_1 \leq \alpha$. Choose $N_1 \in [N_0]$ such that no block basis of 0-averages supported by N_1 is a $c_0^{\beta_1}$-spreading model. We show N_1 is β_1-nice. Because N_0 is assumed to contain no infinite subset which is α-nice, we shall also obtain $\beta_1 < \alpha$.

Let $M \in [N_1]$, $\beta < \beta_1$, $p \in \mathbb{N}$ and $\epsilon > 0$. We shall find a β_1-average supported by M which is (β, p, ϵ)-large. Since $\beta < \beta_1$, there exist $M_1 \in [M]$ and a constant $C > 0$ such that the block basis $(f_m)_{m \in M_1}$ is a c_0^β-spreading model with constant $C > 0$. Let $0 < \delta < \epsilon/(pC)$. Since no block basis of 0-averages supported by M_1 is a $c_0^{\beta_1}$-spreading model, Theorem 5.1 yields a β_1-average u, supported by M_1, positive scalars $(\lambda_i)_{i \in F}$ (where $F = \operatorname{supp} u$) and $I \subset F$ with $I \in S_{\beta_1}$, such that

$$u = \sum_{i \in F} \lambda_i f_i, \max_{i \in I} \lambda_i < \delta, \text{ and } \sum_{i \in F \setminus I} \lambda_i < \delta.$$

Let $t \in K$ and let J be the union of less than, or equal to, p consecutive members of S_β. It follows that

$$\left| \sum_{i \in J \cap F} \lambda_i f_i(t) \right| \leq \left\| \sum_{i \in J \cap F} \lambda_i f_i \right\|$$

$$\leq pC \max_{i \in F} \lambda_i < pC\delta < \epsilon.$$

Thus, u is a β_1-average, (β, p, ϵ)-large, and so N_1 is β_1-nice, as claimed.

We shall now construct, by transfinite induction on $1 \leq \tau < \omega_1$, families $\{N_\tau\}_{1 \leq \tau < \omega_1} \subset [N_0]$ and $\{\beta_\tau\}_{1 \leq \tau < \omega_1} \subset [1, \alpha)$ with the following properties:

(1) $N_{\tau_2} \setminus N_{\tau_1}$ is finite, for all $1 \leq \tau_1 < \tau_2 < \omega_1$.

(2) N_τ is β_τ-nice, for all $1 \leq \tau < \omega_1$.

(3) $\beta_{\tau_1} < \beta_{\tau_2}$, for all $1 \leq \tau_1 < \tau_2 < \omega_1$.

Of course, (3) is absurd since $\alpha < \omega_1$. Hence, our assumption that N_0 contained no infinite subset which is α-nice, was false. The proof of the theorem will be completed, once we give the construction of the above described families, satisfying conditions (1)-(3). N_1 and β_1 have been already constructed. Suppose that $1 < \tau_0 < \omega_1$ and that $\{N_\tau\}_{1 \leq \tau < \tau_0} \subset [N_0]$, $\{\beta_\tau\}_{1 \leq \tau < \tau_0} \subset [1, \alpha)$ have been constructed fulfilling properties (1)-(3), above, with ω_1 being replaced by τ_0.

Assume first that τ_0 is a successor ordinal, say $\tau_0 = \tau_1 + 1$. We know by the inductive construction, that N_{τ_1} is β_{τ_1}-nice. By assumption, N is α-large. Since $\beta_{\tau_1} < \alpha$, there exists $\Gamma \in [N_{\tau_1}]$ such that no block basis of β_{τ_1}-averages supported by Γ is a $c_0^{\eta_{\tau_1}}$-spreading model, where $\beta_{\tau_1} + \eta_{\tau_1} = \alpha$. Lemma 6.14 now implies the existence of $N_{\tau_0} \in [N_{\tau_1}]$ and $1 \leq \zeta_{\tau_1} \leq \eta_{\tau_1}$ such that N_{τ_0} is $(\beta_{\tau_1} + \zeta_{\tau_1})$-nice. Set $\beta_{\tau_0} = \beta_{\tau_1} + \zeta_{\tau_1}$. Necessarily, $\beta_{\tau_0} < \alpha$, by the choice of N_0. It is easy to see that the families $\{N_\tau\}_{1 \leq \tau < \tau_0 + 1}$ and $\{\beta_\tau\}_{1 \leq \tau < \tau_0 + 1}$ satisfy conditions (1)-(3), above, with ω_1 being replaced by $\tau_0 + 1$.

Next assume that τ_0 is a limit ordinal and choose a strictly increasing sequence of ordinals $\tau_1 < \tau_2 < \ldots$ such that $\tau_0 = \lim_n \tau_n$. By the inductive construction we have that $\beta_{\tau_1} < \beta_{\tau_2} < \ldots$ and thus we may define the limit ordinal $\beta_{\tau_0} = \lim_n \beta_{\tau_n}$. In addition to this, $N_{\tau_n} \setminus N_{\tau_m}$ is finite for all integers $m < n$. We deduce from the above, that $\cap_{i=1}^k N_{\tau_i}$ is β_{τ_k}-nice, for all $k \in \mathbb{N}$. Finally, choose $N_{\tau_0} \in [N_0]$ such that $N_{\tau_0} \setminus \cap_{i=1}^k N_{\tau_i}$ is finite, for all $k \in \mathbb{N}$. We infer from Lemma 6.12, that N_{τ_0} is β_{τ_0}-nice. It is easily verified now, that the families $\{N_\tau\}_{1 \leq \tau < \tau_0 + 1}$ and $\{\beta_\tau\}_{1 \leq \tau < \tau_0 + 1}$ satisfy conditions (1)-(3), above, with ω_1 being replaced by $\tau_0 + 1$. This completes the inductive step and the proof of the theorem. $\qquad\square$

Proof of Corollary 6.3. Assume without loss of generality, that (f_n) has no subsequence equivalent to the unit vector basis of c_0. By the Kunen-Martin boundedness principle (see [19], [28]), we may choose an ordinal $1 \leq \gamma < \omega_1$ such that no subsequence of (f_n) is a c_0^γ-spreading model. Set $K_m = \{t \in K : \sum_n |f_n(t)| \leq m\}$, for all $m \in \mathbb{N}$. Clearly, (K_m) is an increasing sequence of closed subsets of K and $K = \cup_m K_m$. We claim that for every $m \in \mathbb{N}$, every $N \in [\mathbb{N}]$, and all $\epsilon > 0$, there exists a γ-average u of (f_n) supported by N and such that $|u|(t) < \epsilon$, for all $t \in K_m$ (if $u = \sum_i a_i f_i$, we define $|u|(x) = \sum_i |a_i||f_i(x)|$, for all $x \in K$).

To see this, let $0 < \delta < \epsilon/m$. Since no subsequence of (f_n) is a c_0^γ-spreading model, Theorem 5.1 allows us choose a γ-average u of (f_n), supported by N and such that there exist non-negative scalars $(\lambda_i)_{i=1}^p$ and $I \subset \{1, \ldots, p\}$ satisfying the following: (1) $u = \sum_{i=1}^p \lambda_i f_i$ and $\max_{i \in I} \lambda_i < \delta$. (2) $(f_i)_{i \in I}$ is S_γ-admissible (i.e. $I \in S_\gamma$) and $\sum_{i \in \{1, \ldots, p\} \setminus I} \lambda_i < \delta$. It is easy to check now that for every $t \in K_m$ we have $|u|(t) < \epsilon$ and thus our claim holds.

Now let (ϵ_n) be a summable sequence of positive scalars and $N \in [\mathbb{N}]$. Successive applications of the previous claim yield a block basis $v_1 < v_2 < \ldots$ of γ-averages of (f_n), supported by N and satisfying $|v_n|(t) < \epsilon_n$ for every $t \in K_n$ and all $n \in \mathbb{N}$. It follows that for all $t \in K$ the set $\{n \in \mathbb{N} : |v_n(t)| \geq \epsilon_n\}$ is a subset of $\{1, \ldots, q_t\}$, where q_t is the least $m \in \mathbb{N}$ such that $t \in K_m$. We deduce from Theorem 6.1, that there exist $\beta < \omega_1$ and a block basis of β-averages of (v_n), equivalent to the unit vector basis of c_0.

In order to get a block basis of averages of (f_i) equivalent to the unit vector basis of c_0, one needs a somewhat more demanding argument which goes as follows. Choose a countable limit ordinal α with $\gamma < \alpha$ and let $(\alpha_j + 1)_{j=1}^\infty$ be the sequence of ordinals associated to α. Let $N \in [\mathbb{N}]$ and choose $n \in N$ with $n \geq 2$, such that $\gamma < \alpha_n$. Let $m \in \mathbb{N}$. Since no subsequence of (f_i) is a $c_0^{\alpha_n}$-spreading model, our preceding argument allows us choose an α_n-average v of (f_i), supported by $\{i \in N : n < i\}$, and such that $|v|(t) < 1/(2n)$, for all $t \in K_m$. Set $u = ((1/n)f_n + v)/\|(1/n)f_n + v\|$. Clearly, u is an α-average of (f_i) supported by N and satisfying $|u|(t) < 3/n$, for all $t \in K_m$. Note that $n = \min \operatorname{supp} u$.

Summarizing, given $N \in [\mathbb{N}]$ we can select a block basis $u_1 < u_2 < \ldots$ of α-averages of (f_i) supported by N and satisfying $|u_n|(t) < 3/m_n$, for all $t \in K_n$ and $n \in \mathbb{N}$. In the above, we have let $m_n = \min \operatorname{supp} u_n$, for all $n \in \mathbb{N}$. It follows that for all $n \in \mathbb{N}$, if $t \in K_n$ and $|u_i|(t) \geq 3/m_i$, then $i < n$. Given $L \in [\mathbb{N}]$, set $l_n = \min \operatorname{supp} \alpha_n^L$, for all $n \in \mathbb{N}$. We now define

$$\mathcal{D} = \{L \in [\mathbb{N}] : \forall n \in \mathbb{N}, \forall t \in K_n, \text{ if } |\alpha_i^L|(t) \geq 3/l_i, \text{ then } i < n\}.$$

$\mathcal{D}$ is closed in the topology of pointwise convergence, thanks to Lemma 4.3. Our preceding discussion and Lemma 4.3, show that every $N \in [\mathbb{N}]$ contains some $L \in \mathcal{D}$ as a subset. We infer from Theorem 4.5, that $[N] \subset \mathcal{D}$ for some $N \in [\mathbb{N}]$.

Next, let $\mathcal{T}_0$ be the collection of those finite subsets E of N that can be written in the form $E = \cup_{i=1}^m \operatorname{supp} \alpha_i^L$, for some $L \in [N]$ (depending on E) for which there exists some $t \in K$ (depending on E and L) such that $|\alpha_i^L|(t) \geq 3/l_i$, for all $i \leq m$.

Let $\mathcal{T}$ be the collection of all initial segments of elements of $\mathcal{T}_0$. We claim that $\mathcal{T}$ is compact in the topology of pointwise convergence. Indeed, were this false, there would exist $M \in [N]$, $M = (m_i)$, such that $\{m_1, \ldots, m_n\} \in \mathcal{T}$, for all $n \in \mathbb{N}$. Let $n \in \mathbb{N}$. It follows that $\cup_{i=1}^n \operatorname{supp} \alpha_i^M \in \mathcal{T}$. Hence, there exist $L_n \in [N]$, $k_n \in \mathbb{N}$ and $t_n \in K$ such that $\cup_{i=1}^n \operatorname{supp} \alpha_i^M$ is an initial segment of

$\cup_{i=1}^{k_n}\operatorname{supp}\alpha_i^{L_n}$ and $|\alpha_i^{L_n}|(t_n) \geq 3/d_i$, for all $i \leq k_n$, where $d_i = \min\operatorname{supp}\alpha_i^{L_n}$, for all $i \in \mathbb{N}$. We now deduce from Lemma 4.3, that $n \leq k_n$ and that $\alpha_i^M = \alpha_i^{L_n}$, for all $i \leq n$. Therefore, $|\alpha_i^M|(t_n) \geq 3/m_i$, for all $i \leq n$, where $m_i = \min\operatorname{supp}\alpha_i^M$, for all $i \in \mathbb{N}$. The compactness of K now implies that there is some $t \in K$ satisfying $|\alpha_i^M|(t) \geq 3/m_i$, for all $i \in \mathbb{N}$. This is a contradiction, as $M \in \mathcal{D}$. Thus, our claim holds and so $\mathcal{T}$ is indeed compact.

We next apply Theorem 4.2 of [34] to obtain $P \in [N]$ such that $\mathcal{T}[P]$ is a hereditary and compact family. The result in [24] now yields $Q \in [P]$ and a countable ordinal $\eta > \alpha$, such that $\mathcal{T}[Q] \subset \mathcal{S}_\eta$. It follows that for every $L \in [Q]$ and all $n \in \mathbb{N}$ such that there exists some $t \in K$ satisfying $|\alpha_i^L|(t) \geq 3/l_i$, for all $i \leq n$, we have $\cup_{i=1}^n\operatorname{supp}\alpha_i^L \in \mathcal{S}_\eta$.

We now claim that $\xi^Q \leq \eta$ (see Definition 6.4). If this is not the case, we may choose $R \in [Q]$, $R = (r_i)$, which is ζ-large, for some countable ordinal ζ with $\eta < \zeta$. Let $\epsilon > 0$. We shall assume, as we clearly may, that $\sum_i(1/r_i) < \epsilon$. Since $\alpha < \eta$, we may choose an ordinal β with $\alpha + \beta = \zeta$. By passing to an infinite subset of R, if necessary, we may assume without loss of generality, thanks to Proposition 5.9, that every ζ-average of (f_i) supported by R admits an $(\epsilon, \alpha, \beta)$-decomposition.

Because R is ζ-large, it is also ζ-nice, by Theorem 6.7. We may thus select a ζ-average u of (f_i), supported by R, which is $(\eta, 1, \epsilon)$-large. We infer from Proposition 5.9 that there exist normalized blocks $u_1 < \cdots < u_n$, positive scalars $(\lambda_i)_{i=1}^n$ and $I \subset \{1, \ldots, n\}$ such that $u = \sum_{i=1}^n \lambda_i u_i$ and u_i is an α-average for all $i \in I$, while $\|\sum_{i \notin I} \lambda_i u_i\|_{\ell_1} < \epsilon$.

Now let $t \in K$ and define $H = \{i \in I : |u_i|(t) \geq 3/q_i\}$, where $q_i = \min\operatorname{supp}u_i$, for all $i \in I$. Let $\{i_1 < \ldots, < i_k\}$ be an enumeration of H. Lemma 4.3 yields some $L \in [R]$ such that $u_{i_j} = \alpha_j^L$, for all $j \leq k$. Set $J = \cup_{i \in H}\operatorname{supp}u_i$. Since $L \in [Q]$, it follows that $J \in \mathcal{S}_\eta$. Writing $u_i = \sum_s a_s^i f_s$, for all $i \leq n$, we conclude, as u is $(\eta, 1, \epsilon)$-large, that

$$\Big|\sum_{i \in H} \lambda_i \sum_s a_s^i f_s(t)\Big| \leq \epsilon + \sum_{i \notin H} \lambda_i \sum_s a_s^i |f_s(t)|.$$

We now have the estimates

$$|u(t)| = \Big|\sum_{i \in H} \lambda_i \sum_s a_s^i f_s(t) + \sum_{i \notin H} \lambda_i \sum_s a_s^i f_s(t)\Big|$$

$$\leq \Big|\sum_{i \in H} \lambda_i \sum_s a_s^i f_s(t)\Big| + \Big|\sum_{i \notin H} \lambda_i \sum_s a_s^i f_s(t)\Big|$$

$$\leq \epsilon + 2 \sum_{i \notin H} \lambda_i \sum_s a_s^i |f_s(t)|$$

$$\leq \epsilon + 2 \sum_{i \in I \setminus H} \lambda_i \sum_s a_s^i |f_s(t)| + 2 \sum_{i \notin I} \lambda_i \sum_s a_s^i |f_s(t)|$$

$$\leq \epsilon + 6 \sum_{i \in I \setminus H} |u_i|(t) + 2\Big\|\sum_{i \notin I} \lambda_i u_i\Big\|_{\ell_1}$$

$$< \epsilon + 18 \sum_{i \in I \setminus H} (1/q_i) + 2\epsilon$$

$$< 21\epsilon.$$

Since $\|u\| = 1$, we reach a contradiction for ϵ small enough. Therefore, $\xi^Q \leq \eta$. Proposition 6.5 now yields a block basis of ξ-averages of (f_i), for some $\xi \leq \eta$, equivalent to the unit vector basis of c_0. $\qquad\square$

REFERENCES

[1] D. Alspach and S.A. Argyros, *Complexity of weakly null sequences*, Dissertationes Mathematicae, **321**, (1992), 1–44.

[2] D. Alspach and E. Odell, *Averaging weakly null sequences*, Lecture Notes in Math. Vol. 1332, Springer, Berlin, 1988, 126–144.

[3] D. Alspach, R. Judd and E. Odell, *The Szlenk index and local ℓ_1-indices*, Positivity **9** (2005), 1–44.

[4] G. Androulakis, *A subsequence characterization of sequences spanning isomorphically polyhedral Banach spaces*, Studia Math. **127** (1998), no.1, 65–80.

[5] G. Androulakis and E. Odell, *Distorting mixed Tsirelson spaces*, Israel J. Math. **109** (1999), 125–149.

[6] S.A. Argyros and I. Deliyanni, *Examples of asymptotic ℓ_1 Banach spaces*, Trans. Amer. Math. Soc. **349** (1997), 973–995.

[7] S.A. Argyros and I. Gasparis, *Unconditional structures of weakly null sequences*, Trans. Amer. Math. Soc. **353** (2001), 2019–2058.

[8] S.A. Argyros, G. Godefroy and H.P. Rosenthal, *Descriptive set theory and Banach spaces*, Handbook of the geometry of Banach spaces, Vol.2, (W.B. Johnson and J. Lindenstrauss eds.), North Holland (2003), 1007–1069.

[9] S.A. Argyros and V. Kanellopoulos, *Determining c_0 in $C(\mathcal{K})$ spaces*, preprint.

[10] S.A. Argyros, S. Mercourakis and A. Tsarpalias, *Convex unconditionality and summability of weakly null sequences*, Israel J. Math. **107** (1998), 157–193.

[11] A.D. Arvanitakis, *Weakly null sequences with an unconditional subsequence*, Proc. Amer. Math. Soc., in press.

[12] B. Beauzamy and J.T. Lapreste, *Modeles etales des espaces de Banach. Travaux en cours*, Hermann, Paris, 1984.

[13] Y. Benyamini, *An extension theorem for separable Banach spaces*, Israel J. Math. **29** (1978), 24–30.

[14] C. Bessaga and A. Pelczynski, *On bases and unconditional convergence of series in Banach spaces*, Studia Math. **17** (1958), 151–164.

[15] C. Bessaga and A. Pelczynski, *A generalization of results of R.C. James concerning absolute bases in Banach spaces*, Studia Math. **17** (1958), 165–174.

[16] C. Bessaga and A. Pelczynski, *Spaces of continuous functions IV*, Studia Math. **19** (1960), 53–62.

[17] A. Brunel and L. Sucheston, *On B-convex Banach spaces*, Math. Systems Theory **7** (1974), 294–299.

[18] D.W. Dean, I. Singer and L. Sternbach, *On shrinking basic sequences in Banach spaces*, Studia Math. **40** (1971), 23–33.

[19] C. Dellacherie, *Les derivations en theorie descriptive des ensembles et le theoreme de la borne* (French), Seminaire de Probabilites, XI (Univ. Strasbourg, Strasbourg, 1975/76) p. 34–46, Lecture Notes in Math., Vol. 581, Springer, Berlin, 1977.

[20] E. Ellentuck, *A new proof that analytic sets are Ramsey*, J. Symbolic Logic **39** (1974), 163–165.

[21] J. Elton, *Thesis*, Yale University (1978).

[22] J. Elton, *Extremely weakly unconditionally convergent series*, Israel J. Math. **40** (1981), no.3-4, 255–258.

[23] V. Fonf, *A property of Lindenstrauss-Phelps spaces*, Functional Anal. Appl. **13** (1979), no.1, 66–67.

[24] I. Gasparis, *A dichotomy theorem for subsets of the power set of the natural numbers*, Proc. Amer. Math. Soc. **129** (2001), 759–764.

[25] R. Haydon, E. Odell and H. Rosenthal, *On certain classes of Baire-1 functions with applications in Banach space theory*, Functional analysis (Austin, TX, 1987/89) 1–35, Lecture Notes in Math., Vol. 1470, Springer, Berlin, 1991.

[26] R.C. James, *Bases and reflexivity of Banach spaces*, Ann. of Math. (2) **52** (1950), 518–527.

[27] R. Judd, *A dichotomy on Schreier sets*, Studia Math. **132** (1999), 245–256.

[28] A. Kechris, *Classical descriptive set theory*, Graduate Texts in Mathematics, Vol. 156, Springer-Verlag, NY, 1995.

[29] D.H. Leung, *Symmetric sequence subspaces of $C(\alpha)$*, J. London Math. Soc. (2) **59** (1999), 1049–1063.

[30] J. Lindenstrauss and L. Tzafriri, *Classical Banach spaces I*, Springer-Verlag, New York (1977).

[31] B. Maurey and H.P. Rosenthal, *Normalized weakly null sequence without unconditional subsequence*, Studia Math. **61** (1977), 77–98.

[32] S. Mazurkiewicz and W. Sierpinski, *Contributions a la topologie des ensembles denomrables*, Fund. Math. **1** (1920), 17–27.

[33] A. Miljutin, *Isomorphism of the spaces of continuuous functions over compact sets of the cardinality of the continuum* (Russian), Teor. FunkciĭFunkcional. Anal. i Priložen. Vyp **2** (1966), 150–156.

[34] E. Odell, *Applications of Ramsey theorems to Banach space theory*, Notes in Banach spaces, (H.E. Lacey, ed.), Univ. Texas Press (1980), 379–404.

[35] E. Odell, *On Schreier unconditional sequences*, Contemp. Math. **144** (1993), 197–201.

[36] E. Odell, N. Tomczak-Jaegermann and R. Wagner, *Proximity to ℓ_1 and distortion in asymptotic ℓ_1 spaces*, J. Funct. Anal. **150** (1997), 101–145.

[37] A. Pelczynski, *Linear extensions, linear averagings, and their applications to linear topological classification of spaces of continuuous functions*, Dissertationes Math. Rozpraway Mat. **58** (1968).

[38] A. Pelczynski and Z. Semadeni, *Spaces of continuous functions III. Spaces $C(\Omega)$ for Ω without perfect subsets*, Studia Math. **18** (1959), 211–222.

[39] J. Rainwater, *Weak convergence of bounded sequences*, Proc. Amer. Math. Soc. **14** (1963), 999.

[40] J. Schreier, *Ein Gegenbeispiel zur theorie der schwachen konvergenz*, Studia Math. **2**, (1930), 58–62.

[41] B. Wahl, *Thesis*, The University of Texas (1993).

DEPARTMENT OF MATHEMATICS, ARISTOTLE UNIVERSITY OF THESSALONIKI, THESSA-LONIKI 54124, GREECE.
E-mail address: ioagaspa@auth.gr

DEPARTMENT OF MATHEMATICS, THE UNIVERSITY OF TEXAS AT AUSTIN, AUSTIN, TEXAS 78712, U.S.A.
E-mail address: odell@math.utexas.edu

DEPARTMENT OF MATHEMATICS, HANOVER COLLEGE, P.O. BOX 890, HANOVER, INDIANA 47243, U.S.A.
E-mail address: wahl@hanover.edu

YET ANOTHER PROOF OF SOBCZYK'S THEOREM

FÉLIX CABELLO SÁNCHEZ

As everybody knows Sobczyk theorem states that c_0, the space of null sequences with the maximum norm, is 2-complemented in any separable Banach space containing it. Equivalently, if Y is a subspace of a separable Banach space X, then every operator $T : Y \to c_0$ has an extension $\tilde{T} : X \to c_0$ with $\|\tilde{T}\| \leq 2\|T\|$.

Since the publication of [14] many proofs of Sobczyk theorem appeared. The survey [4] contains every proof whose authors (or some of them) knew when the paper was completed.

After that at least three papers extending Sobczyk theorem to the vector valued setting appeared, namely [12, 10, 5]. In my opinion, Rosenthal's proof is very close in spirit to the original arguments of Sobczyk (see [4, 2.7]). The proof given by Johnson and Oikhberg reminds me of the proof of Werner [4, 2.9]. The proof by Castillo and Moreno has something to do with the argument in [4, 2.10].

Before going further, let us say, following Rosenthal, that E has the separable extension property (SEP) if every operator $T : Y \to E$ from a subspace of a separable Banach space X can be extended to an operator $\tilde{T} : X \to E$. If this can be achieved with $\|\tilde{T}\| \leq \lambda\|T\|$ we say that E has the λ-SEP. It is a deep result of Zippin [15, 16] that c_0 is the only separable space with the SEP, but there are nonseparable spaces with the SEP: the most obvious ones are injective spaces such as ℓ_∞. It follows from results by Aronszajn and Panitchpakdi [1] that $C(K)$ has the 1-SEP if and only if K is an F-space, that is, a compact space where disjoint cozero sets are completely separated (see also [8, 11, 13]). Thus, for instance, $\ell_\infty/c_0 = C(\beta\mathbb{N}\backslash\mathbb{N})$ has the 1-SEP. The 'isomorphic part' also follows from recent work by Castillo, Moreno and Suárez [6]: the quotient of a space with the SEP by a subspace with the SEP has the SEP. Finally, extensions of two spaces with the SEP have the SEP (see [3]). This includes the Johnson-Lindenstrauss space JL_∞ or the space contructed by Benyamini in [2].

Each of the papers [12, 10, 5] just mentioned contains a proof of the fact that if (E_n) is a sequence of Banach spaces having the λ-SEP then $c_0(E_n)$ has the Λ-SEP, where $\Lambda = \Lambda(\lambda)$ (but to be true, each paper considers a different property equivalent to the SEP and proves much more than this).

In this short note I present a new proof of this result based on a kind of open mapping theorem for the strong operator topology. Some preparation

Supported in part by DGICYT project MTM2004—02635.
Key words and phrases: Extension, Operator, Separably injective Banach space.

is necesary, however, to uncover the core of the argument. First of all, note that it suffices to prove the corresponding result when all the E_n are the same space, say E. For if E_n has the λ-SEP for all n, then $E = \ell_\infty(E_n)$ has it and if $c_0(E)$ has the Λ-SEP then so do all its 1-complemented subspaces, in particular $c_0(E_n)$. Thus our result is as follows:

Theorem 1. *It E has the SEP then so does $c_0(E)$.*

Proof

Notice we have not quantified the SEP in the statement. This is unnecessary because of the following typical result —which answers a question curiously posed in [12]. The proof is due to J.M.F. Castillo.

Proposition 1. *If E has the SEP then it has the λ-SEP for some λ.*

Proof. A set of real numbers is bounded if it contains bounded sequences only, hence it suffices to prove that if X_n is a sequence of separable Banach spaces with subspaces Y_n and $T_n : Y_n \to E$ are operators with $\|T_n\| \leq 1$, then there are extensions $\tilde{T}_n : X_n \to E$ such that $\|\tilde{T}_n\| \leq \lambda$, for some finite λ. Fortunately direct sums do exist in the category of Banach spaces: consider the sum operator $T : \ell_1(Y_n) \to E$ given by $T((y_n)) = \sum_n T_n y_n$ and extend it to $\ell_1(X_n)$. $\square$

We pass to the proof of Theorem 1. Let us begin with the observation that an operator $T : Y \to c_0(E)$ is given by a sequence $T_n : Y \to E$ such that $T_n(y) \to 0$ for all $y \in Y$. This just means that the sequence (T_n) converges to zero in the strong operator topology we briefly describe. Let A and B be Banach spaces and $L(A, B)$ the corresponding space of operators. The closed ball of radius r in that space will be denoted $L(A, B)_r$. The strong operator topology (SOT) in $L(A, B)$ is the smallest linear topology for which the sets

$$U = \{T \in L(A, B) : \|Tx\| \leq 1\} \qquad (x \in A)$$

are neighborhoods of the origin. The SOT is never metrizable (unless A is finite dimensional: in this case the SOT agrees with the usual norm topology). However, it is metrizable on bounded sets provided A is separable. Indeed, if (x_n) is bounded and spans a dense subspace of A, then the norm

$$T \longmapsto \sum_{n=1}^{\infty} \frac{\|Tx_n\|}{2^n}$$

induces the (relative) SOT on bounded sets of $L(A, B)$. It is clear that a sequence $T_n : A \to B$ is convergent to zero in the SOT if and only if $T_n x \to 0$ for all $x \in A$ (this already implies that (T_n) is uniformly bounded, by the uniform boundedness principle). Hence we can restate Theorem 1 as follows: if E has the SEP and Y is a subspace of a separable space X, then every SOT-null sequence $T_n : Y \to E$ extends to a SOT-null sequence $\tilde{T}_n : X \to E$. Note that $c_0(E)$ has the λ-SEP iff and only if this can be done with $\|\tilde{T}_n\| \leq \lambda$ provided $\|T_n\| \leq 1$.

The following simple lemma gives a simple criterion for the existence of such extensions.

Lemma 1. *Let $R : S \to T$ be a continuous mapping between topological spaces such that $R(s) = t$. Let S' (resp. T') be a metrizable subset of S (resp. T) containing s (resp. t). The following are equivalent:*

(a) *Every sequence (t_n) converging to t in T' is the image under R or some sequence (s_n) converging to s in S'.*

(b) *R is relatively open at s: if U is a neighborhood of s in S, then $R(U \cap S')$ contains a neighborhood of t in T'.*

(c) *R admits a section (relatively) continuous at t: there is a mapping $\varrho : T' \to S'$ continuous at t and such that $\varrho(t) = s$ and $R \circ \varrho = 1_{T'}$.*

Proof. Let us see that (a) implies (b). If (b) fails there is a neighborhood U of s relative to S' such that t is not interior to $R(U) \cap T'$. Hence there is a sequence $t_n \to t$ with t_n outside $R(U) \cap T'$ for all n. It is clear that (t_n) cannot be the image of any sequence converging to s in S'.

Now we prove that (b) implies (c). This part of the proof borrows from [7, Proof of Lemma 2.2]. Let (U_n) be a decreasing base for the topology of S' at s, with $U_1 = S'$. Since R is continuous at s (b) implies that $V_n = R(U_n) \cap T'$ is also a decreasing base of neighborhoods of t in T'. For each n, let $\varrho_n : V_n \to U_n$ any map such that $R \circ \varrho_n = 1_{V_n}$, with $\varrho_n(t) = s$. Now, for $y \neq t$ in T', put $n(y) = \max\{n : y \in V_n\}$. Finally, define $\varrho : T' \to S'$ taking

$$\varrho(y) = \varrho_{n(y)}(y)$$

(and $\varrho(t) = s$). It is clear that $R \circ \varrho = 1_{T'}$ and also that ϱ is continuous at t.

That (c) implies (a) is obvious: if $t_n \to t$ in T', then $s_n = \varrho(t_n)$ is a sequence converging to s whose image under R is the starting sequence. $\square$

Of course this applies to restriction operators (they are always SOT-continuous):

Lemma 2. *Let Y be a subspace of a separable Banach space X and let E be another Banach space. If $R : L(X, E) \to L(Y, E)$ denotes the restriction map, then the following are equivalent:*

(d) *Every sequence of operators $T_n : Y \to E$ converging to zero in the SOT with $\|T_n\| \leq 1$ for all n is the restriction of some SOT null sequence $\tilde{T}_n : X \to E$ with $\|\tilde{T}_n\| \leq \kappa$.*

(e) *Every operator $T : Y \to c_0(E)$ has an extension $\tilde{T} : X \to c_0(E)$ with $\|\tilde{T}\| \leq \kappa\|T\|$.*

(f) *If U is a neighborhood of the origin in the SOT of $L(X, E)$, then $R(U \cap L(X, E)_\kappa)$ contains a neighborhood of the origin in the relative SOT of $L(Y, E)_1$.* $\square$

We are now ready to prove Theorem 1. We shall show that if E has the λ-SEP, then $c_0(E)$ has the $(3\lambda^2 + \varepsilon)$-SEP for all $\varepsilon > 0$. It suffices to verify that (f) holds true for $\kappa = 3\lambda^2 + \varepsilon$. The proof is accomplished in two steps. Once U has been fixed we choose an intermediate $\tilde{X}$ (depending on U) such that $Y \subset \tilde{X} \subset X$ with $\tilde{X}/Y$ finite dimensional. This induces a decomposition of R as the composition

$$L(X, E) \xrightarrow{\ R_1\ } L(\tilde{X}, E) \xrightarrow{\ R_2\ } L(Y, E),$$

where R_i are the corresponding restriction operators. That $R_1(U)$ is large enough will follow from the choice of $\tilde{X}$. That $R_2(R_1(U))$ is large enough will follow from very easy finite dimensional considerations, thanks to the implication (e) $\Rightarrow$ (f) in Lemma 2.

First step. Let U be a neighborhood of the origin in the SOT of $L(X, E)$. Without loss of generality we may assume that, for some $x_1, \ldots, x_k \in X$, one has

$$U = \{T \in L(X, E) : \|Tx_i\| \leq 1 \text{ for } 1 \leq i \leq k\}$$

Let $\tilde{X}$ denote the least subspace of X containing $x_1, \ldots, x_k$ and Y. By the very definition of λ-SEP we see that for each $r > 0$ the set $R_1(U \cap L(X, E)_{\lambda r})$ contains $\tilde{U} \cap L(\tilde{X}, E)_r$, where

$$\tilde{U} = \{T \in L(\tilde{X}, E) : \|Tx_i\| \leq 1 \text{ for } 1 \leq i \leq k\}.$$

This was the first step. I emphasize that we have not proved that operators $\tilde{X} \to c_0(E)$ extend to X since $\tilde{X}$ depends on U. $\qquad\square$

Second step. Finally, we prove that every operator $T : Y \to c_0(E)$ extends to an operator $\tilde{T} : \tilde{X} \to c_0(E)$ with $\|\tilde{T}\| \leq (3 + \varepsilon)\lambda\|T\|$. Let Z denote $\tilde{X}/Y$ and $\pi : \tilde{X} \to Z$ the natural quotient map. Since Z is finite dimensional there is a finite dimensional $F \subset \tilde{X}$ such that for every $z \in Z$ there is $f \in F$, with $\|f\| \leq (1 + \varepsilon)\|z\|$ such that $\pi(f) = z$. Writing $G = Y \cap F$, we have a commutative diagram

$$
\begin{array}{ccccccccc}
0 & \longrightarrow & G & \longrightarrow & F & \xrightarrow{\ \pi\ } & Z & \longrightarrow & 0 \\
 & & \imath_G \downarrow & & \imath_F \downarrow & & \| & & \\
0 & \longrightarrow & Y & \longrightarrow & \tilde{X} & \xrightarrow{\ \pi\ } & Z & \longrightarrow & 0
\end{array}
$$

with exact rows. Notice that (the restriction) $T \circ \imath_G$ can be easily extended to an operator $\tilde{T} : F \to c_0(E)$ with $\|\tilde{T}\| \leq \lambda\|T\|$: indeed, as G is finite dimensional, the corresponding sequence $T_n \circ \imath_G : G \to E$ converges to zero in norm! So, just extend in each coordinate, with $\|\tilde{T}_n\| \leq \lambda\|T_n \circ \imath_G\|$. The rest is straightforward once one realizes that we have a push-out diagram (we refer the reader to [9] or [3] for explanations). Whatever this means the relevant information is that, due to the 'form' of the above diagram, $\tilde{X}$ is well isomorphic to a certain quotient of $Y \oplus F$ from where T extends easily. In any case some control of the constants is needed and so we verify these facts by hand.

So consider the direct sum space $Y \oplus F$ (with the sum norm) and the map $S : Y \oplus F \to \tilde{X}$ sending (y, f) into $y + f$. Clearly $\|S\| \leq 1$. On the other hand, given $x \in \tilde{X}$ we can write $x = (x - f) + f$, where f is any element of F such that $\pi(f) = \pi(x)$, with $\|f\| \leq (1 + \varepsilon)\|x\|$. It follows that $x - f$

belongs to Y and $S(x - f, f) = x$, in particular S is onto, and we see that $\|(x - f, f)\| \leq (3 + \varepsilon)\|x\|$. Let $\Delta = \ker S$, that is:

$$\Delta = \{(y, f) \in Y \times F : y + f = 0\} = \{(g, -g) : g \in G\}.$$

The push-out space associated to the pair of operators (actually inclusions) $G \to Y$ and $G \to F$ is just $\mathrm{PO} = (Y \oplus F)/\Delta$. We have seen that it is $(3 + \varepsilon)$-isomorphic to $\tilde{X}$ via the sum map. Consider the map $L : Y \oplus F \to c_0(E)$ given by $L(y, f) = T(y) + \tilde{T}(f)$. It is clear that $\|L\| \leq \|\tilde{T}\| \leq \lambda \|T\|$ and also that L vanishes on Δ. Hence it defines a operator $\tilde{L} : \mathrm{PO} \to c_0(E)$ and composition with $S^{-1} : \tilde{X} \to \mathrm{PO}$ gives an extension of T to $\tilde{X}$ of norm at most $(3 + \varepsilon)\lambda\|T\|$. $\qquad\qquad\square$

CONCLUDING REMARKS

Theorem 1 suggests some questions.

- Must $D \otimes_\varepsilon E$ have the SEP if both D and E have it? By Theorem 1 this is so when $D = c_0$ since $c_0 \otimes_\varepsilon E = c_0(E)$, but I do not know even if $\ell_\infty \otimes_\varepsilon \ell_\infty$ has the SEP!
- Let Y be a subspace of a separable space X such that every operator $Y \to E$ extends to X. Does every operator $Y \to c_0(E)$ extend to X? By the main result in [5] the answer is 'yes' if X/Y has the BAP.
- Let Y be a subspace of a separable space X such that every operator $Y \to E_i$ extends to X $(i = 1, 2)$. Does every operator $Y \to E_1 \otimes_\varepsilon E_2$ extend to X?

REFERENCES

[1] N. Aronszajn and P. Panitchpakdi, Extension of uniformly continuous transformations and hyperconvex metric spaces. Pacific Journal of Mathematics, 6 (1956) 405–439.

[2] Y. Benyamini, An M-space which is not isomorphic to a $C(K)$-space, Israel Journal of Mathematics 28 (1977) 98–104.

[3] F. Cabello Sánchez and J.M.F. Castillo, The long homology sequence for quasi-Banach spaces, with applications, Positivity 8 (2004) 379–374.

[4] F. Cabello Sánchez, J.M.F. Castillo and D.T. Yost, Sobczyk's theorems from A to B, Extracta Mathematicæ 15 (2000) 391–420.

[5] J.M.F. Castillo and Y. Moreno, Sobczyk cohomology for separable Banach spaces, preprint (2005).

[6] J.M.F. Castillo, Y. Moreno and J. Suárez, On Lindenstrauss-Pelczyński spaces, Studia Mathematica (to appear).

[7] P. Domański, On the splitting of twisted sums and the three space problem for local convexity, Studia Mathematica 82 (1985) 155–189.

[8] M. Henriksen, Some remarks on a paper of Aronszajn and Panitchpakdi. Pacific Journal of Mathematics 7 (1957) 1619–1621.

[9] W.B. Johnson, Extensions of c_0, Positivity 1 (1997) 55–74.

[10] W.B. Johnson and T. Oikhberg, Separable lifting property and extensions of local reflexivity, Illinois Journal of Mathematics 45 (2001) 123–137.

[11] J. Lindenstrauss, On the extension of operators with range in a $C(K)$ space, Proceedings of the American Mathematical Society 15 (1964) 218–225.

[12] H.P. Rosenthal, The complete separable extension property, Journal of Operator Theory 43 (2000) 329–374.

[13] G.L. Seever, Measures on F-spaces, Transactions of the American Mathematical Society 133 (1968) 269–280.

[14] A. Sobczyk, On the extension of linear transformations, Transactions of the American Mathematical Society 55 (1944) 153–169.

[15] M. Zippin, The separable extension problem, Israel Journal of Mathematics 26 (1977) 372–387.

[16] M. Zippin, Extension of bounded linear operators, Handbook of the Geometry of Banach Spaces, Vol. 2, W.B. Johnson and J. Lindenstrauss, eds., Elsevier, Amsterdam 2003.

DEPARTAMENTO DE MATEMÁTICAS, UNIVERSIDAD DE EXTREMADURA, 06071–BADAJOZ, SPAIN

E-mail address: fcabello@unex.es, http://kolmogorov.unex.es/~fcabello

THE CATEGORY OF EXACT SEQUENCES
OF BANACH SPACES

JESÚS M. F. CASTILLO AND YOLANDA MORENO

To Atenea.

> *Consider nature's magnificent foresight*
> *spreading seeds of madness everywhere.*
> *If mortals would refrain from no matter which contact*
> *with wisdom, even the old age would not exist.*
> *Life is not different from a dreaming game*
> *whose greatest gifts come to us through craziness.*
> *Consider nature's magnificent foresight*
> *in making the heart be always right.*

The purpose of this paper is to lay the foundations for the construction of the category of exact sequences of Banach spaces; the construction for quasi-Banach spaces is analogous and thus we omit it. The construction of a category associated to a theory, in addition to its intrinsic value, provides the right context to study, among others, isomorphic and universal objects. In our particular case, let us describe a couple of phenomena often encountered when working with exact sequences of Banach spaces for which the categorical approach provides rigorous explanations.

If one "multiplies" an exact sequence $0 \to Y \to X \to Z \to 0$ by the left (resp. right) by a given space E, the resulting exact sequence $0 \to E \oplus Y \to E \oplus X \to Z \to 0$ (resp. $0 \to Y \to X \oplus E \to Z \oplus E \to 0$) is "the same". And this holds despite the fact that the original and the "multiplied" sequences are not equivalent under any known definition. The categorical approach provides the simplest explanation: the two sequences are isomorphic objects in the category. Another aspect we will consider is the following: given a family of exact sequences $0 \to Y_n \to X_n \to Z_n \to 0$ it is rather standard to use their l_∞-amalgam: $0 \to l_\infty(Y_n) \to l_\infty(X_n) \to l_\infty(Z_n) \to 0$; but also their c_0- or even l_p-amalgam. For instance, the c_0-amalgam is used in [12] to obtain a separably injective space not isomorphic to a complemented subspace of any $C(K)$-space; while the l_2-amalgam is at the basis of the Enflo-Lindenstrauss-Pisier solution to the 3-space problem for Hilbert spaces [14]. One could expect, as we will show it occurs, that amalgams correspond to products in the category.

1991 *Mathematics Subject Classification.* Primary: 46M15,46M18,46B07,46B20.
Supported in part by DGICYT project MTM2004-02635.

The organization of this paper is as follows: Section 2 contains the construction of the category 3; Section 3 describes the isomorphisms of the category; Sections 4 and 5 describe the universal constructions and their associated diagrams. The last two sections contain applications of these tools to classical Banach space theory: in section 6 there is a review of Pełczyński c-amalgam [24] and Stegall's application to obtain a counterexample about the Dunford-Pettis property [26]; while in Section 7 (Theorem 7.1) we give a method to embedd every separable Banach space E into a separable $\mathcal{L}_\infty$-space $\mathcal{L}_\infty(E)$ in such a way that every operator from E into an isometric L_1-predual can be extended to $\mathcal{L}_\infty(E)$ and, moreover, the quotient $\mathcal{L}_\infty(E)/E$ has the BAP. This example combines the Bourgain-Pisier construction [3] with Zippin's construction [29].

1. BACKGROUND

We assume from the reader some acquaintance with categories, pull-back and push-out constructions [16, 22], exact sequences of Banach and quasi-Banach spaces [7] and their representations by means of quasi-linear and z-linear maps [17, 18, 7, 4, 6, 11].

From now on **Ban** denotes the category of Banach spaces and linear continuous operators. An exact sequence $0 \to Y \to X \to Z \to 0$ in **Ban** is a diagram in which the kernel of each arrow coincides with the image of the preceding. Two exact sequences $0 \to Y \to X \to Z \to 0$ and $0 \to Y \to X_1 \to Z \to 0$ are said to be equivalent if there exists an operator $T : X \to X_1$ making commutative the diagram

$$
\begin{array}{ccccccccc}
0 & \longrightarrow & Y & \longrightarrow & X & \longrightarrow & Z & \longrightarrow & 0 \\
 & & \| & & \downarrow{\scriptstyle T} & & \| & & \\
0 & \longrightarrow & Y & \longrightarrow & X_1 & \longrightarrow & Z & \longrightarrow & 0.
\end{array}
$$

An exact sequence is said to split if it is equivalent to the trivial sequence $0 \to Y \to Y \oplus Z \to Z \to 0$. There is a correspondence (see [17, 18, 7]) between exact sequences $0 \to Y \to X \to Z \to 0$ of Banach spaces and the so-called z-linear maps, which are homogeneous maps $F : Z \curvearrowright Y$ (we use this notation to stress the fact that these are not linear maps) with the property that there exists some constant $C > 0$ such that for all finite sets $x_1, \dots, x_n \in Z$ one has $\|F(\sum_{n=1}^N x_n) - \sum_{n=1}^N F(x_n)\| \le C \sum_{n=1}^N \|x_n\|$. The infimum of those constants C is called the z-linearity constant of F and denoted $Z(F)$. Precisely, given $F : Z \curvearrowright Y$ one can construct the quasi-Banach space $Y \oplus_F Z = (Y \times Z, \| \cdot \|_F)$ where $\|(y, z)\|_F = \|y - Fz\| + \|z\|$; then form its Banach envelope $co(Y \oplus_F Z)$, to finally obtain an exact sequence $0 \to Y \to co(Y \oplus_F Z) \to Z \to 0$. Conversely, given an exact sequence $0 \to Y \to X \to Z \to 0$, if b denotes a homogeneous bounded selection for the quotient map and ℓ a linear selection then $F = b - \ell$ is a z-linear map whose associated sequence $0 \to Y \to co(Y \oplus_{b-\ell} Z) \to Z \to 0$ is equivalent to the starting one.

Two z-linear maps $F, G : Z \curvearrowright Y$ are said to be equivalent, and we write $F \equiv G$, if the induced exact sequences are equivalent; which happens if and only if the difference $F - G$ can be written as $B + L$, where $B : Z \to Y$ is a homogeneous bounded map and $L : Z \to Y$ a linear map. A z-linear map $F : Z \curvearrowright Y$ is said to be trivial if $F \equiv 0$. We will use the notation

$$0 \longrightarrow Y \longrightarrow X \longrightarrow Z \longrightarrow 0 \equiv F$$

to mean that the exact sequence and the z-linear map correspond one to the other.

1.1. Pull-back and push-out.

In a category $\mathfrak{C}$, the push-out of two arrows $S : Y \to M$ and $j : Y \to X$ is an object PO and two arrows $u_S : X \to PO$ and $u_j : M \to PO$ such that $u_S j = u_j S$, and with the universal property that given two operators $\alpha : M \to E$ and $\beta : X \to E$ such that $\alpha S = \beta T$ there exists a unique arrow $\gamma : PO \to E$ such that $\gamma u_j = \alpha$ and $\gamma u_S = \beta$. In **Ban**, their push-out is the quotient space $PO = (M \oplus_1 X)/\overline{\Delta}$ where $\Delta = \{(Sy, -jy) \in M \oplus_1 X\}$. The operators: $u_S : X \to PO$ and $u_j : M \to PO$ are the restrictions to M and X of the quotient map $M \oplus_1 X \to PO$. It has the universal property that given two operators $\alpha : M \to E$ and $\beta : X \to E$ such that $\alpha S = \beta j$ there exists a unique operator $\gamma : PO \to E$ such that $\gamma u_j = \alpha$, $\gamma u_S = \beta$, and $\|\gamma\| \leq \max\{\|\alpha\|, \|\beta\|\}$. If, moreover, j is an embedding, there exists a commutative diagram

$$
\begin{array}{ccccccccc}
0 & \longrightarrow & Y & \xrightarrow{\ j\ } & X & \xrightarrow{\ q\ } & Z & \longrightarrow & 0 \equiv F \\
 & & \downarrow{\scriptstyle S} & & \downarrow{\scriptstyle u_S} & & \| & & \\
0 & \longrightarrow & M & \xrightarrow[u_j]{} & PO & \longrightarrow & Z & \longrightarrow & 0 \equiv SF.
\end{array}
$$

Observe that the z-linear map associated with the lower push-out sequence is SF. It is a standard result in algebra [16, 22] that the lower push-out sequence splits if and only if the operator S can be extended to an operator $X \to M$. Thus, given another operator $S' : Y \to M$ one has that $SF \equiv S'F$ if and only if $S - S'$ can be extended to X.

The dual notion is that of pull-back. In **Ban**, given operators $q : X \to Z$ and $T : W \to Z$ the pull-back space of $\{q, T\}$ is $PB = \{(x, w) : qx = Tw\} \subset X \oplus_\infty W$ endowed with the relative product topology. If, moreover, q is a quotient, then there exists a commutative diagram

$$
\begin{array}{ccccccccc}
0 & \longrightarrow & Y & \xrightarrow{\ j\ } & X & \xrightarrow{\ q\ } & Z & \longrightarrow & 0 \equiv F \\
 & & \| & & \uparrow & & \uparrow{\scriptstyle T} & & \\
0 & \longrightarrow & Y & \xrightarrow[i]{} & PB & \longrightarrow & W & \longrightarrow & 0 \equiv FT
\end{array}
$$

in which the operators $PB \to X$ and $PB \to W$ are the restrictions of the canonical projections of $X \oplus_\infty W$ into, respectively, X and W, and $i(y) = (jy, 0)$. Observe that the z-linear map associated with the lower pull-back sequence is FT. Since, again by folklore result in algebra [16, 22], the lower pull-back sequence splits if and only if the operator T can be lifted to an operator $W \to X$, it turns out that if $T' : W \to Z$ is another operator, one has that $FT \equiv FT'$ if and only if $T - T'$ can be lifted to X.

2. The Category 3 of z-linear maps

Let $\equiv$ denote the classical equivalence relation for either exact sequences or z-linear maps.

Definition. The category 3 of z-linear maps between Banach spaces has as objects $\equiv$-equivalence classes of z-linear maps. Given a pair of objects F, G, the set of morphisms $\mathrm{Hom}_3(F, G)$ is formed by equivalence classes of couples of operators (α, γ) such that $\alpha F \equiv G\gamma$, with respect to the equivalence relation

$$(\alpha', \gamma') \asymp (\alpha, \gamma) \iff (\alpha - \alpha')F \equiv 0 \iff G(\gamma - \gamma') \equiv 0.$$

The composition is defined as $(\alpha, \gamma)(\alpha', \gamma') = (\alpha\alpha', \gamma\gamma')$, and the identity morphism is therefore (id, id).

2.1. Equivalence of categories.
The category 3 and its counterpart, the category of exact sequences of Banach spaces $\mathfrak{S}$ (see the appendix), are, as one might expect, equivalent in any reasonable sense, and thus we shall identify them. "Equivalent categories", following [22], means that there exist two covariant functors $\mathcal{S} : 3 \to \mathfrak{S}$ and $\mathcal{F} : \mathfrak{S} \to 3$ such that for each pair of objects $\heartsuit$ of $\mathfrak{S}$ and $\spadesuit$ of 3 there is a bijection

$$\mathrm{Hom}_{\mathfrak{S}}(\heartsuit, \mathcal{S}(\spadesuit)) \longleftrightarrow \mathrm{Hom}_{3}(\mathcal{F}(\heartsuit), \spadesuit);$$

(namely, these are adjoint functors, see [16]). In our case, $\mathcal{F}$ and $\mathcal{S}$ establish the biunivoque correspondence between classes of extensions and classes of z-linear maps: $\mathcal{F}$ assigns to the $\equiv$-class of $0 \to Y \to X \xrightarrow{q} Z \to 0$ the $\equiv$-class of the z-linear map $b - l$, where b and l are, respectively, a bounded homogeneous and a linear selection for q. Reciprocally, $\mathcal{S}$ assigns to the $\equiv$-class of a z-linear map F the class of the extension $0 \to Y \to Y \oplus_F Z \to Z \to 0$.

2.2. The categories 3^Z and 3_Y associated to 3.
The basic homological tools can be naturally interpreted inside 3. In particular, a push-out diagram

$$
\begin{array}{ccccccccc}
0 & \longrightarrow & Y & \longrightarrow & X & \longrightarrow & Z & \longrightarrow & 0 \equiv F \\
& & \alpha\downarrow & & \downarrow & & \| & & \\
0 & \longrightarrow & Y' & \longrightarrow & X' & \longrightarrow & Z & \longrightarrow & 0 \equiv G,
\end{array}
$$

corresponds to a morphism (α, id), which we shall denote $F \xrightarrow{\alpha} G$; and a pull-back diagram

$$
\begin{array}{ccccccccc}
0 & \longrightarrow & Y & \longrightarrow & X & \longrightarrow & Z & \longrightarrow & 0 \equiv F \\
& & \| & & \uparrow & & \uparrow\gamma & & \\
0 & \longrightarrow & Y & \longrightarrow & X' & \longrightarrow & Z' & \longrightarrow & 0 \equiv G,
\end{array}
$$

corresponds to a morphism (id, γ) which we shall denote $F \xleftarrow{\gamma} G$. Since those are specially interesting morphisms it shall be helpful to consider the associated categories 3^Z and 3_Y in which those are the only morphisms: Let

Z be a Banach space; the category $\mathbf{3}^Z$ shall have as objects $\equiv$-classes of z-linear maps $F : Z \curvearrowright \heartsuit$; and given a pair of objects F y G, a morphism $F \longrightarrow G$ shall be the $\asymp$-class formed by all operators α such that $\alpha F \equiv G$. Analogously, given a Banach space Y, the category $\mathbf{3}_Y$ shall have as objects $\equiv$-classes of z-linear maps with range in Y; while given a pair of objects F, G, a morphism $G \longleftarrow F$ from F to G is the $\asymp$-class of all operators γ such that $G\gamma \equiv F$. It is easy to check that the category $\mathbf{3}$ is additive, while $\mathbf{3}^Z$ and $\mathbf{3}_Y$ are not.

3. Isomorphisms of $\mathbf{3}$

Exact sequences of Banach spaces is one of those mathematical objects for which the notion of equality is not well defined. Some attempts have been made in the literature to determine when two exact sequences are "equal". On one hand there stands the classical notion of equivalent sequences; moreover Kalton and Peck [18] also consider the notion of "projectively equivalent" sequences. On the other hand, metric considerations suggest to the authors of [8] to work with "isometricaly equivalent" sequences. The papers [9] introduces and studies the notion of "isomorphically equivalent" sequences, which turns out to be quite useful (see [4, 10]), as follows: Two exact sequences $0 \to Y \to X \to Z \to 0$ and $0 \to Y_1 \to X_1 \to Z_1 \to 0$ of Banach spaces are said to be *isomorphically equivalent* if there exist isomorphisms $\alpha : Y \to Y_1$, $\beta : X \to X_1$ and $\gamma : Z \to Z_1$ making commutative the diagram

$$
\begin{array}{ccccccccc}
0 & \longrightarrow & Y & \longrightarrow & X & \longrightarrow & Z & \longrightarrow & 0 \\
& & \downarrow{\scriptstyle\alpha} & & \downarrow{\scriptstyle\beta} & & \downarrow{\scriptstyle\gamma} & & \\
0 & \longrightarrow & Y_1 & \longrightarrow & X_1 & \longrightarrow & Z_1 & \longrightarrow & 0.
\end{array}
$$

Moreover, it is clear that this notion contains all the previous ones as particular cases. In terms of z-linear maps, F and G are isomorphically equivalent if and only if there exist isomorphisms (in **Ban**) α and γ such that $\alpha F \equiv G\gamma$. So, we see that isomorphically equivalent z-linear maps are a particular case of isomorphic objects in $\mathbf{3}$. Further examples of isomorphic objects can be obtained as follows: Given an exact sequence $0 \to Y \xrightarrow{j} X \xrightarrow{q} Z \to 0 \equiv F$ and a space E we shall write $0 \to E \oplus Y \to E \oplus X \to Z \to 0 \equiv E \oplus F$ to denote the obvious "multiplied" sequence with embedding $(e, y) \to (e, jy)$. Analogously, we write $0 \to Y \to X \oplus E \to Z \oplus E \to 0 \equiv F \oplus E$ to represent the sequence with quotient map $(x, e) \to (qx, e)$. The sequences $F, E \oplus F$ and $F \oplus E$ shall be called multiples. One has:

Proposition 3.1. *Two objects in $\mathbf{3}$ are isomorphic if and only if they have isomorphically equivalent multiples.*

Proof. Let $F : Z \curvearrowright Y$ be a z-linear map and E a Banach space. It is obvious that the sequences $E \oplus F$ and $F \oplus E$ are isomorphic objects (although, in general, they are not isomorphically equivalent since, say, Z and $Z \oplus E$ need not be isomorphic).

Conversely let $0 \to Y \to X \to Z \to 0 \equiv F$ and $0 \to Y' \to X' \to Z' \to 0 \equiv F'$ be isomorphic objects in $\mathfrak{Z}$. Let $(\alpha, \gamma) : F \to G$ be an isomorphism, which means that there is another morphism $(\alpha', \gamma') : G \to F$ such that $(\alpha', \gamma')(\alpha, \gamma) \asymp (id, id) \asymp (\alpha, \gamma)(\alpha', \gamma')$. Thus, $F \equiv \alpha' \alpha F$. We use now the first diagonal principle developed in [10] applied to F and αF to obtain that $PO \oplus F$ and $X \oplus \alpha F$ are isomorphically equivalent (here PO is of course the middle push-out space in the sequence αF). Therefore $PO \oplus F \oplus X'$ and $X \oplus \alpha F \oplus X'$ are isomorphically equivalent. Since $(\alpha, \gamma) : F \to F'$ is a morphism, $\alpha F \equiv F' \gamma$ and thus $X \oplus \alpha F$ and $X \oplus F' \gamma$ are still equivalent. Applying the second diagonal principle [10] to F' and $F' \gamma$ we obtain that $F' \oplus PB$ and $F' \gamma \oplus X'$ are isomorphically equivalent (here, of course, PB is the middle space in the pull-back sequence $F' \gamma$). Their multiples $X \oplus F' \oplus PB$ and $X \oplus F' \gamma \oplus X'$ are isomorphically equivalent as well.

Summing up, $PO \oplus F \oplus X'$ and $X \oplus \alpha F \oplus X' \equiv X \oplus F' \gamma \oplus X'$ are isomorphically equivalent; and this last object is isomorphically equivalent to $X \oplus F' \oplus PB$. $\qquad\square$

4. UNIVERSAL CONSTRUCTIONS

In a category $\mathfrak{C}$, the product of objects $(A_i)_i$ is an object ΠA_i plus a family of morphisms $\pi_j : \Pi A_i \to A_j$ such that for each object B they induce a bijection $\mathrm{Hom}_{\mathfrak{C}}(B, \prod A_i) \longrightarrow \prod_i \mathrm{Hom}_{\mathfrak{C}}(B, A_i)$ in the form $T \to (\pi_i \circ T)_i$. The coproduct is the (categorically) dual notion: an object $\bigoplus A_i$ plus a family of morphisms $\iota_j : A_j \to \bigoplus A_i$ such that for each object B they induce a bijection $\mathrm{Hom}_{\mathfrak{C}}(\bigoplus A_i, B) \longrightarrow \prod_i \mathrm{Hom}_{\mathfrak{C}}(A_i, B)$ in the form $T \to (T\iota_i)_i$.

Banach spaces obviously admit finite products; they also exist in $\mathfrak{Z}$ since, as one might guess, the product in $\mathfrak{Z}$ of a finite family $F_i : Z_i \curvearrowright Y_i$ of objects of $\mathfrak{Z}$ is the class generated by the z-linear map $\Pi F_i : \Pi Z_i \curvearrowright \Pi Y_i$ that takes at $(z_i) \in \Pi Z_i$ the value $(F_i z_i)$, together with the canonical projections. We do not know if $\mathfrak{Z}_Y$ admits finite products, or if $\mathfrak{Z}^Z$ admits finite coproducts.

Regarding infinite products, the phenomenon which happens in Banach spaces (as well as in many other categories) is that, given an infinite family (A_i), there exists an object $\mathcal{P}$ and a collection of morphisms $\pi_i : \mathcal{P} \to A_i$ such that the natural induced correspondence $\mathrm{Hom}_{\mathfrak{C}}(C, \mathcal{P}) \longrightarrow \prod_i \mathrm{Hom}_{\mathfrak{C}}(C, A_i)$ is a bijection only onto a region $\mathcal{D}(C) \subset \prod_i \mathrm{Hom}_{\mathfrak{C}}(C, A_i)$. We shall say that the couple $(\mathcal{P}, (\pi_i))$ is a restricted product or, more precisely, a $\mathcal{D}$-product of (A_i). The same considerations can be made about the coproduct. A type of categories in which the notion of restricted product or coproduct is necessary are those termed *augmented categories*; namely, categories in which there exists an additional structure on the sets $\mathrm{Hom}_{\mathfrak{C}}(A, B)$ (we hasten to remark that this is just a vague notion; see [23]). Banach spaces is the perfect example for us of augmented category.

A restricted product can be as "good" as a standard product, provided one additional condition is satisfied: that the map $\mathcal{D}(\cdot) : \mathfrak{C} \to \mathbf{Set}$ assigning to each object C the region $\mathcal{D}(C) \subset \prod_i \mathrm{Hom}_{\mathfrak{C}}(C, A_i)$ defines a functor. In that case, the functor $\mathcal{D}(\cdot)$ is *representable*. Recall that a contravariant functor

$\mathbf{F} : \mathfrak{C} \longrightarrow \mathbf{Set}$ is said to be representable if there exists an object M de $\mathfrak{C}$ such that $\mathbf{F}$ and $\mathrm{Hom}_{\mathfrak{C}}(\,\cdot\,, M)$ are naturally equivalent. A covariant functor $\mathbf{F} : \mathfrak{C} \longrightarrow \mathbf{Set}$ is said to be representable if there is some object M of $\mathfrak{C}$ such that $\mathbf{F}$ and $\mathrm{Hom}_{\mathfrak{C}}(M, \,\cdot\,)$ are naturally equivalent. The object M is called in both cases the representative of $\mathbf{F}$. Since a way of thinking is to understand that universal properties are nothing else but the affirmation that a certain functor is representable, what we have got is that when $\mathcal{D}$ is a representable functor the $\mathcal{D}$-product defines a *universal property*, exactly as the standard product.

The simplest example of the necessity of this approach is what happens in the category of Banach spaces, where "standard" products of infinite families do not exist while it is the natural notion of l_∞-amalgam which plays that role. And the fact is that it is indeed a restricted product in the previous sense: given a family (A_i) of Banach spaces the space $l_\infty(A_i)$ is the representative, together with the canonical projections, of the contravariant functor $l_\infty\left[\mathcal{L}(\,\cdot\,, A_i)\right] : \mathbf{Ban} \to \mathbf{Set}$ that assigns to each object X of $\mathbf{Ban}$ the set $l_\infty\left[\mathcal{L}(X, A_i)\right]$ of uniformly bounded families of operators $X \to A_i$. Less trivial is the c_0-amalgam: still a restricted product since $c_0(A_i)$ is the representative of the functor $c_0^{SOT}\left[\mathcal{L}(\,\cdot\,, A_i)\right] : \mathbf{Ban} \to \mathbf{Set}$ that assigns to each object X the set $c_0^{SOT}\left[\mathcal{L}(X, A_i)\right]$ of pointwise convergent to 0 families of operators $X \to A_i$. The restricted coproduct in Banach spaces is the l_1-coproduct: given a family (A_i) of Banach spaces the space $l_1(A_i)$ is the representative, together with the canonical injections, of the covariant functor $l_\infty\left[\mathcal{L}(A_i, \,\cdot\,)\right] : \mathbf{Ban} \to \mathbf{Set}$ that assigns to each object X of $\mathbf{Ban}$ the set $l_\infty\left[\mathcal{L}(A_i, X)\right]$ of uniformly bounded families of operators $A_i \to X$.

4.1. What amalgams of z-linear maps conceal. The existence of restricted products (resp. coproducts) in $\mathfrak{Z}$ analogous to those of $\mathbf{Ban}$ does not appear to be impossible. We start observing that $\mathfrak{S}$ admits some constructions corresponding to the amalgams of $\mathbf{Ban}$: given an *adequate* family of sequence $0 \to Y_i \to X_i \to Z_i \to 0$ it is customary to construct the *amalgams* $0 \to l_p(Y_i) \to l_p(X_i) \to l_p(Z_i) \to 0$ for $1 \le p \le +\infty$ and $p = 0$. Are those objects the corresponding l_∞, l_p, c_0-products/coproducts in $\mathfrak{S}$? Do those objects depend, and in which sense, on the exact sequences chosen as representatives?

Intuition suggests that given a collection $0 \to Y_i \to X_i \to Z_i \to 0 \equiv F_i$, the l_∞-product extension should be the l_∞-amalgam $0 \to l_\infty(Y_i) \to l_\infty(X_i) \to l_\infty(Z_i) \to 0$ in such a way that the l_∞-product of the z-linear maps F_i would be the z-linear map associated to that extension. An elementary observation is that the family of extensions must satisfy some "uniformity" assumption so that the l_∞-amalgam turn out to be an exact sequence in $\mathbf{Ban}$. This "uniformity" information can be found in the associated family (F_i) of z-linear maps as follows: if we form the canonical extensions

$$0 \longrightarrow Y_i \longrightarrow Y_i \oplus_{F_i} Z_i \longrightarrow Z_i \longrightarrow 0 \equiv F_i,$$

we find that the constant of concavity C_i for the quasi-norm $\|\cdot\|_{F_i}$ can be a problem to make the amalgam $l_\infty(Y_i \oplus_{F_i} Z_i)$ a quasi-Banach space. Since we have replaced the quasi-Banach twisted sums $Y_i \oplus_{F_i} Z_i$ by their Banach envelopes $\mathrm{co}(Y_i \oplus_{F_i} Z_i)$ we just have to worry about the Banach-Mazur distances $\mathrm{d}(Y_i \oplus_{F_i} Z_i, \mathrm{co}(Y_i \oplus_{F_i} Z_i))$. It is not hard to calculate that that distance is $Z(F_i)$, so the condition to amalgamate families of z-linear maps (F_i) is that $\sup Z(F_i) < +\infty$. This suggests that to amalgamate a family of objects (F_i) of $\mathfrak{Z}$ requires to have a $Z(\cdot)$-bounded family (F_i) of representatives.

Keeping these remarks in mind, it is easy to show that the l_∞-amalgam of the sequences (F_i) has as associated z-linear map "essentialy" the product map $\prod F_i$: fix a normalized Hamel basis $(e_\alpha)_\alpha$ of $l_\infty(Z_i)$ and then construct the linear map $L_{\prod F_i} : l_\infty(Z_i) \to \prod Y_i$ that coincides with $\Pi_i F_i$ at the elements of the Hamel basis. Since the family (F_i) is $Z(\cdot)$-bounded it is easy to deduce that the map $G = \Pi_i F_i - L_{\Pi_i F_i}$ is z-linear and has its range contained in $l_\infty(Y_i)$. Finally, there is a commutative diagram

$$
\begin{array}{ccccccccc}
0 & \longrightarrow & l_\infty(Y_i) & \longrightarrow & l_\infty(Y_i \oplus_{F_i} Z_i) & \longrightarrow & l_\infty(Z_i) & \longrightarrow & 0 \\
 & & \| & & T\uparrow & & \| & & \\
0 & \longrightarrow & l_\infty(Y_i) & \longrightarrow & l_\infty(Y_i) \oplus_G l_\infty(Z_i) & \longrightarrow & l_\infty(Z_i) & \longrightarrow & 0,
\end{array}
$$

in which T is the operator sending $((y_i),(z_i))$ to $(y_i, z_i)_i$. The same type of construction can be made for other amalgams. The question now is:

Are amalgams of z-linear maps restricted products (coproducts) in $\mathfrak{Z}$?

The question is not entirely innocent because if we accept that the product (say) of trivial objects should be trivial, we are forced to discard the previous constructions. The reason is that it is easy to give examples in which $l_\infty(F_n)$ is not trivial although (F_n) is a $Z(\cdot)$-bounded family of trivial objects: choose $B_n : Z_n \to Y_n$ bounded maps increasingly far from linear maps while $Z(B_n) \le 1$ for all $n \in \mathbb{N}$. The l_∞-amalgam of (trivial) extensions $l_\infty(B_n)$ is quite clearly not trivial. This makes it impossible to continue with the idea of making products of z-linear maps in terms of their equivalence classes.

4.2. Restricted products of z-linear maps. Let us replace the category $\mathfrak{Z}$ by a new one $\mathbf{Z}$ whose objects are z-linear maps (not equivalence classes), and whose morphisms $F \rightrightarrows G$ are couples of operators (α, γ) such that $\alpha F \equiv G\gamma$. Let $l_\infty(Z(X_i, Y_i))$ denote the space formed by all uniformly $Z(\cdot)$-bounded collections of z-linear maps. We introduce the equivalence relation:

$$(F_i) \bowtie (G_i) \iff \forall i, \quad F_i - G_i = B_i + L_i, \quad \text{with } \sup_i \|B_i\| < +\infty,$$

here each $B_i : X_i \to Y_i$ denotes a bounded homogeneous map and $L_i : X_i \to Y_i$ a linear map. We shall say that a collection of morphisms $(\alpha_i, \gamma_i) : F_i \rightrightarrows G_i$ of $\mathbf{Z}$ is *uniformly representable* if $(\alpha_i), (\gamma_i)$ are uniformly bounded families of operators such that

$$(\alpha_i F_i) \bowtie (G_i \gamma_i).$$

We set now a new category $\mathbf{Z}^{\bowtie}$ whose objects are $\bowtie$-equivalence classes of uniformly $Z(\cdot)$-bounded families of z-linear maps, and whose morphisms are uniformly representable families of morphisms $(\alpha_i, \gamma_i) : F_i \rightrightarrows G_i$ in $\mathbf{Z}$. Consider the diagonal functor $\Delta : \mathbf{Z} \to \mathbf{Z}^{\bowtie}$ defined by $\Delta(F) = (F)$. The representative, if it exists, of the functor

$$\mathbf{Z} \xrightarrow{\Delta} \mathbf{Z}^{\bowtie} \xrightarrow{\mathrm{Hom}_{\mathbf{Z}^{\bowtie}}(\cdot, (F_i))} \mathbf{Set}$$

shall be called the l_∞-product of the family (F_i) of z-linear maps. One has.

Proposition 4.1. *The functor* $\mathrm{Hom}_{\mathbf{Z}^{\bowtie}}(\Delta(\cdot), (F_i)) : \mathbf{Z} \longrightarrow \mathbf{Set}$ *is representable and its representative is the amalgam* $l_\infty(F_i)$ *together with the canonical projections* $\pi_j : l_\infty(Y_i) \to Y_j$ *and* $\eta_j : l_\infty(Z_i) \to Z_j$.

This notion of l_∞-product in combination with the l_∞-product in **Ban** allows the construction of a bounded or l_∞-cohomology for Banach spaces. The considerably more difficult case of a c_0-comohomology is considered in [11]. The first step, namely, the existence of a c_0-product for z-linear maps can be obtained as follows. Let $F : Z \curvearrowright Y$ be a z-linear map. We consider the subset $\mathrm{Hom}_{\mathbf{Z}^{\bowtie}}(\Delta F, (F_i))_0$ of families of morphisms $(\alpha_i, \gamma_i) : F \rightrightarrows F_i$ such that:

(1) (α_i) y (γ_i) are pointwise convergent to 0.
(2) There is a $\| \cdot \|$-bounded family of homogeneous maps $B_i : Z \to Y_i$ pointwise convergent to 0 such that $\alpha_i F - F_i \gamma_i = B_i + L_i$, where $L_i : Z \to Y_i$ are linear.

The representative, if it exists, of the functor $\mathbf{Z} \xrightarrow{\Delta} \mathbf{Z}^{\bowtie} \xrightarrow{\mathrm{Hom}_{\mathbf{Z}^{\bowtie}}(\cdot, (F_i))_0} \mathbf{Set}$ shall be called the c_0-product of the family (F_i). One has

Proposition 4.2. *Let* $(F_i : Z_i \curvearrowright Y_i)$ *be a* $Z(\cdot)$-*bounded family of* z-*linear maps. The functor* $\mathrm{Hom}_{\mathbf{Z}^{\bowtie}}(\Delta(\cdot), (F_i))_0$ *is representable and its representative is* $c_0(F_i)$ *together with the canonical projections.*

This method of restricted products is quite flexible and can be used to obtain other products. Working in an entirely dual form it can be shown, with just a few technical difficulties, that $l_1(F_i)$, the z-linear map that corresponds to the l_1-amalgam of sequences- yields their natural coproduct.

4.3. Products in 3^Z and coproducts in 3_Y. Let us consider now the situation in the associated categories 3^Z and 3_Y.

Products in 3, products and pull-back in **Ban**, and products in this associated category are connected as follows: Let $0 \to Y_i \to X_i \xrightarrow{q_i} Z \to 0 \equiv F_i$ be a family of objects in 3^Z. Their l_∞-product can be obtained from the corresponding l_∞-product in 3 just making pull-back with the diagonal operator $D : Z \to l_\infty(Z)$ as it is shown in the diagram

$$
\begin{array}{ccccccccc}
0 & \longrightarrow & l_\infty(Y_i) & \longrightarrow & l_\infty(X_i) & \longrightarrow & l_\infty(Z) & \longrightarrow & 0 \equiv l_\infty(F_i) \\
 & & \| & & \uparrow & & \uparrow{\scriptstyle D} & & \\
0 & \longrightarrow & l_\infty(Y_i) & \longrightarrow & PB_\infty & \longrightarrow & Z & \longrightarrow & 0 \equiv l_\infty(F_i)D.
\end{array}
$$

Moreover, the space PB_∞ in sequence $0 \to l_\infty(Y_i) \to PB_\infty \to Z \to 0$ is nothing different from the restricted l_∞-pull-back in **Ban** of the quotients $X_i \to Z$. Indeed, there is no reason why pull-back and push-out should be defined for just two arrows. In fact, since pull-backs and push-outs are just products in an associated category (see [16]), having at our disposal restricted products allows us to define arbitrary pull-back and push-outs in the same restricted sense. All this means that the category of Banach spaces which admits l_∞-products, also admits l_∞-pull-backs; and since it admits l_1-coproducts, it also admits l_1-push-outs.

Dually, co-products in 3, in 3_Y and co-products and push-out in **Ban** are connected as follows: Let $0 \to Y \overset{j_i}{\to} X_i \to Z_i \to 0 \equiv F_i$ be a family of objects in 3_Y. Their l_1-coproduct in 3_Y is obtained from the corresponding l_1-coproduct in 3 making push-out with the sum-operator $\Sigma : l_1(Y) \to Y$ as it is shown in the diagram:

$$
\begin{array}{ccccccccc}
0 & \longrightarrow & l_1(Y) & \longrightarrow & l_1(X_i) & \longrightarrow & l_1(Z_i) & \longrightarrow & 0 \equiv l_1(F_i) \\
& & \Sigma \downarrow & & \downarrow & & \| & & \\
0 & \longrightarrow & Y & \longrightarrow & PO_\infty & \longrightarrow & l_1(Z_i) & \longrightarrow & 0 \equiv \Sigma l_1(F_i).
\end{array}
$$

The space PO_∞ is the restricted push-out of the family of embeddings $Y \to X_i$.

Other restricted products and coproducts need a careful case-by-case examination. For instance, although the existence of the l_∞-product in 3 guarantees the existence of the l_∞-product in 3^Z, the same cannot be said of the c_0-products, since no diagonal operator $D : Z \to c_0(Z)$ exists. The interested reader is addressed to [11].

4.4. **Pull-back in 3_Y.** The problem of the existence of pull-back is genuinely interesting in 3_Y since no product is known in this case. If one has a diagram $F \overset{\gamma_i}{\leftarrow} F_i$ in 3_Y, namely,

$$
\begin{array}{ccccccccc}
0 & \longrightarrow & Y & \longrightarrow & X & \overset{q}{\longrightarrow} & Z & \longrightarrow & 0 \equiv F \\
& & \| & & \uparrow \beta_i & & \uparrow \gamma_i & & \\
0 & \longrightarrow & Y & \longrightarrow & X_i & \overset{q_i}{\longrightarrow} & Z_i & \longrightarrow & 0 \equiv F_i
\end{array}
$$

then it is not difficult to verify the existence of an exact sequence

$$
0 \longrightarrow Y \overset{\varsigma}{\longrightarrow} PB(\beta_i) \overset{\eta}{\longrightarrow} PB(\gamma_i) \longrightarrow 0 \equiv FB.
$$

The sequence FB describes the pull-back in 3_Y (this is a bit trying to prove since the universal property of the pull-back spaces $PB(\gamma_i)$ in **Ban** cannot be directly used). The following 3D diagram connects products and pull-backs in **Ban**, products in 3^Z and pull-backs in 3_Y

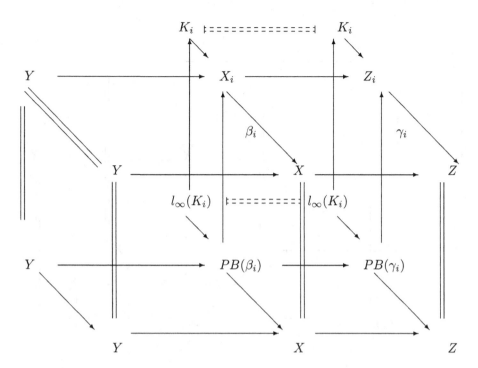

4.5. Push-out in 3^Z. There is an entirely dual situation starting with a diagram $F \xrightarrow{\alpha_i} F_i$ in 3^Z, namely,

$$
\begin{array}{ccccccccc}
0 & \longrightarrow & Y & \xrightarrow{\ j\ } & X & \longrightarrow & Z & \longrightarrow & 0 \equiv F \\
& & {\scriptstyle \alpha_i}\downarrow & & {\scriptstyle \beta_i}\downarrow & & \| & & \\
0 & \longrightarrow & Y_i & \longrightarrow & X_i & \longrightarrow & Z_i & \longrightarrow & 0 \equiv F_i.
\end{array}
$$

It is not difficult to verify the existence of an exact sequence

$$
0 \longrightarrow PO(\alpha_i) \xrightarrow{\ \varsigma\ } PO(\beta_i) \xrightarrow{\ \eta\ } Z \longrightarrow 0 \equiv FO.
$$

The sequence FO describes the push-out in 3^Z (this is, once more, bit trying to prove). The following 3D diagram connects coproducts and push-out in **Ban**, coproducts in 3_Y and push-out in 3^Z in a diagram

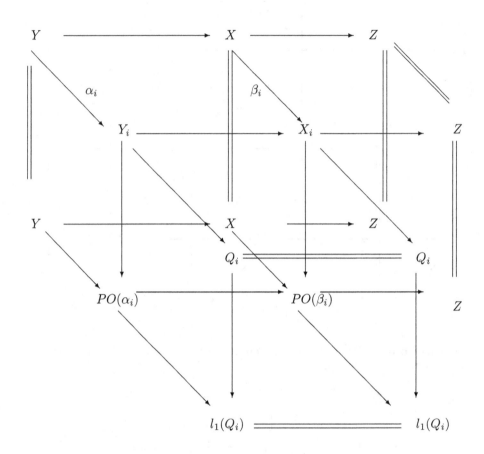

5. LIMITS AND COLIMITS

Products are a particular instance of limits, and coproducts are particular instances of co-limits. Let us recall the definition of co-limit in a category. Let I be a partially ordered set. An inductive system $\langle A_i, \phi_{ij} \rangle$, with indices in I, of objects of a category $\mathfrak{C}$ is a family of objects (A_i) and morphisms $\phi_{ij} : A_i \to A_j$, $i \leq j$ such that $\phi_{ik} = \phi_{jk} \circ \phi_{ij}$ for each ordered triple $i \leq j \leq k$. The co-límit of the system $\langle A_i, \phi_{ij} \rangle$ in $\mathfrak{C}$ is an object $\lim_{\to} A_i$ endowed with a collection of morphisms $\psi_j : A_j \to \lim_{\to} A_i$ such that for all $i < j$ one has $\psi_j \phi_{ij} = \psi_i$; and such that it is universal with respect to that property, that is, if there is another object G and another collection of morphisms $\psi'_j : A_j \to G$ such that for each $i < j$ one has $\psi'_j \phi_{ij} = \psi'_i$, then there is a unique morphism $\Phi : \lim_{\to} A_i \to G$ verifying that for each $i \in$ I, $\Phi \psi_i = \psi'_i$. Obviously, the co-limit of an inductive system is unique in the category, up to isomorphisms.

5.1. **Inductive limits of Banach spaces.** Tradition says that co-limits with respect to a countable totally ordered set in **Ban** are called inductive limits (or simply limits). Banach spaces admit very restricted limits, as we show next. Through this and the following section we will work inside the category **Ban₁** of Banach spaces and operators having norm at most 1.

So, let $\langle E_n, i_n \rangle$ be an inductive system in **Ban₁**. Let us consider the vector space

$$X = \{(x_n)_n \in l_\infty(E_n) : \exists \mu \in \mathbb{N} : i_n(x_n) = x_{n+1}, \forall n > \mu\},$$

endowed with the seminorm $\|(x_n)_n\| = \lim \|x_n\|$. If $K = \ker \|\cdot\|$ the inductive limit $\mathfrak{X}$ of $\langle E_n, i_n \rangle$ is the completion of the quotient X/K together with the family of isometries $I_n : E_n \to \mathfrak{X}$ defined as $I_n(x) = [(0, 0, \ldots, x, x, \ldots)]$. The inductive limit can be described as $\mathfrak{X} = \overline{\cup I_n(E_n)}$. It is important to observe that the object $\mathfrak{X}$ scan only be universal with respect to the property "$\forall n \in \mathbb{N}, I_{n+1}i_n = I_n$" for uniformly bounded families of operators $f_n : E_n \to X$. Thus, if for all $n \in \mathbb{N}$ it occurs that $f_{n+1}i_n = f_n$ then there is an operator $T : \mathfrak{X} \to X$ defined by $T((x_n) + K) = f_N x_N$, where N is the least index for which $i_j x_j = x_{j+1}$ for all $j \geq N$. That operator verifies $TI_n = f_n$ and is unique. If each f_n is an into isometry then also T is an into isometry as well: $\|T((x_n) + K)\| = \|f_N x_N\| = \|x_N\| = \lim \|x_n\| = \|(x_n) + K\|$. In this case, the limit of the system will be called $\lim E_n$ and it can be identified with $\overline{\cup_n E_n}$.

5.2. **Inductive limits of exact sequences.** As we have already said at the introduction, the categories $\mathfrak{Z}$ of z-linear maps and $\mathfrak{S}$ of short exact sequences (see the Appendix) are "equivalent". This does not mean however that all constructions are equally easy to obtain in the two categories. We construct now the restricted inductive limits first in $\mathfrak{S}$ and then in $\mathfrak{Z}$.

Given an inductive system $(*)$ $[0 \to Y_n \to X_n \to Z_n \to 0, (\alpha_n, \beta_n, \gamma_n)]$ in which the operators $\alpha_n, \beta_n, \gamma_n$ are into isometries, it is not hard to verify that the sequence

$$0 \longrightarrow \lim Y_n \longrightarrow \lim X_n \longrightarrow \lim Z_n \longrightarrow 0$$

is exact and acts as its restricted inductive limit.

If one moreover has $F_n : Z_n \curvearrowright Y_n$ so that $F_{n+1}\gamma_n = \gamma_n F_n$ then the inductive system $[F_n, (\alpha_n, \gamma_n)]$ admits a z-linear map $F : \lim Z_n \curvearrowright \lim Y_n$ inducing the limit exact sequence. Standard folklore results reduce the problem to define $F : \cup J_n Z_n \curvearrowright \cup I_n Y_n$; in each point $z \in \cup_n J_n Z_n$ it takes the value $F(z) = I_N F_N(z_N)$ for $N = \min\{n \in \mathbb{N} : z \in J_n Z_n\}$ and $J_N(z_N) = z$. We check that F is z-linear: given $z_i \in \cup J_n Z_n$, take N_i such that $z_i = I_{N_i}(z_{N_i})$.

Let $M = \max\{N_i\}$.

$$\|F(\sum_i z_i) - \sum_i F z_i\| = \|I_M F_M(\gamma_M(\sum z_{N_i}) - \sum I_{N_i} F_{N_i}(z_{N_i})\| =$$
$$= \|I_M F_M(\gamma_M(\sum z_{N_i}) - \sum I_M \alpha_{M-1} \cdots \alpha_{N_i} F_{N_i}(z_{N_i})\|$$
$$\leq Z(F_M) \sum_i \|z_i\|.$$

And F verifies that for all $n \in \mathbb{N}$, $F J_n(z_n) = I_n F_n(z_n)$. To see that F defines the limit sequence we just have to define the operator:

$$T((y_n, z_n) + K_0) = ((y_n) + K_1, (z_n) + K_2),$$

(where $K_0 = \ker(\lim \| \cdot \|_{F_n})$, $K_1 = \ker(\lim \| \cdot \|_{Y_n})$ and $K_2 = \ker(\lim \| \cdot \|_{Z_n})$ and observe that it makes commutative the diagram

$$
\begin{array}{ccccccccc}
0 & \longrightarrow & \lim Y_n & \longrightarrow & \lim Y_n \oplus_{F_0} \lim Z_n & \longrightarrow & \lim Z_n & \longrightarrow & 0. \\
& & \| & & T \uparrow & & \| & & \\
0 & \longrightarrow & \lim Y_n & \longrightarrow & \lim [Y_n \oplus_{F_n} Z_n] & \longrightarrow & \lim Z_n & \longrightarrow & 0.
\end{array}
$$

Remark. This limit map is the limit of the sequence F_n in the smaller category $\mathbf{Z_1}$ (like $\mathbf{Z}$ but only into isometries are allowed as operators). One should not however to expect that F behaves even as a restricted inductive limit in 3. The problem of the existence of limits in 3 thus remains open.

6. THE c-PRODUCT OF PEŁCZYŃSKI AND LUSKY; AND THE STEGALL PRODUCT

Let us discuss now other products appearing in the literature such as l_1 or c-products. On one hand, it is plain that there is no c-product: this notion needs an ambient space. If, however, one has a sequence $X_n \subset X$ of subspaces of a given Banach space X then their c-product should be

$$c(X_n) = \{(x_n) : x_n \in X_n, \ \exists \lim x_n\}.$$

There exists a natural exact sequence

$$0 \longrightarrow c_0(X_n) \longrightarrow c(X_n) \xrightarrow{\lim \cdot} X \longrightarrow 0.$$

When the spaces X_n are finite-dimensional, this sequence was first considered by Pełczyński in [24]. Lusky [21] and Johnson-Oikhberg [15] show that this sequence splits if and only if X has the BAP (see also [11]). It is not difficult to prove that this sequence always locally splits (i.e. its dual sequence splits).

If we set $l_1^0(X_n) = \{(x_n) : x_n \in X_n, \sum \|x_n\| < +\infty, \sum x_n = 0\}$, then we can consider the sequence associated to their "l_1-product"

$$0 \longrightarrow l_1^0(X_n) \longrightarrow l_1(X_n) \xrightarrow{\Sigma} X \longrightarrow 0.$$

Let us show that when X has the BAP this sequence locally splits. Indeed, let (B_n) be a sequence of finite rank operators pointwise convergent to the identities; assuming that $B_n(X) \subset X_n$, the dual sequence

$$0 \longrightarrow X^* \longrightarrow l_\infty(X_n^*) \longrightarrow l_1^0(X_n)^* \longrightarrow 0$$

splits since a linear continuous projection can be easily obtained taking a free ultrafilter $\mathcal{U}$ on $\mathbb{N}$ and setting: $P((x_n^*)) = w^* - \lim_\mathcal{U} B_n^* x_n^*$. In particular, X^* is complemented in $l_\infty(X_n^*)$. Stegall [26] was the first to use a similar construction, see also [13], to provide the first example $l_1(l_2^n)$ of a Banach space with the Dunford-Pettis property whose dual has not the Dunford-Pettis property: the key of his argumentation is to show that its dual $l_\infty(l_2^n)$ contains complemented copies of l_2, something we have already proved.

7. CONSTRUCTION OF EXOTIC $\mathcal{L}_\infty$-SPACES

Recall that a Banach space X is said to be a $\mathcal{L}_{\infty,\lambda}$ space if every finite dimensional subspace of X is contained in a finite dimensional subspace λ-isomorphic to some l_∞^n. A Banach space that is a $\mathcal{L}_{\infty,\lambda}$ space for all $\lambda > 1$ is called a Lindenstrauss space. As it is well-known, a Banach space X is a Lindenstrauss space if and only of X^* is isometric to some L_1-space. Our purpose in this section is to prove the following result.

Theorem 7.1. *Given a separable Banach space E there exists an exact sequence*

$$0 \longrightarrow E \longrightarrow \mathcal{L}_\infty(E) \longrightarrow Q \longrightarrow 0$$

in which

(1) *$\mathcal{L}_\infty(E)$ is a separable $\mathcal{L}_\infty$-space*
(2) *Every operator from E into a Lindenstrauss space can be extended to an operator on $\mathcal{L}_\infty(E)$.*
(3) *The space Q has the BAP.*

This construction is intermediate between the Bourgain-Pisier construction in [3], in which every separable Banach space is embedded into some separable $\mathcal{L}_{\infty,\lambda}$ space in such a way that the quotient space has the Schur and Radon-Nikodym properties; and Zippin's construction in [29], in which every separable Banach space E is embedded into a separable Banach space $Z(E)$ with an FDD such that $Z(E)/E$ has an FDD and moreover every $C(K)$-valued operator defined on E can be extended to $Z(E)$. Using the result of Lusky [21] asserting that if X has BAP and $c_\infty = c_0(F_n)$ for a dense sequence of finite dimensional spaces (see below) then $X \oplus c_\infty$ has basis then the sequence $0 \to E \to \mathcal{L}_\infty(E) \oplus c_\infty \to Q \oplus c_\infty \to 0$ improves Zippin's construction.

Proof. Let $\varepsilon > 0$ and consider a sequence (ε_n) such that $\prod(1 + \varepsilon_n) \leq 1 + \varepsilon$. Relabelling without mercy, let $F_n \to l_\infty^{m(n)}$ be a dense (in the Banach-Mazur distance) sequence of subspaces of l_∞^n-spaces. Let $c_1 = l_1(F_n)$ be their coproduct. Let I_0 be a countable set of finite rank operators such that every norm-one finite-rank operator $c_1 \to E$ has an element of I_0 at distance at most

ε_0. Let $\phi_0 : l_1(I_0, c_1) \to E$ be the coproduct operator $\phi_0((s_j)) = \sum_{j \in I_0} j(s_j)$. We form the push-out diagram

$$
\begin{array}{ccccccccc}
0 & \longrightarrow & l_1(I_0, c_1) & \longrightarrow & l_1(I_0, l_1(l_\infty^{m(n)})) & \longrightarrow & S_1 & \longrightarrow & 0 \\
& & \phi_0 \downarrow & & \downarrow & & \| & & \\
0 & \longrightarrow & E & \underset{i_1}{\longrightarrow} & P_1 & \longrightarrow & S_1 & \longrightarrow & 0
\end{array}
$$

to obtain an isometric enlargement $i_1 : E \to P_1$ such that every norm one operator $F_n \to E$ can be extended to an operator $l_\infty^{m(n)} \to P_1$ with norm at most $1 + \varepsilon_0$.

For the inductive step, assume that I_{n-1}, ϕ_{n-1} and P_n have already been constructed. We set now I_n a countable set of finite rank operators to be added to I_{n-1} in such that every norm-one finite-rank operator $c_1 \to P_n$ has an element of $\cup_{j=1}^n I_j$ at distance at most ε_n. We form the operator $\phi_n((s_j)) = \sum_{k \leq n} \sum_{j \in I_k} j(s_j)$ and the push-out diagram

$$
\begin{array}{ccccccccc}
0 & \longrightarrow & l_1(\cup_{j=1}^n I_j, c_1) & \longrightarrow & l_1(\cup_{j=1}^n I_j, l_1(l_\infty^{m(n)})) & \longrightarrow & S_n & \longrightarrow & 0 \\
& & \phi_n \downarrow & & \downarrow & & \| & & \\
0 & \longrightarrow & P_n & \underset{i_{n+1}}{\longrightarrow} & P_{n+1} & \longrightarrow & S_n & \longrightarrow & 0
\end{array}
$$

to obtain an isometric enlargement $i_{n+1} : P_n \to P_{n+1}$ such that every norm one operator $F_k \to P_n$ can be extended to an operator $l_\infty^{m(k)} \to P_{n+1}$ with norm at most $\prod_{j=1}^n 1 + \varepsilon_j$.

Let $\lim P_n$ be the inductive limit of the sequence

$$
0 \longrightarrow E \overset{i_1}{\longrightarrow} P_1 \overset{i_2}{\longrightarrow} P_2 \longrightarrow \cdots
$$

It is clear that every norm-one operator $F_n \to \lim P_n$ admits an extension to $l_\infty^{m(n)}$ with norm at most $\prod(1 + \varepsilon_n) \leq 1 + \varepsilon$; from which it immediately follows, by the arguments in [19] (see also [12]), that $\lim P_n$ is an $\mathcal{L}_\infty$-space. The sequence we are looking for is

$$
0 \longrightarrow E \longrightarrow \lim P_n \overset{q}{\longrightarrow} Q \longrightarrow 0
$$

We pass to show (2). In what follows, let $\mathcal{L}_{\infty, \lambda}$ denote a unspecified $\mathcal{L}_{\infty, \lambda}$-space. Let us show that every norm one operator $\tau : E \to \mathcal{L}_{\infty, \lambda}$ can be extended to P_1 with norm at most $(1 + \varepsilon_0)\lambda$. To see this, consider the restriction of $\tau \phi_0$ to each F_n. It can be assumed that its range lies in a λ-isomorphic copy of a finite dimensional l_∞^n space, and therefore can be extended to $l_\infty^{m(n)}$ with norm at most $(1 + \varepsilon_0)\lambda$. The coproduct property yields thus an extension of $\tau \phi_0$ to $l_1(I_0, l_1(l_\infty^{m(n)}))$ with norm at most $(1 + \varepsilon_0)\lambda$, while the push-out property yields therefore an extension of τ to P_1 with norm at most $(1 + \varepsilon_0)\lambda$. An analogous reasoning shows that every operator $P_n \to \mathcal{L}_{\infty, \lambda}$ can be extended to P_{n+1} with norm at most $(1 + \varepsilon_n)\lambda$. In particular, the sequence $0 \to P_n \to P_{n+1} \to P_{n+1}/P_n \to 0$ splits if and only if P_n is an $\mathcal{L}_\infty$-space.

To obtain an extension of τ to $\lim P_n$ we need to proceed inductively, which is why it is necessary to restrict the range to be a Lindenstrauss space, namely an $\mathcal{L}_{\infty,\lambda}$-space for all $\lambda > 1$. So, let $\mathcal{L}$ be a unspecified Lindenstrauss space and $\tau : E \to \mathcal{L}$ be a norm one operator. Choosing a sequence (λ_n) such that $\prod(1+\varepsilon_n)\lambda_n < \infty$, and obtaining then successive extensions τ_1 of τ to P_1 with norm $(1 + \varepsilon_0)\lambda_1$; then τ_2 of τ_1 to P_2 with norm $(1 + \varepsilon_1)(1 + \varepsilon_0)\lambda_2\lambda_1$, and so on, one obtains in the end an extension $T : \lim P_n \to \mathcal{L}$ of τ with norm at most $\prod(1 + \varepsilon_n)\lambda_n$ (which can be made as close to 1 as one wishes).

It remains to prove that Q has the BAP. We start with two observations: the first one is that $Q = (\lim P_n)/E$; the second is that the inductive construction allows us to assume without loss of generality that the space S_1 is finite dimensional, and thus that also each P_n/E is finite dimensional. In this form, $\lim(P_n/E)$ has the BAP if and only if its associated c-product sequence $0 \to c_0(P_n/E) \to c(P_n/E) \to \lim(P_n/E) \to 0$ splits, which is what we are going to prove.

Let us define the space $c[P_n; E] = \{(p_n) : p_n \in P_n : \exists \ \lim p_n \in E\}$. Recalling that $q_n : P_n \to P_n/E$ denotes the natural quotient, there is a push-ut diagram

$$
\begin{array}{ccccccccc}
0 & \longrightarrow & c[P_n; E] & \longrightarrow & c(P_n) & \xrightarrow{\lim q_n(\cdot)} & \lim(P_n/E) & \longrightarrow & 0 \\
& & \downarrow{\scriptstyle (q_n(\cdot))} & & \downarrow & & \| & & \\
0 & \longrightarrow & c_0(P_n/E) & \longrightarrow & c(P_n/E) & \xrightarrow[\lim \cdot]{} & \lim(P_n/E) & \longrightarrow & 0.
\end{array}
$$

Let us add a new line to obtain a commutative diagram (warning: the left column is not exact!):

$$
\begin{array}{ccccccccc}
0 & \longrightarrow & c_0(P_n) & \xrightarrow{i} & c(P_n) & \longrightarrow & \lim P_n & \longrightarrow & 0 \\
& & \downarrow{\scriptstyle j} & & \| & & \downarrow{\scriptstyle q} & & \\
0 & \longrightarrow & c[P_n; E] & \xrightarrow{\bar{j}} & c(P_n) & \xrightarrow{\lim q_n(\cdot)} & \lim(P_n/E) & \longrightarrow & 0 \\
& & \downarrow{\scriptstyle (q_n(\cdot))} & & \downarrow & & \| & & \\
0 & \longrightarrow & c_0(P_n/E) & \longrightarrow & c(P_n/E) & \xrightarrow[\lim \cdot]{} & \lim(P_n/E) & \longrightarrow & 0.
\end{array}
$$

The new upper sequence splits: it is a c-product sequence and $\lim P_n$ is a separable $\mathcal{L}_\infty$ space, so it has basis. If $\pi : c(P_n) \to c_0(P_n)$ is a projection through i

$$
\lim q_n(\cdot)j\pi : c(P_n) \longrightarrow c_0(P_n/E)
$$

is an extension for $\lim q_n(\cdot)$ through $\bar{j}$, and the lower sequence splits.

$\square$

Comment. Consider the family of embeddings $E \xrightarrow{i_n \cdot i_1} P_n$ and form their push-out PO. The coproduct yields an exact sequence

$$0 \longrightarrow E \longrightarrow PO \longrightarrow l_1(P_n/E) \longrightarrow 0,$$

such that every norm one operator $E \to \mathcal{L}_{\infty,\lambda}$ can be extended to an operator $PO \to \mathcal{L}_\infty$ with norm at most λ. This sequence is close to the Bourgain-Pisier construction since the push-out diagram

$$
\begin{array}{ccccccccc}
 & & E & = & E & & & & \\
 & & \downarrow & & \downarrow & & & & \\
0 & \longrightarrow & P_n & \longrightarrow & P_{n+1} & \longrightarrow & S_n & \longrightarrow & 0 \\
 & & {\scriptstyle q_n}\downarrow & & \downarrow & & \| & & \\
0 & \longrightarrow & P_n/E & \longrightarrow & P_{n+}/E & \longrightarrow & S_n & \longrightarrow & 0 \\
 & & \downarrow & & \downarrow & & & & \\
 & & 0 & = & 0, & & & &
\end{array}
$$

and the fact that S_n enjoys the Schur and RN properties plus a 3-space argument (see [7]) show that P_n/E and P_{n+1}/E enjoy or not the Schur and RN properties simultaneously. Since $P_1/E = S_1$ has both the Schur and RN properties, it follows that $l_1(P_n/E)$ has the Schur and RN properties. Unfortunately, PO cannot be an $\mathcal{L}_\infty$-space since it has l_1 as a quotient.

8. APPENDIX. THE CATEGORY OF EXACT SEQUENCES

When working with short exact sequences of Banach spaces, one does not want to distinguish between equivalent sequences. This motivates the necessity of constructing "the" category $\mathfrak{S}$ having as objects equivalence classes of short exact sequences. The choice of morphisms comes now determined: if it would seem natural to act as in the case of complexes (see [22, 16]) and choose as morphisms of $\mathfrak{S}$ triples of operators (α, β, γ) making commutative a diagram

(1)
$$
\begin{array}{ccccccccc}
0 & \longrightarrow & Y & \xrightarrow{i} & X & \xrightarrow{q} & Z & \longrightarrow & 0 \\
 & & \downarrow{\scriptstyle \alpha} & & \downarrow{\scriptstyle \beta} & & \downarrow{\scriptstyle \gamma} & & \\
0 & \longrightarrow & Y' & \xrightarrow[i']{} & X' & \xrightarrow[q']{} & Z' & \longrightarrow & 0,
\end{array}
$$

those morphisms are not well-defined between equivalence classes. We then make use of the notion of homotopy. Although the notion makes sense for arbitrary complexes, we will use it just to work with short exact sequences. Two triples of operators (α, β, γ) and $(\alpha', \beta', \gamma')$ making a commutative diagram like (1) are said to be homotopic if and only if there exist operators $S : X \to Y'$ and $T : Z \to X'$ such that: i) $Si = \alpha - \alpha'$; ii) $q'T = \gamma - \gamma'$; and iii) $i'S + Tq = \beta - \beta'$. In the case of exact sequences, it is easy to check that (α, β, γ) and $(\alpha', \beta', \gamma')$ are homotopic if and only if just i) and ii) hold.

On the other hand, the existence of a diagram like (1) is equivalent, modulus homotopy, to the fact that the pull-back and push-out extensions in the diagram

$$
\begin{array}{ccccccccc}
0 & \longrightarrow & Y & \longrightarrow & X & \longrightarrow & Z & \longrightarrow & 0 \\
 & & \downarrow \alpha & & \downarrow \hat{\alpha} & & \| & & \\
0 & \longrightarrow & Y' & \longrightarrow & PO & \longrightarrow & Z & \longrightarrow & 0 \\
 & & \| & & & & \| & & \\
0 & \longrightarrow & Y' & \longrightarrow & PB & \longrightarrow & Z & \longrightarrow & 0 \\
 & & \| & & \downarrow \hat{\gamma} & & \downarrow \gamma & & \\
0 & \longrightarrow & Y' & \longrightarrow & X' & \longrightarrow & Z' & \longrightarrow & 0
\end{array}
$$

are equivalent. Thus, $i)$ and $ii)$ hold if and only if either $i)$ or $ii)$ holds. It is now clear which should be the definition of the morphisms of $\mathfrak{S}$: given two (equivalence classes of) exact sequences E and E', a morphism $E \rightrightarrows E'$ shall be the equivalence class of the operators (α, γ) for which there is a third operator (the arrow without name below) making commutative the diagram

$$
\begin{array}{ccccccccc}
0 & \longrightarrow & Y & \longrightarrow & X & \longrightarrow & Z & \longrightarrow & 0 \in E \\
 & & \downarrow \alpha & & \downarrow & & \downarrow \gamma & & \\
0 & \longrightarrow & Y' & \longrightarrow & X' & \longrightarrow & Z' & \longrightarrow & 0 \in E',
\end{array}
$$

with respect to the equivalence relation

$$(\alpha, \gamma) \asymp (\alpha', \gamma') \iff \alpha - \alpha' \text{ extends to } X \iff \gamma - \gamma' \text{ lifts to } X'.$$

We shall write $\mathrm{Hom}_{\mathfrak{S}}(E, E')$ to denote the set of all morphisms of $\mathfrak{S}$ from E to E'. To simplify notation we shall write (α, γ) to denote an element of $\mathrm{Hom}_{\mathfrak{S}}(E, E')$, although it has to be understood that it represents the $\asymp$ −equivalence class of the couple (α, γ). The composition of morphisms comes defined in a natural way: $(\alpha, \gamma) \circ (\alpha', \gamma') = (\alpha \circ \alpha', \gamma \circ \gamma')$ and the identity morphism shall be (id, id).

REFERENCES

[1] Y. Benyamini and J. Lindenstrauss, *Geometric nonlinear functional analysis, Vol. 1*, Amer. Math. Soc., Providence (1999).

[2] J. Bourgain and F. Delbaen, *A class of special $\mathcal{L}_\infty$-spaces*, Acta Math. 145 (1980) 155-176.

[3] J. Bourgain and G. Pisier, *A construction of $\mathcal{L}_\infty$-spaces and related Banach spaces*, Bol. Soc. Brasil. Mat. 14 (1983) 109-123.

[4] F. Cabello Sánchez and J.M.F. Castillo, *Duality and twisted sums of Banach spaces*, J. Funct. Anal. 175 (2000) 1-16.

[5] F. Cabello Sánchez, J.M.F. Castillo, N.J. Kalton and D.T. Yost, *Twisted sums with $C(K)$ spaces,*, Trans. Amer. Math. Soc. 355 (2003) 4523-4541.

[6] F. Cabello Sánchez and J.M.F. Castillo, *The long homology sequence for quasi-Banach spaces, with applications*, Positivity 8 (2004) 379-394.

[7] J.M.F. Castillo and M. González, *Three-space problems in Banach space theory*, Springer LNM 1667 (1997).

[8] J.M.F. Castillo and Y. Moreno, *König-Wittstock quasi-norms on quasi-Banach spaces* Extracta Math. 17 (2002) 273-280.

[9] J.M.F. Castillo and Y. Moreno, *On isomorphically equivalent extensions of quasi-Banach spaces*, in Recent Progress in Functional Analysis, (K.D. Bierstedt, J. Bonet, M. Maestre, J. Schmets (eds.)), North-Holland Math. Studies 187 (2000) 263-272.

[10] J.M.F.Castillo and Y. Moreno, *On the Lindenstrauss-Rosenthal theorem*, Israel J. Math. 140 (2004) 253-270.

[11] J.M.F.Castillo and Y. Moreno, *Sobczyk cohomology for separable Banach spaces*, preprint 2006.

[12] J.M.F. Castillo, Y. Moreno, J. Suárez, *On Lindenstrauss-Pełczyński spaces*, Studia Math. (to appear).

[13] J. Diestel, *A survey of results related to the Dunford-Pettis property*, Contemp. Math. 2 (1980) 15-60.

[14] P. Enflo, J. Lindenstrauss and G. Pisier, *On the "three space problem"*, Math. Scand. **36** (1975), 199–210.

[15] W. B. Johnson and T. Oikhberg, *Separable lifting property and extensions of local reflexivity*, Ill. J. of Math. 45 (2001) 123-137.

[16] E. Hilton and K. Stammbach, *A course in homological algebra*, Springer GTM 4.

[17] N. Kalton, *The three-space problem for locally bounded F-spaces*, Compo. Math. 37 (1978), 243–276.

[18] N. Kalton and N.T. Peck, *Twisted sums of sequence spaces and thee three-space problem*, Trans. Amer. Math. Soc. 255 (1979), 1–30.

[19] J. Lindenstrauss, *On the extension of compact operators*, Mem. Amer. Math. Soc. 48 (1964).

[20] J. Lindenstrauss and L. Tzafriri, *Classical Banach spaces I, sequence spaces*, Ergeb. Math. 92, Springer-Verlag 1977.

[21] W. Lusky, *A note on Banach spaces containing c_0 and c_∞*, J. Funct. Anal. 62 (1985) 1-7.

[22] S. Mac Lane, *Homology*, Grund. math. Wiss. 114, Springer-Verlag 1975.

[23] Scott Osborne, M. *Basic homological algebra* Springer GTM 196.

[24] A. Pełczyński, *A separable Banach space with the bounded aproximation property is a complemented subspace of a Banach space with a basis*, Studia Math. 40 (1971) 239-243.

[25] Z. Semadeni, *Banach spaces of continuous functions*, Monografie Matematyczne 55, PWN, 1971.

[26] Ch. Stegall, *Duals of certain spaces with the Dunford-Pettis property*, Notices AMS 19 (1972) 799

[27] Ch. Stegall, *Banach spaces whose duals contain $_1(\Gamma)$ with applications to the study of dual $L_1(\mu)$ spaces*, Trans. AMS 176 (1973)463-477.

[28] Ch. A. Weibel, *An introduction to homological algebra*, Cambridge studies 38

[29] M. Zippin, *The embedding of Banach spaces into spaces with structure*, Ill. J. Math. 34 (1990) 586–606.

DEPARTAMENTO DE MATEMÁTICAS, UNIVERSIDAD DE EXTREMADURA, AVDA. DE ELVAS S/N 06071-BADAJOZ, SPAIN

E-mail address: castillo@unex.es, ymoreno@unex.es

EXTENSION PROBLEMS FOR $\mathcal{C}(K)$–SPACES AND TWISTED SUMS

N. J. KALTON

1. INTRODUCTION

This article can be regarded as an update on the handbook article by Zippin [27]. In this article Zippin drew attention to problems surrounding extensions of linear operators with values in $\mathcal{C}(K)$-spaces. The literature on this subject may be said to start with the work of Nachbin, Goodner and Kelley on the case when K is extremally disconnected around 1950. Thus the subject is over fifty years old, but it still seems that comparatively little is known in the general case. We are particularly interested in extending operators on separable Banach spaces when we can assume the range is $\mathcal{C}(K)$ for K a compact metric space. In this article we will sketch some recent progress on these problems.

2. LINEAR EXTENSION PROBLEMS

It is, by now, a very classical result that a Banach space X is 1-injective if and only if X is isometric to a space $\mathcal{C}(K)$ where K is extremally disconnected; this is due to Nachbin, Goodner and Kelley [22], [11] and [17]. For a general compact Hausdorff space K the space $\mathcal{C}(K)$ is usually not injective (and, in particular, never if K is metrizable). However it is a rather interesting question to determine conditions when linear operators into arbitrary $\mathcal{C}(K)$–spaces can be extended. This problem was first considered in depth by Lindenstrauss in 1964 [18].

Let us introduce some notation. Suppose X is a Banach space and E is a closed subspace. Then, for $\lambda \geq 1$, we will say that the pair (E, X) has the $(\lambda, \mathcal{C})$-extension property if whenever $T_0 : E \to \mathcal{C}(K)$ is a bounded operator then there is an extension $T : X \to \mathcal{C}(K)$ with $\|T\| \leq \lambda\|T_0\|$. We say that X has the $(\lambda, \mathcal{C})$-extension property if (E, X) has the $(\lambda, \mathcal{C})$-extension property for every closed subspace E. We will use the term $\mathcal{C}$-extension property to

1991 *Mathematics Subject Classification*. Primary: 46B03, 46B20.
The author was supported by NSF grant DMS-0244515.

denote the $(\lambda, \mathcal{C})$-extension property for some $\lambda \geq 1$.

Usually we will want to suppose that X is separable and in this case it obviously suffices to take K metrizable; indeed since every $\mathcal{C}(K)$ for K metrizable is a contractively complemented subspace of $\mathcal{C}[0, 1]$ we may even take $K = [0, 1]$. Notice that c_0 is separably injective by Sobczyk's theorem [23]. This implies that if one chooses K to be the one-point compactification of $\mathbb{N}$ so that $\mathcal{C}(K) = c \approx c_0$ then one always has extensions when X is separable. A deep result of Zippin [25] shows that c_0 is the unique separably injective separable Banach space.

The spaces $\mathcal{C}(K)$ are $\mathcal{L}_\infty$–spaces, which means that locally they behave like ℓ_∞ and so are injective in a local sense. In 1964, Lindenstrauss [18] showed that if we restrict the operator T_0 to be compact then indeed an extension always exists and we can choose $\lambda = 1 + \epsilon$ for any $\epsilon > 0$. However the extension of bounded operators is more delicate. Indeed consider the Cantor set $\Delta = \{0, 1\}^{\mathbb{N}}$ and $\varphi : \Delta \to [0, 1]$ be the canonical surjection

$$\varphi((t_n)_{n=1}^\infty) = \sum_{n=1}^\infty \frac{t_n}{2^n}.$$

Then $\mathcal{C}[0, 1]$ can be isometrically embedded into $\mathcal{C}(\Delta)$ via the embedding $f \to f \circ \varphi$. For this embedding, $\mathcal{C}[0, 1]$ is uncomplemented in $\mathcal{C}(\Delta)$ (much more is true, cf. [3] p. 21). Thus the identity map on $\mathcal{C}[0, 1]$ cannot be extended to $\mathcal{C}(\Delta)$ i.e. $\mathcal{C}(\Delta)$ fails the $\mathcal{C}$-extension property. The existence of this counterexample already implies that ℓ_1 fails the $\mathcal{C}$-extension property. Indeed let $Q : \ell_1 \to \mathcal{C}(\Delta)$ be a quotient map and let $E = Q^{-1}\mathcal{C}[0, 1]$. Then the map $Q : E \to \mathcal{C}[0, 1]$ cannot be extended to an operator $T : E \to \mathcal{C}[0, 1]$. Indeed if such an extension exists then $T = SQ$ where $S : \mathcal{C}(\Delta) \to \mathcal{C}(\Delta)$ is a bounded operator, which is a projection of $\mathcal{C}(\Delta)$ onto $\mathcal{C}[0, 1]$. Thus any space that contains ℓ_1 fails the $\mathcal{C}$-extension property. However in 1971, Lindenstrauss and Pełczyński [19] gave a positive result:

Theorem 2.1. *The space c_0 has the $(1 + \epsilon, \mathcal{C})$-extension property for every $\epsilon > 0$.*

For a discussion of which spaces can replace $\mathcal{C}(K)$-spaces in this theorem see [8]. Twenty years later Zippin [26] gave a stronger result for ℓ_p when $p > 1$.

Theorem 2.2. *For $p > 1$ the spaces ℓ_p have the $(1, \mathcal{C})$-extension property.*

The characterization of spaces with the $\mathcal{C}$-extension property remains mysterious. It is for example not known if L_p for $1 < p < \infty$ has the $\mathcal{C}$-extension property but it is known that if $p \neq 2$ then L_p fails the $(1,\mathcal{C})$–extension property [12]. Recently the author [15] has characterized separable Banach spaces with the $(1+\epsilon,\mathcal{C})$-extension property in terms of properties of types. We will not discuss this in detail, but we note the following application:

Theorem 2.3. *Let X be a separable Orlicz sequence space not containing ℓ_1. Then X has the $\mathcal{C}$-extension property.*

Note that we do not claim the $(1,\mathcal{C})$ or $(1+\epsilon,\mathcal{C})$-extension property; this is a renorming theorem, so that X can be renormed to have the $(1+\epsilon,\mathcal{C})$-extension property.

3. EXTENSIONS BY $\mathcal{C}(K)$-SPACES

An *extension* of a Banach space X by a space Y is a short exact sequence:

$$0 \longrightarrow Y \longrightarrow Z \longrightarrow X \longrightarrow 0.$$

More informally we refer to Z as an extension of X by Y if Z is a Banach space with a subspace isometric to Y so that Z/Y is isometric to X. Such an extension *splits* if it reduces to a direct sum, i.e. Y is complemented in Z. We write Ext $(X,\mathcal{C}) = \{0\}$ if *every* extension of X by a $\mathcal{C}(K)$–space splits.

Now suppose $T_0 : Y \to \mathcal{C}(K)$ is a bounded linear operator. Then we can construct an extension of X by $\mathcal{C}(K)$ by the pushout construction. Then T_0 has a bounded extension $T : Z \to \mathcal{C}(K)$ if and only if this extension splits:

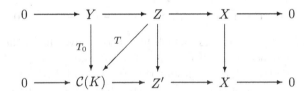

This means that Ext $(X,\mathcal{C}) = \{0\}$ if and only if whenever Y is a Banach space and E is a subspace of Y with $Y/E \approx X$ then (E, Y) has the $\mathcal{C}$-extension property. For the special case when $Y = \ell_1$ one obtains a complete classification of subspaces of ℓ_1 with the $\mathcal{C}$-extension property (first noted by Johnson and Zippin [12]):

Theorem 3.1. *Let E be a subspace of ℓ_1; then (E, ℓ_1) has the $\mathcal{C}$-extension property if and only if Ext $(\ell_1/E, \mathcal{C}) = \{0\}$.*

Johnson and Zippin [12] went on to prove:

Theorem 3.2. *Let E be a subspace of ℓ_1 which is weak*-closed as a subspace of c_0^*; then (E, ℓ_1) has the $\mathcal{C}$-extension property.*

Recently the author [15] refined their arguments to show that in fact under these hypotheses (E, ℓ_1) has the $(1+\epsilon, \mathcal{C})$-extension property for every $\epsilon > 0$. (The original argument yielded only $3 + \epsilon$ in general.)

In terms of extensions this means:

Corollary 3.3. *If X is the dual of a subspace of c_0 then Ext $(X,\mathcal{C}) = \{0\}$.*

This suggests a natural problem:

Problem 1. *Let X be a separable Banach space. Is it true that $Ext\,(X,\mathcal{C}) = \{0\}$ if and only if X is isomorphic to the dual of a subspace of c_0?*

This is equivalent (via the automorphism results of Lindenstrauss and Rosenthal [20]) to asking if whenever (E,ℓ_1) has the $\mathcal{C}$-extension property if and only if there is an automorphism $U : \ell_1 \to \ell_1$ such that $U(E)$ is weak*-closed. There is some evidence for a positive answer to Problem 1. The author proved the following results in [14]:

Theorem 3.4. *Let X be a separable Banach space such that $Ext\,(X,\mathcal{C}) = \{0\}$. Then*
(i) X has the Schur property.
(ii) If X has a (UFDD) then X is isomorphic to the dual of a subspace of c_0.

The method of proof revolved around taking one non-trivial extension of $\mathcal{C}(K)$ namely the example created in §2, and performing a pullback construction for an arbitrary operator $T : X \to c_0$:

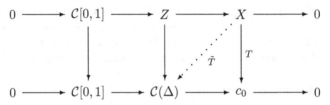

The existence of $\tilde{T} : X \to \mathcal{C}(\Delta)$ which lifts T is equivalent to the splitting of the pullback sequence. Thus if $Ext\,(X,\mathcal{C}) = \{0\}$ one can always lift T and this allows us to make deductions about the structure of X.

For a characterization of spaces X such that $Ext\,(X,\mathcal{C}) = \{0\}$ see [7]. We also remark that if K is a fixed countable compact metric space one may expect that $Ext\,(X,\mathcal{C}(K)) = \{0\}$ more often. The first non-trivial case is $K = \omega^\omega$ which is discussed in [5].

4. UNIVERSAL EXTENSIONS AND AUTOMORPHISMS

In a recent paper Castillo and Moreno [6] related extension properties with the Lindenstrauss-Rosenthal automorphism theorems [20]. In their paper, Lindenstrauss and Rosenthal showed that if E and F are two isomorphic subspaces of c_0 of infinite codimension then there is an automorphism $U :$ $c_0 \to c_0$ so that $U(E) = F$. In somewhat less exact language, one can say that there is at most one embedding, up to automorphism, of a separable Banach space into c_0. This is related to Sobczyk's theorem. They also investigated embeddings of separable spaces into ℓ_∞ and proved dual results for ℓ_1 (which we have already mentioned).

Now by Miljutin's theorem [21] all $\mathcal{C}(K)$-spaces are isomorphic for K uncountable and compact metric. The classical Banach-Mazur theorem states that every separable Banach space embeds into $\mathcal{C}[0,1]$ isometrically. The problem of obtaining automorphism results in $C(K)$-spaces is clearly related to the extension problem; we will now make this relationship precise.

Let us say that a separable Banach space has the *separable universal C-extension property* if (X, Y) has the C-extension property whenever Y is a separable Banach space containing X. In effect one may always suppose that $Y = C[0, 1]$. The following result is a more precise version of a theorem of Castillo-Moreno [6] (see [16]):

Theorem 4.1. *Let X be a separable Banach space. The following conditions on X are equivalent:*
(i) X has the separable universal C-extension property.
(ii) If X_1 and X_2 are two subspaces of $C[0, 1]$ with $X \approx X_1 \approx X_2$ then there is an automorphism $U : C[0, 1] \to C[0, 1]$ with $U(X_1) = X_2$.

Given this it becomes rather interesting to determine which spaces have the separable universal C-extension property. It is a trivial consequence of Sobczyk's theorem [23] that c_0 has this property; in fact it has the separable universal $(2, C)$-extension property with the obvious meaning. It is then a consequence of Theorem 2.1 that every subspace of c_0 has the separable universal $(2 + \epsilon, C)$-extension property. To see this we observe that if X is a subspace of c_0 which is also a subspace of a separable Banach space Y then we can form a separable superspace Z so that $X \subset c_0 \subset Z$ and $X \subset Y \subset Z$; just let Z be the quotient of $c_0 \oplus_1 Y$ by the subspace $\{(x, -x) : x \in X\}$ and identify X with the subspace of Z spanned by the cosets of $\{(x, 0) : x \in X\}$.

The obvious place to start looking for more spaces is to consider space with the separable universal $(1, C)$-extension property. However, in 1964, Lindenstrauss [18] showed that these spaces are exactly the finite-dimensional polyhedral spaces. There are no infinite-dimensional examples. The next obvious try is to consider the separable universal $(1 + \epsilon, C)$-extension property for every $\epsilon > 0$. This was first done by Speegle [24], whose main result is that such a space cannot have a uniformly smooth norm. Speegle also asked whether ℓ_1 has this property.

In fact we have recently shown that the answer to Speegle's question is positive:

Theorem 4.2. *Let X be almost isometric to the dual of a subspace of c_0. Then X has the separable universal $(1 + \epsilon, C)$-extension property for every $\epsilon > 0$.*

This extends Theorem 3.2 because a weak*-closed subspace of ℓ_1 is the dual of a quotient of c_0, and Alspach [1] showed that a quotient of c_0 is almost isometric to a subspace of c_0. Let us notice here a connection with Problem 1. It is a result of Bourgain [4] that ℓ_1 contains an uncomplemented copy X of ℓ_1. Now (X, ℓ_1) has the C-extension property and so Ext $(\ell_1/X, C) = \{0\}$.

Problem 2. *Suppose X is a subspace of ℓ_1 which is isomorphic to ℓ_1; is ℓ_1/X the dual of a subspace of c_0?*

Unfortunately, Bourgain's construction in local in nature and so if one creates the example in a natural way the space ℓ_1/X is simply an ℓ_1-sum of finite-dimensional spaces. This Problem asks for a global construction.

Theorem 4.2 is not the complete answer to the characterization of spaces with the $(1 + \epsilon, \mathcal{C})$-extension property for every $\epsilon > 0$. We also have:

Theorem 4.3. *Let X be a subspace of $L_1(0, 1)$ whose unit ball is compact for the topology of convergence in measure. Then X has the separable universal $(1 + \epsilon, \mathcal{C})$-extension property for every $\epsilon > 0$.*

In [10] an example is given of a subspace of L_1 where the unit ball is compact for convergence in measure and yet X is not almost isometric to the dual of a subspace of c_0. Another example constructed in [15] is a Nakano space ℓ_{p_n} where $\lim_{n \to \infty} p_n = 1$.

We now have a fairly rich class of spaces for which the equivalent conditions of Theorem 4.1 hold; this class includes all weak*-closed subspaces of ℓ_1 and all subspaces of c_0. It is not hard to see we can expand the class by taking direct sums (e.g. $c_0 \oplus \ell_1$) and with slightly more work, extensions. Thus any extension of c_0 by ℓ_1 satisfies Theorem 4.1. The fact that there are non-trivial extensions of c_0 by ℓ_1 is proved in [5].

5. HOMOGENEOUS ZIPPIN SELECTORS

Suppose E is a subspace of a Banach space X. Then Zippin [26, 27] introduced a criterion for (E, X) to have the $(\lambda, \mathcal{C})$-extension property. We say that a map $\Phi : B_{E^*} \to X^*$ is a *Zippin selector* of Φ is weak*-continuous and $\Phi(e^*)|_E = e^*$ for every $e^* \in B_{E^*}$. Then [26] (E, X) has the (λ, C) extension property if and only if there is a Zippin selector $\Phi : B_{E^*} \to \lambda B_{X^*}$.

In certain special cases one can find a homogeneous selector, i.e. one can choose Φ so that $\Phi(\alpha e^*) = \alpha \Phi(e^*)$ for every $e^* \in E^*$. Indeed suppose $X = \ell_p$ where $1 < p < \infty$ and E is any closed subspace. Define $\Phi(e^*)$ to be the unique norm-preserving extension of e^* to ℓ_p. Then Φ is homogeneous and weak*-continuous. To see this suppose (e_n^*) is a sequence in B_{E^*} so that e_n^* converges weak* to e^*. To show that $(\Phi(e_n^*))_{n=1}^{\infty}$ converges weak* to $\Phi(e^*)$ it suffices to show this for some subsequence. We therefore select $e_n \in B_E$ so that $e_n^*(e_n) = \|e_n^*\|$ and suppose, by passing to a subsequence that (e_n) weakly converges to some $e \in E$. Then using the special properties of ℓ_p it is quite clear that $\Phi(e_n^*)$ converges weak* to some x^* so that $\|x^*\| = \|e\|^{p-1}$ and $x^*(e) = \|e\|^p$. Now $x^*|_E = e^*$ and, if $e^* = 0$ we have $e = 0$ and $x^* = 0$; if ! not $\|e^*\| \geq e^*(e)/\|e\| = \|e\|^{p-1} = \|x^*\|$ so that $\Phi(e^*) = x^*$.

If we have a homogeneous Zippin selector for (E, X) we can extend Φ to be defined on homogeneous on E^* and continuous for the bounded weak*-topology (equivalently weak*-continuous on bounded sets). We define

$$\|\Phi\| = \sup\{\|\Phi(e^*)\| : \|e^*\| \leq 1\}.$$

Now consider the embedding of ℓ_1 into $\mathcal{C}(B_{\ell_\infty})$. It is shown in [16] that there is a homogeneous Zippin selector $\|\Phi\|$ selector for $(\ell_1, \mathcal{C}(B_{\ell_\infty}))$ with $\|\Phi\| = 1$. It follows that:

Theorem 5.1. *Suppose X is a separable Banach space containing ℓ_1. Then for any $\epsilon > 0$ there is a homogeneous Zippin selector Φ so that $\|\Phi\| < 1 + \epsilon$.*

To see this, use Theorem 4.2. There is a linear operator $T : X \to \mathcal{C}(B_{\ell_\infty})$ with $\|T\| < 1 + \epsilon$ and $Tx = x$ for $x \in \ell_1$. Define $\Psi : \ell_1^* \to X^*$ by $\Psi = T^* \circ \Phi$.

In the general the $(\lambda, \mathcal{C})$-extension property on a pair (E, X) does not imply the existence of a homogeneous Zippin selector Φ with $\|\Phi\| = \lambda$. In fact if E is non-separable there are examples where no homogeneous Zippin selector exists [9]. However Castillo and Suarez [9] recently applied an old result of Benyamini [2] to obtain:

Theorem 5.2. *If E is a separable subspace of a Banach space X so that (E, X) has the $\mathcal{C}$-extension property then there is a homogeneous Zippin's selector $\Phi : E^* \to X^*$.*

It turns out that the existence of homogeneous Zippin selectors is important for c_0–products. In fact we can now prove [16]:

Theorem 5.3. *If X has the separable universal $\mathcal{C}$-extension property then $c_0(X)$ also has the separable universal $\mathcal{C}$-extension property. The space $c_0(\ell_1)$ has the separable universal $(2 + \epsilon, \mathcal{C})$-extension property.*

The space $c_0(\ell_1)$ is the space with the most complicated structure that we know satisfies Theorem 4.1. We now turn to the question raised by Castillo and Moreno [6]: does a separable Hilbert space satisfy this Theorem? Indeed Speegle's theorem [24] shows that ℓ_2 fails the separable universal $(1 + \epsilon, \mathcal{C})$-extension property but, as the example of c_0 shows, this cannot resolve the question in general. Let us start by considering the canonical inclusion $\ell_2 \subset \mathcal{C}(B_{\ell_2^*})$. Theorem 5.2 implies:

Theorem 5.4. *Suppose $1 < p < \infty$. Then there is a homogeneous Zippin selector for $(\ell_p, \mathcal{C}(B_{\ell_p^*}))$.*

This does not seem to immediately help us decide whether ℓ_p has the separable universal $\mathcal{C}$-extension property. However if ℓ_p is embedded in some X so that (ℓ_p, X) has the $\mathcal{C}$-extension property then the argument of Theorem 5.1 shows that (ℓ_p, X) has a homogeneous Zippin selector Φ. This allows to make a renorming of X by setting, for example:

$$|x| = \sup\{|\langle x, \Phi(e^*)\rangle| : \ e^* \in B_{\ell_p^*}\}$$

and then

$$\|x\|_1 = (\tfrac{1}{2}\|x\|^p + \tfrac{1}{2}|x|^p)^{1/p}.$$

Thus $\|\cdot\|_1$ is an equivalent norm on X which agrees with the original norm on ℓ_p. However it has an additional property. There exists a constant $c > 0$ so that

$$\lim_{n \to \infty} \|x + u_n\|_1^p \geq \|x\|_1^p + c^p \lim_{n \to \infty} \|u_n\|_1^p$$

whenever $x \in X$, $(u_n)_{n=1}^\infty$ is a weakly null sequence in ℓ_p and all the limits exist. This condition as it turns out is also sufficient for the $\mathcal{C}$-extension property:

Theorem 5.5. *Suppose $1 < p < \infty$. Suppose $\ell_p \subset X$ where X is a separable Banach space. In order that (ℓ_p, X) has the $\mathcal{C}$-extension property it is necessary and sufficient that there is an equivalent norm $\|\cdot\|_1$ on X ℓ_p so that for some $c > 0$,*

$$\lim_{n\to\infty} \|x + u_n\|_1^p \geq \|x\|_1^p + c^p \lim_{n\to\infty} \|u_n\|_1^p$$

whenever $x \in X$, $(u_n)_{n=1}^\infty$ is a weakly null sequence in ℓ_p and all the limits exist.

Thus our problem is reduced to a renorming question. Let us note here that we do not require that the new norm $\|\cdot\|_1$ coincides with the original norm on ℓ_p. To see what this means let us suppose we have $1 < p < \infty$ and H is an Hilbertian subspace of L_p. Then if $p \geq 2$, H is complemented by a result of Kadets and Pełczyński [13] and so (H, L_p) has the $\mathcal{C}$-extension property. On the other hand, if $1 < p < 2$ then the hypothesis of Theorem 5.5 holds for the original norm on L_p. To see this, observe first that for a suitable constant $a > 0$ we have an inequality

$$|1 + t|^p \geq 1 + pt + a\min(|t|^p, |t|^2) \qquad -\infty < t < \infty.$$

Then suppose $\|f\|_p = 1$ and $(g_n)_{n=1}^\infty$ is a weakly null sequence with $\|g_n\|_p \leq 1$. Let $\operatorname{sgn} t = t/|t|$ if $t \neq 0$ and let $\operatorname{sgn} 0 = 0$.

$$\int |f + g_n|^p dt$$

$$\geq 1 + p\int |f|^{p-1}(\operatorname{sgn} f)g_n \, dt + a\int_{|g_n|<|f|} |f|^{p-2}|g_n|^2 dt + a\int_{|g_n|\geq|f|} |g_n|^p dt$$

$$\geq 1 + p\int |f|^{p-1}(\operatorname{sgn} f)g_n \, dt + a\left(\int_{|g_n|<|f|} |g_n|^p dt\right)^{2/p} + a\left(\int_{|g_n|\geq|f|} |g_n|^p dt\right)^{2/p}$$

$$\geq 1 + p\int |f|^{p-1}(\operatorname{sgn} f)g_n \, dt + \frac{a}{2}\|g_n\|_p^2.$$

Note that

$$\lim_{n\to\infty} \int |f|^{p-1}(\operatorname{sgn} f)g_n \, dt = 0.$$

From this it follows easily that the norm on L_p has the property that for some $c > 0$ we have

$$\lim_{n\to\infty} \|f + g_n\|_p^2 \geq \|f\|_p^2 + c^p \lim_{n\to\infty} \|g_n\|_p^2$$

whenever $f \in L_p$, $(g_n)_{n=1}^\infty$ is a weakly null sequence in L_p and all the limits exist. In fact it easy to see via renorming that any Banach space with a 2-concave unconditional basis satisfies a similar condition. These considerations are, however, a form of overkill. We do not require a condition on every weakly null sequence; instead we need the conditions for weakly null sequences in the given Hilbertian subspace.

Surprisingly when one attempts a more delicate analysis one finds that the (UMD)-property of Burkholder begins to play a role. Recall that a Banach space X has the (UMD)-property if for some (respectively, every) $1 < p < \infty$

there is a constant $C = C(p)$ so that for any finite X-valued martingale $(M_n)_{n=0}^N$ one has an estimate

$$(\mathbb{E}\|\sum_{j=1}^N \epsilon_j dM_j\|^p)^{1/p} \le C\mathbb{E}\|M_N\|^p)^{1/p} \qquad \epsilon_j = \pm 1, \ j = 1, 2, \ldots, N$$

where $dM_j = M_j - M_{j-1}$.

The connection is expressed in the following theorem:

Theorem 5.6. *Suppose $1 < p < \infty$ and $\ell_p \subset X$ where X is a Banach space with (UMD). Then X can be given an equivalent norm so that*

$$\lim_{n\to\infty} (\tfrac{1}{2}\|x + u_n\|^p + \tfrac{1}{2}\|x - u_n\|^p) \ge \|x\|^p + c^p \lim_{n\to\infty} \|u_n\|^p$$

whenever $x \in X$, $(u_n)_{n=1}^\infty$ is a weakly null sequence in ℓ_p and all the limits exist.

This is not quite what we need but one quickly gets:

Theorem 5.7. *Suppose $1 < p < \infty$ and $\ell_p \subset X$ where X is a Banach space with (UMD). If X has an unconditional basis (or even a $(UFDD)$) then (ℓ_p, X) has the C-extension property.*

This result applies when $X = L_r$ for some $1 < r < \infty$ or is a reflexive Schatten ideal. We do not know whether the result remains true if one removes the (UFDD) hypothesis: however the (UMD) hypothesis is necessary:

Theorem 5.8. *If $1 < p < \infty$ there is a super-reflexive Banach space X with unconditional basis containing ℓ_p so that (ℓ_p, X) fails to have the C-extension property.*

This answers the question of Castillo and Moreno negatively. There must be at least two non-automorphic embeddings of a Hilbert space into $C[0, 1]$. However the methods are very specific to ℓ_p-spaces and it is natural to ask:

Problem 3. *Is there any super-reflexive example of a separable Banach space with the separable universal C-extension property?*

Of course we can eliminate any space which contains a complemented copy of ℓ_p for $1 < p < \infty$ (such as L_p). The example in Theorem 5.8 proves that there are super-reflexive spaces failing the C-extension property (thus answering a question of Zippin [27]). However we may still ask:

Problem 4. *Does every separable Banach space with (UMD) have the C-extension property?*

We do not even know the answer for L_p when $1 < p < \infty$. See also [27].

REFERENCES

[1] D. E. Alspach, *Quotients of c_0 are almost isometric to subspaces of c_0*, Proc. Amer. Math. Soc. **76** (1979), 285–288.

[2] Y. Benyamini, *Separable G spaces are isomorphic to $C(K)$ spaces*, Israel J. Math. **14** (1973), 287–293.

[3] Y. Benyamini and J. Lindenstrauss, *Geometric nonlinear functional analysis. Vol. 1*, American Mathematical Society Colloquium Publications, vol. 48, American Mathematical Society, Providence, RI, 2000.

[4] J. Bourgain, *A counterexample to a complementation problem*, Compositio Math. **43** (1981), 133–144.

[5] F. Cabello Sánchez, J. M. F. Castillo, N. J. Kalton, and D. T. Yost, *Twisted sums with $C(K)$ spaces*, Trans. Amer. Math. Soc. **355** (2003), 4523–4541 (electronic).

[6] J. M. F. Castillo and Y. Moreno, *On the Lindenstrauss-Rosenthal theorem*, Israel J. Math. **140** (2004), 253–270.

[7] ———, *Extensions by spaces of continuous functions*, to appear.

[8] J. M. F. Castillo, Y. Moreno, and J. Suárez, *On Lindenstrauss-Pełczyński spaces*, to appear in Studia Mathematica.

[9] J. M. F. Castillo and J. Suárez, *Extending operators into Lindenstrauss spaces*, to appear.

[10] G. Godefroy, N. J. Kalton, and D. Li, *On subspaces of L^1 which embed into l_1*, J. Reine Angew. Math. **471** (1996), 43–75.

[11] D. B. Goodner, *Projections in normed linear spaces*, Trans. Amer. Math. Soc. **69** (1950), 89–108.

[12] W. B. Johnson and M. Zippin, *Extension of operators from weak*-closed subspaces of l_1 into $C(K)$ spaces*, Studia Math. **117** (1995), 43–55.

[13] M. I. Kadets and A. Pełczyński, *Bases, lacunary sequences and complemented subspaces in the spaces L_p*, Studia Math. **21** (1961/1962), 161–176.

[14] N. J. Kalton, *On subspaces of c_0 and extension of operators into $C(K)$-spaces*, Q. J. Math. **52** (2001), 313–328.

[15] ———, *Extension of linear operators and Lipschitz maps into $C(K)$-spaces*, to appear.

[16] ———, *Automorphisms of $C(K)$-spaces and extension of linear operators*, to appear.

[17] J. L. Kelley, *Banach spaces with the extension property*, Trans. Amer. Math. Soc. **72** (1952), 323–326.

[18] J. Lindenstrauss, *Extension of compact operators*, Mem. Amer. Math. Soc. No. **48** (1964), 112.

[19] J. Lindenstrauss and A. Pełczyński, *Contributions to the theory of the classical Banach spaces*, J. Functional Analysis **8** (1971), 225–249.

[20] J. Lindenstrauss and H. P. Rosenthal, *Automorphisms in c_0, l_1 and m*, Israel J. Math. **7** (1969), 227–239.

[21] A. A. Miljutin, *Isomorphism of the spaces of continuous functions over compact sets of the cardinality of the continuum*, Teor. Funkciĭ Funkcional. Anal. i Priložen. Vyp. **2** (1966), 150–156. (1 foldout). (Russian)

[22] L. Nachbin, *On the Hahn-Banach theorem*, Anais Acad. Brasil. Ci. **21** (1949), 151–154.

[23] A. Sobczyk, *Projection of the space (m) on its subspace (c_0)*, Bull. Amer. Math. Soc. **47** (1941), 938–947.

[24] D. M. Speegle, *Banach spaces failing the almost isometric universal extension property*, Proc. Amer. Math. Soc. **126** (1998), 3633–3637.

[25] M. Zippin, *The separable extension problem*, Israel J. Math. **26** (1977), 372–387.

[26] ———, *A global approach to certain operator extension problems*, Functional Analysis (Austin, TX, 1987/1989), LNM, vol. 1470, Springer, Berlin, pp. 78–84.

[27] ———, *Extension of bounded linear operators*, Handbook of the Geometry of Banach Spaces, Vol. 2, North-Holland, Amsterdam, 2003, pp. 1703–1741.

DEPARTMENT OF MATHEMATICS, UNIVERSITY OF MISSOURI-COLUMBIA, COLUMBIA, MO 65211.

E-mail address: nigel@math.missouri.edu

PALAMODOV'S QUESTIONS FROM HOMOLOGICAL METHODS IN THE THEORY OF LOCALLY CONVEX SPACES

JOCHEN WENGENROTH

ABSTRACT. In his seminal work from 1971, V. P. Palamodov introduced methods from category theory and homological algebra to the theory of locally convex spaces. These methods shed new light on many classical topics and led to many applications in analysis, e.g. for partial differential operators. The final section of Palamodov's article posed eight open problems. We will try to explain the motivation for these questions as well as their solutions which show that Palamodov's problems had been a bit too optimistic.

1. THE EMPEROR'S NEW CLOTHES?

Usually, homological methods do not solve a problem at once, but they may tell rather precisely what has to be done for the solution. Once knowing the solution it is often possible to give a presentation which avoids the abstract homological methods and which may then look even ingenious. Perhaps, this is one of the reasons why seemingly many mathematicians dislike homological tools: it looks as if they were superfluous. Another reason certainly is that these general tools, which found applications in all parts of mathematics from algebra and topology to algebraic geometry as well as – and this is the concern of the present article – to functional analysis, have to be formulated in a very general abstract language. One might get the feeling that only trivialities can be true in such a generality, and like in Andersen's "The emperor's new clothes" one waits for a child telling that he is naked. Indeed, in his book on algebra [11, page 105], Serge Lang posed the exercise: "Take any book on homological algebra, and prove all the theorems without looking at the proofs given in that book".

We want to describe what homolgical methods can do for analytical problems by considering the list of questions posed by Palamodov [12, section 12] who was the first to use homological ideas systematically in connection with problems arising from partial differential equations. In order to make the presentation readable for analysts we try to avoid as much of the language from category theory as possible (although the resulting mixture of homological and functional analytical language is probably not very satisfying).

Before entering into our subject let us look at three fundamental results of complex analysis in one variable. It is almost a commonplace that Runge's approximation theorem, the Mittag-Leffler theorem about meromorphic functions with prescribed poles, and the surjectivity of the $\overline{\partial}$-operator on $\mathscr{C}^\infty(\Omega)$ are "essentially equivalent". As all true theorems are logical tautologies, this statement has no precise meaning. We suppose that on the one hand, it means that is more or less "easy" to deduce one result from either of the others and, on the other hand, that there is some kind of common source for the proofs of the results. And indeed, we shall see that there is one *particular property* of the space $\mathscr{H}(\Omega)$ of analytic functions which implies the results. We admit that such kind of "meta-result" would be a poor consequence of an elaborated theory. What makes it worthwhile are evaluable conditions for the "particular property" which can be applied to less classical situations. For instance, they can be translated into Hörmander's convexity condtions for supports and singular supports to characterize surjectivity of linear partial differential operators (with constant coefficients) on spaces of functions or distributions, see [6, 7].

2. Localizing and pasting

Perhaps one may say that category theory formalizes one of the most fundamental aspect of mathematics: to search for invariances and relations of objects in some enviroments. To do so one considers a class of objects X and for each pair of objects a set $\mathrm{Mor}(X,Y)$ of admissible morphisms of X to Y such that whenever f "deforms" X to Y and g "deforms" Y to Z one has a morphism gf of X to Z. Except for associativity of this "composition" the nature of the morphisms does not play any role for category theory, what counts are additional properties of the sets $\mathrm{Mor}(X,Y)$ for instance being abelian groups such that the composition distributes with the group operations.

Many familiar properties of functions between sets, say, can be formulated in category language (to give just one fancy example: a set X is empty if and only if $|\mathrm{Mor}(X,Y)| = 1$ for every set Y, in category terms X is called an initial object) and it is a matter of taste whether one finds it amusing or stupid to give such translations. Instead of developing category language and "applying" it to a particular case, we directly turn to the category $\mathscr{LCS}$ which is one of the most useful to deal with problems in analysis.

We consider locally convex spaces (l.c.s. for short) over a fixed field $\mathbb{K} \in \{\mathbb{R}, \mathbb{C}\}$ as objects and the sets $L(X,Y)$ of continuous linear operators from X to Y as morphisms. Even if one is primarily interested in Banach spaces this larger category quickly enters the game since it is indispensable to deal with weak and weak-$*$-topologies. The composition of morphisms is just the composition of operators, and with respect to pointwise addition, the sets of morphisms are abelian groups (of course, even vector spaces) with the zero operator as neutral element.

Virtually every "linear problem" in analysis can be stated as the question whether a particular operator is surjective and $\mathscr{LCS}$ is a good environment for this question. To get a taste of what is meant by this, remember that for

Fréchet spaces Y and Z the open mapping theorem tells us that $g : Y \to Z$ is surjective if g is continuous and open onto its dense range. To prove density one has to solve only very few equations $g(y) = z$ and for the requirement g being open onto its range one must prove "continuity properties" of $g(y)$.

Still this has nothing to do with homological methods which enter the game if one does not stick to a given problem whether a particular $g : Y \to Z$ is surjective but instead, one first studies a *model* for this problem. To visualize this strategy one does not only consider single operators but *diagrams* of operators, the most simple being $X \xrightarrow{f} Y \xrightarrow{g} Z$ (meaning that f and g are morphisms from X to Y and Y to Z, respectively).

This diagram is called a *complex* if $gf = 0$, or equivalently $\mathrm{Im}(f) \subseteq \mathrm{Kern}(g)$ where Im and Kern denote the image and kernel of an operator. A large part of homological algebra is concerned with the quotients $\mathrm{Kern}(g)/\mathrm{Im}(f)$, and for the moment we are interested in the case where this quotient is trivial, i.e. $\mathrm{Im}(f) = \mathrm{Kern}(g)$. Then the complex is called *acyclic* (at Y). In particular, this notion gives a very short way of stating surjectivity or injectivity of an operator by saying that $Y \to Z \to 0$ or $0 \to X \to Y$, respectively, is an acyclic complex.

Of course, what happened is a triviality, but the conciseness of the notion allows an easy formulation of a general principle, how to prove surjectivity of a given map: Assume the following diagram of linear spaces and maps to be commutative such that each row and each column is an acyclic complex.

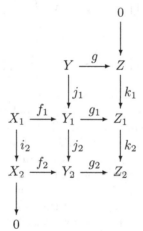

Then $\mathrm{Im}(k_1) \subseteq \mathrm{Im}(g_1)$ implies that g is surjective.

The proof is like a pin-ball machine where we put a ball $z \in Z$ in the upper-right corner, it flips through the diagram and comes out as some $y \in Y$ solving $g(y) = z$ (officially, this method is called diagram chase):

Since $\mathrm{Im}(k_1) \subseteq \mathrm{Im}(g_1)$ there is $y_1 \in Y_1$ with $g_1(y_1) = k_1(z)$. As $k_2 \circ k_1 = 0$ we obtain $g_2 \circ j_2(y_1) = k_2 \circ g_1(y_1) = 0$ from commutativity. Acyclicity of the lower row gives an $x_2 \in X_2$ with $f_2(x_2) = j_2(y_1)$, and by surjectivity of i_2 we find $x_1 \in X_1$ with $i_2(x_1) = x_2$. Commutativity then implies $j_2(y_1 - f_1(x_1)) = j_2(y_1) - f_2(i_2(x_1)) = 0$. Finally, acyclicity of the middle column yields $y \in Y$

with $j_1(y) = y_1 - f_1(x_1)$, and hence $k_1 \circ g(y) = g_1 \circ j_1(y) = g_1(y_1) - g_1 \circ f_1(x_1) = k_1(z)$ which implies $g(y) = z$ since k_1 is injective.

Diagrams as above appear naturally if one attemps to prove surjectivity of an operator by "localizing and pasting". Assume that Y and Z are spaces determined by countably many conditions requiring e.g. differentiability, integrability or other growth conditions of elements in Y and Z. We then describe these conditions by linear spaces Y_n such that they become stronger if n increases and we have natural *spectral maps* $\varrho_{n+1}^n : Y_{n+1} \to Y_n$ (which in many cases are inclusions or restrictions if Y_{n+1} describes conditions on larger sets). If we succeeded in describing all conditions for Y we can identify this space with $\{(y_n)_{n\in\mathbb{N}} \in \prod_{n\in\mathbb{N}} Y_n : \varrho_{n+1}^n(y_{n+1}) = y_n\}$. If we have a similar description for Z and an operator $g : Y \to Z$, it usually happens that g also acts on Y_n, i.e. we have operators $g_n : Y_n \to Z_n$ commuting with the spectral maps. Denoting by X_n and X the kernels of g_n and g, respectively, we obtain a commutative diagram

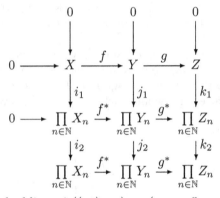

where j_1 is the embedding, $j_2((y_n)_{n\in\mathbb{N}}) = (y_n - \varrho_{n+1}^n y_{n+1})_{n\in\mathbb{N}}$ makes the middle column acyclic and i_1, i_2 and k_1, k_2 are defined in the same way.

To apply the pin-ball machine successfully we now need two things: First we have to solve locally, which means $\operatorname{Im}(k_1) \subseteq \operatorname{Im}(g^*)$ where $g^*((y_n)_{n\in\mathbb{N}}) = (g_n(y_n))_{n\in\mathbb{N}}$. This step is usually much easier than the original problem since z satisfies all conditions simultaneously and we need solutions which only satisfy finitely many. The second step then requires surjectivity of $i_2 : \prod_{n\in\mathbb{N}} X_n \to \prod_{n\in\mathbb{N}} X_n$, $(x_n)_{n\in\mathbb{N}} \mapsto (x_n - \varrho_{n+1}^n x_{n+1})$. If this is so we can "paste" the local solutions to a global solution. Palamodov described what we did here by a "functor" acting on the category of "countable spectra" (X_n, ϱ_{n+1}^n) as above and obtained surjectivity of i_2 as a characterization when the "derivative" Proj^1 of this functor vanishes.

We do not want to explain this right now, but to make these remarks comparable with Palamodov's work let us just define for $\mathscr{X} = (X_n, \varrho_{n+1}^n)$ the linear space $\operatorname{Proj}^1 \mathscr{X} = \prod_{n\in\mathbb{N}} X_n / \operatorname{Im}(i_2)$.

We want to emphasize that one should consider the calculation of $\operatorname{Proj}^1 \mathscr{X}$ as a *model* or a *test* for the original problem whether $g : Y \to Z$ is surjective. What both problems have in common is the kernel $X = \operatorname{Kern}(g) = \operatorname{Kern}(i_2)$

and the model problem is simpler since the operator i_2 as well as the spaces between which it acts are very simple. We have seen that $\text{Proj}^1 \mathscr{X} = 0$ is sufficient for the surjecitivity of g whenever we can solve the local problems. Looking at the transposed diagram above we get that $\text{Proj}^1 \mathscr{X} = 0$ is also necessary whenever $\text{Im}(f^*) \subseteq \text{Im}(j_2)$ which is always the case if $\text{Proj}^1 \mathscr{Y} = 0$ for $\mathscr{Y} = (Y_n, \varrho_{n+1}^n)$ holds.

To see a concret situation let us consider the $\bar{\partial}$-operator acting on $Y = \mathscr{C}^\infty(\Omega)$ for some open set $\Omega \subseteq \mathbb{C} \cong \mathbb{R}^2$. If $(\Omega_n)_{n \in \mathbb{N}}$ is an exhaustion of Ω by open and relatively compact sets $\Omega_n \supset \bar{\Omega}_{n-1}$ we consider $Y_n = \mathscr{C}^\infty(\Omega_n)$ and the restrictions ϱ_{n+1}^n. The local problems require solutions of $\bar{\partial} f = g$ only in Ω_n where f is in $\mathscr{C}^\infty(\Omega)$. These problems are easy to solve: We multiply with a $\mathscr{C}^\infty$-function having compact support and being equal to one in $\bar{\Omega}_n$ and form the convolution with the Cauchy-kernel $\frac{1}{\pi z}$. The kernels X and X_n become $X = \mathscr{H}(\Omega)$, the space of holomorphic functions on Ω, and $X_n = \mathscr{H}(\Omega_n)$, respectively, and it remains to show $\text{Proj}^1 \mathscr{X} = 0$.

Before considering this, let us look at the Mittag-Leffler theorem about the existence of meromorphic functions with prescribed principle parts. In this case we denote by Y_n the space of meromorphic functions with finitely many prescribed poles in Ω_n and by Z_n some countable dimensional space describing the principle parts. The local problems are then trivial and we obtain exactly the same kernels as for the $\bar{\partial}$-case.

The elementary algebraic considerations thus tell us that both results, surjectivily of $\bar{\partial}$ and the Mittag-Leffler theorem, will follow from $\text{Proj}^1 \mathscr{X} = 0$. Stated slightly differently, the following problem is a model for these results:

Given $f_n \in \mathscr{H}(\Omega_n)$ find $g_n \in \mathscr{H}(\Omega_n)$ with $f_n = g_n - g_{n+1}|_{\Omega_n}$ for all $n \in \mathbb{N}$.

The first idea to solve this is to sum up $g_n = \sum\limits_{k=n}^{\infty} f_k$ and to make these series convergent on needs Runges approximation theorem to achieve small summands. Knowing this classical device, the following theorem of Palamodov [12, §5] is not too surprising. We denote $\varrho_m^n = \varrho_{n+1}^n \circ \cdots \circ \varrho_m^{m-1}$ for $n < m$ and $\varrho_n^n = id_{X_n}$.

Theorem 1. *If X_n are Fréchet-spaces and $\varrho_{n+1}^n : X_{n+1} \to X_n$ are linear and continous then $\text{Proj}^1 \mathscr{X} = 0$ if and only if*

$$\forall \, n \in \mathbb{N}, \, U \in \mathscr{U}_0(X_n) \quad \exists \, m \geq n \quad \forall \, k \geq m \quad \varrho_m^n(X_m) \subseteq \varrho_k^n(X_k) + U,$$

i.e. the images $\varrho_k^n(X_k)$ remain dense in $\varrho_m^n(X_m)$ with respect to the seminorm having unit ball U.

Runge's approximation theorem implies the "Mittag-Leffler condition" of Theorem 1 and hence $\text{Proj}^1 \mathscr{H} = 0$ for the spectrum $\mathscr{H}$ consisting of the Fréchet space $\mathscr{H}(\Omega_n)$ and the restrictions (one has to choose $m \geq n$ in such a way that Ω_m contains all "holes" of Ω_n).

In the situation described above for the $\bar{\partial}$-operator Theorem 1 can be applied to prove $\text{Proj}^1 \mathscr{Y} = 0$ (one even has $\varrho_{n+1}^n(Y_{n+1}) \subseteq \varrho_k^n(Y_k)$ for all $k \geq n + 1$) and since $\text{Proj}^1 \mathscr{X} = 0$ is thus necessary for the surjecitivity of $\bar{\partial}$ the necessity part of Theorem 1 implies a rough version of Runge's theorem:

Every function which is holomorphic on a neighbourhood of some compact set can be approximated by functions which are holomorphic on much larger sets.

Theorem 1 is just the beginning of a very successful story about conditions ensuring $\operatorname{Proj}^1 \mathscr{X} = 0$ and thus giving tools to solve linear problems in analysis. We do not want to go further in this direction now (instead, we refer to Domanski's survey [1] about (PLS)-spaces and to [19, chapter 3] and the references therein), but we would like to emphasize that the abstract algebraic considerations were indispensable for the development of these tools.

$$\boxed{\text{Question 2}}$$

It is high time to remember the title of this article and to come to one of Palamodov's question. If $\operatorname{Proj}^1 \mathscr{X} = 0$ holds for a spectrum of Fréchet spaces and continuous linear spectral maps then the "test operator" $i_2 : \prod_{n\in\mathbb{N}} X_n \to \prod_{n\in\mathbb{N}} X_n$ is a continous surjection between Fréchet spaces and hence open by the open mapping theorem.

Although stated differently, Palamodov asked whether this is always true if X_n are replaced by locally convex spaces. If this were true, one would get from the abstract machinery not only solvability of a linear problem but also some "continuity estimates" for the solutions.

Since the open mapping theorem holds for very large classes of l.c.s. it is not surprising that the answer is yes, if X_n satisfy a suitable condition (having "strict ordered webs"), see [19, theorem 3.3.3]. But since the open mapping principle may fail, it is neither surprising that the answer in general is no: Playing with the finest locally convex topology and arguing with Hamel bases an example is constructed in [19, 3.3.2] destroying Palamodov's hope.

3. DERIVED FUNCTORS

Every mathematical object X lives in a natural environment which should be described as a category where the morphisms describe the invariances of the object. Then every reasonable operation on X should also respect its environment, i.e. it should transform also the morphisms of the category. Thus, a *functor* is a rule (or a mapping – although there are some set theoretical problems which arise from the fact that usually the objects of a category are too many to form a set) which transforms objects to objects and morphisms to morphisms and distributes with the composition of morphisms.

A functor between two categories where the sets of morphisms are abelian groups is said to be *additive* if it respects these group structures.

Let us consider two examples which are important with respect to Palamodov's work. For the first one we consider the category whose objects are projective spectra $\mathscr{X} = (X_n, \varrho_{n+1}^n)$ of vector spaces, and morphisms $f : \mathscr{X} \to \mathscr{Y}$ consist of sequences $f_n : X_n \to Y_n$ of linear maps with $f_n \circ \varrho_{n+1}^n = \sigma_{n+1}^n \circ f_{n+1}$. Forming the projective limit $\operatorname{Proj}\mathscr{X} = \{(x_n)_{n\in\mathbb{N}} \in \prod_{n\in\mathbb{N}} X_n : \varrho_{n=1}^n x_{n+1} = x_n\}$

with $\mathrm{Proj}(f)((x_n)_{n\in\mathbb{N}}) = (f_n x_n)_{n\in\mathbb{N}}$ is then an additive functor from the category of projective spectra to the category of vector spaces. As explained in the second section this functor is used to solve problems by localization and pasting.

For the second example we fix any category and an object E and transform objects X to the sets $\mathrm{Mor}(E, X)$ of morphisms from E to X and morphisms $f : X \to Y$ to the mapping $f^* : \mathrm{Mor}(E, X) \to \mathrm{Mor}(E, Y)$ defined by composition $g \mapsto fg$.

Using this functor one can, for instance, describe the existence of a solution operator for a continuous linear operator $T \in L(Y, Z)$. T has a continuous linear right inverse $R \in L(Z, Y)$, which means $T \circ R = id_Z$, if and only if id_Z belongs to the range of $T^* : L(Z, Y) \to L(Z, Z)$ (then T^* is even surjective). Thus, considering the functor $L(Z, \cdot)$ allows a formulation of the original question as an acyclicity problem, namely whether

$$0 \to L(Z, X) \xrightarrow{J^*} L(Z, Y) \xrightarrow{T^*} L(Z, Z) \to 0$$

is acyclic, where $J : X \to Y$ is the embedding of $X = \mathrm{Kern}(T)$ into Y.

As in section 2, homological algebra enters the game if one does not apply the functor $L(Z, \cdot)$ to the complex $0 \to X \xrightarrow{J} Y \xrightarrow{T} Z \to 0$ (which is the ultimate goal) but instead to a "test complex" $0 \to X \to \cdots$ with simpler spaces and operators.

There is always "canonical" method to construct a test complex $0 \to X \to \cdots$ which uses the notion of *injective objects*.

An object I in $\mathscr{LCS}$ is called injective if it has the following extension property: Whenever $f : X \to I$ is an operator and X is a topological subspace of some Y there is an extension $g : Y \to I$ with $g|_X = f$.

If in our original complex $0 \to X \xrightarrow{J} Y \xrightarrow{T} Z \to 0$ the kernel X were injective we could extend $J^{-1} : J(X) \to X$ to $g : Y \to X$ and would obtain the desired right inverse by the formula $R(T(y)) = y - J(g(y))$.

The first example of an injective object is supplied by the Hahn-Banach theorem: The field $\mathbb{K}$ is injective. Knowing this, one also gets the injectivity of the Banach spaces spaces $\ell_\infty(M)$ of bounded functions on sets M. Since moreover, products of injective objects are easily seen to be injective and finally, every vector space V with the trivial topology $\{\emptyset, V\}$ is injective one arrives at the fundamental result that every l.c.s. X is isomorphic to a subspace of an injective one, hence there is an exact complex $(*)$ $0 \to X \to I \to Q \to 0$ where Q is the quotient I/X. This complex can always be used as a test for the original problem, and the *first derivative* $\mathscr{F}^1(X) = \mathscr{F}(Q)/\mathscr{F}(I)$ of an additive functor $\mathscr{F}$ from $\mathscr{LCS}$ to the category of abelian groups describes the "success" of the test, i.e. the acyclicity of the complex $(+)$ $0 \to \mathscr{F}(X) \to \mathscr{F}(I) \to \mathscr{F}(Q) \to 0$.

As stated, this simple definition (which differs slightly from the usual one which uses so-called injective resolutions) depends on the injective object I. However, it is not too hard to prove that that the quotient groups depend on I in an isomorphic way.

Although we are mainly interested in the first derivative, let us define the higher derivatives $\mathscr{F}^{k+1}(X) = \mathscr{F}^k(Q)$.

The following fundamental theorem of Palamodov (see [12] or [19]) shows that our test indeed gives information about the action of the functor on general complexes. We give the formulation only for *injective* functors from $\mathscr{LCS}$ to the abelian groups which are defined to be additive such that $(+)$ is acyclic at $\mathscr{F}(X)$.

Theorem 2. *Let $\mathscr{F}$ be an injective functor on $\mathscr{LCS}$ and $0 \to X \xrightarrow{J} Y \xrightarrow{T} Z \to 0$ an* exact *complex, this means that it is acyclic and that J and T are not only continuous but also open onto the range. Then there is an acyclic complex*

$$0 \to \mathscr{F}(X) \xrightarrow{J^*} \mathscr{F}(Y) \xrightarrow{T^*} \mathscr{F}(Z) \to \mathscr{F}^1(X) \to \mathscr{F}^1(Y) \to \mathscr{F}^1(Z) \to \mathscr{F}^2(X) \to \cdots$$

In particular we get that $0 \to \mathscr{F}(X) \to \mathscr{F}(Y) \to \mathscr{F}(Z) \to 0$ is always acyclic if $\mathscr{F}^1(X) = 0$. If $\mathscr{F}^1(Y) = 0$ this condition is also necessary.

Let us now see what this theorem means for the functor $L(E, \cdot)$ which we want to use to find solution operators. The derivatives of $L(E, \cdot)$ are usually denoted by $\operatorname{Ext}^k(E, \cdot)$:

Corollary 3. *1. For two locally convex spaces E and X we have $\operatorname{Ext}^1(E, X) = 0$ if and only if for every exact complex $0 \to X \xrightarrow{J} Y \xrightarrow{T} E \to 0$ there exists a right inverse for T.*

2. $\operatorname{Ext}^{k+1}(E, X) = 0$ holds if and only if the "k-th acyclicity class" $\{Y \in \mathscr{LCS} : \operatorname{Ext}^k(E, Y) = 0\}$ is stable with respect to forming quotients by subspaces isomorphic to X.

The first part explains the notion $\operatorname{Ext}^1(E, \cdot)$ for the derivative of $L(E, \cdot)$: An exact complex $0 \to X \xrightarrow{J} Y \xrightarrow{T} E \to 0$ is called an *extension* of (E, X) and $\operatorname{Ext}^1(E, X) = 0$ means that "up to isomorphism of complexes" there is only the trivial one with $Y = E \times X$.

To apply the corollary in concrete cases one must be able to compute $\operatorname{Ext}^1(E, X)$, and typically one of the difficulties is to find a good description for the quotient $Q = I/X$.

For nuclear Fréchet spaces X this problem can be solved with the aid of Theorem 1. In this case there is always a representation $X = \operatorname{Proj}\mathscr{X}$ with a projective spectrum $\mathscr{X} = (X_n, \varrho_{n+1}^n)$ of Banach spaces X_n isomorphic to the injective space $\ell_\infty = \ell_\infty(\mathbb{N})$ and nuclear spectral maps ϱ_{n+1}^n such that $\operatorname{Proj}^1 \mathscr{X} = 0$ holds. Hence, $0 \to X \to \prod_{n\in\mathbb{N}} X_n \xrightarrow{i_2} \prod_{n\in\mathbb{N}} X_n \to 0$ is an exact complex (with very simple quotient Q) permitted for the definition of the derived functors.

Applying $L(E, \cdot)$ yields $0 \to L(E, X) \to \prod L(E, X_n) \to \prod L(E, X_n)$ and therefore the calculation of $\operatorname{Ext}^1(E, X)$ is again a question about Proj^1.

If E is such that $L(E, X_n)$ have canonical Fréchet topologies (which is the case if E is a so-called (DF)-space and in particular if E is a normed space or the strong dual of a Fréchet space) one can solve the Proj^1-question once more with Palamodov's Theorem 1. Using nuclearity it is not too hard to prove the Mittag-Leffler condition in Theorem 1 (see [19, lemma 5.6.1]) and one can thus conclude $\mathrm{Ext}^1(E, X) \cong \mathrm{Proj}^1 L(E, X_n) = 0$.

The same holds true if X is any Fréchet space and instead E is a nuclear (DF)-space (instead of the definition of the first derivative one then has to use Theorem 2). We thus arrive at the following result of Palamodov [12, theorem 9.1]:

Corollary 4. *Let E be (DF)-space and X a Fréchet space such that one of them is nuclear. Then $\mathrm{Ext}^k(E, X) = 0$ for all $k \in \mathbb{N}$.*

As an example, we consider a surjective operator $T : Y \to Z$ between Fréchet spaces and the induced opertator $\tilde{T} : \mathscr{C}^\infty(\Omega, Y) \to \mathscr{C}^\infty(\Omega, Z)$ between the corresponding spaces of infinitely differentiable vector valued function on an open subset of $\Omega \subseteq \mathbb{R}^n$. Using Schwartz's ε-products we have $\mathscr{C}^\infty(\Omega, E) = \mathscr{C}^\infty(\Omega)\varepsilon E = L(\mathscr{C}^\infty(\Omega)'_\beta, E)$ where $\mathscr{C}^\infty(\Omega)'_\beta$ is a nuclear (DF)-space. Since $\mathrm{Ext}^1(\mathscr{C}^\infty(\Omega)'_\beta, X) = 0$ for $X = \mathrm{Kern}(T)$ we conclude that $\tilde{T}$ is again surjective.

One could attack this result more directly and would probably use certain approximations and corrections which are behind the Mittag-Leffler condition in Theorem 1 in a more concrete form. But then analytical aspects (which yield nuclearity of $\mathscr{C}^\infty(\Omega)$) and topological consideration would be mixed up which makes the proof much less transparent than the use of Corollary 4.

Corollary 4 and its little application are again only the beginning of a long story about the functors $\mathrm{Ext}^k(E, \cdot)$. In particular, there is a complete characterization of $\mathrm{Ext}^1(E, X) = 0$ if E and X are both Fréchet spaces and one of them is nuclear or a Köthe sequence space (this is due to Vogt [15] and Frerick [4] and Frerick and the author [5]), and there are very strong results of Domański and Vogt [3] and of the present author in [18] about $\mathrm{Ext}^1(E, X)$ for subspaces and quotients of the space of distributions $\mathscr{D}'$.

Comparing the applications of those results to distributional complexes in [3] with a more classical approach of Palamodov [13] using Fourier analysis shows rather impressivly the power of functional analytic techniques inspired by homological ideas.

We have now enough information to deal with several of Palamodov's questions related to derived functors.

$$\boxed{\text{Question 6}}$$

We have seen that every nuclear Fréchet space X admits an exact complex $0 \to X \to \ell_\infty^\mathbb{N} \to \ell_\infty^\mathbb{N} \to 0$ which yields $\mathscr{F}^k(X) = 0$ for $k \geq 2$ and every injecitve functor $\mathscr{F}$, in particular $\mathrm{Ext}^k(E, X) = 0$ for $k \geq 2$ and every l.c.s. E. Question 6 asks whether this is true for every Fréchet space X. Using only classical facts from Banach space theory (which can be found e.g. in [21]) we will now show that there are even Banach spaces E and X for which this fails.

Using Khinchin's inequality, we can find the Hilbert space ℓ_2 as a topological subspace of $L_1([0,1])$ and thus obtain an exact complex $0 \to \ell_2 \xrightarrow{i} L_1([0,1]) \xrightarrow{q} E \to 0$. Taking duals we get an exact complex $0 \to X \to \ell_\infty \to \ell_2 \to 0$ with $X = E'$. Since ℓ_∞ is injective we have $\mathrm{Ext}^1(E, \ell_\infty) = 0$ and assuming $\mathrm{Ext}^2(E, X) = 0$ we obtain $\mathrm{Ext}^1(E, \ell_2) = 0$ from the second part of corollary 3. Whence its first part gives a right inverse $R : E \to L_1([0,1])$ for q above which then yields a left inverse $T : L_1([0,1]) \to \ell_2$ for the embedding $i : \ell_2 \to L_1([0,1])$, i.e. $T \circ i = id_{\ell_2}$. A theorem of Grothendieck (see e.g. [21, III.F7]) implies that T is absolutely summing and then id_{ℓ_2} would be absolutely summing, too. This contradiction shows $\mathrm{Ext}^2(E, X) \neq 0$.

Question 7

In connection with corollary 4 Palamodov writes "it is natural to expect that the following proposition 'dual' to [corollary 4] is valid: $\mathrm{Ext}^i(E, X) = 0, i \geq 1$, if E is metric, X is a complete dual metric space, and one of them is a nuclear space". It follows form results of Grothendieck [9, chap. 2, 10.3] that this is true if X is Banach and E is any nuclear space.

This question is solved under the assumption of the continuum hypothesis in [20]: For every infinite-dimensional nuclear (DF)-space X we have:

(1) $\mathrm{Ext}^k(E, X) = 0$ for $k \geq 3$ and every l.c.s. E.
(2) $\mathrm{Ext}^1(\mathbb{K}^\mathbb{N}, X) = 0$ and $\mathrm{Ext}^2(\mathbb{K}^\mathbb{N}, X) \neq 0$.
(3) There is a normed space E with $\mathrm{Ext}^1(E, X) \neq 0$.

Although this answers Palamodov's question as stated, it remains an open problem whether for instance $\mathrm{Ext}^1(s, s') = 0$ where s is the nuclear Fréchet space of all rapidly decreasing scalar sequences.

Question 8

The methods used to solve the conjecture above also led to an answer to a question about the completeness of quotient spaces: If $\mathscr{C}(X)$ denotes to vector space which is obtained as the Hausdorff completion of a l.c.s. X we get an injective functor from $\mathscr{LCS}$ to the category of vector spaces. Then $\mathscr{C}^1(X) = 0$ means that Y/X is complete for every complete l.c.s. Y containing X as a subspace. Except for the formulation it is a classical result that $\mathscr{C}^1(X) = 0$ holds for every metrizable l.c.s. X [12, proposition 10.2], and Palamodov asked whether this is still true for the space $X = \mathscr{D}'$ of distributions or for the space φ of all finite scalar sequences endowed with the finest locally convex topology. The methods in [20] yield that (under the continuum hypothesis) $\mathscr{C}^1(X) \neq 0$ for each l.c.s. X containing a complemented copy of some infinite dimensional nuclear (DF)-space (the case $X = \varphi$ was already solved by F.C. Schmerbeck [14]).

Question 5

Yet another problem solved by the methods in [20] is concerned with du-ality, namely whether for an exact complex $0 \to E \xrightarrow{j} F \xrightarrow{q} G \to 0$ in $\mathscr{LCS}$ the complex of the strong duals $0 \to G'_\beta \xrightarrow{q^t} F'_\beta \xrightarrow{j^t} E'_\beta \to 0$ is exact (the Hahn-Banach theorem implies that it is acyclic). Instead of repeating Palam-odov's construction (which is not as smooth as the one leading to Ext^k and Proj^k) we define $D^1(E) = 0$ and $D^+(E) = 0$ if for every exact complex $0 \to E \xrightarrow{j} F \xrightarrow{q} G \to 0$ the dual maps q^t or j^t, respectively, are open onto their range. Based on results of Grothendieck Palamodov [12, §8] proved that for a Fréchet space E we have $D^1(E) = 0$ if and only if E is quasinormable, i.e.

$$\forall\, U \in \mathscr{U}_0(E) \quad \exists\, V \in \mathscr{U}_0(E) \quad \forall\, \varepsilon > 0 \quad \exists\, B \subseteq E \text{ bounded} \quad V \subseteq \varepsilon U + B$$

whereas $D^+(E) = 0$ if and only if E'_β is barrelled. Moreover $D^1(E) = 0$ implies $D^+(E) = 0$.

Motivated by these results (and perhaps a theorem of Schwartz which im-plies $D^+(E) = 0$ for each complete quasinormable space where all bounded sets are relatively compact – according to Grothendieck such spaces are called Schwartz spaces), Palamodov asked whether $D^1(E) = 0$ and $D^+(E) = 0$ for every quasinormable l.c.s. The answer to these questions is again negative [19, p. 125], and under the continuum hypothesis we even have $D^1(E) \neq 0$ for every infinite-dimensional nuclear (DF)-space E.

$$\boxed{\text{Question 1}}$$

We finish this section with the Palamodov's first question which refers to the possibility whether one can construct a "dual" theory of derived functors where one investigates the action of a functor on a complex $0 \to X \to Y \to Z \to 0$ not by keeping X and replacing Y and Z by simpler spaces but by keeping Z and replacing X and Y. The notion which would allow such a construction is that of a projective object P which means that whenever $T : P \to Z$ is an operator into a quotient Z of some l.c.s. Y with quotient map $q : Y \to Z$ then there is a "lifting" $S : P \to Y$ with $q \circ S = T$. In the category of Banach spaces the space ℓ_1 of absolutely summable sequences is projective (if e_n are the unit vectors and y_n is a bounded sequence in Y with $q(y_n) = T(e_n)$ one can just define S by $S(e_n) = y_n$). Palamodov suspected that there are very few projective objects in $\mathscr{LCS}$ and this conjecture was confirmed by V.A. Geĭler [8] who showed that projective objects in $\mathscr{LCS}$ necessarily carry the finest locally convex topology.

On the other hand, Domański, Krone, and Vogt [2] gave a kind of sub-stitute for the non-existent projetive resolutions in the category of (nuclear) Fréchet spaces. This approach gives a somehow positive answer to Palam-odov's question and allows to carry out similar constructions as in categories having many projective objects.

4. INDUCTIVE LIMITS

Inductive spectra are in a sense dual to projective ones. We have a sequence of l.c.s. X_n and continuous linear mappings $i_{n,n+1} : X_n \to X_{n+1}$ which are assumed to be injective (thus we have $X_n \subseteq X_{n+1}$). Then $X = \text{ind } X_n = \bigcup_{n \in \mathbb{N}} X_n$ is a vector space which we endow with the finest locally convex topology such that all inclusions $X_n \hookrightarrow X$ are continuous. A seminorm p on X is thus continuous with respect to that topology if and only if all its restrictions $p|_{X_n}$ are continuous on X_n. X is always a quotient of the locally convex direct sum $\bigoplus_{n \in \mathbb{N}} X_n = \text{ind } \prod_{k \leq n} X_k \times \prod_{k > n} \{0\}$ where the quotient map is just the sum $\sigma((x_n)_{n \in \mathbb{N}}) = \sum_{n \in \mathbb{N}} x_n$. This observation yields an acyclic complex $(*)$ $0 \to \bigoplus X_n \xrightarrow{d} \bigoplus X_n \xrightarrow{\sigma} X \to 0$ where d is defined by $(x_n)_{n \in \mathbb{N}} \mapsto (x_n - x_{n-1})_{n \in \mathbb{N}}$ with $x_0 = 0$.

Palamodov defined an inductive spectrum $(X_n)_{n \in \mathbb{N}}$ to be *(weakly) acyclic* if the complex above is exact (exact if all spaces are endowed with their weak topologies), i.e. d is a (weak) isomorphism onto its range.

This notion is suitable to decide which subspaces L of X are limit subspaces (or well-located) which means that the relative topology coincides with the inductive limit topology of $\text{ind } L \cap X_n$ (or if they have at least the same continuous linear functionals). In fact a diagram chase as described in section 2 yields:

> If $(X_n)_{n \in \mathbb{N}}$ is (weakly) acyclic then L is a limit subspace (well-located) if and only if $(X_n/L \cap X_n)_{n \in \mathbb{N}}$ is (weakly) acyclic.

This means again that the complex $(*)$ for the spectrum $(X_n/L \cap X_n)_{n \in \mathbb{N}}$ is a test for the exactness of $0 \to \text{ind } L \cap X_n \to X \to X/L \to 0$.

The typical example of an acyclic inductive limit is $X = \mathscr{D}(\Omega)$ the space of $\mathscr{C}^\infty$-functions on an open set $\Omega \subseteq \mathbb{R}^N$ with compact support. Then a continuous linear operator $T : \mathscr{D}'(\Omega) \to \mathscr{D}'(\Omega)$ on the space of distributions is surjective if and only if the kernel of the transposed operator is a well-located subspace of $\mathscr{D}(\Omega)$. To apply this one needs evaluable conditions to ensure (weak) acyclicity and there is indeed a very general result [17], [19, theorem 6.1]:

Theorem 5. *1. An inductive spectrum $(X_n)_{n \in \mathbb{N}}$ is acyclic if for every $n \in \mathbb{N}$ there is $m \geq n$ such that for every $k \geq m$ the topologies of X_m and X_k coincide on some 0-neighbourhood of X_n.*

2. If all X_n are metrizable this condition is also necessary for acyclicity.

Question 3 and 4

Palamodov's questions about acyclic spectra ask for necessary conditions. Inspired by his result that inductive limits of acyclic spectra of Fréchet spaces are complete and regular (i.e. every bounded subset of $\text{ind} X_n$ is contained and bounded in some "step" X_n) he asked whether this is true for complete l.c.s. X_n instead of Fréchet spaces. We do not know the answer in general. However, if we define $X = \text{ind}X_n$ to have local closed neighbourhoods if

$$\forall\, n \in \mathbb{N} \quad \exists\, m \geq n \quad \forall\, U \in \mathscr{U}_0(X_m) \quad \exists\, V \in \mathscr{U}_0(X_n) \quad \overline{V}^X \cap X_n \subseteq U$$

(this notion had been introduced by D. Vogt [16], and it is satisfied by all spectra verifying the condition in Theorem 5) we have [19, theorem 6.10]:

Theorem 6. *The inductive limit of an acyclic spectrum of complete separated l.c.s. with local closed neighbourhoods is complete and regular.*

Let us finally mention that there is no hope for a positive answer if one only assumes weak acyclicity instead of acyclicity. In fact, using Palamodov's characterization [12, theorem 6.3] one can check that G. Köthe's example [10, §31.6] of a non-regular and non-complete inductive limit of Banach spaces is indeed weakly acyclic.

REFERENCES

[1] P. Domański: *Classical PLS-spaces: spaces of distributions, real analytic functions and their relatives*, Orlicz centenary volume, 51–70, Banach Center Publ., 64, Polish Acad. Sci., Warsaw, 2004.

[2] P. Domański, J. Krone, D. Vogt: *Standard exact projective resolutions relative to a countable class of Fréchet spaces*, Studia Math. **123** (1997), 275-290.

[3] P. Domański, D. Vogt: *Distributional complexes split for positive dimensions*, J. Reine Angew. Math. **522** (2000), 63 - 79.

[4] L. Frerick: *A splitting theorem for nuclear Fréchet spaces*, pp 163-167 in: Functional Analysis, Proceedings Int. Workshop Trier (S. Dierolf, S. Dineen, P. Domański, eds.), de Gruyter, Berlin, 1996.

[5] L. Frerick, J. Wengenroth: *A sufficient condition for the vanishing of the derived projective limit functor*, Arch. Math. (Basel) **67** (1996), 296-301.

[6] L. Frerick, J. Wengenroth: *Surjective convolution operators on spaces of distributions*. Rev. R. Acad. Cien. Seria A. Mat. **97** (2003), 263-272.

[7] L. Frerick, J. Wengenroth: *Convolution equations for ultradifferentiable functions and ultradistributions*. J. Math. Anal. Appl. **297** (2004), 506-517.

[8] V.A. Geĭler: *The projective objects in the category of locally convex spaces*, Funkcional. Anal. i Priložen **6** (1972), 79-80.

[9] A. Grothendieck: *Produits Tensoriels Topologiques et Espaces Nucléaires*, Mem. Amer. Math. Soc. **16**, 1955.

[10] G. Köthe: *Topological Vector Spaces I*, Springer, New-York, 1969.

[11] S. Lang: *Algebra*, Addison-Wesley, Reading, 1965.

[12] V.P. Palamodov: *Homological methods in the theory of locally convex spaces*. Uspekhi Mat. Nauk **26**, 3 - 65 (1971) (in Russian); English transl.: Russian Math. Surveys **26**, 1-64 (1971)

[13] V.P. Palamodov: *A criterion for splitness of differential complexes with constant coefficients*, in: Geometrical and algebraical aspects in several complex variables, C.A. Berenstein, D. Struppa (eds), EditEl, Rende, 1991, 265–291.

[14] F.C. Schmerbeck: *Über diejenigen topologischen Vektorräume N, für die X/N bei vollständigem topologischen Vektorraum X stets vollständig ausfällt*, Dissertation Ludwig-Maximilians-Univ., München, 1986.

[15] D. Vogt: *On the functors* $\text{Ext}^1(E, F)$ *for Fréchet spaces*, Studia Math. **85** (1987), 163-197.

[16] D. Vogt: *Regularity properties of (LF)-spaces*, in: Progress in Functional Analysis, North-Holland, Amsterdam,1992, 57-84.

[17] J. Wengenroth: *Acyclic inductive spectra of Fréchet spaces*, Studia Math. **120**, 247-258 (1996)

[18] J. Wengenroth: *A splitting theorem for subspaces and quotients of $\mathscr{D}'$*, Bull. Polish Acad. Sci. Math. **49** (2001), 349-354.

[19] J. Wengenroth: *Derived Functors in Functional Analysis*, Lecture Notes in Mathematics 1850, Springer, Berlin, 2003.

[20] J. Wengenroth: *A conjecture of Palamodov about the functors Ext^k in the category of locally convex spaces*, J. Func. Anal. **201** (2003), 561-571.

[21] P. Wojtaszczyk: *Banach Spaces for Analysts*, Cambridge University Press, Cambridge, 1991.

INSTITUT DE MATHÉMATIQUE, GRANDE TRAVERSE, 12, BÂTIMENT B37, B-4000 LIÈGE (SART-TILMAN), BELGIUM

E-mail address: J.Wengenroth@ulg.ac.be

ORDINAL REPRESENTABILITY IN BANACH SPACES

M. J. CAMPIÓN, J. C. CANDEAL, A. S. GRANERO, AND E. INDURÁIN

ABSTRACT. The objective of this note is to investigate the role of ordinal representability in the theory of Banach spaces. Necessary and sufficient conditions have been achieved for the norm or weak topologies of a Banach space to have the continuous (respectively, semicontinuous) representability property.

1. Introduction

In the mathematical theory of *ordered structures*, a topology τ defined on a nonempty set X is said to satisfy the *continuous representability property* (CRP) if every continuous total preorder $\precsim$ defined on X admits a numerical representation by means of a continuous real-valued isotony. (Topologies satisfying CRP have been also called *"useful topologies"* in the literature, see Herden [1991]).

Examples of topologies satisfying CRP are the connected and separable (see Eilenberg [1941]) and the second countable ones (see Debreu [1954]). Also, on metrizable spaces, the topology (metric) satisfies CRP if and only if it is separable. (See Candeal et al. [1998]). On non-metrizable spaces the analogous property is no longer true: For instance, the Hilbert space $\ell_2(\mathbb{R})$ endowed with the weak topology ω is non-separable but satisfies CRP, as we shall prove later.

The structure of topologies satisfying CRP was fully stated in Herden and Pallack [2000]. However, the conditions that characterize CRP topologies are difficult to check, so that this suggests *to analyze directly the satisfaction of CRP in some important cathegories of topological spaces.*

Continuing a way initiated in Candeal et al. [1999], in the present paper we pay attention to the main topologies in *Banach spaces,* so that given a Banach space X, we shall study the ordinal representability of the norm and weak topologies on X and the norm, weak and weak-star topologies on its dual X^*. Hopefully, if X or X^* satisfy good properties related to its ordinal substructures, they should also have some extra good additional properties

1991 *Mathematics Subject Classification.* (2000) Primary: 46 B 40. Secondary: 06 A 06, 46 B 25, 54 F 05.

Key words and phrases. Banach spaces, weak topologies, total preorders, continuous and semicontinuous isotonies.

related with their own Banach space structures, so that we can think about the possibility of characterizing classical properties of Banach spaces (e.g. separability) by means of properties related with *order*.

It is known (see Monteiro [1987]) that *the weak-star topology of a dual Banach space satisfies CRP*. In the present paper we will complete the panorama by proving also that *the weak topology of a Banach space also satisfies CRP*. These facts are important not only from the point of view of Functional Analysis, but also because they carry powerful consequences in applications. For instance, in Economic Theory, when dealing with problems of General Equilibrium related to "large economies", several models that strongly depend on the structure of a Banach space are encountered.

In those models, "preferences" (usually understood as continuous total preorders) are defined, and continuous "utility functions", or in other words, numerical representations by means of continuous real-valued isotonies, are required to prove the existence of an equilibrium. In that framework it is usual to deal with *nonseparable Banach spaces* (see e.g. Mas-Colell [1986]) where weak and weak-star topologies are more adequate because the metric (norm) topology does not satisfy CRP.

Another classical property in the topological theory of ordered structures is the *semicontinuous representability property* (SRP). A topology τ defined on a nonempty set X is said to satisfy SRP if every upper (respectively lower) semicontinuous total preorder $\precsim$ defined on X admits a numerical representation by means of an upper (respectively lower) semicontinuous real-valued isotony. Topologies satisfying SRP have been also called *"completely useful topologies"*. SRP implies CRP but the converse is not true in general. (See Bosi and Herden [2002]).

It is known that some topological conditions, as the satisfaction of the second axiom of countability (see Isler [1997]), are sufficient for a given topology to satisfy SRP. The main structure of topologies accomplishing SRP has also been established in Bosi and Herden [2002], but as in the case of CRP, the conditions that characterize SRP topologies are not easy to check.

In what concerns Banach spaces, matching some well-known results it is straightforward to see that *the norm topology of a Banach space satisfies SRP if and only if it satisfies CRP*. Thus, it remains to characterize the satisfaction of SRP by the weak topology of a Banach space X and the weak-star topology of a dual Banach space X^*.

In the present paper we fill these gaps, showing that *the weak topology of a Banach space X satisfies SRP if and only if it is separable* and, finally, *the weak-star topology of a dual Banach space X^* satisfies SRP if and only if the predual X is separable in norm*.

2. Previous concepts and results

Let X be a nonempty set. Let $\precsim$ be a *total preorder* (i.e.: a reflexive, transitive and complete binary relation) on X. (If $\precsim$ is also antisymmetric, it is said to be a *total order*). We denote $x \prec y$ instead of $\neg(y \precsim x)$. Also $x \sim y$ will stand for $(x \precsim y) \wedge (y \precsim x)$ for every $x, y \in X$.

The total preorder $\precsim$ is said to be *representable* if there exists a real-valued order-preserving isotony $f : X \longrightarrow \mathbb{R}$. Thus $x \precsim y \iff f(x) \leq f(y)$ $(x, y \in X)$. This fact is characterized (see e.g. Bridges and Mehta [1995], p. 23) by equivalent conditions of *"order-separability"* that the preorder $\precsim$ must satisfy. Thus, the total preorder $\precsim$ is said to be *order-separable in the sense of Debreu* if there exists a countable subset $D \subseteq X$ such that for every $x, y \overset{\cdot}{\in} X$ with $x \prec y$ there exists an element $d \in D$ such that $x \precsim d \precsim y$. Such subset D is said to be *order-dense* in $(X, \precsim)$.

If X is endowed with a topology τ, the total preorder $\precsim$ is said to be *continuously representable* if there exists an isotony f that is continuous with respect to the topology τ on X and the usual topology on the real line $\mathbb{R}$. The total preorder $\precsim$ is said to be τ-continuous if the sets $U(x) = \{y \in X \ , \ x \prec y\}$ and $L(x) = \{y \in X \ , \ y \prec x\}$ are τ-open, for every $x \in X$.

In this case, the topology τ is said to be *natural* or *compatible* with the preorder $\precsim$. (See Bridges and Mehta [1995], p. 19). The coarsest natural topology is the *order topology* θ whose subbasis is the collection $\{L(x) \ : \ x \in X\} \cup \{U(x) \ : x \in X\}$.

The topology τ on X is said to have the *continuous representability property* (CRP) if every continuous total preorder $\precsim$ defined on X admits a numerical representation by means of a continuous real-valued isotony.

A powerful tool to obtain *continuous* representations of an order-separable totally preordered set $(X, \precsim)$ endowed with a natural topology τ is the so-called *Debreu's open gap lemma*. (See Debreu [1954], or Ch. 3 in Bridges and Mehta [1995]). To this extent, let T be a subset of the real line $\mathbb{R}$. A *lacuna* L corresponding to T is a nondegenerate interval of $\mathbb{R}$ that has both a lower bound and upper bound in T and that has no points in common with T. A maximal lacuna is said to be a *Debreu gap*. Debreu's open gap lemma states that if S is a subset of the extended real line $\overline{\mathbb{R}}$, then there exists a strictly increasing function $g : S \longrightarrow \mathbb{R}$ such that all the Debreu gaps of $g(S)$ are open. Using Debreu's open gap lemma, the classical process to get a *continuous* real-valued isotony goes as follows: First, one can easily construct a (non necessarily continuous!) isotony f representing $(X, \precsim)$ when $\precsim$ is order-separable (see e.g. Birkhoff [1967], Theorem 24 on p. 200, or else Bridges and Mehta [1995], Theorem 1.4.8 on p. 14). Once we have an isotony f, Debreu's open gap is applied to find a strictly increasing function $g : f(X) \longrightarrow \mathbb{R}$ such that all the *Debreu gaps* of $g(f(X))$ are open. Consequently, the composition $F = g \circ f : X \longrightarrow \mathbb{R}$ is also an isotony representing $(X, \precsim)$, but now F is continuous with respect to any given natural topology τ on X.

A topological space (X, τ) is said to be *separably connected* if for every $a, b \in X$ there exists a connected and separable subset $C(a, b) \subseteq X$ such that $a, b \in C(a, b)$. Separably connected implies connected, but the converse is not true in general. (See Candeal et al. [1998]).

Let $A = [0, \Omega)$ where Ω denotes the first uncountable ordinal. Between each ordinal $\alpha \in A$ and its follower $\alpha + 1$ we insert a copy of the interval $(0, 1) \subset \mathbb{R}$. The space L that we got in this way is called *the long line*. (See Steen and Seebach [1970], pp. 71-72). L is ordered in the obvious way. Denote by $\precsim_L$ the usual total order on L. Let $Y = L \times \{0, 1\}$. Endow Y with the total preorder $\precsim_Y$ defined as follows:

i) $(a, 0) \prec_Y (b, 1)$ for every $a, b \in L$ such that either $a \neq 0$ or $b \neq 0$,

ii) $(0, 0) \sim_Y (0, 1)$,

iii) $(a, 0) \precsim_Y (b, 0) \iff b \precsim_L a$ for every $a, b \in L$,

iv) $(a, 1) \precsim_Y (b, 1) \iff a \precsim_L b$ for every $a, b \in L$.

$(Y, \precsim_Y)$ is called *the double long line*.

Separably connected topological spaces have a property of continuous ordinal representability in the double long line, as the next lemma states.

LEMMA 2.1. :   *Let (X, τ) be a separably connected topological space. Let $\precsim$ be a τ-continuous total preorder on X. Let θ_Y denote the order topology on the double long line Y. There exists a continuous map $F : (X, \tau) \longrightarrow (Y, \theta_Y)$ such that*

$$x \precsim y \iff F(x) \precsim_Y F(y) \quad (x, y \in X).$$

PROOF: We can suppose without loss of generality that $\precsim$ is actually a *total order*, because since $\precsim$ is τ-continuous, the quotient space $(X/ \sim) = \{[x] : x \in X\}$, where $[x] = \{y \in X : y \sim x\}$ is also separably connected with respect to the quotient topology induced by τ on $(X/ \sim)$. Also, again because $\precsim$ is τ-continuous, it is enough to find a continuous isotony $F : (X, \theta_X) \to (Y, \theta_Y)$ where θ_X stands for the order topology on X. Since θ_X is coarser than τ, (X, θ_X) is also separably connected. Indeed, it is path-connected by Remark 2 (iv) in Candeal et al. [1998]. Two cases may occur now: If $(X, \precsim)$ is representable, there exists a continuous real-valued isotony. But it is clear that the real line with its usual topology and order can be also continuously embedded in the double long line (Y, θ_Y). If otherwise $(X, \precsim)$ fails to be representable, Remark 3.2 (iii) in Beardon et al. [2002] states that it can be also continuously embedded in (Y, θ_Y) through an isotony. (See also Theorem 5 in Monteiro [1987], of which this result is a generalization). This finishes the proof. ∎

Given a topological space (X, τ) a total preorder $\precsim$ is said to be τ-*lower semicontinuous* if the sets $U(x) = \{y \in X , x \prec y\}$ are τ-open, for every $x \in X$. In a similar way, $\precsim$ is said to be τ-*upper semicontinuous* if the sets $L(x) = \{y \in X , y \prec x\}$ are τ-open, for every $x \in X$. The topology τ is said to satisfy the semicontinuous representability property (SRP) if every τ-upper (respectively τ-lower) semicontinuous total preorder $\precsim$ defined on X admits a numerical representation by means of a τ-upper (respectively τ-lower) semicontinuous real-valued isotony.

If $\precsim$ is a τ-upper (respectively τ-lower) semicontinuous total preorder on X, and it is also representable (or, equivalently, order-separable in the sense of Debreu), using again Debreu's open gap lemma we can get a τ-upper (respectively τ-lower) *semicontinuous* real-valued isotony that represents $\precsim$. The process goes as follows: First, we construct an isotony f representing $(X, \precsim)$. Then Debreu's open gap is applied to find a strictly increasing function $g : f(X) \longrightarrow \mathbb{R}$ such that all the *Debreu gaps* of $g(f(X))$ are open. Consequently, the composition $F = g \circ f : X \longrightarrow \mathbb{R}$ is also an isotony representing $(X, \precsim)$, but now F is τ-upper (respectively τ-lower) semicontinuous.

Let us see now the relationship between CRP and SRP.

LEMMA 2.2. : *SRP implies CRP, but the converse is not true.*

PROOF: (See also Proposition 4.4 also Bosi and Herden [2002]). Suppose that (X, τ) is a topological space such that τ satisfies SRP. Let $\precsim$ be a τ-continuous total preorder on X. Since $\precsim$ is continuous, it is in particular τ-upper semicontinuous, so that, by SRP, there exists an upper semicontinuous isotony f representing $(X, \precsim)$. Debreu's open gap is applied immediately to find a strictly increasing function $g : f(X) \longrightarrow \mathbb{R}$ such that all the *Debreu gaps* of $g(f(X))$ are open, so that the composition $F = g \circ f : X \longrightarrow \mathbb{R}$ is also an isotony representing $(X, \precsim)$, but now F is τ-continuous because, being $\precsim$ τ-continuous, the topology τ is, by definition, a natural topology. Therefore τ satisfies CRP.

Some example that proves that the converse does not hold in the general case will be given in next sections. For instance, we will prove that the weak topology of a Banach space X always satisfies CRP, but it satisfies SRP if and only if X is separable. For instance, (ℓ_∞, ω) satisfies CRP but not SRP, where ω stands for the weak topology. ∎

Now we show another important topological consequence of SRP, that shall be used in the sequel.

LEMMA 2.3. : *Let (X, τ) be a topological space such that τ satisfies SRP. Then (X, τ) is hereditarily separable and hereditarily Lindelöf.*

PROOF: (See also Lemma 4.1 and Proposition 4.2 in Bosi and Herden [2002]. For basic definitions and results involving topological concepts, we refer to Dugundji [1966]).

Remember that, being Ω the first uncountable ordinal, an uncountable subset $(x_\alpha)_{\alpha < \Omega}$ of a topological space (X, τ) is said to be *right-separated* (respectively, *left-separated*) if for every ordinal $\alpha < \Omega$ we have that x_α does not belong to the τ-closure $\bar{B}_\alpha$ of the set $B_\alpha = \{x_\beta : \alpha < \beta < \Omega\}$ (respectively, if for every ordinal $\alpha < \Omega$ we have that x_α does not belong to the τ-closure $\bar{C}_\alpha$ of the set $C_\alpha = \{x_\beta : \beta < \alpha\}$.

It is well-known (see e.g. Theorem 3.1. in Roitman [1984]) that:

A topological space (X, τ) is hereditarily separable (respectively, hereditarily Lindelöf) if and only if it has not an uncountable left-separated family (respectively, if and only if it has not an uncountable right-separated family).

Suppose now that (X, τ) has a right-separated family $(x_\alpha)_{\alpha < \Omega}$. Given $\alpha < \Omega$, let $O_\alpha = X \setminus \bar{B}_\alpha$. Let $U_0 = \emptyset$; $U_\alpha = \bigcup_{\gamma < \alpha} O_\gamma$ $(0 < \alpha < \Omega)$. Observe that given $0 \leq \alpha < \beta$ it holds that $U_\alpha \subsetneq U_\beta$. Now consider the preorder $\precsim$ on X defined as follows:

$$x \precsim y \iff (y \in U_\alpha \Rightarrow x \in U_\alpha \text{ , for every } \alpha < \Omega).$$

By construction, the preordered set $(X, \precsim)$ contains a subset isotonic to $([0, \Omega), <)$, hence non-representable. (See Beardon et al. [2002]). Thus, in particular, there is no upper-semicontinuous isotony representing $\precsim$.

Notice, in addition, that $\precsim$ is τ- upper semicontinuous: Indeed, given $x \in X$ we have that

$$L(x) = \{y \in X \ : \ y \prec x\} = \bigcup_{\alpha < \Omega \ , \ x \notin U_\alpha} U_\alpha,$$

which is a τ-open subset of X.

Therefore (X, τ) does not satisfy SRP.

Finally, suppose that (X, τ) has a left-separated family $(x_\alpha)_{\alpha < \Omega}$. Given $\alpha < \Omega$, let $O_\alpha = X \setminus \bar{C}_\alpha$. Let $V_0 = X$; $V_\alpha = \bigcup_{\alpha < \gamma} O_\gamma$ $(0 < \alpha < \Omega)$. Observe that given $0 \leq \alpha < \beta$ it holds that $V_\beta \subsetneq V_\alpha$. Now consider the preorder $\precsim$ on X defined as follows:

$$x \precsim y \iff (x \in V_\alpha \Rightarrow y \in V_\alpha \text{ , for every } \alpha < \Omega).$$

For every $x \in X$ we have that

$$U(x) = \{y \in X \ : \ x \prec y\} = \bigcup_{\alpha < \Omega \ , \ x \notin V_\alpha} V_\alpha,$$

which is a τ-open subset of X. Henceforth, $\precsim$ is τ-lower semicontinuous.

By construction, the preordered set $(X, \precsim)$ contains a subset isotonic to $([0, \Omega), <)$. Hence $\precsim$ is a non-representable total preorder. Thus, in particular, there is no lower-semicontinuous isotony representing $\precsim$.

Consequently, (X, τ) does not satisfy SRP.

This finishes the proof. ∎

We conclude with another key result that involves the Lindelöf property.

LEMMA 2.4. : Let (X, τ) be a Lindelöf and separably connected topological space. Then τ satisfies CRP.

PROOF: Let $\precsim$ be a τ-continuous total preorder on X. By Lemma 2.1, there exists a continuous isotony $F : (X, \tau) \longrightarrow (Y, \theta_Y)$ that maps X to the double long line Y. Observe that F being continuous, the image $F(X)$ must be a connected subset of Y. Moreover, by Lindelöf property, $F(X)$ must be order-bounded in Y, that is, there exist elements $a, b \in Y$ such that

$a \prec_Y x \prec_Y b$ for every $x \in X$. Indeed, if $F(X)$ is not order-bounded, for every ordinal number $\alpha \in (0, \Omega)$, denote $S_\alpha = \{z \in Y : (\alpha, 0) \prec_Y z \prec_Y (\alpha, 1)\}$. It is clear that $\{F^{-1}(S\alpha) : \alpha \in (0, \Omega)\}$ is an open covering of X without a countable subcovering, in contradiction with the assumption of X being Lindelöf. Finally, observe that every order-bounded subset of the double long line is continuously isotonic to a subset of the real line. (See Monteiro [1987] or Beardon et al. [2002]).

This finishes the proof. ∎

3. The continuous representability property (CRP) in Banach spaces

In what follows, let X be a Banach space. (For basic definitions and results involving Functional Analysis concepts, we refer to Aliprantis and Border [1999]).

In Candeal et al. [1998] it has been proved that for a *metric space* (X, d), where d denotes the distance function and metric topology on X, the following statements are equivalent:
i) (X, d) satisfies CRP, ii) (X, d) is separable, iii) (X, d) is second countable.

Also, we know that a second countable topology always satisfies SRP. (This is known as *Rader's theorem*, despite its original proof in Rader [1963] had a flaw, as proved in Mehta [1997]. A correct proof of Rader's theorem appears in Isler [1997]). The following result follows now as a consequence of all these facts.

THEOREM 3.1. : *Let $(X, || \cdot ||)$ be a Banach space endowed with the norm topology. The following statements are equivalent:*

   i) $(X, || \cdot ||)$ *satisfies SRP,*
   ii) (X, ω) *is hereditarily separable and hereditarily Lindelöf (where ω stands for the weak topology on X),*
   iii) $(X, || \cdot ||)$ *is separable.*

PROOF:

   i) $\Rightarrow$ ii)

Let $\precsim$ a total preorder on X. Suppose that $\precsim$ is upper semicontinuous with respect to the weak topology ω. Then $\precsim$ is also upper semicontinuous with respect to the norm topology, which is finer than ω. Since the norm topology satisfies SRP, in particular the preorder $\precsim$ is representable by a real-valued order-preserving isotony f. Moreover, by Debreu's open gap lemma there exists a strictly increasing function $g : f(X) \longrightarrow \mathbb{R}$ such that all the Debreu gaps of $g(f(X))$ are open. Finally, observe that since $\precsim$ is upper semicontinuous with respect to ω, the composition $F = g \circ f : X \longrightarrow \mathbb{R}$ is ω-upper semicontinuous, and it is a real-valued isotony that represents $\precsim$.

The proof for the lower-semicontinuous case is entirely analogous. Thus we conclude that (X, ω) also satisfies SRP, so that we by Lemma 3, (X, ω) is is hereditarily separable and hereditarily Lindelöf.

   ii) $\Rightarrow$ iii) $\Rightarrow$ i)

By hypothesis, (X, ω) is separable, and this is equivalent to say that $(X, ||\cdot||)$ is separable (see e.g. Corson [1961] or Jain et al. [1997], p. 196.) Now, since in a metric space separability is equivalent to second countability, it follows that $(X, ||\cdot||)$ satisfies SRP by Rader's theorem.

This finishes the proof. (Observe that more equivalences, as "$(X, ||\cdot||)$ is second countable" or "$(X, ||\cdot||)$ satisfies CRP" among others, could have been added to the statement). ∎

Now let X be a Banach space and X^* its dual. Endow X^* with the weak-star topology ω^*. Since by Alaoglu's theorem the closed unit ball U of $(X^*, ||\cdot||)$ is compact with respect to the weak-star topology ω^*, and this also happens for nU ($n \in \mathbb{N}$), it follows that X^* is σ-compact (i.e.: countable union of compact subsets) with respect to the weak-star topology. But σ-compact implies Lindelöf, so that applying Lemma 2.4 we arrive to the following fact, that was already known in the literature. (See Example 4 in Monteiro [1987]):

THEOREM 3.2. : *Let X be a Banach space. Then (X^*, ω^*) satisfies CRP.*

REMARK 3.3.: Suppose now that X is a *reflexive* Banach space. Then the weak and weak-star topologies agree on X^*. This fact, jointly with Theorem 3.1 and Theorem 3.2, allows us to put an example of a Banach space that satisfies CRP but not SRP, as announced before. For that purposes, it is enough to consider a reflexive and non-separable Banach space X, and take the weak topology on X^*. For instance, we can consider the non-separable Hilbert space $X = \ell_2(\mathbb{R})$. □

To conclude this section, it remains to study the behaviour of the *weak topology* ω of a Banach space X, with respect to the satisfaction of the continuous representability property (CRP).

To do so, we introduce some necessary definition and lemmas.

DEFINITION: Let X be a nonempty set endowed with a topology τ. The topological space (X, τ) is said to satisfy the *countable chain condition (CCC)* if every family of pairwise disjoint τ-open subsets is countable.

LEMMA 3.4. : *Let $< E, F >$ be real linear spaces in duality. Then $(E, \omega(E, F))$ satisfies the countable chain condition CCC, where $\omega(E, F)$ stands for the weak topology that F induces on E.*

PROOF: (See also Corson [1961] p. 8). Notice that, in the topology $\omega(E, F)$, the space E is homeomorphic in a natural way to a dense subset of a product of copies of the real numbers. The real line $\mathbb{R}$ is separable in its usual Euclidean topology, and a product of separable spaces satisfies CCC in the product topology. (See Engelking [1989], Corollary 2.3.18). Finally, a dense subset of a topological space that satisfies CCC also satisfies CCC in the induced topology. (See Szpilrajn-Marczewski [1941], or Lemma 4 in Corson [1961]). ∎

LEMMA 3.5. : *Let (X, τ) be a separably connected topological space that satisfies the countable chain condition CCC. Then it also satisfies the continuous representability property CRP.*

PROOF: By Lemma 2.1, if $\precsim$ is a τ-continuous total preorder on X, there exists a continuous map $F : (X, \tau) \longrightarrow (Y, \theta_Y)$ such that $x \precsim y \iff F(x) \precsim_Y F(y)$ $(x, y \in X)$ where $(Y, \precsim_Y)$ denotes the double long line. By continuity of F, the subset $F(X) \subseteq Y$ is connected. As in Lemma 2.4, it is enough now to prove that $F(X)$ is order-bounded in Y, that is, there exist elements $a, b \in Y$ such that $a \prec_Y x \prec_Y b$ for every $x \in X$: Suppose, by contradiction, that $F(X)$ is not order-bounded. For a given ordinal number $\alpha < \Omega$ let $\alpha + 1$ be its follower, and call $A_\alpha = \{z \in Y : (\alpha + 1, 0) \prec_Y z \prec_Y (\alpha, 0)\}$, $B_\alpha = \{z \in Y : (\alpha, 1) \prec_Y z \prec_Y (\alpha + 1, 1)\}$. It is clear that the family $\{F^{-1}(A_\alpha)\} \bigcup \{F^{-1}(B_\alpha)\}$ $(\alpha < \Omega)$ is an uncountable family of nonempty and pairwise disjoint open subsets of X, which contradicts the fact of (X, τ) satisfying the countable chain condition CCC. This finishes the proof. ∎

Now we are ready to present the final result of this section.

THEOREM 3.6. : *Let $< E, F >$ be real linear spaces in duality. Then $(E, \omega(E, F))$ satisfies CRP.*

PROOF: Any compatible topology in a topological vector space is separably connected because it is path-connected. Now, by Lemma 3.4 we have that $(E, \omega(E, F))$ satisfies the countable chain condition CCC. And finally, according to Lemma 3.5, $(E, \omega(E, F))$ must satisfy CRP.

Observe, in particular, that if X is a Banach space and X^* is its topological dual, then starting from the duality $< X, X^* >$ we arrive to the fact:

"(X, ω) *satisfies CRP*".

Also, starting from the duality $< X^*, X >$ we arrive to the (already known) fact:

"(X^*, ω^*) *satisfies CRP*".

(See also Theorem 3.2 above). ∎

4. The semicontinuous representability property (SRP) in Banach spaces

Through the results given in Theorem 3.1 of the previous section, it follows that given a Banach space X, it holds that $(X, \|\cdot\|)$ satisfies SRP $\iff$ (X, ω) satisfies SRP $\iff$ $(X, \|\cdot\|)$ is a separable Banach space. It remains now to characterize the satisfaction of SRP by the weak-star topology ω^* of the dual Banach space X^*. To do so, we introduce a previous lemma.

LEMMA 4.1. : *Let X be a non-separable Banach space. Then (X^*, ω^*) is not hereditarily Lindelöf.*

PROOF: Assume, by contradiction, that (X^*, ω^*) is hereditarily Lindelöf. For any element $f \neq 0 \in X^*$ choose some element $x_f \in X$ such that $f(x_f) > 0$. Let $U_f = \{g \in X^* : g(x_f) > 0\}$ and $\mathcal{F} = \{U_f : f \neq 0 \in X^*\}$. The family $\mathcal{F}$ is a covering of $X^* \setminus \{0\}$ by ω^*-open subsets. Since $X^* \setminus \{0\}$ is assumed to be ω^*-Lindelöf, there exists a countable subfamily $\{U_{f_n} : f_n \neq 0 \in X^*$, $n \in \mathbb{N}\} \subset \mathcal{F}$ which also covers $X^* \setminus \{0\}$. And this implies that if $x_n = x_{f_n}$ the sequence $(x_n)_{n \in \mathbb{N}}$ separates points in X^* or, in other words, given $g \neq 0 \in X^*$ we can find $k \in \mathbb{N}$ such that $g(x_k) \neq 0$. Now let D be the norm closure in X of the set of all linear combinations of the elements of the sequence $(x_n)_{n \in \mathbb{N}}$. It is clear that D is norm-separable, so that, being X non-separable, the annihilator $D^0 = \{h \in X^* : h(d) = 0 \text{ for every } d \in D\}$ is not trivial. But this is obviously a contradiction. ∎

Now we are ready to prove the following Theorem 4.2, so characterizing SRP for the weak-star topology of a dual Banach space.

THEOREM 4.2. : *Let X be a Banach space. The following statements are equivalent:*

i) $(X, || \cdot ||)$ *is separable,*

ii) (X^*, ω^*) *satisfies SRP,*

iii) (X^*, ω^*) *is hereditarily Lindelöf.*

PROOF:

i) $\Rightarrow$ ii)

Since $(X, || \cdot ||)$ is separable, the closed unit ball U of $(X^*, || \cdot ||)$ is compact and *metrizable*, with respect to the weak-star topology ω^*. This also happens for nU $(n \in \mathbb{N})$. If $\precsim$ is a total preorder on X^*, and it is upper semicontinuous with respect to ω^*, the restriction of $\precsim$ to nU $(n \in \mathbb{N})$ is also upper semicontinuous with respect to the restriction of ω^* to nU, which is a second countable topology. Now, by Rader's theorem, the restriction of $\precsim$ to nU is, in particular, representable by means of a real-valued isotony, hence, it is *order-separable* (in the sense of Debreu). Since $X^* = \bigcup_{n \in \mathbb{N}} (nU)$, it is straightforward to prove that $\precsim$ is actually order-separable on the whole X^*. Thus, there exists a real-valued order-preserving isotony $f : X^* \longrightarrow \mathbb{R}$ that represents $\precsim$. Moreover, by Debreu's open gap lemma there exists a strictly increasing function $g : f(X^*) \longrightarrow \mathbb{R}$ such that all the Debreu gaps of $g(f(X^*))$ are open. Finally, observe that since $\precsim$ is upper semicontinuous with respect to ω^*, the composition $F = g \circ f : X^* \longrightarrow \mathbb{R}$ is upper semicontinuous, and it is also a real-valued isotony that represents $\precsim$.

The proof for the lower-semicontinuous case is entirely analogous. Thus we conclude that (X^*, ω^*) also satisfies SRP.

ii) $\Rightarrow$ iii) $\Rightarrow$ i)

This follows from Lemma 2.3 and Lemma 4.1.

This finishes the proof. ∎

REMARK 4.3.: A Hausdorff topological space (X, τ) is said to be an *L-space* (see Roitman [1984]) if it is hereditarily Lindelöf but fails to be hereditarily separable. Thus, as an immediate corollary of Theorem 4.2, we obtain the following fact:

In the cathegory of dual Banach spaces endowed with the weak-star topology there is no L-space.

Similarly, a Hausdorff topological space (X, τ) is said to be an *S-space* whenever it is hereditarily separable but not hereditarily Lindelöf. Now observe that, as a consequence of Theorem 3.1, we also have:

i) In the cathegory of Banach spaces endowed with the weak topology there is no S-space.

ii) In the cathegory of Banach spaces endowed with the norm topology there are neither S-spaces nor L-spaces. □

5. Final remarks

1) If we assume the *continuum hipothesis (CH)* , then in Theorem 3.1 we *cannot* add "(X, ω) is hereditarily Lindelöf" to the equivalent conditions of the statement. Also, under CH we *cannot* add "(X^*, ω^*) is hereditarily separable" to the equivalent conditions of the statement of Theorem 4.2. The reason is the following: Let K be a *Kunen compact* (See p. 1128 in Negrepontis [1984]) such that the Banach space $X = C(K)$ is non-separable in norm. Then $(C(K), \omega)$ is hereditarily Lindelöf but not hereditarily separable because $C(K)$ is not separable in norm, hence it is not ω-separable, either. In addition $(C(K)^*, \omega^*)$ is hereditarily separable, but not hereditarily Lindelöf, by Lemma 4.1. In other words, $(C(K), \omega)$ is a L-space, whereas $(C(K)^*, \omega^*)$ is an S-space. (See Negrepontis [1984], Granero [1997] or Granero et al. [2003] for further information).

2) On certain Banach spaces it is still possible to add "(X^*, ω^*) is hereditarily separable" to the equivalent conditions of the statement of Theorem 4.2. In some families of Banach spaces it happens that if the closed unit ball U of $(X^*, || \cdot ||)$ is ω^*-separable then $(X^*, || \cdot ||)$ is also separable, which carries the separability of $(X, || \cdot ||)$. Examples of such families have been given in Dancer and Sims [1979].

3) Once we know that some topology does not satisfy CRP, as it is the case of the norm topology of a non-separable Banach space, we can still ask ourselves about which continuous total preorders (of course, not everyone!) admit a continuous real-valued isotony. The same question appears for topologies that do not satisfy SRP, but on which we want to characterize the semicontinuous total preorders that do admit a semicontinuous real-valued isotony. This kind of studies is still *open*. However, there are some partial results in Candeal et al. [1999].

   To put only one example, it is straightforward to prove that in a Banach space $(X, || \cdot ||)$ endowed with the norm topology, a continuous total preorder $\precsim$ such that the sets $A(x) = \{y \in X \, , \, x \precsim y\}$ and $B(x) = \{y \in X \, , \, y \precsim x\}$ are all *convex* , admits a continuous representation by means of a real-valued isotony. Observe that a norm-continuous total preorder $\precsim$ satisfying that condition is also ω-continuous, because the closed convex sets in $(X, || \cdot ||)$ and (X, ω) coincide. Thus since (X, ω) satisfies CRP, we have in particular that there exists a representation of $\precsim$ through a real-valued isotony $f : X \longrightarrow \mathbb{R}$. Again by Debreu's open gap lemma, there exists a strictly increasing function $g : f(X) \longrightarrow \mathbb{R}$ such that all the Debreu gaps of $g(f(X))$ are open. And, since $\precsim$ is norm continuous, the composition $F = g \circ f : X \longrightarrow \mathbb{R}$ is actually a continuous real-valued isotony that represents $\precsim$.

4) We can ask ourselves why the properties CRP and SRP have been defined considering total *preorders* instead of just *total orders*. One

important reason is that, despite it is easy to define continuous total preorders in the general case, *in some topologies there is no continuous total order.* This indeed happens in some Banach space $(X, \| \cdot \|)$ endowed with the norm topology, even if X is of finite dimension greater or equal than two. (See e.g. Theorem 4 in Candeal and Induráin [1993]). The study and characterization of topological spaces (X, τ) for which every continuous (respectively, semicontinuous) *total order* admits a continuous (respectively, semicontinuous) representation through a real-valued isotony is also an *open problem.* □

REFERENCES

[1] Aliprantis, Ch.D. and K.C. Border: "Infinite Dimensional Analysis". Springer. Berlin. 1999.
[2] Beardon, A.F., Candeal, J.C., Herden, G., Induráin, E. and G.B. Mehta: "The non-existence of a utility function and the structure of non-representable preference relations.". Journal of Mathematical Economics 37, 17-38. (2002).
[3] Birkhoff, G.: "Lattice theory. (Third edition)". American Mathematical Society. Providence, RI. 1967.
[4] Bosi G. and G. Herden: "On the structure of completely useful topologies". Applied General Topology 3 (2), 145-167. (2002).
[5] Bridges, D.S. and G.B. Mehta: "Representations of preference orderings". Springer-Verlag. Berlin. 1995.
[6] Candeal, J.C., Hervés, C. and E. Induráin: "Some results on representation and extension of preferences". Journal of Mathematical Economics 29, 75-81. (1998).
[7] Candeal, J.C. and E. Induráin: "Utility functions on chains". Journal of Mathematical Economics 22, 161-168. (1993).
[8] Candeal, J.C., Induráin, E. and G.B. Mehta: "Order preserving functions on ordered topological spaces". Bulletin of the Australian Mathematical Society 60, 55-65. (1999).
[9] Corson, H.H.: "The weak topology of a Banach space". Transactions of the American Mathematical Society 101, 1-15. (1961).
[10] Dancer, E.N. and B. Sims: "Weak star separability". Bulletin of the Australian Mathematical Society 20, 253-257. (1979).
[11] Debreu, G.: "Representation if a preference ordering by a numerical function". In: Thrall, R., Coombs, C. and R. Davies (eds.), *"Decision Processes"*, pp. 159-166. John Wiley. New York. 1954.
[12] Dugundji, J.: "Topology". Allyn and Bacon. Boston. 1966.
Eilenberg, S.: "Ordered topological spaces" American Journal of Mathematics 63, 39-45. (1941).
[13] Engelking, R.: "General Topology. Revised and completed edition". Heldermann Verlag. Berlin. 1989.
[14] Granero, A.S.: "On nonseparable Banach spaces and S- and L-spaces". Universidad Complutense de Madrid. Departamento de Análisis Matemático. 1997.
[15] Granero, A.S., Jiménez, M., Montesinos, A., Moreno, J.P. and A. Plichko: "On the Kunen-Shelah properties in Banach spaces". Studia Mathematica 157(2), 97-120. (2003).
[16] Herden, G.: "Topological spaces for which every continuous total preorder can be represented by a continuous utility function". Mathematical Social Sciences 22, 123-136. (1991).
[17] Herden, G. and A. Pallack: "Useful topologies and separable systems". Applied General Topology 1 (1), 61-82. (2000).
[18] Isler, R.: "Semicontinuous utility functions in topological spaces". Rivista di Matematica per le Scienze economiche e sociali 20 (1), 111-116. (1997).

[19] Jain, P.K., Ahuja, O.P. and K. Ahmed: "Functional Analysis". New Age Publishers. New Delhi. 1997.

[20] Mas-Colell, A.: "The price equilibrium existence problem in topological vector lattices". Econometrica 54 (5), 1039-1053. (1986).

[21] Mehta, G.B.: "A remark on a utility representation theorem of Rader". Economic Theory 9, 367-370. (1997).

Monteiro, P.K.: "Some results on the existence of utility functions on path connected spaces". Journal of Mathematical Economics 16, 147-156. (1987).

[22] Negrepontis, S.: "Banach spaces and Topology". In: Kunen, K. and J.E. Vaughan (eds.), "Handbook of set-theoretical topology", pp. 1045-1142. Norht Holland. Amsterdam. 1984.

[23] Rader, T.: "The existence of a utility function to represent preferences". Review of Economic Studies 30, 229-232. (1963).

[24] Roitman, J.: "Basic S and L". In: Kunen, K. and J.E. Vaughan (eds.), "Handbook of set-theoretical topology", pp. 295-326. Norht Holland. Amsterdam. 1984.

[25] Steen, L.A. and J.A. Seebach Jr.: "Counterexamples in Topology". Holt, Rinehart and Winston. New York. 1970.

[26] Szpilrajn-Marczewski, E. : "Remarque sur les produits cartésiens d'espaces topologiques". Comptes Rendus de l'Academie des Sciences de l' U.R.S.S. 31, 525-527. (1941).

Acknowledgement: Thanks are given to an anonymous referee for his/her helpful suggestions and comments.

Departamento de Matemática e Informática, Universidad Pública de Navarra, Campus Arrosadía. E-31006. PAMPLONA (SPAIN).
E-mail address: mjesus.campion@unavarra.es

Departamento de Análisis Económico, Facultad de Ciencias Económicas y Empresariales, Universidad de Zaragoza, c/ Doctor Cerrada 1-3. E-50005. ZARAGOZA (SPAIN).
E-mail address: candeal@unizar.es

Departamento de Análisis Matemático, Facultad de Matemáticas, Universidad Complutense de Madrid, E-28040. MADRID (SPAIN).
E-mail address: antonio_suarez@mat.ucm.es

Departamento de Matemática e Informática, Universidad Pública de Navarra, Campus Arrosadía. E-31006. PAMPLONA (SPAIN).
E-mail address: steiner@unavarra.es

OVERCLASSES OF THE CLASS OF RADON-NIKODÝM COMPACT SPACES

MARIÁN FABIAN[†]

Dedicated to the memory of Simon Fitzpatrick

ABSTRACT. We map the state of arts of the class of Radon-Nikodým compact spaces and some overclasses of it occurring in the study of a widely open problem whether a continuous image of a Radon-Nikodým compact space is such. We consider classes of quasi-Radon-Nikodým, strongly fragmentable, and countably lower fragmentable compact spaces and we show that they all coincide. A compact space which is simultaneously Corson and quasi-Radon-Nikodým must be Eberlein. An almost totally disconnected, quasi-Radon-Nikodým compact space must be Radon-Nikodým. Quasi-Radon-Nikodým compact spaces whose density are not too big are shown to be Radon-Nikodým. The question whether the union of two Radon-Nikodým compact spaces is such is discussed. Banach space counterparts of some of the above results are given.

INTRODUCTION

A (Hausdorff) compact space K is called *Radon-Nikodým* if there exists a continuous injection $K \hookrightarrow (X^*, w^*)$ for a suitable Asplund space X. This concept crystallized in the break of seventies and eighties of the past century in minds of several mathematicians: Fitzpatrick, Namioka, Reinov, Stegall, ... While Fitzpatrick and Namioka in their mutual correspondence used the name Asplund compact space, Reinov [27] preferred the name Radon-Nikodým. Some other circles think that a right name should refer to Grothendieck and Šmulyan. Nowadays the name suggested by Reinov is commonly used. The Radon-Nikodým compact spaces have been intensively studied by Namioka [21, 22, 23], Stegall [30, 31, 32] (who originated from Grothendieck's memoir [15]), and several other authors [25, 4, 12, 20, 17, 6, 7, ...].

We recall that several other classes – metric, (uniform) Eberlein, Gul'ko, Corson, scattered, ..., compact spaces – are stable under making continuous images. It was I. Namioka who first articulated, in his paper [21], the following

Problem. *Is a continuous image of a Radon-Nikodým compact space Radon-Nikodým?*

The aim of this survey is to collect a relevant material and thus encourage a reader to investigate this so far still widely open and interesting question.

[†] Supported by grants A 1019301, IAA 100190610, and by the Institutional Research Plan of Academy of Sciences of Czech Republic No. AV0Z 101 905 03. We thank Departamento de Matemáticas de Universidad Politécnica de Valencia for its generous hospitality.

2000 *Mathematics Subject Classification.* Primary: 46B50. Secondary 46B22.

Key words. Radon-Nikodým, quasi-Radon-Nikodým, fragmentable, and Corson compact spaces, weakly compactly generated space, ε−Asplund set, regular averaging operator, low density.

We introduce and study several overclasses of the class of Radon-Nikodým compact spaces, we present positive answers under additional assumptions, and also we outline some ways how to approach the problem. We also study this topic from the perspective of Banach space theory.

PRELIMINARIES

Let $(X, \| \cdot \|)$ be a Banach space with its dual space X^*. For $x \in X$ and $x^* \in X^*$ we write $\langle x, x^* \rangle$ instead of $x^*(x)$. The symbol B_X denotes the closed unit ball in X. The weak* topology on X^* is denoted by w^*. Given a nonempty set $A \subset X$, we say that a set $M \subset X^*$ is A-*separable* if it is separable in the topology of uniform convergence on A. ¿From among tens of various equivalent possibilities how to define an Asplund space we shall prefer the following: A Banach space is called *Asplund* if the dual of each separable subspace of it is separable. We mention that one equivalent condition for the Asplund property of X roughly says that, in the dual space X^*, a "vector" variant of the well known Radon-Nikodým theorem holds. This is where the name of Radon-Nikodým compact space comes from. The Banach space X is called *Asplund generated* if there are an Asplund space Y and a linear continuous mapping $T : Y \to X$ so that TY is dense in X. Let K be a compact space. $C(K)$ denotes the Banach space of continuous functions on K, provided with the supremum norm. Given a function $\rho : K \times K \to [0, +\infty)$, we say that ρ is *lower semicontinuous* if the set $\{(k_1, k_2) \in K \times K; \ \rho(k_1, k_2) \le a\}$ is closed for every $a \ge 0$, and we say that ρ *fragments* K, or K is *fragmented* by ρ, if for every $\varepsilon > 0$ and every nonempty set $M \subset K$ there is an open set $\Omega \subset K$ such that the intersection $M \cap \Omega$ is nonemty and has ρ-diameter less than ε, that is, ρ-diam $M \cap \Omega := \sup\{\rho(k_1, k_2); \ k_1, k_2 \in M \cap \Omega\} < \varepsilon$. A compact space K is called *fragmentable* if it admits a function $\rho : K \times K \to [0, 1]$ which fragments it and separates distinct points of K. It should be noted that the original definition of the fragmentability required that ρ was a metric. That these two notions coincide was observed by Ribarska [10, page 86]; for another proof of this we can also use Theorem 4 (i) below. If K is a compact space and $\rho : K \times K \to [0, +\infty)$ is a function, we say that $f \in C(K)$ is ρ-*Lipschitz* if there is a constant $c > 0$ such that $f(k_1) - f(k_2) \le c\rho(k_1, k_2)$ for all $k_1, k_2 \in K$. A set $E \subset C(K)$ is called *equimeasurable* if for every regular Borel probability measure μ on K and for every $\varepsilon > 0$ there is a closed set $H \subset K$ such that $\mu(H) > 1 - \varepsilon$ and the set of restrictions $E_{|H} := \{f_{|H}; \ f \in E\}$ is relatively norm-compact in $C(H)$. The class of all Radon-Nikodým compact spaces will be denoted by $\mathcal{RN}$. The symbol $\mathcal{IRN}$ will mean the family of all continuous images of Radon-Nikodým compact spaces. Thus we can write the above problem as "$\mathcal{IRN} = \mathcal{RN}$?".

RADON-NIKODÝM COMPACT SPACES

For better understanding of the concept of Radon-Nikodým compact space we list several equivalent characterizations of it.

Theorem 1. *For a compact space K the following assertions are equivalent:*

(i) $K \in \mathcal{RN}$.

(ii) *There are a Banach space X and a continuous injection $\iota : K \hookrightarrow (X^*, w^*)$ such that $(\iota(K), w^*)$ is fragmented by the metric generated by the norm.*

(iii) *There are a Banach space X and a continuous injection $\iota : K \hookrightarrow (X^*, w^*)$ such that for every countable set $A \subset B_X$ the set $\iota(K)$ is A–separable.*

(ii') *There are a set Γ and a continuous injection $\iota : K \hookrightarrow [-1,1]^\Gamma$ such that $\iota(K)$ is fragmented by the metric of uniform convergence on Γ.*

(iii') *There are a set Γ and a continuous injection $\iota : K \hookrightarrow [-1,1]^\Gamma$ such that for every countable set $A \subset \Gamma$ the set $\iota(K)$ is separable in the topology of uniform convergence on A.*

(iv) *There exists a lower semicontinuous metric $d : K \times K \to [0,1]$ which fragments K.*

(v) *The Banach space $C(K)$ is Asplund generated.*

(vi) $(B_{C(K)^*}, w^*) \in \mathcal{RN}$.

(vii) *There is a function $\rho : K \times K \to [0,1]$ which fragments K and is such that the set $E = \{f \in C(K);\ f \text{ is } \rho - \text{Lipschitz}\}$ separates the points of K.*

(viii) *There is an equimeasurable set $E \subset C(K)$ which separates the points of K.*

Let us say a few words about the proofs.

(i)$\Rightarrow$(ii) is trivial.

(ii)$\Rightarrow$(i) is [31, Corollary 1.16] or [22, Theorem 3.1]; the proofs are based on a variant of the interpolation technique of Davis, Figiel, Johnson and Pełczyński.

(ii)$\Leftrightarrow$(iii) is [22, Theorem 3.4]. The argument had a predecessor in [24].

(ii)$\Leftrightarrow$(ii') and (iii)$\Rightarrow$(iii') are easy to prove.

(iii')$\Rightarrow$(iii) is not difficult, see the proof of [22, Theorem 3.6].

(i)$\Leftrightarrow$(v)$\Leftrightarrow$(vi) goes back to [30]. For a proof see [22], eventually [10, Theorem 1.5.4]. It is based on the Stone-Weierstrass theorem and a variant of the interpolation technique mentioned above.

(ii)$\Rightarrow$(iv). Assuming the situation $\iota : K \hookrightarrow (X^*, w^*)$ from (ii), put $d(k, k') = \|\iota(k) - \iota(k')\|$, $k, k' \in K$.

(iv)$\Rightarrow$(ii) was first proved by Namioka [22, Corollary 6.7]. An alternative proof, using [16], was suggested by Ghoussoub. In [18, Theorem 2.1], there is an even better result: *If a compact space K admits a lower semicontinuous metric $d : K \times K \to [0,1]$, then there exists a Banach space X and a continuous injection $\iota : K \hookrightarrow (B_{X^*}, w^*)$ such that $d(k, k') = \|\iota(K) - \iota(k')\|$ for all $k, k' \in K$.*

(ii)$\Rightarrow$(vii). Put $\rho(k, k') = \|\iota(k) - \iota(k')\|$, $k, k' \in K$. Further, for $x \in X$ define $\hat{x}(k) = \langle x, \iota(k)\rangle$, $k \in K$; thus $\hat{x} \in C(K)$. Now, put $E = \{\hat{x};\ x \in X\}$.

(vii)$\Rightarrow$(viii) follows from [20, Lemma 3.3] and the theorem of Arzela-Ascoli.

(viii)$\Leftrightarrow$(iii') is due to Stegall, see [20, Theorem 5.1] or [8, pages 144–148].

(vii)$\Rightarrow$(iv). Put $d(k, k') = \sup\{|f(k) - f(k')|;\ f \in E \cap B_{C(K)}\}$, $k, k' \in K$. ∎

Questions. 1. *Let K be a compact space admitting a lower semicontinuous metric $d : K \times K \to [0,1]$ fragmenting it (hence $K \in \mathcal{RN}$). Does there exist an Asplund space X and a continuous injection $\iota : K \hookrightarrow (B_{X^*}, w^*)$ such that $d(k, k') = \|\iota(k) - \iota(k')\|$ for all $k, k' \in K$?* It seems that this question is open even for metrizable compact spaces.

2. (Argyros [6]) *Does every $K \in \mathcal{RN}$ continuously inject into the space $\mathcal{P}(S) (\subset C(S)^*)$ of regular Borel probability measures, endowed with the weak* topology, for a suitable scattered compact space S?* Note that $C(S)$ is an Asplund space whenever S is a scattered compact space.

Examples of Radon-Nikodým compact spaces are: metric compact spaces (trivially from (iv)), closed balls in a reflexive Banach space provided with the weak topology (as reflexive spaces are Asplund), Eberlein compact spaces (as they can be embedded into a reflexive space), intervals $[0, \lambda]$ for any ordinal number λ, more generally, all scattered compact spaces (as the "discrete" metric shows), dual balls (B_{X^*}, w^*), where X is an Asplund generated space, ... Note also that the implication (i)$\Rightarrow$(vi) in Theorem 1 serves for producing further Radon-Nikodým compact spaces.

It remains to mention some compact spaces which are not Radon-Nikodým. For sure $\{0,1\}^{\aleph}$, where $\aleph$ is any uncountable cardinal, is such. This can be seen, for instance from the condition (iv). Actually, this space is not even fragmentable. Further, all Corson compact spaces which are not Eberlein are not Radon-Nikodým, according to Theorem 9 below, see, say [10, Sections 4.3, 7.3, 8.4]. There are also several specially constructed compact spaces not belonging to $\mathcal{RN}$, see [22].

We can easily see that our problem whether $\mathcal{IRN} = \mathcal{RN}$ has a Banach space equivalent:

Problem'. *For a Banach space X and its subspace $Y \subset X$, does $(B_{X^*}, w^*) \in \mathcal{RN}$ imply that $(B_{Y^*}, w^*) \in \mathcal{RN}$?*

Obviously, if X is an Asplund generated Banach space, then $(B_{X^*}, w^*) \in \mathcal{RN}$. The converse is false. Actually, every non-weakly compactly generated subspace of a weakly compactly generated space is such. To check this, we need, say, [10, Theorem 8.3.4] and a well known fact that a continuous image of an Eberlein compact space is Eberlein, hence Radon-Nikodým. ¿From (i)$\Rightarrow$(vi) in Theorem 1 we however get that *a Banach space is a subspace of an Asplund generated space if and only if $(B_{X^*}, w^*) \in \mathcal{IRN}$.* Theorem 1 also gives that $K \in \mathcal{IRN}$ *(if and) only if $(B_{C(K)}, w^*) \in \mathcal{IRN}$.*

It should be noted that the question whether $\mathcal{IRN} = \mathcal{RN}$ is open even for the union of two Radon-Nikodým compact spaces, that is: *If a compact space K can be written as $K = K_1 \cup K_2$, with $K_1, K_2 \in \mathcal{RN}$, does this imply that $K \in \mathcal{RN}$?* Note that such a K is a continuous image of the (obviously Radon-Nikodým) compact space which is a "disjoint" union of K_1 and K_2, and hence $K \in \mathcal{IRN}$. At the end of the paper, we present some partial positive results related to this subclass of $\mathcal{IRN}$.

OVERCLASSES OF THE RADON-NIKODÝM COMPACT SPACES

A quite natural way how to approach the class $\mathcal{IRN}$ is as follows. Consider the situation $\varphi : L \twoheadrightarrow K$, where $L \in \mathcal{RN}$, φ is a continuous surjection, and let $d : L \times L \to [0, 1]$ be a lower semicontinuous metric which fragments L. Define then

$$\rho(k_1, k_2) = \inf d(\varphi^{-1}(k_1), \varphi^{-1}(k_2)), \quad k_1, k_2 \in K.$$

It is easy to check that ρ separates the points of K, i.e., $\rho(k_1, k_2) > 0$ whenever $k_1, k_2 \in K$ are distinct, that ρ is lower semicontinuous, and Zorn's lemma guarantees that ρ fragments K. The only trouble is the triangle inequality – we do not see any way how to prove it for this ρ in general. The situation described here leads to the following concept introduced by Arvanitakis [6]. We say that a compact space K is *quasi-Radon-Nikodým* if there exists a function $\rho : K \times K \to [0, 1]$ which separates distinct points of K, is lower semicontinuous, and fragments K. (Unlike of the original Arvanitakis' definition we omit the requirement of symmetry for ρ since it is redundant.) The class of all quasi Radon-Nikodým compact spaces will be denoted by $\mathcal{QRN}$. We have

$$\mathcal{RN} \subset \mathcal{IRN} \subset \mathcal{QRN}$$

while the validity of both reverse inclusions is open. Trivially, every quasi-Radon-Nikodým compact space is fragmentable. This is a sharp inclusion: Every Gul'ko non-Eberlein compact space shows this; see Theorem 9. Note also an easily verifiable fact that *a continuous image of a quasi Radon-Nikodým compact space is such.*

There is another generalization of the class $\mathcal{RN}$, due to Rezničenko [3, page 104]: A compact space K is called *strongly fragmentable* if there exists a metric $d : K \times K \to [0, 1]$ which fragments K and is such that for every two distinct $k, h \in K$ there are neighbourhoods $U \ni k$, $V \ni h$ such that $\inf \{ d(u, v); \ u \in U, \ v \in V \} > 0$. The class of all strongly fragmentable compact spaces will be denoted by $\mathcal{SF}$. It is easy to check from (iv) in Theorem 1 that $\mathcal{RN} \subset \mathcal{SF}$.

Theorem 3. (Namioka [23]) $\mathcal{SF} = \mathcal{QRN}$.

Proof. $\mathcal{SF} \subset \mathcal{QRN}$: Consider any $K \in \mathcal{SF}$, with a metric d witnessing for that. For $n \in \mathbb{N}$ let C_n be the closure of the set $\{(k, h) \in K \times K; \ d(k, h) < \frac{1}{n}\}$, and define a function $\rho_n : K \times K \to \{0, 1\}$ by

$$\rho_n(k, h) = \begin{cases} 0, & \text{if } (k, h) \in C_n, \\ 1, & \text{if } (k, h) \in K \times K \setminus C_n. \end{cases}$$

Clearly, each ρ_n is lower semicontinuous. Put then $\rho = \sum_{n=1}^{\infty} 2^{-n} \rho_n$; this function will also be lower semicontinuous. ρ separates the points of K. Indeed, assume that this is not so. Then there are distinct points $k, h \in K$ with

$\rho(k, h) = 0$. From the strong fragmentability find $n \in \mathbb{N}$ and neighbourhoods $U \ni k$, $V \ni h$ such that $\inf \{d(u, v); \ u \in U, \ v \in V\} > \frac{1}{n}$. Now, $\rho(k, h) = 0$ implies that $\rho_n(k, h) = 0$, and so $(k, h) \in C_n$. Then, from the definition of C_n, we can find $k' \in U$, $h' \in V$ such that $d(k', h') < \frac{1}{n}$, a contradiction.

It remains to show that ρ fragments K. Consider any $\emptyset \neq M \subset K$ and any $\varepsilon > 0$. Find $m \in \mathbb{N}$ so large that $2^{-m} < \varepsilon$. From the strong fragmentability we find an open set $\Omega \subset K$ so that the set $M \cap \Omega$ is nonempty and has d–diameter less than $\frac{1}{m}$. Then for every $k, h \in M \cap \Omega$ we have $(k, h) \in C_m$ $(\subset C_{m-1} \subset \cdots \subset C_1)$, and hence $\rho(k, h) = \sum_{n=m+1}^{\infty} \rho_n(k, h) \leq \sum_{n=m+1}^{\infty} 2^{-n} = 2^{-m}$ $(< \varepsilon)$. Therefore $K \in \mathcal{QRF}$. We have proved that $\mathcal{SF} \subset \mathcal{QRN}$.

$\mathcal{QRN} \subset \mathcal{SF}$: Consider any $K \in \mathcal{QRN}$, with $\rho : K \times K \to [0, 1]$ witnessing for that. Fix any $n \in \mathbb{N}$.

Claim. *There exists an ordinal λ_n and a pairwise disjoint family $\{K_\alpha^n; \ 0 \leq \alpha < \lambda_n\}$ of nonempty subsets of K such that*

(i) $\bigcup \{K_\alpha^n; \ 0 \leq \alpha < \lambda_n\} = K$,

(ii) for every $\beta < \lambda_n$ the set $\bigcup \{K_\alpha^n; \ 0 \leq \alpha \leq \beta\}$ is open, and

(iii) $\rho - \operatorname{diam} K_\alpha^n < \frac{1}{n}$ *for every $\alpha < \lambda_n$.*

We prove the claim. Let $K_0^n \subset K$ be a nonempty open set such that $\rho - \operatorname{diam} K_0^n < \frac{1}{n}$ Consider any ordinal λ and assume that the sets $K_\alpha^n \subset K$ have been constructed for all $\alpha < \lambda$ such that (ii) and (iii) are valid when λ_n is replaced by our λ. Put $U = \bigcup \{K_\alpha^n; \ 0 \leq \alpha < \lambda\}$; this will be, by (ii), an open set. If $U = K$, put $\lambda_n = \lambda$ and finish the process. Further assume that $U \neq K$. As $K \in \mathcal{QRN}$, there is an open set $\Omega \subset K$ such that $\Omega \backslash U \neq \emptyset$ and $\rho - \operatorname{diam} (\Omega \backslash U) < \frac{1}{n}$. Put then $K_\lambda^n = \Omega \backslash U$. Thus $\bigcup \{K_\alpha^n; \ 0 \leq \alpha \leq \lambda\} = \Omega \cup U$ and this will be an open set. Moreover, $\rho - \operatorname{diam} K_\lambda^n < \frac{1}{n}$. This finishes the induction step and proves the claim.

Now, define $d_n : K \times K \to \{0, 1\}$ by

$$d_n(k, h) = \begin{cases} 0 & \text{if } (k, h) \in K_\alpha^n \times K_\alpha^n \ \text{ and } \ 0 \leq \alpha < \lambda_n, \\ 1 & \text{otherwise,} \end{cases}$$

this is a symmetric function. Let us check that d_n satisfies the triangle inequality. Consider any $k, k', k'' \in K$. We have to verify that $d_n(k, k'') \leq d_n(k, k') + d_n(k', k'')$. The only case which may cause troubles is when $d_n(k, k') = d_n(k', k'') = 0$. But then $k, k' \in K_\alpha^n$ and also $k', k'' \in K_\alpha$ for the same $\alpha < \lambda_n$, and hence $k, k'' \in K_\alpha^n$, i.e., $d_n(k, k'') = 0$ as well. The inequality is thus verified.

Having done the above for every $n \in \mathbb{N}$, define $d = \sum_{n=1}^{\infty} 2^{-n} d_n$; this d will also satisfy the triangle inequality. Let us check that d fragments K. So fix any $\emptyset \neq M \subset K$ and any $\varepsilon > 0$. Find $m \in \mathbb{N}$ so large that $2^{-m} < \varepsilon$. Let $\beta_1 < \lambda_1$ be the first $\beta < \lambda_1$ such that $M \cap K_\beta^1 \neq \emptyset$. Put $\Omega_1 = \bigcup \{K_\alpha^1; \ 0 \leq \alpha \leq \beta_1\}$; this is an open set. Moreover, $M \cap \Omega_1 = M \cap K_{\beta_1}^1$ and so d_1–diameter of $M \cap \Omega_1$ is 0 ... Assume we have already found ordinals

$\beta_1 < \lambda_1, \ \beta_2 < \lambda_2, \ldots, \ \beta_{m-1} < \lambda_{m-1}$ and open sets $\Omega_1, \ldots, \Omega_{m-1}$ in K such that the set $M \cap \Omega_1 \cap \cdots \cap \Omega_{m-1}$ is nonemty and has d_1- diameter, ..., $d_{m-1}-$ diameter equal to 0. Let $\beta_m < \lambda_m$ be the first $\beta < \lambda_m$ such that $M \cap \Omega_1 \cap \cdots \cap \Omega_{m-1} \cap K_\beta^m \neq \emptyset$. Define the (open) set $\Omega = \Omega_1 \cap \cdots \cap \Omega_{m-1} \cap \bigcup \{K_\alpha^m; \ 0 \le \alpha \le \beta_m\}$. Then the set $M \cap \Omega$ is nonempty and

$$d_1 - \mathrm{diam}\,(M \cap \Omega) = d_2 - \mathrm{diam}\,(M \cap \Omega) = \cdots = d_m - \mathrm{diam}\,(M \cap \Omega) = 0,$$

and so $d - \mathrm{diam}\,(M \cap \Omega) \le \sum_{n=m+1}^{\infty} 2^{-n} = 2^{-m} \ (< \varepsilon)$. Therefore d fragments K.

What remains to prove is that d "strongly fragments" K. Consider any distinct points $k, h \in K$. ¿From the lower semicontinuity of ρ we find $n \in \mathbb{N}$ and neighbourhoods $U \ni k$, $V \ni h$ so that $\inf \{\rho(u, v); \ u \in U, \ v \in V\} > \frac{1}{n}$. Then for sure $d_n(u, v) = 1$ for all $u \in U$ and $v \in V$, and thus $\inf\{d(u, v); \ u \in U, \ v \in V\} \ge 2^{-n} \ (> 0)$. ∎

In the proof of the inclusion $\mathcal{SF} \subset \mathcal{QRN}$, we did not use that d satisfied the triangle inequality neither did we need that d was symmetric. Thus we get the following

Corollary. *A compact space K belongs to $\mathcal{SF}$ if (and only if) there exists a function $\rho : K \times K \to [0,1]$ which fragments K and is such that for every two distinct $k, h \in K$ there are neighbourhoods $U \ni k$, $V \ni h$ such that $\inf \{\rho(u, v); \ u \in U, \ v \in V\} > 0$.*

In the proof of Theorem 3, the following concept can be traced. Let K be a compact space and put $\Delta_K = \{(k, k); \ k \in K\}$. We say that $C \subset K \times K$ is an *almost neighbourhood* of Δ_K if for every nonempty set $M \subset K$ there is an open set $\Omega \subset K$ such that the set $M \cap \Omega$ is nonempty and $(M \cap \Omega) \times (M \cap \Omega) \subset C$ [22]. Using this concept, Namioka characterized the following three classes of compact spaces.

Theorem 4. (Namioka) *For a compact space K we have:*
  (i) ([22]) *K is fragmentable if and only if it admits a family C_n, $n \in \mathbb{N}$, of almost neighbourhoods of Δ_K such that $\bigcap_{n=1}^{\infty} C_n = \Delta_K$.*
  (ii) ([22]) *$K \in \mathcal{QRN}$ if and only if it admits a family C_n, $n \in \mathbb{N}$, of closed almost neighbourhoods of Δ_K such that $\bigcap_{n=1}^{\infty} C_n = \Delta_K$.*
  (iii) ([23]) *$K \in \mathcal{RN}$ if and only if it admits a family C_n, $n \in \mathbb{N}$, of closed almost neighbourhoods of Δ_K such that $\bigcap_{n=1}^{\infty} C_n = \Delta_K$, and $C_{n+1} \circ C_{n+1} \subset C_n$, i.e., $(k, k') \in C_n$ whenever $(k, k'') \in C_{n+1}$ and $(k'', k') \in C_{n+1}$, for every $n \in \mathbb{N}$.*

The proof of (i) and (ii) can be found in the proof of Theorem 3. (i) also yields Ribarska's result mentioned in [10, page 86].

We have also a generalization of Radon-Nikodým compact spaces defined via Banach spaces [12]. Given a compact space K, $\varepsilon > 0$, and a nonempty set $A \subset C(K)$, we say that K is $\varepsilon - A$*-fragmentable* if for every $\emptyset \neq M \subset K$ there is an open set $\Omega \subset K$ so that $M \cap \Omega \neq \emptyset$ and $\sup\{f(k) - f(h); \ k, h \in M \cap \Omega, \ f \in A\} \leq \varepsilon$. A compact space K is called *countably lower fragmentable* if the conditions of Proposition 5 below hold. The class of all countably lower fragmentable compact spaces will be denoted by $\mathcal{CLF}$.

Proposition 5. *For a compact space K the following assertions are equivalent:*

(i) *For every $\varepsilon > 0$ there exist nonempty sets $A_n^\varepsilon \subset C(K)$, $n \in \mathbb{N}$, such that K is $\varepsilon - A_n^\varepsilon$-fragmentable for every $n \in \mathbb{N}$ and $\bigcup_{n=1}^{\infty} A_n^\varepsilon = C(K)$.*

(ii) *There exist nonempty sets $A_n \subset C(K)$, $n \in \mathbb{N}$, such that K is $1 - A_n$-fragmentable for every $n \in \mathbb{N}$ and $\bigcup_{n=1}^{\infty} A_n = C(K)$.*

(iii) *For every $\varepsilon > 0$ there exist nonempty sets $A_n^\varepsilon \subset B_{C(K)}$, $n \in \mathbb{N}$, such that K is $\varepsilon - A_n^\varepsilon$-fragmentable for every $n \in \mathbb{N}$ and $\bigcup_{n=1}^{\infty} A_n^\varepsilon = B_{C(K)}$.*

(iv) *For every $\varepsilon > 0$ there exist nonempty sets $A_n^\varepsilon \subset C(K)$, $n \in \mathbb{N}$, such that K is $\varepsilon - A_n^\varepsilon$-fragmentable for every $n \in \mathbb{N}$ and the set $\bigcup_{n=1}^{\infty} A_n^\varepsilon$ separates points of K.*

Proof. (i)$\Rightarrow$(ii) and (i)$\Rightarrow$(iii)$\Rightarrow$(iv) are trivial.
(ii)$\Rightarrow$(i). ([7]) For $\varepsilon > 0$ and $n \in \mathbb{N}$ put $A_n^\varepsilon = \varepsilon A_n$, where the A_n's are from (ii).
(iv)$\Rightarrow$(i) needs more work. It is based on the Stone-Weierstrass theorem. For details see the proof of [12, Lemma 4]. ∎

We get easily that $\mathcal{RN} \subset \mathcal{CLF}$. We even have $\mathcal{IRN} \subset \mathcal{CLF}$. Indeed, consider $L \in \mathcal{RN}$ and a continuous surjection $\varphi : L \twoheadrightarrow K$. Define $\varphi^\circ : C(K) \to C(L)$ by $\varphi^\circ(f) = f \circ \varphi$, $f \in C(K)$. Find an Asplund space Y and a linear continuous mapping $T : Y \to C(L)$, with TY dense in $C(L)$. Then it is easy to check that the sets $A_n^\varepsilon := \left(nT(B_Y) + \frac{\varepsilon}{3}B_{C(L)}\right) \cap \varphi^\circ(C(K))$, $n \in \mathbb{N}$, $\varepsilon > 0$, satisfy (i) in Proposition 5.

Next we shall consider the class $\mathcal{CLF}$ in a Banach space framework. Let X be Banach space and let $\varepsilon > 0$. A set $A \subset X$ is called ε*-Asplund* if it is non-empty, bounded, and for every $\emptyset \neq M \subset B_{X^*}$ there is a weak* open set $\Omega \subset X^*$ such that the set $M \cap \Omega$ is nonempty and has A-diameter at most ε, that is, $\sup\{\langle x, x_1^* - x_2^* \rangle; \ x_1^*, x_2^* \in M \cap \Omega, \ x \in A\} \leq \varepsilon$ [13]. Obviously, a nonempty bounded set $A \subset X$ is ε-Asplund if and only if the compact space (B_{X^*}, w^*) is $\varepsilon - \kappa(A)$-fragmentable, where $\kappa : X \hookrightarrow C(B_{X^*}, w^*)$ is the canonical injection. More information about ε-Asplund sets can be found in [13]. Of course, if a subset of X is ε-Asplund for every $\varepsilon > 0$, we get a concept of *Asplund set* [10, Section 1.4]. It should be noted that *for a compact space K the Asplund sets in $C(K)$ are exactly the sets which are equimeasurable.* For the proof of this combine, for instance, [20, Theorem 5.1], [12, Lemma 2], and [13, Theorem 2]. For a more detailed account of the Asplund sets and related concepts see [8, Chapter 5]. Now, we can follow the pattern of Proposition 5 and get a Banach space analogue of it.

Proposition 6. *For a Banach space X the following assertions are equivalent:*

(o) $(B_{X^*}, w^*) \in \mathcal{CLF}$.

(i) *For every $\varepsilon > 0$ there exist $\varepsilon-$Asplund sets $A_n^\varepsilon \subset X$, $n \in \mathbb{N}$, so that $\bigcup_{n=1}^\infty A_n^\varepsilon = X$.*

(ii) *There exist $1-$Asplund sets $A_n \subset X$, $n \in \mathbb{N}$, such that $\bigcup_{n=1}^\infty A_n = X$.*

(iii) *For every $\varepsilon > 0$ there are $\varepsilon-$Asplund sets $A_n^\varepsilon \subset B_X$, $n \in \mathbb{N}$, so that $\bigcup_{n=1}^\infty A_n^\varepsilon = B_X$.*

(iv) *For every $\varepsilon > 0$ there exist $\varepsilon-$Asplund sets $A_n^\varepsilon \subset X$, $n \in \mathbb{N}$, so that the set $\bigcup_{n=1}^\infty A_n^\varepsilon$ is linearly dense in X.*

Again, the only nontrivial implication (iv)$\Rightarrow$(i) can be proved as [12, Lemma 4]. ∎

Theorem 7. (Avilés [7]) $\mathcal{QRN} = \mathcal{CLF}$.

Proof. $\mathcal{QRN} \subset \mathcal{CLF}$: Assume that $K \in \mathcal{QRN}$, with $\rho : K \times K \to [0,1]$ witnessing for this. For $n \in \mathbb{N}$ we define

$$A_n = \{ f \in C(K); \ |f(k) - f(h)| < 1 \text{ whenever } h, k \in K \text{ and } \rho(k,h) < \tfrac{1}{n} \}.$$

Assume that there exists $f \in C(K) \backslash \bigcup_{n=1}^\infty A_n$. For $n \in \mathbb{N}$ find $k_n, h_n \in K$ so that $\rho(k_n, h_n) < \tfrac{1}{n}$ and $|f(k_n) - f(h_n)| \geq 1$. ¿From compactness of K we find cluster points $k, h \in K$ of the sequences (k_n), (h_n) respectively. Then $|f(k) - f(h)| \geq 1$ and so $k \neq h$. However the lower semicontinuity of ρ yields $(0 \leq) \ \rho(k,h) \leq \limsup_{n\to\infty} \rho(k_n, h_n) \leq \lim_{n\to\infty} \tfrac{1}{n} = 0$, a contradiction. Therefore $\bigcup_{n=1}^\infty A_n = C(K)$.

Fix any $n \in \mathbb{N}$. It remains to show that K is $1 - A_n-$fragmentable. So fix any nonempty set $M \subset K$. Since $K \in \mathcal{QRN}$, there is an open set $\Omega \subset K$ such that $M \cap \Omega \neq \emptyset$ and for every $k, h \in M \cap \Omega$ we have $\rho(k,h) < \tfrac{1}{n}$. Thus, for every $f \in A_n$ and for every $k, h \in M \cap \Omega$ we get $|f(k) - f(h)| < 1$, which means that K is $1 - A_n-$fragmentable. Therefore, $K \in \mathcal{CLF}$.

$\mathcal{CLF} \subset \mathcal{QRN}$: From Proposition 5 (ii) we find sets $A_n \subset C(K)$, $n \in \mathbb{N}$, witnessing that $K \in \mathcal{CLF}$. For $n \in \mathbb{N}$ define $B_n = \{tf; \ 0 \leq t \leq 1, \ f \in A_1 \cup \cdots \cup A_n\}$; note that K is also $1 - B_n-$fragmentable. For $f \in C(K)$ and $p \in \mathbb{N}$ put

$$\sigma(f, p) - \min\{m \subset \mathbb{N}; \ (p+1)f \in B_m\}$$

and further denote $\sigma(f) = (\sigma(f,1), \sigma(f,2), \ldots)$; this will be a non-decreasing sequence. Now, for any sequence $\tau \in \mathbb{N}^\mathbb{N}$ we define a function $H^\tau : [0, +\infty) \to [0,1]$ by

$$H^\tau(t) = \begin{cases} 0 & \text{if } t = 0 \\ \frac{1}{\tau(n)} & \text{if } \frac{1}{n+1} < t \leq \frac{1}{n}, \ n \in \mathbb{N}, \\ \frac{1}{\tau(1)} & \text{if } t > 1. \end{cases}$$

Finally, we define $\rho : K \times K \to [0,1]$ by

$$\rho(k,h) = \sup_{f \in C(K)} H^{\sigma(f)}(|f(k) - f(h)|), \quad k, h \in K.$$

Since for a non-decreasing sequence $\tau \in \mathbb{N}^{\mathbb{N}}$ the function H^τ is lower semicontinuous, so is the assignment $(k, h) \mapsto H^{\sigma(f)}(|f(k) - f(h)|)$, $(k, h) \in K \times K$, for every fixed $f \in C(K)$. Therefore ρ is also lower semicontinuous. Further, if points $k, h \in K$ are distinct, there is $f \in C(K)$ which separates them, and so $\rho(k, h) > 0$.

It remains to show that our ρ fragments K. So fix any $\emptyset \neq M \subset K$ and any $\varepsilon > 0$. Find $n \in \mathbb{N}$ so that $\frac{1}{n} < \varepsilon$. Since $K \in \mathcal{CLF}$, there is an open set $\Omega \subset K$, with $M \cap \Omega \neq \emptyset$, such that $\sup \{f(k) - f(h); \ k, h \in M \cap \Omega, \ f \in B_n\} \leq 1$. We shall show that $\sup \{\rho(k, h); \ k, h \in M \cap \Omega\} \leq \frac{1}{n} \ (< \varepsilon)$. For doing so, it is enough to prove that for every $f \in C(K)$ we have $H^{\sigma(f)}(|f(k) - f(h)|) < \frac{1}{n}$ for all $k, h \in M \cap \Omega$. So fix any $0 \neq f \in C(K)$. First, assume that $f \notin B_n$. Then $2f \notin B_n$ as well and so $\sigma(f, 1) = \min \{m \in \mathbb{N}; \ 2f \in B_m\} > n$. Thus $H^{\sigma(f)}(t) \leq \frac{1}{\sigma(f,1)} < \frac{1}{n}$ for all $t \geq 0$, and, in particular, $H^{\sigma(f)}(|f(k) - f(h)|) < \frac{1}{n}$ for all $k, h \in K$. Second, assume $f \in B_n$. Put $q = \max \{p \in \mathbb{N}; \ pf \in B_n\}$; we have $q < +\infty$ as B_n is bounded. Then for every $k, h \in M \cap \Omega$ we have $|qf(k) - qf(h)| < 1$, $|f(k) - f(h)| < \frac{1}{q}$, and so

$$H^{\sigma(f)}(|f(k) - f(h)|) \leq H^{\sigma(f)}(\tfrac{1}{q}) = \frac{1}{\sigma(f, q)} \ .$$

Finally, as $(q + 1)f \notin B_n$, we have $\sigma(f, q) > n$. Therefore $H^{\sigma(f)}(|f(k) - f(h)|) < \frac{1}{n}$.   ∎

Theorem 8. (Avilés [7]) *Given an (uncountable) set Γ and a closed set $K \subset [0, 1]^\Gamma$, then $K \in \mathcal{CLF}$ if and only if for every $\varepsilon > 0$ there are $\Gamma_n^\varepsilon \subset \Gamma$, $n \in \mathbb{N}$, so that $\bigcup_{n=1}^\infty \Gamma_n^\varepsilon = \Gamma$ and K is ε-fragmented by the pseudometric of uniform convergence on Γ_n^ε for every $n \in \mathbb{N}$.*

Proof. **Necessity.** Let sets A_n^ε, $n \in \mathbb{N}$, $\varepsilon > 0$, from (i) in Proposition 5 witness that $K \in \mathcal{CLF}$. For $\gamma \in \Gamma$ put $\pi_\gamma(k) = k(\gamma)$, $k \in K$; then $\pi_\gamma \in C(K)$. We claim that the sets

$$\Gamma_n^\varepsilon = \{\gamma \in \Gamma; \ \pi_\gamma \in A_n^\varepsilon\}, \quad \varepsilon > 0, \quad n \in \mathbb{N},$$

do the job. Indeed, fix any $n \in \mathbb{N}$ and any $\varepsilon > 0$. Let $\emptyset \neq M \subset K$ be any set. Find an open set $\Omega \subset K$ so that $M \cap \Omega \neq \emptyset$ and that for every $k, h \in M \cap \Omega$ we have $\sup \{f(k) - f(h); \ f \in A_n^\varepsilon\} \leq \varepsilon$. Then $\sup\{k(\gamma) - h(\gamma); \ k, h \in M \cap \Omega, \ \gamma \in \Gamma_n^\varepsilon\} \leq \varepsilon$.

Sufficiency. For $\varepsilon > 0$ we define new sets $\Gamma_2^\varepsilon := \Gamma_1^\varepsilon \cup \Gamma_2^\varepsilon$, $\Gamma_3^\varepsilon := \Gamma_1^\varepsilon \cup \Gamma_2^\varepsilon \cup \Gamma_3^\varepsilon$, ... It is easy to check that these sets also satisfy Condition of our theorem. Moreover $\Gamma_1^\varepsilon \subset \Gamma_2^\varepsilon \subset \cdots$ for every $\varepsilon > 0$. Fix any $f \in C(K)$. We claim that *there are $m, n \in \mathbb{N}$ such that $|f(k) - f(h)| < 1$ whenever $k, h \in K$ and $\sup \{|k(\gamma) - h(\gamma)|; \ \gamma \in \Gamma_n^{1/m}\} \leq \frac{1}{m}$.* Let

$$U = \{(k, h) \in K \times K; \ |f(k) - f(h)| < 1\}$$

and

$$C_{\gamma,j} = \left\{(k,h) \in K \times K;\ |k(\gamma) - h(\gamma)| \le \tfrac{1}{j}\right\} \text{ for } \gamma \in \Gamma \text{ and } j \in \mathbb{N}.$$

Then U is open, each $C_{\gamma,j}$ is closed in $K \times K$, and $\bigcap\{C_{\gamma,j};\ (\gamma,j) \in \Gamma \times \mathbb{N}\} \subset U$. Hence, by compactness, there is a finite subset $\Phi \subset \Gamma \times \mathbb{N}$ such that $\bigcap\{C_{\gamma,j};\ (\gamma,j) \in \Phi\} \subset U$. Find a finite set $F \subset \Gamma$ and $m \in \mathbb{N}$ so large that $\Phi \subset F \times \{1,\dots,m\}$. Let $n \in \mathbb{N}$ be so large that $F \subset \Gamma_n^{1/m}$. Then it is easy to check that these m, n satisfy the claim.

Now, for $n, m \in \mathbb{N}$ we put

$$A_{n,m} = \big\{f \in C(K);\ |f(k) - f(h)| < 1 \text{ whenever}$$
$$k, h \in K \text{ and } \sup\big\{|k(\gamma) - h(\gamma)|;\ \gamma \in \Gamma_n^{1/m}\big\} \le \tfrac{1}{m}\big\}.$$

The claim guarantees that $\bigcup\{A_{n,m};\ n, m \in \mathbb{N}\} = C(K)$. And, since for every $n, m \in \mathbb{N}$ the space K is $\frac{1}{m}$−fragmented by the pseudometric of uniform convergence on $\Gamma_n^{1/m}$, we have that K is $1 - A_{n,m}$−fragmentable. We thus verified the condition (ii) from Proposition 5. ∎

Corollary. ([4, 12, 6]) *Let $K \subset \{0,1\}^\Gamma$ be a countably lower fragmentable compact space. Then $K \hookrightarrow \Pi_{n=1}^\infty S_n$, where each S_n is a scattered compact space, and hence $K \in \mathcal{RN}$.*

Proof. Theorem 8 yields the sets $\Gamma_n^{1/2} \subset \Gamma$, $n \in \mathbb{N}$. Then each $K_{|\Gamma_n^{1/2}} := \{k_{|\Gamma_n^{1/2}};\ k \in K\}$ is $\frac{1}{2}-$ fragmented by $\|\cdot\|_\infty$; and hence scattered. Now, we observe that the mapping $k \mapsto \big(k_{|\Gamma_n^{1/2}};\ n \in \mathbb{N}\big)$ injects K continuously into $\Pi_{n=1}^\infty K_{|\Gamma_n^{1/2}}$. ∎

It is worth to compare Theorem 8 with a result of Talagrand-Argyros-Farmaki [14, 11]: *Given a closed subset $K \subset [-1,1]^\Gamma \cap \Sigma(\Gamma)$, then K is an Eberlein compact (if and) only if for every $\varepsilon > 0$ there are sets $\Gamma_n^\varepsilon \subset \Gamma$, $n \in \mathbb{N}$, so that $\bigcup_{n=1}^\infty \Gamma_n^\varepsilon = \Gamma$ and for every $n \subset \mathbb{N}$ and every $k \in K$ the set $\{\gamma \in \Gamma_n^\varepsilon;\ |k(\gamma)| > \varepsilon\}$ is finite.* Up to our knowledge, we do not know of any analogous characterization of Radon-Nikodým compact spaces which are **already sitting** in $[-1,1]^\Gamma$, see the conditions (ii'), (iii') in Theorem 1.

THE FRAMEWORK OF CORSON COMPACT SPACES

Here we investigate what happens with the above classes if we confine to the class of Corson compact spaces. A compact space is called *Corson* if it can be continuously injected into $\Sigma(\Gamma)$, for a suitable (uncountable) set Γ. Here

$$\Sigma(\Gamma) = \big\{x \in \mathbb{R}^\Gamma;\ \#\{\gamma \in \Gamma;\ x(\gamma) \ne 0\} \le \omega\big\}$$

and it is endowed with the topology of pointwise convergence on Γ. A compact space is called *Eberlein* if it is homeomorphic to a weakly compact set of a Banach space. A Banach space X is called *weakly compactly generated* if it contains a weakly compact set whose linear span is dense in it. X is called *weakly Lindelöf determined* if (B_{X^*}, w^*) is a Corson compact space. We have

Theorem 9. (Arvanitakis [6, 23]). *A compact space is Eberlein if (and only if) it is Corson and belongs to the class $\mathcal{QRN}$ $(= \mathcal{CLF} = \mathcal{SF})$.*

This statement had the following prehistory. Let K be a compact space. Alster proved that *K is Eberlein if it is Corson and moreover scattered.* Then Orihuela, Schachermayer and Valdivia [25], and, independently, Stegall [32] proved that *K is Eberlein if (and only if) it is Corson and $K \in \mathcal{RN}$.* According to [3, page 104], this result has already been proved by Rezničenko. However, it is a general belief that Rezničenko's proof has never been published. Then, Stegall extended this equivalence replacing $\mathcal{RN}$ by $\mathcal{IRN}$ [10, Theorem 8.3.6]. And, recently, Arvanitakis replaced $\mathcal{RN}$ in the above equivalence by $\mathcal{QRN}$ [6]. Also, it is asserted in [3, page 104], without any reference or proof, that *every Corson compact space which belongs to $\mathcal{SF}$ is Eberlein.* Of course, by Theorem 3, this is equivalent with Arvanitakis' result.

We do not know of any statement saying that $K \in \mathcal{QRN}$, under some additional conditions, implies $K \in \mathcal{IRN}$. Indeed, all extra conditions on K (known to us) imply already that $K \in \mathcal{RN}$.

Theorem 9 has the following Banach space counterpart.

Theorem 10. ([13]) *For a Banach space X the following assertions are equivalent:*

(i) X is a subspace of a weakly compactly generated space.

(ii) X is weakly Lindelöf determined and $(B_{X^*}, w^*) \in \mathcal{CLF}$.

(ii') X is weakly Lindelöf determined and for every $\varepsilon > 0$ there are $\varepsilon-$Asplund sets $A_n^\varepsilon \subset B_X$, $n \in \mathbb{N}$, such that $B_X = \bigcup_{n=1}^\infty A_n^\varepsilon$.

(iii) (B_{X^*}, w^*) is an Eberlein compact space.

Proof. (i)$\Rightarrow$(ii) can be found in [13]. It is based on the facts that (i) implies that B_X can be written as a countable union of $\varepsilon-$weakly compact sets, and that each $\varepsilon-$weakly compact set is $5\varepsilon-$Asplund.
(ii)$\Leftrightarrow$(ii') is guaranteed by Proposition 6.
(ii)$\Rightarrow$(iii) follows from Theorem 9; this is the main implication.
(iii)$\Rightarrow$(i) is a standard fact due to Amir and Lindenstrauss [2].   ∎

Remarks. 1. Theorem 10 easily yields a well known result that a continuous image of an Eberlein compact space is Eberlein. Indeed, let $\varphi : K \twoheadrightarrow L$ be a continuous mapping from an Eberlein compact space K onto a compact space L. Then $C(L)$ can be understood as a subspace of $C(K)$, which is weakly compactly generated [2]. Thus (i) in Theorem 10 is satisfied. Then, by (iii), $\left(B_{C(L)^*}, w^*\right)$ is an Eberlein compact, and hence, so is its subspace L.

2. In [13], there is a functional analytic proof of Theorem 10, independent of Theorem 9. It uses an elaborated machinery of projectional resolutions of the identity (no wonder), Simons' inequality, an ε-variant of a Jayne-Rogers selection theorem, and a separable reduction.

3. It is also possible to deduce Theorem 9 from Theorem 10. The only fine point we need to check is that if $K \in \mathcal{CLF}$ and K is Corson, then every regular Borel probability measure on K has a separable support. For details see again [13].

There is one more statement in the spirit of Theorem 9. A compact space K is called *almost totally disconnected* if it can be continuously injected into $[0,1]^\Gamma$, with some (uncountable) set Γ, in such a way that for every $k \in K$ the set $\{\gamma \in \Gamma;\ 0 < k(\gamma) < 1\}$ is at most countable. Clearly, every Corson compact is such as well as every closed subspace of $\{0,1\}^\Gamma$ (i.e., every *totally disconnected* compact space). An example giving a right for life for this concept is Namioka's extended long line L obtained from the ordinal interval $[0,\omega_1]$ by inserting a copy of the interval $(0,1)$ in between α and $\alpha+1$ for each ordinal $\alpha < \omega_1$ [21, 22]; here ω_1 means the first uncountable ordinal. Clearly, L is not totally disconnected. Neither is it Corson because it contains the long interval $[0,\omega_1]$. The promised result sounds as:

Theorem 11 (Arvanitakis [6]). *Every almost totally disconnected quasi-Radon-Nikodým compact space K can be continuously injected into the space $\mathcal{P}(S)$ of regular Borel probability measures, endowed with the weak* topology, for a suitable scattered compact space S, and therefore K is a Radon-Nikodým compact space.*

By combining this result with Theorem 7, we get the corollary to Theorem 8.

LOW WEIGHTS AND LOW DENSITIES

Here we shall show a result of Avilés [7] that a $K \in \mathcal{CLF}$ is a Radon-Nikodým compact space provided that the *weight* of K, that is, the least possible cardinal equal to the cardinality of a base of the topology of K, is not "too big". This has also a Banach space analogue. We introduce a special cardinal $\mathbf{b}$ as follows. Consider the product $\mathbb{N}^{\mathbb{N}}$ and endow it by the order that for $\sigma, \tau \in \mathbb{N}^{\mathbb{N}}$ we write $\sigma \prec \tau$ if and only if $\sigma(n) \leq \tau(n)$ for every $n \in \mathbb{N}$. A set $M \subset \mathbb{N}^{\mathbb{N}}$ is called *bounded* if there is $\sigma \in \mathbb{N}^{\mathbb{N}}$ so that $\sigma \succ \tau$ for every $\tau \in M$. The set M is called σ-*bounded* if it can be written as $M = \bigcup_{n=1}^{\infty} M_n$ where each set M_n is bounded. By $\mathbf{b}$ we denote the least possible cardinal equal to the cardinality of a subset of $\mathbb{N}^{\mathbb{N}}$ which is not σ-bounded. It is consistent that $\mathbf{b} > \aleph_1$— the first uncountable cardinal. In fact, Martin's axiom and the negation of the continuum hypothesis imply that $2^{\aleph_0} = \mathbf{b} > \aleph_1$. Now we are ready to formulate

Theorem 12. (Avilés [7]) *Let $K \in \mathcal{CLF}$ and assume that its weight is less than $\mathbf{b}$. Then $K \in \mathcal{RN}$.*

Proof. Let $A_n^\varepsilon \subset B_{C(K)}$, $n \in \mathbb{N}$, $\varepsilon > 0$, be sets from Proposition 5 (iii) witnessing that $K \in \mathcal{CLF}$. For $\varepsilon > 0$ we introduce new sets $A_2^\varepsilon := A_1^\varepsilon \cup A_2^\varepsilon$, $A_3^\varepsilon := A_1^\varepsilon \cup A_2^\varepsilon \cup A_3^\varepsilon$, ... For $\sigma = (\sigma(1), \sigma(2), \ldots) \in \mathbb{N}^{\mathbb{N}}$ put $\varphi(\sigma) = \bigcap_{p=1}^\infty A_{\sigma(p)}^{1/p} \cup \{0\}$. For $\sigma \in \mathbb{N}^{\mathbb{N}}$ the space K is $\varepsilon - \varphi(\sigma)$–fragmentable for every $\varepsilon > 0$. We note that the *density character* of $C(K)$ – the smallest cardinal equal to the cardinality of a dense subset of $C(K)$ – is equal to the weight of the compact space K. Therefore, as $\bigcup_{\sigma \in \mathbb{N}^{\mathbb{N}}} \varphi(\sigma) = B_{C(K)}$, there is a set $\Sigma' \subset \mathbb{N}^{\mathbb{N}}$, with $\#\Sigma' < \mathbf{b}$, such that $\overline{\varphi(\Sigma')} = B_{C(K)}$. By the very definition of the cardinal $\mathbf{b}$, the set Σ' is σ–bounded. Hence, there are bounded sets $\Sigma_m \subset \Sigma'$, $m \in \mathbb{N}$, such that $\bigcup_{m=1}^\infty \Sigma_m = \Sigma'$. For $m \in \mathbb{N}$ find $\sigma_m \in \mathbb{N}^{\mathbb{N}}$ such that $\sigma_m \succ \tau$ for every $\tau \in \Sigma_m$. By the "monotonicity" of φ we get that $\bigcup_{m=1}^\infty \varphi(\sigma_m) \supset \varphi(\Sigma')$. Thus the set $A := \bigcup_{m=1}^\infty \frac{1}{m}\varphi(\sigma_m)$ will still separate the points of K. It then follows that the function $d : K \times K \to [0, 1]$ defined by

$$d(k, h) = \sup \big\{ |f(k) - f(h)|; \ f \in A \big\}, \quad k, h \in K,$$

is a metric on K. Clearly, d is lower semicontinuous. Finally, it fragments K. To prove this, take any $\emptyset \neq M \subset K$ and any $\varepsilon > 0$. Find $m \in \mathbb{N}$ such that $\frac{1}{m+1} < \varepsilon$. Since K is $\varepsilon - \varphi(\sigma_1)$–fragmentable, there is an open set $\Omega_1 \subset K$, with $M \cap \Omega_1 \neq \emptyset$, such that $\sup \big\{ f(k) - f(h); \ k, h \in M \cap \Omega_1, \ f \in \varphi(\sigma_1) \big\} \leq \varepsilon \ldots$ Since K is $\varepsilon - \varphi(\sigma_m)$–fragmentable, there is an open set $\Omega_m \subset K$, with $(M \cap \Omega_1 \cap \cdots \cap \Omega_{m-1}) \cap \Omega_m \neq \emptyset$, such that $\sup \big\{ f(k) - f(h); \ k, h \in M \cap \Omega_1 \cap \cdots \cap \Omega_m, \ f \in \varphi(\sigma_m) \leq \varepsilon \big\}$. Put $\Omega = \Omega_1 \cap \cdots \cap \Omega_m$. Then $M \cap \Omega \neq \emptyset$ and $d - \mathrm{diam}\,(M \cap \Omega) \leq \varepsilon$. Therefore $K \in \mathcal{RN}$ by Theorem 1. ∎

A Banach space counterpart of the above theorem also exists. By combining Theorem 12 with Proposition 6 we immediately get

Theorem 13. (Avilés [7]) *Let X be a Banach space of density character less than $\mathbf{b}$. Assume that X is a subspace of an Asplund generated space, or more generally, that $(B_{X^*}, w^*) \in \mathcal{CLF}$. Then X is Asplund generated.*

Avilés' approach yields several other results: *If X is a subspace of a weakly compactly generated space, or more generally, a weakly $\mathcal{K}$–analytic Banach space, and has density character $< \mathbf{b}$, then X is weakly compactly generated.* This implies a result of Mercourakis, that *under Martin's axiom, weakly $\mathcal{K}$–analytic Banach spaces of density character $< 2^{\aleph_0}$ are already weakly compactly generated.* In the same vein we get a purely topological statement: *Any $\mathcal{K}$–analytic topological space of density character $< \mathbf{b}$ contains a dense σ–compact subset.* By "diminishing" a Talagrand's counterexample [10, Section 4.3] we obtain a *compact space K of weight exactly $\mathbf{b}$ such that $C(K)$ is weakly $\mathcal{K}$–analytic and not a (subspace of a) weakly compactly generated space, i.e., K is not Eberlein.* Likewise, by "diminishing" a Talagrand-Argyros counterexample [10, Section 1.6], we get a *Banach space* of density character **b** which is a subspace of a weakly compactly generated space and is not weakly compactly generated, neither is it Asplund generated.

REGULAR AVERAGING OPERATORS

Though, in solving our problem, there are so far no direct applications of this powerful tool, we think that we should at least mention it briefly. Consider a situation $\varphi : L \twoheadrightarrow K$ where L, K are compact spaces and φ is a continuous surjection. Define $\varphi^\circ : C(K) \to C(L)$ by $\varphi^\circ(f) = f \circ \varphi$, $f \in C(K)$; this is an isometry into. We say that φ admits a *regular averaging operator* (*RAO*) if there exists a linear positive operator $u : C(L) \to C(K)$ such that $u(1_L) = 1_K$ and $u(\varphi^\circ(f)) = f$ for every $f \in C(K)$. We note that in such a case the recipe $Pf = \varphi^\circ(u(f))$, $f \in C(L)$, immediately gives a linear norm one projection $P : C(L) \to C(L)$, with range $\varphi^\circ(C(K))$. The theory of RAO, and of a "dual" concept of regular extension operators, has been broadly developed in Pełczyński memoir [26]. Then, Ditor, using the topological tool of inverse limits, proved the following quite general statement.

Theorem 14. (Ditor [9]) *For every infinite compact space K, with weight $\aleph$, there exist a closed subspace $L \subset \{0,1\}^\aleph$ and a continuous surjection $\varphi : L \twoheadrightarrow K$ which admits a RAO.*

Let us focus for a while on the situation when L is the Cantor discontinuum $\{0,1\}^{\mathbb{N}}$ and $K = [0,1]$. It was Milutin who, in his thesis from 1952, for the first time found, by a clever construction, a surjection $\varphi : \{0,1\}^{\mathbb{N}} \to \to [0,1]$ admitting a RAO. It should be noted that the (surjective) mapping $\{0,1\}^{\mathbb{N}} \ni \varepsilon = (\varepsilon_1, \varepsilon_2, \ldots) \mapsto \frac{1}{2} \sum_{n=1}^{\infty} \left(\frac{1}{2}\right)^{n-1} \varepsilon_n \in [0,1]$ *does not admit any* RAO. On the other hand, for every $\frac{1}{2} < \Delta < 1$, the surjection

$$\{0,1\}^{\mathbb{N}} \ni \varepsilon = (\varepsilon_1, \varepsilon_2, \ldots) \mapsto (1 - \Delta) \sum_{n=1}^{\infty} \Delta^{n-1} \varepsilon_n \in [0,1]$$

does admit a RAO. Both these facts have been proved, using a probabilistic approach, in a recent paper of Argyros and Arvanitakis [5]. Starting from a fixed $\varphi : \{0,1\}^{\mathbb{N}} \twoheadrightarrow [0,1]$, admitting a RAO, then, for any cardinal $\aleph$, a corresponding "product" mapping $\{0,1\}^{\mathbb{N} \times \aleph} \twoheadrightarrow [0,1]^\aleph$ also admits a RAO. And realizing that every compact space K can be considered as a closed subspace of the product $[0,1]^\aleph$, some extra work reproves Ditor's theorem. All this can be found in [5].

We included the above stuff because of the following reasoning. Assume we have given $K \in \mathcal{QRN}$. And let L be the space found for our K in Ditor's theorem. If we proved that $L \in \mathcal{QRN}$, then by Corollary to Theorem 8, L would be a Radon-Nikodým compact space. Thus, by Theorem 1, $C(L)$ would be Asplund generated. And, as $\varphi^\circ(C(K))$ is complemented in $C(L)$, the space $C(K)$ would also be Asplund generated, and therefore K would be a Radon-Nikodým compact space. However, the trouble is that we are in general not able to decide if $L \in Q\mathcal{RN}$. Yet we have several "parallel" positive statements supporting this: In [5, Corollary 24], it is shown that if K is an Eberlein, Talagrand, Gul'ko, or Corson compact space, then the corresponding L in Ditor's theorem can also be constructed such.

THE CASE $K = K_1 \cup K_2$

Here we focus on the following

Subproblem: If a compact space K can be written as $K = K_1 \cup K_2$, with $K_1, K_2 \in \mathcal{RN}$, does this imply that $K \in \mathcal{RN}$?

Note that such a K belongs to $\mathcal{IRN}$. Indeed, put $L = K_1 \times \{1\} \cup K_2 \times \{2\}$. That $L \in \mathcal{RN}$ follows easily from the very definition. And the mapping $\varphi : L \twoheadrightarrow K$ defined by $\varphi(k_i, i) = k_i$, $k_i \in K_i$, $i = 1, 2$, is clearly continuous and surjective. Unfortunately, such a φ may not admit any RAO: Take $K = [0,1]$, $K_1 = [0, \frac{1}{2}]$, $K_2 = [\frac{1}{2}, 1]$. Then, according to [5, Theorem 2], the corresponding φ does not have any RAO, and hence we do not know if $\varphi^\circ(C(K))$ is complemented $C(L)$.

Positive results, known to us, are collected in the next

Theorem 15. *Assume that a compact space K can be written as $K = K_1 \cup K_2$, where $K_1, K_2 \in \mathcal{RN}$. Then $K \in \mathcal{RN}$ provided that one of the following conditions is satisfied.*
 (i) $K_1 \cap K_2$ is G_δ in K,
 (ii) K_1 is a retract of K,
 (iii) $K \backslash K_1$ is scattered,
 (iv) $K_1 \cap K_2$ is metrizable,
 (v) $K_1 \cap K_2$ is scattered.

Question. *What if $K_1 \cap K_2$ is an Eberlein compact space?*

The cases (i)–(iv) are due to Matoušková and Stegall [20]. (v) was shown to us by Spurný [29]. In proofs the following lemma is a basic tool.

Lemma 16 ([20, Lemma 6.4]). *Let $K \in \mathcal{RN}$ and let $F \subset K$ be a closed subset. Then there exists an equimeasurable (i.e., Asplund) set $E \subset C(K)$ such that for every $k \in K$ and $h \in K \backslash F$, $k \neq h$, there is $f \in E$ such that $f(h) > 0$, $f(k) = 0$, and $f_{|F} \equiv 0$.*

Its proof begins from the very definition of Radon-Nikodým compact space and then uses a simple compactness argument.

An important role in proofs of some facts above is played also by

Proposition 17. *Assume that a compact space K can be written as $K = K_1 \cup K_2$, where $K_1, K_2 \in \mathcal{RN}$. Then $K \in \mathcal{RN}$ if (and only if) there exist lower semicontinuous fragmenting metrics d_1, d_2 on K_1, K_2 respectively, such that $\min(d_1, d_2)$ restricted to $K_1 \cap K_2$ is minorized by a lower semicontinuous* metric, or, equivalently, $\min(d_1, d_2)-$Lipschitz functions on $K_1 \cap K_2$ separate the points of it.

In order to get a taste, let us *prove* (iv) in Theorem 15. Find a countable family $f_n \in B_{C(K_1 \cap K_2)}$, $n \in \mathbb{N}$, of functions which separate the (separable metrizable) space $K_1 \cap K_2$. Extend each f_n to $\tilde{f}_n \in B_{C(K)}$. Put then $E_0 = \{\frac{1}{n}\tilde{f}_n; \ n \in \mathbb{N}\}$; this is clearly a relatively norm compact, and hence equimeasurable set. Lemma 16 implies that there exists an equimeasurable set $E_1 \subset C(K)$ which separates any two distinct points $k \in K$ and $h \in K \backslash K_2$. Analogously, we find an equimeasurable set $E_2 \subset C(K)$ which separates any two distinct points $k \in K$ and $h \in K \backslash K_1$. Then the set $E_0 \cup E_1 \cup E_2$ can be easily seen to be equimeasurable and separates all distinct points of K. Therefore, by Theorem 1, $K \in \mathcal{RN}$.

Finally, we want to mention a recent paper of Iancu and Watson [17] providing positive answers to our problem for the situation $\varphi : L \twoheadrightarrow K$, $L \in \mathcal{RN}$, when the set $\{k \in K; \ \varphi^{-1}(k)$ is not a singleton$\}$ has some special properties, for instance, when it is included in a closed metrizable subspace.

References

[1] K. Alster, *Some remarks on Eberlein compacts*, Fund. Math. **104** (1979), 43–46.

[2] D. Amir, J. Lindenstrauss, *The structure of weakly compact sets in Banach spaces*, Ann. Math. **88** (1968), 35–46.

[3] A.V. Archangel'skij, *General Topology II*, Encyclopedia of Mathematical Sciences, Vol. 50, Springer Verlag, Berlin 1995 (Russian original VINITI, Moscow 1989).

[4] S. Argyros, *A note on Radon-Nikodým compact sets*, an unpublished manuscript.

[5] S. Argyros, A. Arvanitakis, *A characterization of regular averaging operators and its consequences*, a preprint.

[6] A. Arvanitakis, *Some remarks on Radon-Nikodým compact spaces*, Fund. Math. **172** (2002), 41–60.

[7] A. Avilés, *Radon-Nikodým compact spaces of low weight and Banach spaces*, Studia Math. **166** (2005), 71–82.

[8] R. D. Bourgin, *Geometric Aspects of Convex Sets wit the Radon-Nikodým Property*, Lect. Notes Math. No. 993, Berlin, Springer-Verlag, 1983.

[9] S. Ditor, *On a lemma of Milutin concerning averaging operators in continuous function spaces*, Trans. Amer. Math. Soc. **149** (1970), 443 452.

[10] M. Fabian, *Gâteaux differentiability of convex functions and topology - weak Asplund spaces*, CMB Series, John Wiley & Sons Inc., New York 1997.

[11] M. Fabian, G. Godefroy, V. Montesinos, V. Zizler, *Inner characterizations of weakly compactly generated Banach spaces and their relatives*, J. Math. Anal. Appl. **297** (2004), 419–455.

[12] M. Fabian, M. Heisler, E. Matoušková, *Remarks on continuous images of Radon-Nikodým compacta*, Comment Math. Univ. Carolinae **39** (1998), 59–69.

[13] M. Fabian, V. Montesinos, V. Zizler, *Quantitative versions of Asplund sets and weakly compact sets*, a preprint.

[14] V. Farmaki, *The structure of Eberlein, uniform Eberlein and Talagrand compact spaces*, Fund. Math. **128** (1987), 15-28.

[15] A. Grothendieck, *Produits tensoriels topologiques et espaces nucléaires*, Memoirs Amer. Math. Soc. No. 16, Providence, Rhode Island 1955.

[16] N. Ghoussoub, B. Maurey, $H_\delta-$*embedings in Hilbert space and optimization on* G_δ *sets*, Memoirs of AMS, Vol. 62, Providence 1986.

[17] M. Iancu, S. Watson, *On continuous images of Radon-Nikodým compact spaces through the metric characterization*, Topology Proc. **26** (2001–2002), 677–693.

[18] J.E. Jayne, I. Namioka, C.A. Rogers, *Norm fragmented weak* compact sets*, Collect. Math. **41** (1990), 133–163.

[19] E. Matoušková, *Extensions of continuous and Lipschitz mappings*, Canadian Math. Bull. **43** (2000), 208–217.

[20] E. Matoušková, Ch. Stegall, *Compact spaces with a finer metric topology and Banach spaces*, in T. Banakh, (Editor), General Topology in Banach Spaces, Nova Sci. Publ., Huntington, N.Y. 2001, 81–101.

[21] I. Namioka, *Eberlein and Radon-Nikodým compact spaces*, Lect. Notes, Univ. College London, Autumn term 1985.

[22] I. Namioka, *Radon-Nikodým compact spaces and fragmentability*, Mathematika **34** (1987), 258–281.

[23] I. Namioka, *On generalizations of Radon-Nikodým compact spaces*, Topology Proc. **26** (2001-2002), 741–750.

[24] I. Namioka, R.R. Phelps, *Banach spaces which are Asplund spaces*, Duke Math. J. **42** (1975), 735–750.

[25] J. Orihuela, W. Schachermayer, M. Valdivia, *Every Radon-Nikodým Corson compact is Eberlein compact*, Studia Math. **98** (1991), 157–174.

[26] A. Pełczyński, *Linear extensions, linear averagings, and their applications to linear topological classification of spaces of continuous functions*, Dissertationes Mathematicae, Warszawa, PWN 1968.

[27] O.I. Reinov, *On a class of Hausdorff compacts and GSG Banach spaces*, Studia Math. **71** (1981), 113–126.

[28] M. Raja, *Embeddings ℓ_1 as Lipschitz functions*, Proc. Amer. Math. Soc. **193** (2005), 2395–2400.

[29] J. Spurný, A private communication.

[30] Ch. Stegall, *The Radon-Nikodým property in conjugate Banach spaces*, Trans. Amer. Math. Soc. **206** (1975), 213–223.

[31] Ch. Stegall, *The Radon-Nikodým property in conjugate Banach spaces II*, Trans. Amer. Math. Soc. **264** (1981), 507–519.

[32] Ch. Stegall, *More facts about conjugate Banach spaces with the Radon-Nikodým property, II*, Acta Univ. Carolinae – Math. Phys. **32** (1991), 47–54.

Acknowledgement. We thank several colleagues, in particular Isaac Namioka, for a careful commenting on this text.

MATHEMATICAL INSTITUTE, CZECH ACADEMY OF SCIENCES, ŽITNÁ 25, 115 67 PRAGUE 1, BOHEMIA

E-mail address : `fabian@math.cas.cz`

CONVEXITY, COMPACTNESS AND DISTANCES

ANTONIO S. GRANERO AND MARCOS SÁNCHEZ

ABSTRACT. In this paper we investigate whether the distances $d(\overline{\mathrm{co}}^{w^*}(K), Z)$ are controlled by the distances $d(K, Z)$ (that is, if $d(\overline{\mathrm{co}}^{w^*}(K), Z) \leq M d(K, Z)$ for some $1 \leq M < \infty$), when K is a weak*-compact subset of $\ell_\infty(I)$ and Z belongs to some particular classes of subspaces of $\ell_\infty(I)$.

1. INTRODUCTION AND PRELIMINARIES

If $(X, \|\cdot\|)$ is a Banach space, let $B(X)$ and $S(X)$ be the closed unit ball and unit sphere of X, respectively, and X^* its topological dual. If $u \in X$ and K, Z are two subsets of X, let $d(u, Z) = \inf\{\|u - z\| : z \in Z\}$ be the distance to Z from u, $d(K, Z) = \sup\{d(k, Z) : k \in K\}$ the distance to Z from K, $\mathrm{co}(K)$ the convex hull of K, $\overline{\mathrm{co}}(K)$ the norm-closure of $\mathrm{co}(K)$ and $\overline{\mathrm{co}}^\tau(K)$ the τ-closure of $\mathrm{co}(K)$, τ being a topology on X. The weak*-topology of a dual Banach space X^* is denoted by w*.

If $Z \subset X^*$ is a subspace of a dual Banach space X^* and $K \subset X^*$ a weak*-compact subset of X^*, an interesting problem is whether the distance $d(\overline{\mathrm{co}}^{w^*}(K), Z)$ is controlled by the distance $d(K, Z)$. The subspace $Z \subset X^*$ is said to have M-control inside X^* for some $1 \leq M < \infty$ if $d(\overline{\mathrm{co}}^{w^*}(K), Z) \leq M d(K, Z)$, for every weak*-compact subset K of X^*, and it is said to have control inside X^* provided it has M-control for some $1 \leq M < \infty$. The behavior of the distances $d(\overline{\mathrm{co}}^{w^*}(K), Z)$ with respect to the distances $d(K, Z)$ is varied. If $Z \subset X^*$ is a weak*-closed subspace of X^*, it is very easy to see that $d(\overline{\mathrm{co}}^{w^*}(K), Z) = d(K, Z)$, even when $K \subset X^*$ is an arbitrary subset of X^*. However, if $Z \subset X^*$ is a subspace of X^*, which is not weak* closed, all situations are possible.

It is worth mentioning that every Banach space X has control inside its bidual X^{**} because of the following extension of the Krein-Šmulian Theorem ([8],[9],[10]): if $Z \subset X$ is a closed subspace of the Banach space X and $K \subset X^{**}$ is a weak* compact subset of X^{**}, then: (1) always $d(\overline{\mathrm{co}}^{w^*}(K), Z) \leq 5d(K, Z)$ and, if $Z \cap K$ is weak*-dense in K, then $d(\overline{\mathrm{co}}^{w^*}(K), Z) \leq 2d(K, Z)$; (2) the equality $d(K, X) = d(\overline{\mathrm{co}}^{w^*}(K), X)$ holds in many cases, for instance: (i) if $\ell_1 \not\subseteq X^*$; (ii) if X has weak*-angelic dual unit ball; (iii) if $X = (\ell_1(I), \|\cdot$

1991 *Mathematics Subject Classification.* 46B20, 46B26.

Key words and phrases. Krein-Šmulyan Theorem, convexity, weak*-compatness, distances.

Supported in part by DGICYT grant MTM2004-01308.

$\|_1$); (iv) if X is the Orlicz space $\ell_\varphi(I)$ and φ satisfies the Δ_2 condition at 0. However, not every Banach space X has 1-control inside its bidual X^{**}, because in [10] there are given examples of Banach spaces X and weak*-compact subsets $K \subset X^{**}$ such that $d(\overline{\mathrm{co}}^{w^*}(K), X) \geq 3d(K, X) > 0$.

In [11] it is proved that a Banach space Y fails to contain a copy of $\ell_1(\mathfrak{c})$ if and only if Y has *universally 3-control* (Y has universally 3-control when Z has 3-control inside X^*, provided X is a Banach space and $Z \subset X^*$ is subspace of X^* isomorphic to Y). Moreover, if Y is a Banach space with a copy of $\ell_1(\mathfrak{c})$ and $1 \leq M < \infty$, there exist a Banach space X and a subspace $Z \subset X^*$ isomorphic to Y such that Z fails to have M-control inside X^*.

This paper is devoted to investigate the control in $\ell_\infty(I)(\cong \ell_1(I)^*)$ of some special subspaces of $\ell_\infty(I)$. First, we study the space $C^*(H)$ of bounded continuous real functions on a topological space H, when $C^*(H)$ is considered as a subspace of $\ell_\infty(H)$. Actually, our frame is something more general: we pick an infinite set I, a topological space H and a mapping $\varphi : I \to H$ such that $\overline{\varphi(I)} = H$. Denote

$$C(H) = \{f : H \to \mathbb{R} : f \text{ continuous}\}, \; C(H)_\varphi = \{f \circ \varphi : f \in C(H)\},$$

$$C^*(H) = \{f : H \to \mathbb{R} : f \text{ continuous and bounded}\} \text{ and}$$

$$C^*(H)_\varphi = \{f \circ \varphi : f \in C^*(H)\}.$$

We shall study the control of $C^*(H)_\varphi$ inside $\ell_\infty(I)$. With this strategy we gain in generality, although the more natural and interesting case is when I is a dense subset of the topological space H and $\varphi : I \to H$ is the canonical inclusion. As we shall see later, in some cases the condition $\overline{\varphi(I)} = H$ is sufficient, in order to have control, but there are situations in which we will need $\varphi(I) = H$.

In Section 2 we show the connection between distances and oscillations. In Section 3 we see that for some topological spaces H (for example, when H is metrizable) the space $C^*(H)_\varphi$ has 1-control inside $\ell_\infty(I)$. In Section 4 we study the control inside $\ell_\infty(I)(= C(\beta I))$ of the subspace of $C(\beta I)$ consisting of elements which are constant on a given closed subset $D \subset \beta I$. The Section 5 is devoted to study the control of $C(H)_\varphi$ inside $\ell_\infty(I)$ when H is compact. Finally, in Section 6 we investigate the control in $\ell_\infty(X)$ of the subspace $B_{1b}(X)$ of bounded functions of the first Baire class on X, X being a metric space.

Our notation and terminology are standard. $\mathfrak{c}$ is the cardinal of $\mathbb{R}$. If I is a set, let $|I|$ denote the cardinality of I. Our topological spaces are Hausdorf and completely regular. For undefined notions and properties we refer the reader to the books [5],[6],[7],[13],[16].

2. OSCILLATIONS AND DISTANCES

In this Section we consider some auxiliary facts, that will be used later. Let I be a set, H a topological space and $\varphi : I \to H$ a mapping such that $\overline{\varphi(I)} = H$. Given a function $f \in \ell_\infty(I)$, we show the connection between the distance $d(f, C^*(H)_\varphi)$ and the "oscillation" of f on H. If $k \in H$ let $\mathcal{V}^k$ denote the family of open neighborhoods of k in H and, if $f \in \ell_\infty(I)$, define the functions $f_1, f_2 : H \to [-\infty, +\infty]$ in the following way:

$$\forall k \in H, \ f_1(k) = \inf\{\sup(f(\varphi^{-1}(V))) : V \in \mathcal{V}^k\} \text{ and}$$

$$f_2(k) = \sup\{\inf(f(\varphi^{-1}(V))) : V \in \mathcal{V}^k\}.$$

Obviously, $-\infty \le f_2(k) \le f_1(k) \le +\infty$ for every $k \in H$. Now, we define the φ-oscillation $Osc_\varphi(f, k)$ of f in $k \in H$ as:

$$Osc_\varphi(f, k) = \lim_{V \in \mathcal{V}^k} \left(\sup\{f(i) - f(j) : i, j \in \varphi^{-1}(V)\} \right) =$$

$$= \lim_{V \in \mathcal{V}^k} \left(\sup f(\varphi^{-1}(V)) - \inf f(\varphi^{-1}(V)) \right).$$

Observe that $Osc_\varphi(f, k) = +\infty$, if $f_1(k) = \pm\infty$ or $f_2(k) = \pm\infty$, and $Osc_\varphi(f, k) = f_1(k) - f_2(k)$ otherwise. Finally define the φ-oscillation of f in H as:

$$Osc_\varphi(f) = \sup\{Osc_\varphi(f, k) : k \in H\}.$$

Lemma 2.1. *With the above notation f_1 is upper semicontinuous (for short, usc) and f_2 is lower semicontinuous (for short, lsc) on H.*

Proof. Let us see that f_1 is usc on H. Suppose that $a \in \mathbb{R}$ and prove that $U_a = \{k \in H : f_1(k) < a\}$ is open in H. If $k \in U_a$ then there exists an open neighbourhood $W \in \mathcal{V}^k$ such that $\sup f(\varphi^{-1}(W)) < a$. If $h \in W$ we have $W \in \mathcal{V}^h$, whence we get $f_1(h) \le \sup f(\varphi^{-1}(W)) < a$ and, so, $W \subset U_a$. Hence, U_a is open.

In an analogous way it can be shown that f_2 is lsc. $\qquad\square$

The following proposition describes the connection between oscillations and distances.

Proposition 2.2. *Let I be a set, H be a normal topological space and $\varphi : I \to H$ a mapping such that $\overline{\varphi(I)} = H$. Then for $f \in \ell_\infty(I)$, we have:*

$$d(f, C^*(H)_\varphi) = \frac{1}{2} Osc_\varphi(f).$$

Proof. A) $d(f, C^*(H)_\varphi) \ge \frac{1}{2}Osc_\varphi(f)$. Indeed, let $\epsilon > 0$ and choose $h \in C^*(H)$ such that $\sup\{|f(x) - h \circ \varphi(x)| : x \in I\} \le d(f, C^*(H)_\varphi) + \epsilon$. Then, if $k \in H$ and $V \in \mathcal{V}^k$ satisfies $h(x) - h(y) < \epsilon$ for every pair $x, y \in V$, we have $\sup\{f(i) - f(j) : i, j \in \varphi^{-1}(V)\} \le 2d(f, C^*(H)_\varphi) + 3\epsilon$, and so, $\frac{1}{2}Osc_\varphi(f) \le d(f, C^*(H)_\varphi) + \frac{3}{2}\epsilon$, whence we get $d(f, C^*(H)_\varphi) \ge \frac{1}{2}Osc_\varphi(f)$.

B) $d(f, C^*(H)_\varphi) \le \frac{1}{2}Osc_\varphi(f)$. Indeed, let $\frac{1}{2}Osc_\varphi(f) = \delta < +\infty$ and let f_1, f_2 be the functions introduced above, which in this case are finite and bounded. Moreover, $f_1 - \delta \le f_2 + \delta$ on H. So, by [6, Ex. 1.7.15 (b), p.88] there exists $h \in C^*(H)$ such that $f_1 - \delta \le h \le f_2 + \delta$ on H. Since $f(i) \in [f_2(\varphi(i)), f_1(\varphi(i))]$ for every $i \in I$, we conclude that:

$$f(i) - \delta \le f_1 \circ \varphi(i) - \delta \le h \circ \varphi(i) \le f_2 \circ \varphi(i) + \delta \le f(i) + \delta,$$

that is, $\sup\{|f(i) - h \circ \varphi(i)| : i \in I\} \leq \delta$, and this implies that $d(f, C^*(H)_\varphi) \leq \frac{1}{2}Osc_\varphi(f)$. $\qquad\square$

Proposition 2.3. *Let I be a set, K a normal countably compact space, $\varphi : I \to K$ a mapping such that $\varphi(I) = K$, K_0 a metric space, $q : K \to K_0$ a continuous mapping with $q(K) = K_0$ and $\varphi_1 = q \circ \varphi : I \to K_0$. Consider $C(K)_\varphi$ and $C(K_0)_{\varphi_1}$ as subspaces of the space $\ell_\infty(I)$. Then $C(K_0)_{\varphi_1} \subset C(K)_\varphi \subset \ell_\infty(I)$ and for $f \in \overline{C(K_0)_{\varphi_1}}^{w^*}$ we have:*

$$d(f, C(K)_\varphi) \leq d(f, C(K_0)_{\varphi_1}) \leq 2d(f, C(K)_\varphi).$$

Proof. Since $f \in \overline{C(K_0)_{\varphi_1}}^{w^*}$, f is constant in each subset $\varphi_1^{-1}(k) \subset I$ for every $k \in K_0$. Thus, there exists $h : K_0 \to \mathbb{R}$ such that $f = h \circ \varphi_1$. Clearly, $d(f, C(K)_\varphi) \leq d(f, C(K_0)_{\varphi_1})$ because $C(K_0)_{\varphi_1} \subset C(K)_\varphi$. In order to prove the inequality $d(f, C(K_0)_{\varphi_1}) \leq 2d(f, C(K)_\varphi)$, it is enough to prove that $Osc_{\varphi_1}(f) \leq 2Osc_\varphi(f)$ by Proposition 2.2.

If $Osc_{\varphi_1}(f) = 0$, obviously $Osc_{\varphi_1}(f) \leq 2Osc_\varphi(f)$. Otherwise, take $Osc_{\varphi_1}(f) > \delta > 0$. So, there exist a point $k_0 \in K_0$ and two sequences $i_n, j_n \in I$, $n \geq 1$, such that $\varphi_1(i_n) \to k_0$, $\varphi_1(j_n) \to k_0$ when $n \to \infty$, and $f(i_n) - f(j_n) > \delta$ for every $n \geq 1$. Since K is countably compact, we can find points $x_0, y_0 \in K$ such that

$$x_0 \in q^{-1}(k_0) \cap \overline{\{\varphi(i_n) : n \geq 1\}} \text{ and } y_0 \in q^{-1}(k_0) \cap \overline{\{\varphi(j_n) : n \geq 1\}}.$$

CLAIM. Either $Osc_\varphi(f, x_0) \geq \frac{\delta}{2}$ or $Osc_\varphi(f, y_0) \geq \frac{\delta}{2}$.

Indeed, let $i_0, j_0 \in I$ be such that $\varphi(i_0) = x_0$ and $\varphi(j_0) = y_0$. Since f is constant on $\varphi_1^{-1}(k_0)$ and $i_0, j_0 \in \varphi_1^{-1}(k_0)$, then $f(i_0) = f(j_0)$, whence, for every $n \geq 1$, we have $\delta < f(i_n) - f(j_n)f(i_n) - f(i_0) + f(j_0) - f(j_n)$. Thus either $f(i_n) - f(i_0) > \frac{\delta}{2}$ or $f(j_0) - f(j_n) > \frac{\delta}{2}$. So, there exists an infinite subset $N \subset \mathbb{N}$ such that either $f(i_n) - f(i_0) > \frac{\delta}{2}$, for every $n \in N$, or $f(j_0) - f(j_n) > \frac{\delta}{2}$, for every $n \in N$, and this proves the Claim.

Therefore, $Osc_\varphi(f) \geq \delta/2$ for every $\delta > 0$ such that $Osc_{\varphi_1}(f) > \delta$, whence we get $Osc_{\varphi_1}(f) \leq 2Osc_\varphi(f)$. $\qquad\square$

3. SPACES OF TYPE $C^*(H)_\varphi$ WITH 1-CONTROL INSIDE $\ell_\infty(I)$

In this Section we consider some classes of topological spaces H such that $C^*(H)_\varphi$ has 1-control inside $\ell_\infty(I)$. We say that a topological space H is in the class $\mathfrak{F}$ (for short, $H \in \mathfrak{F}$) if for every $A \subset H \times H$ and every $h \in H$, with $(h, h) \in \overline{A}$, there exist $d \in H$ and a sequence $(\alpha_n)_{n \geq 1}$ in A such that $\alpha_n \to (d, d)$ as $n \to \infty$. So, H is in $\mathfrak{F}$ provided: (1) H is metrizable; (2) H satisfies the first axiom of countability; (3) $H \times H$ is a Fréchet-Urysohn space. We see in the sequel that the behavior of an topological space $H \in \mathcal{F}$, concerning the control of $C^*(I)_\varphi$ inside $\ell_\infty(I)$, is optimum because: (1) we obtain 1-control; (2) it is sufficient for $\varphi(I)$ to be dense in H. We remark the difference between the case of a topological space $H \in \mathfrak{F}$ and the case of general compact spaces H. As we will see in Section 5, for a general compact space H we only have 5-control and we need that $\varphi(I) = H$, in order to have control.

Proposition 3.1. *Let H be a normal topological space with $H \in \mathfrak{F}$, I a set, $\varphi : I \to H$ a mapping such that $\overline{\varphi(I)} = H$ and $W \subset \ell_\infty(I)$ a weak*-compact subset. Then*

$$d(\overline{co}^{w^*}(W), C^*(H)_\varphi) = d(W, C^*(H)_\varphi).$$

Proof. Suppose that there exist a weak*-compact subset $W \subset B(\ell_\infty(I))$ and two real numbers $a, b > 0$ such that

$$d(\overline{co}^{w^*}(W), C^*(H)_\varphi) > b > a > d(W, C^*(H)_\varphi).$$

Pick $f_0 \in \overline{co}^{w^*}(W)$ with $d(f_0, C^*(H)_\varphi) > b$. By Proposition 2.2 there exists a point $k_0 \in H$ such that $\frac{1}{2}Osc_\varphi(f_0, k_0) > b$. So, there exist $\epsilon > 0$ and, for every $V \in \mathcal{V}^{k_0}$, two points $i_V, j_V \in \varphi^{-1}(V)$ such that

$$f_0(i_V) - f_0(j_V) > 2b + \epsilon.$$

In particular, $(k_0, k_0) \in \overline{\{(\varphi(i_V), \varphi(j_V)) : V \in \mathcal{V}^{k_0}\}}$. Since $H \in \mathfrak{F}$ there exist a sequence $\{(i_n, j_n) : n \geq 1\} \subset \{(i_V, j_V) : V \in \mathcal{V}^{k_0}\}$ and a point $h_0 \in H$ such that $(\varphi(i_n), \varphi(j_n)) \to (h_0, h_0)$. Find a Radon Borel probability measure μ on W such that $f_0 = r(\mu)(= \text{barycenter of } \mu)$. For every $n \geq 1$ let $T_n : \ell_\infty(I) \to \mathbb{R}$ be such that $T_n(f) = f(i_n) - f(j_n)$, for all $f \in \ell_\infty(I)$. Clearly, T_n is a linear mapping which is $\|\cdot\|$-continuous and weak*-continuous. So, we have

$$2b + \epsilon < f_0(i_n) - f_0(j_n) = T_n(f_0) = T_n(r(\mu)) = \int_W T_n(f)d\mu.$$

For every $n \geq 1$ let $A_n = \{f \in W : T_n(f) > 2b + \frac{\epsilon}{2}\}$.

   CLAIM. $\mu(A_n) \geq \frac{\epsilon}{4}$ for all $n \geq 1$.

   Indeed, we have

$$2b + \epsilon < \int_W T_n(f)d\mu = \int_{A_n} T_n(f)d\mu + \int_{W \setminus A_n} T_n(f)d\mu \leq 2\mu(A_n) + (2b + \frac{\epsilon}{2}),$$

whence we deduce that $\mu(A_n) \geq \frac{\epsilon}{4}$ for all $n \geq 1$.

Let $B_n := \bigcup_{m \geq n} A_m$ for every $n \geq 1$. The sequence $\{B_n\}_{n \geq 1}$ is decreasing and $\mu(B_n) \geq \frac{\epsilon}{4}$ for every $n \geq 1$. Hence $\mu(\bigcap_{n \geq 1} B_n) \geq \frac{\epsilon}{4}$ and therefore $\bigcap_{n \geq 1} B_n \neq \emptyset$. Choose $g \in \bigcap_{n \geq 1} B_n$ and inductively a sequence $\{A_{m_k}\}_{k \geq 1}$, $m_k < m_{k+1}$, such that $g \in A_{m_k}$ for all $k \geq 1$. Then $g(i_{m_k}) - g(j_{m_k}) > 2b + \frac{\epsilon}{2}$ for all $k \geq 1$, whence we get $Osc_\varphi(g, h_0) \geq 2b + \frac{\epsilon}{2}$, that is, $d(g, C^*(H)_\varphi) > b$ (by Proposition 2.2), which contradicts the fact that $g \in W$. $\qquad \square$

Corollary 3.2. *Let I be an infinite set, K be a scattered compact Hausdorf space such that $K^{(2)} = \emptyset$ and $\varphi : I \to K$ a mapping such that $\overline{\varphi(I)} = K$. Then for every weak*-compact subset $W \subset \ell_\infty(I)$ we have $d(W, C(K)_\varphi) = d(\overline{co}^{w^*}(W), C(K)_\varphi)$.*

Proof. By Proposition 3.1 it is enough to prove that $K \in \mathfrak{F}$. As $K^{(2)} = \emptyset$, then K is the topological sum of a finite number of disjoint clopen subsets, say $K = \oplus_{i=1}^n K_i$, each K_i being the Alexandrov compactification $K_i = \alpha J_i$ of some discrete set J_i. So, K has property $\mathfrak{F}$ if and only if each αJ_i has. Now apply the trivial fact that the Alexandrov compactification αJ of a discrete set J has property $\mathfrak{F}$. $\qquad\square$

4. The control of $C(\beta I, D)$ and $C_0(\beta I, D)$ inside $\ell_\infty(I)$.

If I is an infinite set, let βI denote the Stone-Čcech compactification of I and $I^* := \beta I \setminus I$. For $f \in \ell_\infty(I)$, $\check{f}$ will be the Stone-Čech extension of f to βI. Pick a closed subset $D \subset \beta I$ and put:

$$C(\beta I, D) := \{f \in \ell_\infty(I) : \check{f} \text{ is constant on } D\},$$

and

$$C_0(\beta I, D) := \{f \in \ell_\infty(I) : \check{f}_{\restriction D} \equiv 0\}.$$

Let us study the control of $C(\beta I, D)$ and $C_0(\beta I, D)$ inside $\ell_\infty(I)$. First, we see an elementary fact concerning the distances to $C(\beta I, D)$ and $C_0(\beta I, D)$.

Proposition 4.1. *Let I be an infinite set, $D \subset \beta I$ be a closed non-empty subset and $z \in \ell_\infty(I)$. Then the distances $d(z, C_0(\beta I, D))$ and $d(z, C(\beta I, D))$ satisfy*

(1) $d(z, C_0(\beta I, D)) = \|\check{z}_{\restriction D}\| : \sup\{|\check{z}(d)| : d \in D\}$;

(2) $d(z, C(\beta I, D)) \frac{1}{2} \sup\{\check{z}(d) - \check{z}(d') : d, d' \in D\} = \frac{1}{2}(\check{z}(d_1) - \check{z}(d_2))$, for some $d_1, d_2 \in D$ such that $\check{z}(d_1) = \max\{\check{z}(d) : d \in D\}$ and $\check{z}(d_2) = \min\{\check{z}(d) : d \in D\}$.

Proof. (1) Consider the canonical restriction mapping $r : C(\beta I) \to C(D)$ such that $r(f) = f_{\restriction D}, \forall f \in C(\beta I)$. Observe that r is a quotient mapping such that $r(B(C(\beta I))) = B(C(D))$ (by Tietze's extension theorem) and $Ker(r) = C_0(\beta I, D)$. So, for $f \in C(\beta I)$ we have

$$\|f_{\restriction D}\| = \|r(f)\| = d(f, C_0(\beta I, D)).$$

(2) Let H be the topological space obtained from βI by the identification of D in one point, say d_0. So, $H = (\beta I \setminus D) \cup \{d_0\}$ as a set and $C(H)$ is isometrically isomorphic to $C(\beta I, D)$. Let $\varphi : \beta I \to H$ be the corresponding quotient mapping. Then considering $C(H)_\varphi$ and $C(\beta I)$ as a subspaces of $\ell_\infty(\beta I)$, we have $C(H)_\varphi \subset C(\beta I) \subset \ell_\infty(\beta I)$. Moreover, $C(H)_\varphi$ as subspace of $C(\beta I)$ is exactly $C(\beta I, D)$. By Proposition 2.2, for every $f \in \ell_\infty(\beta I)$ we have $d(f, C(H)_\varphi) = \frac{1}{2} Osc_\varphi(f)$. Take $f \in C(\beta I)$. It is clear that $Osc_\varphi(f, h) = 0$

for every $h \in \beta I \setminus D$, because f is continuous on βI and $\beta I \setminus D$ is open in βI. Thus

$$\tfrac{1}{2} Osc_\varphi(f) = \tfrac{1}{2} Osc_\varphi(f, d_0) = \tfrac{1}{2} \lim_{V \in \mathcal{V}^{d_0}} \left(\sup\{f(i) - f(j) : i, j \in \varphi^{-1}(V)\} \right),$$

where $\mathcal{V}^{d_0}$ denotes the net of open neighborhoods of d_0 in H. Since D is compact and f is continuous on βI, it is very easy to see that

$$\tfrac{1}{2} \lim_{V \in \mathcal{V}^{d_0}} \left(\sup\{f(i) - f(j) : i, j \in \varphi^{-1}(V)\} \right) =$$

$$= \tfrac{1}{2} \sup\{f(d) - f(d') : d, d' \in D\} = \tfrac{1}{2}(f(d_1) - f(d_2)),$$

for some $d_1, d_2 \in D$ such that $f(d_1) = \max\{f(d) : d \in D\}$ and $f(d_2) = \min\{f(d) : d \in D\}$. $\qquad \square$

The following proposition shows an elementary case in which there is 1-control.

Proposition 4.2. *If I is an infinite set and $D \subset \beta I$ is a closed subset such that $D = \overline{A}^{\beta I}$ for some $A \subset I$, then $C_0(\beta I, D)$ and $C(\beta I, D)$ have 1-control inside $\ell_\infty(I)$.*

Proof. As $D = \overline{A}^{\beta I}$ with $A \subset I$, then

$$C_0(\beta I, D) = \{f \in \ell_\infty(I) : f(a) = 0, \ \forall a \in A\} \quad \text{and}$$

$$C(\beta I, D) = \{f \in \ell_\infty(I) : f(a) = f(b), \ \forall a, b \in A\}.$$

So, $C_0(\beta I, D)$ and $C(\beta I, D)$ are weak*-closed subspaces of $\ell_\infty(I)$ and they have 1-control because every weak*-closed subspace does. $\qquad \square$

Now we consider another class of compact subsets $D \subset \beta I$ that yield 1-control. We begin with the definition of a regular subset. If X is a topological space, a subset $K \subset X$ is said to be *regular* in X if $\text{int}(K)$ is dense in K.

Proposition 4.3. *Let I be an infinite set and $D \subset I^*$ $(= \beta I \setminus I)$ a compact subset regular in I^*. Then $C(\beta I, D)$ and $C_0(\beta I, D)$ have 1-control inside $\ell_\infty(I)$.*

Proof. (A) First, consider the subspace $C(\beta I, D)$. Suppose that $C(\beta I, D)$ fails to have 1-control inside $\ell_\infty(I)$. Then there exist a weak*-compact subset $W \subset B(\ell_\infty(I))$ and two real numbers $a, b > 0$ such that:

$$d(W, C(\beta I, D)) < a < b < d(\overline{co}^{w^*}(W), C(\beta I, D)) \leq 1.$$

Choose $f_0 \in \overline{co}^{w^*}(W)$ such that $b < d(f_0, C(\beta I, D))$. By Proposition 4.1 this means that $\check{f}_0(d) - \check{f}_0(e) > 2b + \epsilon$ for some $d, e \in D$ and some $\epsilon > 0$. Since D is regular, there exist $d_0, e_0 \in \text{int}(D)$ (here $\text{int}(D)$ is relative to I^*) such that $\check{f}_0(d_0) - \check{f}_0(e_0) > 2b + \epsilon$. So, by continuity of $\check{f}_0$, we can find in βI two closed neighbourhoods V and W of d_0 and e_0, respectively, such that $V \cap I^* \subset \text{int}(D)$, $W \cap I^* \subset \text{int}(D)$ and $\check{f}_0(v) - \check{f}_0(w) > 2b + \epsilon$, for every $v \in V$ and every $w \in W$. Since $V \cap I$ and $W \cap I$ are infinite sets, we can choose two sequences $\{d_n : n \geq 1\} \subset V \cap I$ and $\{e_n : n \geq 1\} \subset W \cap I$ pairwise different for every $n \geq 1$. Then

(1) $\overline{\{d_n : n \geq 1\}}^{\beta I} \setminus I \subset V \cap I^* \subset D$ and $\overline{\{e_n : n \geq 1\}}^{\beta I} \setminus I \subset W \cap I^* \subset D$,

and

(2) $f_0(d_n) - f_0(e_n) > 2b + \epsilon$ for all $n \geq 1$.

Now we use an argument similar to the one of Proposition 3.1. Fix a Radon Borel probability measure μ on W such that $f_0 = r(\mu) (= \text{barycenter of } \mu)$ and for every $n \geq 1$ let $T_n : \ell_\infty(I) \to \mathbb{R}$ be such that $T_n(f) = f(d_n) - f(e_n)$, for all $f \in \ell_\infty(I)$. Clearly, T_n is a linear mapping which is $\|\cdot\|$-continuous and weak*-continuous and

$$2b + \epsilon < T_n(f_0) = T_n(r(\mu)) = \int_W T_n(f) d\mu(f).$$

So, if we define $A_n = \{f \in W : T_n(f) > 2b + \frac{\epsilon}{2}\}$ for every $n \geq 1$, then $\mu(A_n) \geq \frac{\epsilon}{4}$ for every $n \geq 1$, as in Proposition 3.1.

Let $B_n := \bigcup_{m \geq n} A_m$ for every $n \geq 1$. The sequence $\{B_n\}_{n \geq 1}$ is decreasing and $\mu(B_n) \geq \frac{\epsilon}{4}$ for every $n \geq 1$. Hence $\mu(\bigcap_{n \geq 1} B_n) \geq \frac{\epsilon}{4}$ and therefore $\bigcap_{n \geq 1} B_n \neq \emptyset$. Choose $g \in \bigcap_{n \geq 1} B_n$ and inductively a sequence $\{A_{n_i}\}_{i \geq 1}$, $n_i < n_{i+1}$, such that $g \in A_{n_i}$ for all $i \geq 1$. Then $g(d_{n_i}) - g(e_{n_i}) > 2b + \frac{\epsilon}{2}$, whence we get $\check{g}(u) - \check{g}(v) \geq 2b + \frac{\epsilon}{2}$, for every $u \in \overline{\{d_{n_i} : i \geq 1\}}^{\beta I} \setminus I$ and every $v \in \overline{\{e_{n_i} : i \geq 1\}}^{\beta I} \setminus I$. Since $\emptyset \neq \overline{\{d_{n_i} : i \geq 1\}}^{\beta I} \setminus I \subset D$ and $\emptyset \neq \overline{\{e_{n_i} : i \geq 1\}}^{\beta I} \setminus I \subset D$, we obtain $d(g, C(\beta I, D)) > b$, which contradicts the fact that $g \in W$.

(B) The proof for the subspace $C_0(\beta I, D)$ is similar to the one of case (A).   □

Corollary 4.4. *If I is an infinite set and $D \subset I^*$ is a closed $\mathcal{G}_\delta$ subset, then $C(\beta I, D)$ and $C_0(\beta I, D)$ have 1-control inside $\ell_\infty(I)$.*

Proof. This follows from Proposition 4.3 and the easy fact that every $\mathcal{G}_\delta$ subset of I^* is regular in I^* (see [16, p. 78]).   □

In the following we show examples of compact subsets $D \subset I^*$ such that the closed subspaces $C(\beta I, D)$ and $C_0(\beta I, D)$ fail to have control inside $\ell_\infty(I)$.

Example 4.5. *There exist a closed subset $D \subset \mathbb{N}^*$ and a weak*-compact subset $K \subset \ell_\infty$ such that $\check{f}_{\restriction D} = 0$ for every $f \in K$ (that is, $K \subset C_0(\beta\mathbb{N}, D)$), but $\overline{co}^{w^*}(K) \not\subseteq C(\beta\mathbb{N}, D)$.*

Proof. Consider the Cantor metric compact space $C = \{0, 1\}^\mathbb{N}$ and the set $S = \{0, 1\}^{(\mathbb{N})} = \{0, 1\} \cup \{0, 1\}^2 \cup \{0, 1\}^3 \cup \cdots$. Let λ be the Haar probability measure on $\{0, 1\}^\mathbb{N}$. If $\sigma = (\sigma_1, \sigma_2, \dots) \in C$ and $n \in \mathbb{N}$, we put $\sigma_{\restriction n} =$

$(\sigma_1, \sigma_2, \ldots, \sigma_n) \in S$. If $A \subset \{0,1\}^n$, let $f_A : C \to \{0,1\}$ be the continuous function

$$\forall \sigma \in C, \ f_A(\sigma) \begin{cases} 1, & \text{if } \sigma_{\restriction n} \in A, \\ 0, & \text{if } \sigma_{\restriction n} \notin A. \end{cases}$$

Let

$$I_+ = \{(A, +) : A \subset \{0,1\}^n \text{ with } |A| = 2^n - n \text{ and } n \in \mathbb{N}\},$$
$$I_- = \{(A, -) : A \subset \{0,1\}^n \text{ with } |A| = 2^n - n \text{ and } n \in \mathbb{N}\}$$

and $I = I_+ \cup I_-$, which satisfies $|I| = \aleph_0$. Observe that:

(1) The family $\{f_A : i = (A, +) \in I_+\}$ separates the points of C.

(2) For every $k \in \mathbb{N}$, the set $\{i = (A, +) \in I_+ : \int_C f_A d\lambda \leq 1 - \frac{1}{k}\}$ is finite.

(3) For $\sigma \in C$ let $\mathcal{O}(\sigma) = \{i = (A, \pm) \in I : f_A(\sigma) = 0\}$. Given a finite subset $\sigma^{(1)}, \sigma^{(2)}, \ldots \sigma^{(p)} \in C$, it is easy to see that $|\bigcap_{i=1}^{p} \mathcal{O}(\sigma^{(i)})^{\epsilon_i}| = \aleph_0$, where $\epsilon_i = \pm 1$, $\mathcal{O}(\sigma^{(i)})^{+1} = \mathcal{O}(\sigma^{(i)})$ and $\mathcal{O}(\sigma^{(i)})^{-1} = I \setminus \mathcal{O}(\sigma^{(i)})$.

(4) For every $i = (A, \pm) \in I$ there exists $\sigma \in C$ such that $f_A(\sigma) = 1$, and so, $i \notin \mathcal{O}(\sigma)$.

By (3) the compact subset $D := \bigcap_{\sigma \in C} \overline{\mathcal{O}(\sigma)}^{\beta I}$ satisfies $D \neq \emptyset$. Moreover, $D \subset \beta I \setminus I$ by (4). Then $\emptyset \neq D_+ := D \cap \overline{I_+}^{\beta I}$ and $\emptyset \neq D_- := D \cap \overline{I_-}^{\beta I}$. Consider the function $\psi : C \to \{-1, 0, +1\}^I \subset B(\ell_\infty(I))$ such that

$$\forall i \in I, \ \forall \sigma \in C, \ \psi(\sigma)(i) \begin{cases} f_A(\sigma), & \text{if } i = (A, +), \\ -f_A(\sigma), & \text{if } i = (A, -). \end{cases}$$

Observe that ψ is a continuous injective function, when we consider in $\{-1, 0, +1\}^I$ the weak*-topology of $\ell_\infty(I)$, which coincides with the product topology of $\{-1, 0, +1\}^I$. So, $K := \psi(C) \subset \{-1, 0, +1\}^I$ is a compact space homeomorphic to C such that $\check{k}_{\restriction D} = 0$, for every $k \in K$, that is, $K \subset C_0(\beta I, D)$.

Let $\mu := \psi(\lambda)$ be the Radon Borel probability measure on K, image of λ under the continuous function ψ, and let $r(\mu) =: z_0 \in \overline{\mathrm{co}}^{w^*}(K)$ be the barycenter of μ. By (2) we have $\check{z}_0(p) = +1$, for every $p \in \overline{I_+}^{\beta I} \setminus I$, and $\check{z}_0(p) = -1$, for every $p \in \overline{I_-}^{\beta I} \setminus I$. Thus, $\check{z}_0 \restriction D_+ = +1$ and $\check{z}_0 \restriction D_- = -1$, whence $z_0 \notin C(\beta I, D)$. $\qquad \square$

If I is an uncountable set, let $I^{*c} := \bigcup \{\overline{A}^{\beta I} : A \subset I, |A| = \aleph_0\} \setminus I$ and $I^{*u} := I^* \setminus I^{*c}$. Observe that I^{*c} is dense and open in I^* and hence I^{*u} is closed nowhere dense in I^*. It is easy to show that:

(1) If $D = I^{*u}$, then $C_0(\beta I, D) = \ell_\infty^c(I) := \{f \in \ell_\infty(I) : |\mathrm{supp}(f)| \leq \aleph_0\}$ and,

(2) $d(f, \ell_\infty^c(I)) = \|\check{f}_{\restriction I^{*u}}\| \sup\{|\check{f}(p)| : p \in I^{*u}\}$, for every $f \in \ell_\infty(I)$.

Let us investigate the control of $C_0(\beta I, D)$ and $C(\beta I, D)$ inside $\ell_\infty(I)$, when $D = I^{*u}$ and I is an uncountable set. We need the following result.

Proposition 4.6. *Let I be an infinite set and $D \subset \beta I$ a closed subset. The following are equivalent*
(a) For every weak*-compact subset $K \subset \ell_\infty(I)$ such that $K \subset C_0(\beta I, D)$ we have $\overline{co}^{w^*}(K) \subset C_0(\beta I, D)$.
(b) $C_0(\beta I, D)$ has 1-control inside $\ell_\infty(I)$.

Proof. (b)$\Rightarrow$ (a). This is obvious because $C_0(\beta I, D)$ is a norm-closed subspace of $\ell_\infty(I)$.

(a)$\Rightarrow$ (b). Suppose that there exist a weak*-compact subset $K \subset B(\ell_\infty(I))$ and two real numbers $a, b > 0$ such that:

$$d(K, C_0(\beta I, D)) < a < b < d(\overline{co}^{w^*}(K), C_0(\beta I, D)).$$

Consider the mapping $B(\ell_\infty(I)) \ni f \to \phi(f) \in B(\ell_\infty(I))$ such that $\phi(f) = (f - a) \bigvee 0$, which is w*-w*-continuous. Observe that $\phi(K)$ is weak*-compact and also $\phi(K) \subset B(C_0(\beta I, D))$. Let $z_0 \in \overline{co}^{w^*}(K)$ be such that $d(z_0, C_0(\beta I, D)) > b$. So, for some $p \in D$ we have $|\check{z}_0(p)| > b$, for example, $\check{z}_0(p) > b$. Choose a net $\{z_\alpha\}_{\alpha \in \mathcal{A}} \subset co(K)$, $z_\alpha = \sum_{i=1}^{n_\alpha} \lambda_{\alpha i} f_{\alpha i}$, $f_{\alpha i} \in K$, $\lambda_{\alpha i} \geq 0$, $\sum_{i=1}^{n_\alpha} \lambda_{\alpha i} = 1$, such that $z_\alpha \xrightarrow[\alpha \in \mathcal{A}]{w^*} z_0$. Clearly, $\phi(z_\alpha) \xrightarrow[\alpha \in \mathcal{A}]{w^*} \phi(z_0)$. Let $w_\alpha = \sum_{i=1}^{n_\alpha} \lambda_{\alpha i} \phi(f_{\alpha i}) \in co(\phi(K))$ and observe that $\phi(z_\alpha) \leq w_\alpha$ for all $\alpha \in \mathcal{A}$ (in the natural partial order " $\leq$ " of $\ell_\infty(I)$). By compactness, there exist a subnet $\{w_\beta\}_{\beta \in \mathcal{B}}$ and a vector $w_0 \in \overline{co}^{w^*}(\phi(K))$ such that $w_\beta \xrightarrow[\beta \in \mathcal{B}]{w^*} w_0$ and also $\phi(z_\beta) \xrightarrow[\beta \in \mathcal{B}]{w^*} \phi(z_0)$.

CLAIM. $\check{w}_0(p) \geq b - a > 0$.

Indeed, since $(\phi(z_0))\check{} (p) > b - a$, there exists a neighborhood V of p in βI such that $(\phi(z_0))\check{} (q) > b - a$ for all $q \in V$. Hence for every $i \in V \cap I$, as $\phi(z_\beta)(i) \xrightarrow[\beta \in \mathcal{B}]{} \phi(z_0)(i)$, there exists $\beta_i \in \mathcal{B}$ such that $\phi(z_\beta)(i) > b - a$, for all $\beta \geq \beta_i$, $\beta \in \mathcal{B}$. So, for every $i \in V \cap I$ we have $w_\beta(i) > b - a$, for all $\beta \geq \beta_i$, $\beta \in \mathcal{B}$, whence we get $w_0(i) \geq b - a$. Thus $\check{w}_0(p) \geq b - a$. As $p \in D$ we get $d(w_0, C_0(\beta I, D)) \geq b - a > 0$, a contradiction because $w_0 \in \overline{co}^{w^*}(\phi(K))$ and, as $\phi(K) \subset B(C_0(\beta I, D))$, then $\overline{co}^{w^*}(\phi(K)) \subset C_0(\beta I, D)$ by (a). $\square$

Let CH mean the continuum hypothesis, $\neg CH$ the negation of the continuum hypothesis and MA the Martin Axiom. A compact space K is Corson if there exists a set I such that K is homeomorphic to some compact subset of $\Sigma([-1, 1]^I) := \{x \in [-1, 1]^I : supp(x) \text{ is countable}\}$ (see [1],[7],[5]). Now we prove the following proposition.

Proposition 4.7. *If I is an uncountable set and $D = I^{*u}$, then the subspaces $C_0(\beta I, D)$ and $C(\beta I, D)$ have 1-control inside $\ell_\infty(I)$, under $\neg CH + MA$.*

Proof. (A) First, we consider the subspace $C_0(\beta I, D)$. Let $K \subset C_0(\beta I, D) = \ell_\infty^c(I)$ be a weak*-compact subset. Then K is a Corson compact space. Pick $z_0 \in \overline{\text{co}}^{w^*}(K)$ and μ a Radon probability on K such that $r(\mu) = z_0$. Recall that, under $\neg CH + MA$, every Corson compact space W has property (M), that is, every Radon measure on W has separable support (see [1, p. 215], [4, p. 201 and 205]). So, K has property (M) and hence there exists a countable subset $J \subset I$ such that $\text{supp}(\mu) =: H \subset \ell_\infty^J(I) := \{x \in \ell_\infty(I) : \text{supp}(x) \subset J\}$. Clearly, $z_0 = r(\mu) \in \overline{\text{co}}^{w^*}(H) \subset \ell_\infty^J(I) \subset \ell_\infty^c(I)$. Thus $\overline{\text{co}}^{w^*}(K) \subset C_0(\beta I, D)$. Now it is enough to apply Proposition 4.6.

(B) Consider now the subspace $C(\beta I, D)$. Assume that $C(\beta I, D)$ fails to have 1-control inside $\ell_\infty(I)$, that is, there exist a weak*-compact subset $K \subset B(\ell_\infty(I))$, a point $z_0 \in \overline{\text{co}}^{w^*}(K)$ and two real numbers $a, b > 0$ such that:

$$d(K, C(\beta I, D)) < a < b < d(z_0, C(\beta I, D)).$$

Since $b < d(z_0, C(\beta I, D))$, we can choose two points $d_1, d_2 \in D$ such that $\check{z}_0(d_1) - \check{z}_0(d_2) > 2b + \eta_0$ for some $\eta_0 > 0$. As $d_1 \neq d_2$, there exist in βI neighbourhoods V_i of d_i, $i = 1, 2$, such that $V_1 \cap V_2 = \emptyset$ and $\check{z}_0(p) - \check{z}_0(q) > 2b + \eta_0$ for every $p \in V_1$ and $q \in V_2$. Let ω_1 denote the first uncountable ordinal. Pick $\{i_\xi : \xi < \omega_1\} \subset V_1 \cap I$ and $\{j_\xi : \xi < \omega_1\} \subset V_2 \cap I$ such that $i_\rho \neq i_\xi$ and $j_\rho \neq j_\xi$ for $\rho < \xi < \omega_1$. Observe that these choices can be done because, as $d_i \in D = I^{*u}$, then $|V_i \cap I| > \aleph_0$, $i = 1, 2$.

Let $J = \{(i_\xi, j_\xi) : \xi < \omega_1\}$ and consider the mapping $T : \ell_\infty(I) \to \ell_\infty(J)$ such that $T(f)(i_\xi, j_\xi) = f(i_\xi) - f(j_\xi)$, $f \in \ell_\infty(I)$, $\xi < \omega_1$. Clearly, T is a linear norm-continuous mapping such that $\|T\| \leq 2$. Moreover, T is w*-w*-continuous.

CLAIM 1. $T(C(\beta I, D)) \subset \ell_\infty^c(J) = C_0(\beta J, J^{*u})$.

Indeed, let $f \in C(\beta I, D)$. Then $\check{f}$ is constant on D; denote this constant by t_f. Then there exists a countable subset $I_f \subset I$ such that $f(i) = t_f$ for every $i \in I \setminus I_f$. By construction, there exists $\alpha_0 < \omega_1$ such that $i_\xi, j_\xi \in I \setminus I_f$ for every $\alpha_0 < \xi < \omega_1$. So, for every $\alpha_0 < \xi < \omega_1$, we get

$$T(f)(i_\xi, j_\xi) = f(i_\xi) - f(j_\xi) = t_f - t_f = 0,$$

that is, $T(f) \in \ell_\infty^c(J)$.

Let $W := T(K)$ and $w_0 := T(z_0)$. Clearly, W is a weak*-compact subset of $\ell_\infty(J)$ and $w_0 \in \overline{\text{co}}^{w^*}(W)$.

CLAIM 2. $d(W, \ell_\infty^c(J)) < 2a$.

Indeed, we have

$$d(W, \ell_\infty^c(J)) \leq d(T(K), T(C(\beta I, D))) \leq 2d(K, C(\beta I, D)) < 2a.$$

CLAIM 3. $d(w_0, \ell_\infty^c(J)) > 2b$.

Indeed, for every $\xi < \omega_1$ we have

$$w_0(i_\xi, j_\xi) = z_0(i_\xi) - z_0(j_\xi) > 2b + \eta_0.$$

Thus, $d(w_0, \ell_\infty^c(J)) \geq 2b + \eta_0$.

So, taking into account part (A), we arrive to a contradiction, which completes the proof. $\qquad\square$

We see in the following proposition that under CH the subspaces $C_0(\beta I, D)$ and $C(\beta I, D)$ fail to have control inside $\ell_\infty(I)$ when $D = I^{*u}$.

Proposition 4.8. *Let I be an uncountable set and $D = I^{*u}$. Then under CH neither $C_0(\beta I, D)$ nor $C(\beta I, D)$ have control inside $\ell_\infty(I)$.*

Proof. Without loss of generality, we suppose that $I = \omega_1$. We use a modification of the Argyros-Mercourakis-Negrepontis Corson compact space without property (M) described in [1, p. 219]. Let Ω be the space of Erdös, that is, the Stone space of the quotient algebra M_λ/N_λ, where λ is the Lebesgue measure on $[0, 1]$, M_λ is the algebra of λ-measurable subsets of $[0, 1]$ and N_λ is the ideal of λ-null subsets of $[0, 1]$. Ω is an extremely disconnected compact space (because M_λ/N_λ is complete) and there is an isomorphism between the algebra M_λ/N_λ and the algebra of clopen subsets of Ω. Moreover, there exists a strictly positive regular Borel normal probability measure $\tilde\lambda$ on Ω, determined by the condition $\tilde\lambda(V) = \lambda(U)$, V being any clopen subset of Ω and U a λ-measurable subset of $[0, 1]$ with $\lambda(U) > 0$ such that the class $U + N_\lambda \in M_\lambda/N_\lambda$ is the image of V by the aforementioned isomorphism.

By CH we can write $[0, 1] = \{x_\xi : \xi < \omega_1\}$ and $\mathcal{K}([0, 1]) = \{F_\xi : \xi < \omega_1\}$, where $\mathcal{K}([0, 1])$ is the family of all compact subsets of $[0, 1]$ with strictly positive Lebesgue measure. For each $\xi < \omega_1$ we choose a compact subset $K_\xi \subset F_\xi \cap \{x_\rho : \xi < \rho < \omega_1\}$ fulfilling the following requirements: (i) if $\lambda(F_\xi) = 1$, we pick K_ξ such that $\lambda(K_\xi) > 0.9$; (ii) if $\lambda(F_\xi) < 1$, K_ξ satisfies $0 < \lambda(K_\xi)$ and $\lambda(F_\xi \setminus K_\xi) < 1 - \lambda(F_\xi)$. Since $K_\xi \subset \{x_\rho : \xi < \rho < \omega_1\}$, the family $\{K_\xi : \xi < \omega_1\}$ has caliber $\aleph_1$, that is, $\bigcap_{\xi \in A} K_\xi = \emptyset$, if $A \subset \omega_1$ and $|A| = \aleph_1$.

Let V_ξ be the clopen subset of Ω corresponding to the class $K_\xi + N_\lambda \in M_\lambda/N_\lambda$ in the isomorphism between the quotient algebra M_λ/N_λ and the algebra of clopen subsets of Ω. Observe that $\tilde\lambda(V_\xi) = \lambda(K_\xi) > 0$ for every $\xi < \omega_1$.

CLAIM. The family $\{V_\xi : \xi < \omega_1\}$ has caliber $\aleph_1$, that is, if $A \subset \omega_1$ and $|A| = \aleph_1$, then $\bigcap_{\xi \in A} V_\xi = \emptyset$. Moreover, by the choice of K_ξ we have $|\{\xi < \omega_1 : \tilde\lambda(V_\xi) \leq 0.5\}| = \aleph_1$ and $|\{\xi < \omega_1 : \tilde\lambda(V_\xi) > t\}| = \aleph_1$ for every $0 < t < 1$.

Indeed, suppose that $\bigcap_{\xi \in A} V_\xi \neq \emptyset$ for some subset $A \subset \omega_1$ with $|A| = \aleph_1$. Then, for every finite subset $F \subset A$, $\bigcap_{\xi \in F} V_\xi$ is a non-empty clopen subset of Ω. Moreover, $\tilde\lambda(\bigcap_{\xi \in F} V_\xi) > 0$ because $\tilde\lambda$ is a strictly positive probability on Ω. Now observe that $\bigcap_{\xi \in F} K_\xi + N_\lambda \in M_\lambda/N_\lambda$ is the class corresponding to $\bigcap_{\xi \in F} V_\xi$ in the isomorphism between the quotient algebra M_λ/N_λ and the algebra of clopen subsets of Ω. So, $\lambda(\bigcap_{\xi \in F} K_\xi) = \tilde{\lambda}(\bigcap_{\xi \in F} V_\xi) > 0$ for every finite subset $F \subset A$. Since the sets K_ξ are compact, we get $\bigcap_{\xi < \omega_1} K_\xi \neq \emptyset$, a contradiction because the family $\{K_\xi : \xi < \omega_1\}$ has caliber $\aleph_1$.

Finally, observe that there exists an uncountable family of compact subsets K of $[0,1]$ fulfilling that $0 < \lambda(K) \leq 0.5$. Also, for every $0 < t < 1$, there exists an uncountable family of compact subsets K of $[0,1]$ fulfilling that $t < \lambda(K) < 1$. And this completes the proof of the Claim.

Consider $\mathcal{A} = \{A \subset \omega_1 : \bigcap_{\xi \in A} V_\xi \neq \emptyset\} \cup \{\emptyset\}$. Clearly, each element of $\mathcal{A}$ is a countable subset of ω_1 and $\mathcal{A}$ is adequate (see [14, p. 1116]), that is:

(i) if $B \subset A$ and $A \in \mathcal{A}$, then $B \in \mathcal{A}$;

(ii) $A \in \mathcal{A}$ whenever $A \subset \omega_1$ and $B \in \mathcal{A}$ for every finite subset $B \subset A$, and

(iii) $\{\xi\} \in \mathcal{A}$ for every $\xi < \omega_1$.

Moreover, there are elements $A \in \mathcal{A}$ with $|A| = \aleph_0$. Indeed, given $0 < \delta < 1$, the family $\{\xi < \omega_1 : \tilde{\lambda}(V_\xi) > \delta\}$ is infinite. So, as in the proof of Proposition 3.1, we can extract from this family a countable infinite subset $\{\xi_n : n \geq 1\}$ such that $\bigcap_{n \geq 1} V_{\xi_n} \neq \emptyset$. Thus $\{\xi_n : n \geq 1\}$ is infinite and $\{\xi_n : n \geq 1\} \in \mathcal{A}$.

As $\mathcal{A}$ is adequate and its elements are countable, $K := \{1_A : A \in \mathcal{A}\}$ is a compact subset of $\ell_\infty^c(\omega_1)$. Hence K is a Corson compact space (see [14, p. 1116]) with respect to the weak*-topology $\sigma(\ell_\infty(\omega_1), \ell_1(\omega_1))$. In particular, every element $k \in K \subset \ell_\infty(I)$ (recall that $I = \omega_1$) satisfies $\tilde{k}_{\upharpoonright I^{*u}} \equiv 0$, that is, $K \subset C_0(\beta I, I^{*u})$.

If $x \in \Omega$ and $A_x = \{\xi \in \omega_1 : x \in V_\xi\}$, then $A_x \in \mathcal{A}$. Thus, we can define the mapping $T : \Omega \to K$ so that, for every $x \in \Omega$, $T(x) = 1_{A_x}$. It is easy to see that T is a continuous mapping. Let $L := T(\Omega) \subset K$ and let $\mu := T(\tilde{\lambda})$ be the image of $\tilde{\lambda}$ under T. Let $z_0 = r(\mu) \in \overline{\mathrm{co}}^{w^*}(K)$ be the barycenter of μ. If $\xi \in I = \omega_1$ (that is, $\xi < \omega_1$), define $\pi_\xi : \ell_\infty(I) \to \mathbb{R}$ by $\pi_\xi(f) = f(\xi)$, for all $f \in \ell_\infty(I)$. Observe that π_ξ is a weak*-continuous linear mapping on $\ell_\infty(I)$. So

$$1 \geq z_0(\xi) = \pi_\xi(z_0) = \pi_\xi(r(\mu)) \int_L \pi_\xi(k) d\mu(k) = \int_{T(\Omega)} k(\xi) d(T(\tilde{\lambda}))(k) =$$

$$= \int_\Omega 1_{A_x}(\xi) d\tilde{\lambda}(x) = \tilde{\lambda}(V_\xi) = \lambda(K_\xi).$$

By the Claim $|\{\xi < \omega_1 : \tilde{\lambda}(V_\xi) > t\}| = \aleph_1$, for every $0 < t < 1$, and this implies that there exists a point $p \in I^{*u}$ such that $\check{z}_0(p) = 1$, and so, $1 = d(z_0, \ell_\infty^c(I)) = d(z_0, C_0(\beta I, I^{*u}))$. On the other hand, $|\{\xi < \omega_1 : \tilde{\lambda}(V_\xi) \leq 0.5\}| = \aleph_1$, which implies that there exists another point $q \in I^{*u}$ such that

$\check{z}_0(q) \leq 0.5$. So,

$$d(z_0, C(\beta I, I^{*u})) \geq \tfrac{1}{2}(\check{z}_0(p) - \check{z}_0(q)) \geq \tfrac{1}{2}(1 - 0.5) = 0.25,$$

and this completes the proof.   □

5. The control of $C(K)$ inside $\ell_\infty(K)$ when K is compact

In [3] Cascales, Marciszewski and Raja proved that, if H is a normal countably compact space and W a weak*-compact subset of $\ell_\infty(H)$, then $d(\overline{co}^{w^*}(W), C(H)) \leq 5d(W, C(H))$ and, moreover, if $W \cap C(H)$ is weak*-dense in W, $d(\overline{co}^{w^*}(W), C(H)) \leq 2d(W, C(H))$. They use the technique of double limits. In this Section we generalize this result. We use a different technique based on Proposition 3.1 and the Weierstrass Theorem.

Proposition 5.1. *Let I be a set, H a normal countably compact space and $\varphi : I \to H$ a mapping such that $\varphi(I) = H$. Then every multiplicative sub-algebra Z of $\ell_\infty(I)$ such that $Z \subset C(H)_\varphi$ and $1_I \in Z$ has 5-control inside $\ell_\infty(I)$.*

Proof. Suppose that Z does not have 5-control inside $\ell_\infty(I)$. Then there exist a weak*-compact subset $K \subset B(\ell_\infty(I))$, two real numbers $a, b > 0$ and a point $z_0 \in \overline{co}^{w^*}(K)$ such that

$$5d(K, Z) < 5a < b < d(z_0, Z).$$

Pick $\psi \in S(Z^\perp) \subset S((\ell_\infty(I))^*)$ such that $\psi(z_0) > b$.

Step 1. Since $\psi(z_0) > b$, there exists $x_1^* \in S(\ell_1(I))$ such that $x_1^*(z_0) > b$. So, as $z_0 \in \overline{co}^{w^*}(K)$ we can find $\eta_1 \in co(K)$ with

$$\eta_1 = \sum_{i=1}^{n_1} \lambda_{1i}\eta_{1i}, \ \eta_{1i} \in K, \ \lambda_{1i} \geq 0, \ \sum_{i=1}^{n_1} \lambda_{1i} = 1,$$

such that $x_1^*(\eta_1) > b$. Since $d(\eta_{1i}, Z) < a$ we have the decomposition $\eta_{1i} = \eta_{1i}^1 + \eta_{1i}^2$ with $\eta_{1i}^1 \in Z$ and $\eta_{1i}^2 \in aB(\ell_\infty(I))$. Let $f_{1i} \in C(H)$ be such that $\eta_{1i}^1 = f_{1i} \circ \varphi$, $\mathcal{F}_1 = \{1_H\} \cup \{f_{1i} : 1 \leq i \leq n_1\}$ and $\mathcal{P}_1$ the family of finite products of elements of $\mathcal{F}_1$. Write $\mathcal{P}_1 = \{g_{1n} : n \geq 1\}$.

Step 2. Let $Y_1 = [\{f \circ \varphi : f \in \mathcal{F}_1 \cup \{g_{11}\}\}]$ ([A] means the subspace spanned by A) and observe that $Y_1 \subset Z$. Since $\dim(Y_1) < \infty$, $\psi(z_0) > b$ and $\psi \in Y_1^\perp$, there exists $x_2^* \in S(\ell_1(I))$ such that $x_2^*(z_0) > b$ and $x_{2|Y_1}^* = 0$. So, as $x_i^*(z_0) > b$, $i = 1, 2$, and $z_0 \in \overline{co}^{w^*}(K)$, we can find $\eta_2 \in co(K)$ with

$$\eta_2 = \sum_{i=1}^{n_2} \lambda_{2i}\eta_{2i}, \ \eta_{2i} \in K, \ \lambda_{2i} \geq 0, \ \sum_{i=1}^{n_2} \lambda_{2i} = 1,$$

such that $x_i^*(\eta_2) > b, i = 1, 2$. Since $d(\eta_{2i}, Z) < a$ we have the decomposition $\eta_{2i} = \eta_{2i}^1 + \eta_{2i}^2$ with $\eta_{2i}^1 \in Z$ and $\eta_{2i}^2 \in aB(\ell_\infty(I))$. Let $f_{2i} \in C(H)$ be such that $\eta_{2i}^1 = f_{2i} \circ \varphi$, $\mathcal{F}_2 = \mathcal{F}_1 \cup \{f_{2i} : 1 \leq i \leq n_2\}$ and $\mathcal{P}_2$ the family of finite products of elements of $\mathcal{F}_2$. Write $\mathcal{P}_2 = \{g_{2n} : n \geq 1\}$.

Step 3. Let $Y_2 = [\{f \circ \varphi : f \in \mathcal{F}_2 \cup \{g_{ij} : i, j \leq 2\}\}]$ and observe that $Y_2 \subset Z$. Since $\dim(Y_2) < \infty$, $\psi(z_0) > b$ and $\psi \in Y_2^\perp$, there exists $x_3^* \in S(\ell_1(I))$ such that $x_3^*(z_0) > b$ and $x_{3|Y_2}^* = 0$. So, as $x_i^*(z_0) > b$, $i = 1, 2, 3$, and $z_0 \in \overline{\mathrm{co}}^{w^*}(K)$, we can find $\eta_3 \in \mathrm{co}(K)$ with

$$\eta_3 = \sum_{i=1}^{n_3} \lambda_{3i}\eta_{3i}, \quad \eta_{3i} \in K, \quad \lambda_{3i} \geq 0, \quad \sum_{i=1}^{n_3} \lambda_{3i} = 1,$$

such that $x_i^*(\eta_3) > b, i = 1, 2, 3$. Since $d(\eta_{3i}, Z) < a$ we have the decomposition $\eta_{3i} = \eta_{3i}^1 + \eta_{3i}^2$ with $\eta_{3i}^1 \in Z$ and $\eta_{3i}^2 \in aB(\ell_\infty(I))$. Let $f_{3i} \in C(H)$ be such that $\eta_{3i}^1 = f_{3i} \circ \varphi$, $\mathcal{F}_3 = \mathcal{F}_2 \cup \{f_{3i} : 1 \leq i \leq n_3\}$ and $\mathcal{P}_3$ the family of finite products of elements of $\mathcal{F}_3$. Write $\mathcal{P}_3 = \{g_{3n} : n \geq 1\}$.

Further, we proceed by reiteration.

Let $Y = \overline{\bigcup_{k \geq 1} Y_k}$, $\mathcal{F}_0 = \bigcup_{n \geq 1} \mathcal{F}_n$ and $\mathcal{P}_0 = \bigcup_{n \geq 1} \mathcal{P}_n$. Observe that Y is a closed separable subspace of $\ell_\infty(I)$ and $\mathcal{F}_0$, $\mathcal{P}_0$ are countable subsets of $C(H)$.

Let $H_0 = H/\mathcal{F}_0$ be the quotient topological space obtained from H by means of the following equivalence $\sim$: if $h_1, h_2 \in H$, then $h_1 \sim h_2$ if and only if $f(h_1) = f(h_2)$ for all $f \in \mathcal{F}_0$. H_0 is a metric separable space, because $\mathcal{F}_0$ is countable. Moreover, H_0 is compact because H_0 is countably compact. Let $q : H \to H_0$ denote the quotient mapping and $\varphi_1 = q \circ \varphi$.

CLAIM 1. $\overline{[\mathcal{P}_0]}$ is the subspace $\{g \circ q : g \in C(H_0)\} =: C(H_0)_q$ of $C(H)$ and $Y = \left(\overline{[\mathcal{P}_0]}\right)_\varphi = C(H_0)_{\varphi_1}$.

First, it is clear that if $f \in \mathcal{P}_0$ there exists $g \in C(H_0)$ such that $g \circ q = f$. Let $\mathcal{Q}_0\{g \in C(H_0) : g \circ q \in \mathcal{P}_0\}$ and observe that $\mathcal{P}_0\{g \circ q : g \in \mathcal{Q}_0\} =: (\mathcal{Q}_0)_q$ and $[\mathcal{P}_0] = ([\mathcal{Q}_0])_q$. On the other hand, $[\mathcal{Q}_0]$ is a multiplicative sub-algebra of $C(H_0)$ such that $\mathbf{1}_{H_0} \in \mathcal{Q}_0$ and $[\mathcal{Q}_0]$ separates points in H_0, whence $\overline{[\mathcal{Q}_0]} = C(H_0)$ by the Weierstrass Theorem. Thus

$$C(H_0)_q = \left(\overline{[\mathcal{Q}_0]}\right)_q = \overline{([\mathcal{Q}_0])_q} = \overline{[\mathcal{P}_0]}.$$

Finally, observe that $Y = \overline{[\{f \circ \varphi : f \in \mathcal{P}_0\}]} \left(\overline{[\mathcal{P}_0]}\right)_\varphi$, whence we get $Y = C(H_0)_{\varphi_1}$, because $\varphi_1 = q \circ \varphi$ and $\overline{[\mathcal{P}_0]} = C(H_0)_q$.

Let $K_1 = (K + aB(\ell_\infty(I))) \cap \overline{Y}^{w^*}$, which is a weak*-compact subset of $\overline{Y}^{w^*}$. Observe that

$$\{\eta_{ij}^1 : i \geq 1, 1 \leq j \leq n_i\} \subset K_1 \subset \overline{\mathrm{co}}^{w^*}(K) + aB(\ell_\infty(I)).$$

CLAIM 2. $d(\overline{\mathrm{co}}^{w^*}(K_1), Y) = d(K_1, Y) \leq 4a.$

Indeed, we have $d(K_1, C(H)_\varphi) \leq d(K_1, Z) \leq 2a$. As $Y = C(H_0)_{\varphi_1}$ and $K_1 \subset \overline{Y}^{w^*}$, by Proposition 2.3, we get $d(K_1, Y) \leq 4a$. Finally, by Proposition 3.1 we obtain $d(\overline{co}^{w^*}(K_1), Y) = d(K_1, Y)$.

Let $\eta_0 \in B(\ell_\infty(I))$ be a weak* cluster point of $\{\eta_k\}_{k \geq 1}$.

CLAIM 3. $d(\eta_0, Y) \leq 5a$.

Indeed, it is clear that
$$\eta_0 \in \overline{co}^{w^*}(\{\eta_{ij} : i \geq 1, 1 \leq j \leq n_i\}) \subset \overline{co}^{w^*}(K_1) + aB(\ell_\infty(I)),$$
whence we get $d(\eta_0, Y) \leq d(\overline{co}^{w^*}(K_1), Y) + a \leq 5a$.

CLAIM 4. $d(\eta_0, Y) \geq b$.

Indeed, let $\phi \in B(\ell_\infty(I)^*)$ be a weak* cluster point of $\{x_n^*\}_{n \geq 1}$. Since $x_n^*(\eta_k) > b$, if $k \geq n$, then $x_n^*(\eta_0) \geq b$, for all $n \geq 1$, whence $\phi(\eta_0) \geq b$. Moreover, $\phi \in Y^\perp$ because $x_{n+1|Y_n}^* = 0$ and $Y_n \subset Y_{n+1}$. Hence, $d(\eta_0, Y) \geq \phi(\eta_0) \geq b$.

Since $b > 5a$ we get a contradiction and this completes the proof. $\square$

Proposition 5.2. *Let I be a set, H a normal countably compact space and $\varphi : I \to H$ a mapping such that $\varphi(I) = H$. Then, if Z is a multiplicative sub-algebra of $\ell_\infty(I)$ such that $Z \subset C(H)_\varphi$ and $\mathbf{1}_I \in Z$ and K is a weak*-compact subset of $\ell_\infty(I)$ with $K \cap Z$ weak*-dense in K, we have $d(\overline{co}^{w^*}(K), C(H)_\varphi) \leq 2d(K, C(H)_\varphi)$.*

Proof. Otherwise, there exist a weak*-compact subset $K \subset B(\ell_\infty(I))$, with $Z \cap K$ weak*-dense in K, a point $z_0 \in \overline{co}^{w^*}(K)$ and two real numbers $a, b > 0$ such that $d(z_0, Z) > b > 2a > 2d(K, Z)$. Pick $\psi \in S(Z^\perp)$ such that $\psi(z_0) > b$. We follow the argument of Proposition 5.1 with the following changes:

(i) As $Z \cap K$ is weak*-dense in K, we can choose η_k in $co(Z \cap K)$ with $\eta_k = \sum_{i=1}^{n_k} \lambda_{ki} \eta_{ki}$, $\eta_{ki} \in Z \cap K$ and $\lambda_{ki} \geq 0, \sum_{i=1}^{n_k} \lambda_{ki} = 1$. We also pick $\eta_{ki} = \eta_{ki}^1$ and $\eta_{ki}^2 = 0$.

(ii) Define
$$K_1 = \overline{\{\eta_{ij} : i \geq 1, 1 \leq j \leq n_i\}}^{w^*} \subset \overline{Y}^{w^*} \cap K.$$

Clearly, $d(K_1, Z) \leq d(K, Z) < a$, whence
$$d(K_1, Y) \leq 2d(K_1, C(H)_\varphi) \leq 2d(K_1, Z) \leq 2a.$$

Since $Y = C(H_0)_{\varphi_1}$ and H_0 is a separable metric compact space, we have $d(\overline{co}^{w^*}(K_1), Y) = d(K_1, Y) \leq 2a$. Finally, every weak* cluster point η_0 of $\{\eta_k\}_{k \geq 1}$ in $\ell_\infty(I)$ satisfies $\eta_0 \in \overline{co}^{w^*}(K_1)$, $d(\eta_0, Y) \leq 2a$ and $d(\eta_0, Y) \geq b$, a contradiction. $\square$

Counterexample. If I is a set, H a compact Hausdorf space and $\varphi : I \to H$ a mapping such that $\overline{\varphi(I)} = H$ but $\varphi(I) \neq H$, then $C(H)_\varphi$ may not have control inside $\ell_\infty(I)$. Let us construct a counterexample of this fact. Consider the construction of Example 4.5 and let H be the quotient obtained from βI by identification of D in one point, say d_0. So, $H = (\beta I \setminus D) \cup \{d_0\}$ as a set. Let $\phi : I \to H$ be the identity on I. Then $\overline{\varphi(I)} = H$ but $\varphi(I) \neq H$. Also $C(H)_\varphi = C(\beta I, D)$. Now recall that we have proved in Example 4.5 that $C(\beta I, D)$ does not have control inside $\ell_\infty(I)$.

6. FUNCTIONS OF THE FIRST BAIRE CLASS AND DISTANCES

Let $B_{1b}(X)$ be the space of real bounded functions of the first Baire class considered as a subspace of the space $\ell_\infty(X)$. In [2] it is proved that if X is a polish space and K is a weak*-compact subset of $\ell_\infty(X)$ such that $K \subset B_{1b}(X)$, then $\overline{co}^{w^*}(K) \subset B_{1b}(X)$. So, the subspace $B_{1b}(X)$ is a good candidate to have control inside $\ell_\infty(X)$. In this Section we characterize the distance $d(f, B_{1b}(X))$, $f \in \ell_\infty(X)$, by means of a suitable fragmentation index $Frag(f, X)$ and, using the techniques of [2], we prove that $B_{1b}(X)$ has 2-control inside $\ell_\infty(X)$.

Let (X, τ) be a topological space, $f : X \to \mathbb{R}$ a real function and $\epsilon \geq 0$. We say that (X, τ) is ϵ-*fragmented* by f if for every $\eta > \epsilon$ and every non-empty subset $F \subset X$ there exists an open subset $V \subset X$ such that $V \cap F \neq \emptyset$ and $\text{diam}(f(V \cap F)) \leq \eta$. Define the *fragmentation index* $Frag(f, X)$ of f with respect to (X, τ) as

$$Frag(f, X) = \inf\{\epsilon \geq 0 : (X, \tau) \text{ is } \epsilon\text{-fragmented by } f\}.$$

Denote by

(1) $B_1^{(\epsilon)}(X) = \{f : X \to \mathbb{R} : Frag(f, X) \leq \epsilon\}$.

(2) $B_1(X) \subset \mathbb{R}^X$ will be the space of real functions of the first Baire class, that is, $B_1(X) = \{f : X \to \mathbb{R} : f^{-1}(T_\mathbb{R}) \subset \Sigma_2^0(X)\}$, where $\Sigma_2^0(X)$ is the class of $\mathcal{F}_\sigma$ subsets of X (see [13, pg. 190],[2, 1A]).

Observe that

(a) If $\epsilon_1, \epsilon_2 \geq 0$, then $B_1^{(\epsilon_1)}(X) + B_1^{(\epsilon_2)}(X) \subset B_1^{(\epsilon_1+\epsilon_2)}(X)$.

(b) For every $\lambda > 0$ and $\epsilon \geq 0$ we have $\lambda B_1^{(\epsilon)}(X) = B_1^{(\lambda\epsilon)}(X)$. So, $B_1^{(\epsilon)}(X)$ is a convex subset of $\mathbb{R}^X$.

(c) It is well known [2, 1E, 1C] that, if X is a complete metric space, then $B_1(X) = B_1^{(0)}(X)$.

A topological space (X, τ) is said to be *hereditarily Baire* if every closed subset of X is Baire in its relative topology. If $f \in \mathbb{R}^X$ and $\alpha \in \mathbb{R}$, we denote $\{f \leq \alpha\} = \{x \in X : f(x) \leq \alpha\}$, $\{f \geq \alpha\}\{x \in X : f(x) \geq \alpha\}$, etc.

Proposition 6.1. *Let (X, τ) be a hereditarily Baire space, $\epsilon \geq 0$ and $f \in \mathbb{R}^X$. The following are equivalent:*

(1) $f \in B_1^{(\epsilon)}(X)$.

(2) For every non-empty closed subset $F \subset X$ and every pair of real numbers $s < t$ such that $t - s > \epsilon$ we have either $\overline{F \cap \{f \leq s\}} \neq F$ or $\overline{F \cap \{f \geq t\}} \neq F$.

(3) For every non-empty closed subset $F \subset X$ and every pair of real numbers $s < t$ such that $t - s > \epsilon$ we have $int_F(\overline{F \cap \{f \leq s\}}) \cap int_F(\overline{F \cap \{f \geq t\}}) = \emptyset$.

Proof. (1) $\Rightarrow$ (2). Otherwise, for some non-empty closed subset $F \subset X$ and real numbers $s < t$ with $t - s > \epsilon$ we have $\overline{F \cap \{f \leq s\}} = F = \overline{F \cap \{f \geq t\}}$. By (1) find an open set $V \subset X$ so that $V \cap F \neq \emptyset$ and $\operatorname{diam}(f(V \cap F)) < t - s$. Find then $u \in V \cap F \cap \{f \leq s\}$ and $v \in V \cap F \cap \{f \geq t\}$. Then $f(v) - f(u) \geq t - s > \epsilon$, a contradiction.

(2) $\Rightarrow$ (3). Otherwise, for some non-empty closed subset $F \subset X$ and real numbers $s < t$ with $t - s > \epsilon$ we have $int_F(\overline{F \cap \{f \leq s\}}) \cap int_F(\overline{F \cap \{f \geq t\}}) \neq \emptyset$. If we put

$$U = int_F(\overline{F \cap \{f \leq s\}}) \cap int_F(\overline{F \cap \{f \geq t\}})$$

and $H = \overline{U}$, then we have $\overline{H \cap \{f \leq s\}} = H = \overline{H \cap \{f \geq t\}}$, a contradiction to (2).

(3) $\Rightarrow$ (1). Let $F \subset X$ be a non-empty closed subset of X and $\eta > \epsilon$. We try to find an open subset $V \subset X$ such that $V \cap F \neq \emptyset$ and $\operatorname{diam}(f(V \cap F)) \leq \eta$. For every pair $s < t$ of real numbers with $t - s > \epsilon$ let

$$G_{st} = int_F(F \cap \{f > s\}) \cup int_F(F \cap \{f < t\}).$$

Then each G_{st} is dense in F, because

$$G_{st} = int_F(F \setminus F \cap \{f \leq s\}) \cup int_F(F \setminus F \cap \{f \geq t\}),$$

whence $\overline{G_{st}} = F \setminus [int_F(\overline{F \cap \{f \leq s\}}) \cap int_F(\overline{F \cap \{f \geq t\}})] = F$.

Let $Y = \bigcap\{G_{st} : s, t \in \mathbb{Q}, s < t, t - s > \epsilon\}$. Then Y is dense in F, because F is Baire. For $y \in Y$ define

$$U(y) := \inf\{t > f(y) : y \in int_F(F \cap \{f < t\})\}, \quad \inf(\emptyset) = +\infty,$$

and

$$L(y) := \sup\{s < f(y) : y \in int_F(F \cap \{f > s\})\}, \quad \sup(\emptyset) = -\infty.$$

CLAIM. For every $y \in Y$, we have $-\infty < L(y) \leq f(y) \leq U(y) < +\infty$ and $U(y) - L(y) \leq \epsilon$.

Fix $y \in Y$. It is clear that

$$-\infty \leq L(y) \leq f(y) \leq U(y) \leq +\infty.$$

Let us prove that $U(y) - L(y) \leq \epsilon$. Suppose that $U(y) - L(y) > \epsilon$. Then we can find $s, t \in \mathbb{Q}$ such that $L(y) < s < t < U(y)$ and $t - s > \epsilon$. Then $y \notin int_F(F \cap \{f > s\})$ and $y \notin int_F(F \cap \{f < t\})$, and hence $y \notin G_{st}$, a contradiction. Finally, observe that the fact $U(y) - L(y) \leq \epsilon < \infty$ implies that $-\infty < L(y)$ and $U(y) < +\infty$. This proves the Claim.

Pick $x \in Y$ arbitrarily, $\eta > \epsilon$ and $s, t \in \mathbb{Q}$ such that $s < L(x) \leq U(x) < t$ and $\epsilon < t - s \leq \eta$. Then $x \in int_F(F \cap \{f > s\}) \cap int_F(F \cap \{f < t\})$. Let

$V \subset X$ be an open subset such that

$$V \cap F = int_F(F \cap \{f > s\}) \cap int_F(F \cap \{f < t\}).$$

Then $V \cap F \neq \emptyset$ and $\operatorname{diam}(f(V \cap F)) \leq t - s \leq \eta$. □

Corollary 6.2. *Let (X, τ) be a hereditarily Baire space and $f \in \mathbb{R}^X$. Then:*

$$\begin{aligned} Frag(f, X) &= \inf\{\epsilon > 0 : \forall F \subset X \ closed\ non\text{-}empty,\ \forall s, t \in \mathbb{R}\ with \\ &\qquad t - s > \epsilon,\ either\ \overline{F \cap \{f \leq s\}} \neq F\ or\ \overline{F \cap \{f \geq t\}} \neq F\} \\ &= \inf\{\epsilon > 0 : \forall F \subset X\ closed\ non\text{-}empty,\ \forall s, t \in \mathbb{R}\ with \\ &\qquad t - s > \epsilon,\ then\ int_F(F \cap \{f \leq s\}) \cap int_F(F \cap \{f \geq t\}) = \emptyset\}. \end{aligned}$$

Proposition 6.3. *Let (X, d) be a complete metric space, $\epsilon \geq 0$ and $f \in \mathbb{R}^X$. The following are equivalent:*

(1) $f \in B_1^{(\epsilon)}(X)$.

(2) For every compact subset $K \subset X$ we have $f_{\restriction K} \in B_1^{(\epsilon)}(K)$.

Proof. $(1) \Rightarrow (2)$. This follows from the definition of the space $B_1^{(\epsilon)}(X)$.

$(2) \Rightarrow (1)$. Suppose that $f \notin B_1^{(\epsilon)}(X)$. By Proposition 6.1 there is a non-empty closed subset $F \subset X$ and two real numbers $s < t$ with $t - s > \epsilon$ such that

$$\overline{F \cap \{f \leq s\}} = F = \overline{F \cap \{f \geq t\}}.$$

Now we choose an increasing sequence $\{A_n : n \geq 1\}$ of finite subsets $A_n \subset (F \cap \{f \leq s\}) \cup (F \cap \{f \geq t\})$ such that for every $n \geq 1$: (i) $A_n \neq \emptyset$ and $A_n \subset A_{n+1}$; (ii) if $x \in A_{n+1}$, there is an $y \in A_n$ such that $d(x, y) \leq 2^{-n}$; (iii) if $x \in A_n$, there is an $y \in A_{n+1}$ such that $d(x, y) \leq 2^{-n}$ and $|f(x) - f(y)| \geq t - s$. In the construction of the sequence $\{A_n : n \geq 1\}$ we use induction:

Step 1. Choose arbitrarily $x \in F \cap \{f \leq s\}$, $y \in F \cap \{f \geq t\}$ and put $A_1 = \{x, y\}$.

Step 2. Suppose that we have constructed the sequence $\{A_i : i = 1, 2, ..., n\}$ fulfilling the above conditions. For every $u \in A_n$ we choose an element $z(u) \in F$ in the following manner. If $u \in F \cap \{f \leq s\}$, we choose $z(u) \in F \cap \{f \geq t\}$ with $d(u, z(u)) \leq 2^{-n}$. And vice versa, if $u \in F \cap \{f \geq t\}$, we choose $z(u) \in F \cap \{f \leq s\}$ with $d(u, z(u)) \leq 2^{-n}$. Let $A_{n+1} = A_n \cup \{z(u) : u \in A_n\}$.

It is now easy to see that, if $K = \overline{\bigcup_{n \geq 1} A_n}$, then K is compact (because (X, d) is complete), and $\overline{K \cap \{f \leq s\}} = K = \overline{K \cap \{f \geq t\}}$, so that $f_{\restriction K} \notin B_1^{(\epsilon)}(K)$, a contradiction to (2). □

If X is a set and $f, g \in \mathbb{R}^X$, we define $d(f, g) = \sup\{|f(x) - g(x)| : x \in X\}$, so that $d(f, g)$ can be $+\infty$. In the following proposition we characterize the distance $d(f, B_{1b}(X))$, $f \in \ell_\infty(X)$, by means of the fragmentation index $Frag(f, X)$. This proposition is very much like Proposition 2.2.

Proposition 6.4. *Let (X,d) be a metric space and $f \in \mathbb{R}^X$. Then:*

(I) $d(f, B_1(X)) \leq \frac{1}{2}Frag(f,X)$, if (X,d) is separable.

(II) $d(f, B_1(X)) \geq \frac{1}{2}Frag(f,X)$, if (X,d) is a complete metric space.

(III) $d(f, B_1(X)) = \frac{1}{2}Frag(f,X)$, if (X,d) is a polish space.

Proof. (I) Take any $Frag(f,X) < \eta < \infty$. Since (X,d) is a separable metric space η-fragmented by f, we can find a countable ordinal ξ and a sequence of open subsets $U_\alpha \subset X$, $\alpha < \xi$, of X such that $X = \cup_{\alpha < \xi} U_\alpha$ and diam$(f(U_\beta \setminus \bigcup_{\alpha < \beta} U_\alpha)) \leq \eta$ for every $\beta < \xi$. Observe that the subsets $G_\beta := U_\beta \setminus \bigcup_{\alpha < \beta} U_\alpha$, $\beta < \xi$, are $\mathcal{F}_\sigma$ and pairwise disjoint. Let $t_\beta \in \mathbb{R}$ be the middle point of $f(G_\beta)$ for $\beta < \xi$ (if $G_\beta = \emptyset$ we do not pick t_β). Then for every $x \in G_\beta$ we have $|f(x) - t_\beta| \leq \frac{1}{2}\eta$. Let $g = \sum_{\beta < \xi} t_\beta \cdot \mathbf{1}_{G_\beta}$. Then $g \in B_1(X)$ because for every open subset $V \subset \mathbb{R}$ we have $g^{-1}(V) = \bigcup\{G_\beta : t_\beta \in V\}$, which is an $\mathcal{F}_\sigma$. Obviously, $d(f,g) \leq \frac{1}{2}\eta$.

(II) Let $\eta > d(f, B_1(X))$ and $g \in B_1(X)$ be such that $d(f,g) < \eta$. Fix some non-empty closed subset $F \subset X$. Since (X,d) is a complete metric space, by [2, 1C,1E], $g_{\upharpoonright F}$ is continuous in some point $x_0 \in F$. So, there exists an open neighborhood $V \subset X$ of x_0 such that diam$(g(V \cap F)) < \eta - d(f,g)$. Thus, for every $x,y \in V \cap F$ we have

$$
\begin{aligned}
|f(x) - f(y)| &\leq |f(x) - g(x)| + |g(x) - g(y)| + |g(y) - f(y)| \leq \\
&\leq 2d(f,g) + \text{diam}(g(V \cap F)) < 2\eta.
\end{aligned}
$$

So, diam$(f(V \cap F)) \leq 2\eta$ and $\frac{1}{2}Frag(f,X) \leq \eta$, whence we get $\frac{1}{2}Frag(f,X) \leq d(f, B_1(X))$.

(III) This follows from (I) and (II). $\qquad\qquad\qquad\qquad\qquad\qquad\square$

Let A and X be sets and $f : A \times X \to \mathbb{R}$ a mapping. For every $a \in A$ and $x \in X$, let $f(a, \cdot) : X \to \mathbb{R}$ and $f(\cdot, x) : A \to \mathbb{R}$ be the mappings such that $f(a, \cdot)(y) = f(a,y)$, $\forall y \in X$, and $f(\cdot, x)(b) = f(b,x)$, $\forall b \in A$.

Lemma 6.5. *Let (A, Σ, μ) be a complete probability space, $\epsilon \geq 0$, (X,d) a complete metric space and $f : A \times X \to \mathbb{R}$ a bounded function such that:*

(1) for every $x \in X$, the mapping $f(\cdot, x) : A \to \mathbb{R}$ is Σ-measurable, and

(2) for every $a \in A$, the mapping $f(a, \cdot) : X \to \mathbb{R}$ belongs to the space $B_1^{(\epsilon)}(X)$.

Then, if $z(x) = \int_A f(a,x)d\mu(a)$, $x \in X$, we have $z \in B_1^{(2\epsilon)}(X)$.

Proof. By Proposition 6.3 it is enough to prove that $z_{\upharpoonright K} \in B_1^{(2\epsilon)}(K)$, for every compact subset $K \subset X$. So, without loss of generality, we suppose that X is compact and, fixed a number $\eta_0 > \epsilon$, we have to find an open subset $V \subset X$ such that $V \neq \emptyset$ and diam$(z(V)) \leq 2\eta_0$.

Since f is bounded, adding a constant if necessary, we can suppose that $0 \leq f \leq M < \infty$ for some $M < \infty$. Let $\epsilon < \eta < \eta_0$, $m = [\frac{M}{\eta}]$ (where $[s]$ means the integer part of $s \in \mathbb{R}$) and choose $\delta > 0$ such that $2\eta + m\eta\delta \leq 2\eta_0$. Since $f(a, \cdot) \in B_1^{(\epsilon)}(X)$ for every $a \in A$, then for every non-empty closed subset $F \subset X$ and $r = 0, 1, \cdots, m$, either $\overline{F \cap \{f(a, \cdot) \leq r\eta\}} \neq F$ or $\overline{F \cap \{f(a, \cdot) \geq (r+1)\eta\}} \neq F$.

Now we apply the following result proved in [2, 5C. Lemma]:

"*Let (A, Σ, μ) be a complete probability space and (X, ρ) a complete metric space. Let S and T be two subsets of $A \times X$ such that: (i) $\{a \in A : (a, x) \in S\}$ and $\{a \in A : (a, x) \in T\}$ belong to Σ for every $x \in X$; (ii) for every $a \in A$ and every closed subset $F \subset X$, either $\overline{F \cap \{x \in X : (a, x) \in S\}} \neq F$ or $\overline{F \cap \{x \in X : (a, x) \in T\}} \neq F$. Then, for every $d > 0$ and any non-empty open set $U \subset X$, there is a non-empty open set $V \subset U$ such that*

$$\mu(\{a \in A : (a, x) \in S\}) + \mu(\{a \in A : (a, y) \in T\}) \leq 1 + 3d, \ \forall x, y \in V.$$"

Using this result, an induction yields non-empty open subsets

$$V_m \subset V_{m-1} \subset \cdots \subset V_1 \subset V_0 = X,$$

such that for $r = 0, 1, \cdots, m - 1$

$$\mu(\{f(\cdot, x) \leq r\eta\}) + \mu(\{f(\cdot, y) \geq (r+1)\eta\}) \leq 1 + \delta, \quad \text{for every } x, y \in V_{r+1}.$$

Let $V = V_m$ and $x, y \in V$. Then

$$z(y) = \int_A f(a, y) d\mu(a) \leq \sum_{r=0}^{m} \eta \cdot \mu(\{f(\cdot, y) \geq r\eta\}) \leq$$

$$\leq \eta + \sum_{r=0}^{m-1} \eta \cdot \mu(\{f(\cdot, y) \geq (r+1)\eta\}),$$

while also

$$z(x) = \int_A f(a, x) d\mu(a) \geq \sum_{r=1}^{m} \eta \cdot \mu(\{f(\cdot, x) \geq r\eta\}) \geq$$

$$\geq \sum_{r-1}^{m} \eta \cdot \mu(\{f(\cdot, x) > r\eta\}) = \sum_{r=1}^{m} \eta[1 - \mu(\{f(\cdot, x) \leq r\eta\})] \geq$$

$$\geq \sum_{r=1}^{m} \eta[\mu(\{f(\cdot, y) \geq (r+1)\eta\}) - \delta] = \sum_{r=1}^{m-1} \eta \cdot \mu(\{f(\cdot, y) \geq (r+1)\eta\}) - m\eta\delta,$$

because $\{f(\cdot, y) \geq (m+1)\eta\} = \emptyset$. So

$$z(y) - z(x) \leq \eta + \eta \cdot \mu(\{f(\cdot, y) \geq \eta\}) + m\eta\delta \leq 2\eta + m\eta\delta \leq 2\eta_0.$$

Thus $\text{diam}(z(V)) \leq 2\eta_0$ and, so, $z \in B_1^{(2\epsilon)}(X)$. $\square$

Proposition 6.6. *Let (X, d) be a metric space, $B_{1b}(X)$ the space of real bounded functions of the first Baire class considered as a subspace of $\ell_\infty(X)$ and $K \subset \ell_\infty(X)$ a weak*-compact subset. Then:*

(1) If (X,d) is a complete metric space and $K \subset B_1^{(\epsilon)}(X)$ for some $\epsilon \geq 0$, we have $\overline{co}^{w^*}(K) \subset B_1^{(2\epsilon)}(X)$.

(2) If (X,d) is a polish space, then $d(\overline{co}^{w^*}(K), B_{1b}(X)) \leq 2d(K, B_{1b}(X))$.

Proof. (1) Let $z \in \overline{co}^{w^*}(K)$. As K is a weak*-compact subset of $\ell_\infty(X)$, there exists a Radon probability measure μ on K such that $z = r(\mu)$, that is, z is the barycenter of μ. So, for every $x \in X$ we have

$$z(x) = \pi_x(z) = \pi_x(r(\mu)) = \int_K \pi_x(a)d\mu(a) = \int_K a(x)d\mu(a).$$

Consider the function $f : K \times X \to \mathbb{R}$ given by $f(a,x) = a(x)$, for all $a \in K$ and $x \in X$, which satisfies the conditions of Lemma 6.5. So $z \in B_1^{(2\epsilon)}(X)$ as required.

(2) It is enough to apply (1) and Proposition 6.4. $\qquad\square$

Remark. In Lemma 6.5 the boundedness of the function $f : A \times X \to \mathbb{R}$ is fundamental. This fact justifies that in Proposition 6.6 we deal with w*-compact subsets K of $\ell_\infty(X)$, that is, with τ_p-compact subsets of $\mathbb{R}^X$ (τ_p = topology of pointwise convergence on X) which are uniformly bounded. If K is not uniformly bounded, then Proposition 6.6 can fail, as states the following easy counterexample given in [2, 5G. Example]. Take $X = [0,1]$ and $K := \{2^{n+1}1_{\{q_n\}} : n \geq 1\} \cup \{0\} \subset \mathbb{R}^X$, where $\{q_n : n \geq 1\} = \mathbb{Q} \cap [0,1]$. Clearly, K is a not uniformly bounded τ_p-compact subset of $B_1(X)$ and it is easy to see that $1_{\mathbb{Q}\cap[0,1]} \in \overline{co}^{\tau_p}(K)$ but $1_{\mathbb{Q}\cap[0,1]} \notin B_1(X)$. Actually (see [2, 5H. Proposition]), if X is a polish space and $K \subset \mathbb{R}^X$ is a τ_p-compact subset with $K \subset B_1(X)$, then $\overline{co}^{w^*}(K) \subset B_2(X)$, where $B_2(X)$ is the space of functions $f : X \to \mathbb{R}$ which are τ_p-limits of sequences in $B_1(X)$.

Acknowledgements. The authors are very grateful to the referee who made numerous remarks and suggestions which helped to improve this paper.

REFERENCES

[1] S. ARGYROS, S. MERCOURAKIS AND S. NEGREPONTIS, Functional-analytic properties of Corson-compact spaces, *Studia Math.* 89 (1988), 197-229.

[2] J. BOURGAIN, D. H. FREMLIN AND M. TALAGRAND, Pointwise compact sets of Baire-measurable functions, *Amer. J. of Math.* 100 (1978), 845-886.

[3] B. CASCALES, W. MARCISZEWSKI AND M. RAJA, Distance to spaces of continuous functions, to appear.

[4] W. W. COMFORT AND S. NEGREPONTIS, *Chain Conditions in Topology*, Cambridge Tracts in Math. 79, Cambridge Univ. Press, 1982.

[5] R. DEVILLE, G. GODEFROY AND V. ZIZLER, *Smoothness and renormings in Banach spaces*, Longman Scientific and Technical, Harlow, 1993.

[6] R. ENGELKING, *General Topology*, PWN-Polish Scientific Publishers, Mathematical Monographs Vol. 60, Warsaw, 1977.

[7] M. FABIAN, *Gâteaux differentiability of convex functions and topology. Weak Asplund Spaces*, John Wiley and Sons, New-York, 1997.

[8] M. FABIAN, P. HÁJEK, V. MONTESINOS AND W. ZIZLER, A quantitative version of Krein's Theorem, *Rev. Mat. Iberoam.*, 21(1) (2005), 237-248.

[9] A. S. GRANERO, An extension of the Krein-Smulian Theorem, to appear in *Rev. Mat. Iberoam.*, 22 (2006).

[10] A. S. GRANERO, P. HÁJEK AND V. MONTESINOS, Convexity and w^*-compactness in Banach spaces, *Math. Ann.*, 328 (2004), 625-631.

[11] A. S. GRANERO AND M. SÁNCHEZ, The class of universally Krein-Šmulian Banach spaces, to appear.

[12] R. HAYDON, Some more characterizations of Banach spaces containing ℓ_1, *Math. Proc. Cambridge Phil. Soc.* 80 (1976), 269-276.

[13] A. S. KECHRIS, *Classical Descriptive Set Theory*, Springer-Verlag, New-York,1995.

[14] S. NEGREPONTIS, Banach spaces and topology, *Handbook of Set-Theoretic Topology*, K. Kunen and J. Vaughan (eds.), North-Holland, Amsterdam, 1984, 1045-1142.

[15] M. TALAGRAND,Sur les espaces de Banach contenant $\ell_1(\tau)$, *Israel J. Math.*, 40 (1981), 475-483.

[16] R. C. WALKER, *The Stone-Čech compactification*, Springer-Verlag, Berlin, 1974.

DEPARTAMENTO DE ANÁLISIS MATEMÁTICO, FACULTAD DE MATEMÁTICAS, UNIVERSIDAD COMPLUTENSE DE MADRID, 28040-MADRID, SPAIN.

E-mail address: **AS_ granero@mat.ucm.es , msam0003@encina.pntic.mec.es**

WEYL'S AND BROWDER'S THEOREMS THROUGH THE QUASI-NILPOTENT PART OF AN OPERATOR

PIETRO AIENA AND MARIA TERESA BIONDI

ABSTRACT. In this paper we give a survey on bounded linear operators T on Banach space satisfying Browder's theorem, Weyl's theorem and their approximate point spectrum variants: a-Browder's theorem and a-Weyl's theorem. All these theorems are related to the form assumed by the quasi-nilpotent part $H_0(\lambda I - T)$ as λ ranges in certain subsets of $\mathbb{C}$. Browder's theorems and Weyl's theorems are also related to a property which has a fundamental role in local spectral theory: the single-valued extension property.

1. PRELIMINARIES

Let X be an infinite-dimensional complex Banach space X and let $T \in L(X)$ denote a bounded linear operator on X. Set $\alpha(T) := \dim \ker T$, the dimension of the kernel of T, and $\beta(T) := \operatorname{codim} T(X)$, the codimension of the range $T(X)$. The class of all *upper semi-Fredholm* operators is defined as the set $\Phi_+(X)$ of all $T \in L(X)$ such that $\alpha(T) < \infty$ and $T(X)$ is closed. The class of all *lower semi-Fredholm* operators is defined as the set $\Phi_-(X)$ of all $T \in L(X)$ such that $\beta(T) < \infty$. The class of all *semi-Fredholm operators* is denoted by $\Phi_\pm(X)$, while by $\Phi(X) := \Phi_+(X) \cap \Phi_-(X)$ we shall denote class of all *Fredholm operators*. The *index* of $T \in \Phi_\pm(X)$ is defined by ind $(T) := \alpha(T) - \beta(T)$.

The class of *upper semi-Weyl's operators* is defined by

$$W_+(X) := \{T \in \Phi_+(X) : \operatorname{ind} T \leq 0\},$$

the class of *lower semi-Weyl's operators* is defined by

$$W_+(X) := \{T \in \Phi_-(X) : \operatorname{ind} T \geq 0\},$$

while, the class of *Weyl operators* is defined by

$$W(X) := W_+(X) \cap W_-(X) = \{T \in \Phi(X) : \operatorname{ind} T = 0\}.$$

This paper also concerns with two other classical quantities associated with an operator T: the *ascent* $p := p(T)$, i.e. the smallest non-negative integer p, if it does exist, such that ker $T^p = $ ker T^{p+1}, and the *descent* $q := q(T)$, i.e the smallest non-negative integer q, if it does exists, such that $T^q(X) = T^{q+1}(X)$. It is well-known that if $p(\lambda I - T)$ and $q(\lambda I - T)$ are both finite

[1]1991 *Mathematics Reviews* Primary 47A10, 47A11. Secondary 47A53, 47A55.
Key words and phrases: Single valued extension property, Fredholm operators, Weyl's theorems and Browder's theorems.

then $p(\lambda I - T) = q(\lambda I - T)$, and if $\lambda \in \sigma(T)$ then $p(\lambda I - T) = q(\lambda I - T) > 0$ precisely when λ is a pole of the function resolvent $\lambda \to (\lambda I - T)^{-1}$, see Proposition 38.3 and Proposition 50.2 of Heuser [27]. The class of all *upper semi-Browder operators* is defined

$$B_+(X) := \{T \in \Phi_+(X) : p(T) < \infty\},$$

the class of all *lower semi-Browder operators* is defined

$$B_-(X) := \{T \in \Phi_-(X) : q(T) < \infty\},$$

while the class of *Browder operators* is defined by

$$B(X) := B_+(X) \cap B_-(X) = \{T \in \Phi(X) : p(T) = q(T) < \infty.\}$$

We have

$$B_+(X) \subseteq W_+(X), \quad B_-(X) \subseteq W_-(X), \quad B(X) \subseteq W(X),$$

see [1, Chapter 3]. To each of these classes of operators is associated a spectrum: the *upper semi-Weyl spectrum* of $T \in L(X)$ is defined by

$$\sigma_{uw}(T) := \{\lambda \in \mathbb{C} : \lambda I - T \notin W_+(X)\},$$

the *lower semi-Weyl spectrum* of $T \in L(X)$ is defined by

$$\sigma_{lw}(T) := \{\lambda \in \mathbb{C} : \lambda I - T \notin W_-(X)\},$$

the *Weyl spectrum* is defined by

$$\sigma_w(T) := \{\lambda \in \mathbb{C} : \lambda I - T \notin W(X)\}.$$

Analogously, the *upper semi-Browder spectrum* of $T \in L(X)$ is defined by

$$\sigma_{ub}(T) := \{\lambda \in \mathbb{C} : \lambda I - T \notin B_+(X)\},$$

the *lower semi-Browder spectrum* of $T \in L(X)$ is defined by

$$\sigma_{lb}(T) := \{\lambda \in \mathbb{C} : \lambda I - T \notin B_-(X)\},$$

while the *Browder spectrum* of $T \in L(X)$ is defined by

$$\sigma_b(T) := \{\lambda \in \mathbb{C} : \lambda I - T \notin B(X)\}.$$

In the sequel we shall denote by T^* the dual of $T \in L(X)$, X a Banach space. From the classical Fredholm theory we have

$$\sigma_{uw}(T) = \sigma_{lw}(T^*), \quad \sigma_{lw}(T) = \sigma_{uw}(T^*),$$

and

$$\sigma_{ub}(T) = \sigma_{lb}(T^*), \quad \sigma_{lb}(T) = \sigma_{ub}(T^*).$$

Moreover,

(1) $$\sigma_w(T) = \sigma_w(T^*), \quad \sigma_b(T) = \sigma_b(T^*),$$

and

$$\sigma_w(T) \subseteq \sigma_b(T) = \sigma_w(T) \cup \text{acc } \sigma(T),$$

where we write acc K for the accumulation points of $K \subseteq \mathbb{C}$, see [1, Chapter 3].

The single valued extension property was introduced in the early years of local spectral theory by Dunford [20], [21] and plays an important role in spectral theory, see the recent monographs by Laursen and Neumann [29], or by Aiena [1]. We shall consider the following local version of this property introduced by Finch [22] and successively studied by several authors, see [6], [8], and [9].

Definition 1.1. *Let X be a complex Banach space and $T \in L(X)$. The operator T is said to have* the single valued extension property *at $\lambda_0 \in \mathbb{C}$ (abbreviated SVEP at λ_0), if for every open disc U centered at λ_0, the only analytic function $f : U \to X$ which satisfies the equation $(\lambda I - T)f(\lambda) = 0$ for all $\lambda \in U$ is the function $f \equiv 0$.*
An operator $T \in L(X)$ is said to have SVEP if T has SVEP at every point $\lambda \in \mathbb{C}$.

Evidently, an operator $T \in L(X)$ has SVEP at every point of the resolvent $\rho(T) := \mathbb{C} \setminus \sigma(T)$. From the identity theorem for analytic function it easily follows that $T \in L(X)$ has SVEP at every point of the boundary $\partial\sigma(T)$ of the spectrum $\sigma(T)$. In particular, both T and T^* have SVEP at every isolated point of $\sigma(T) = \sigma(T^*)$. Moreover, see [1, Theorem 3.8],

$$(2) \qquad\qquad p(\lambda I - T) < \infty \Rightarrow T \text{ has SVEP at } \lambda,$$

and dually

$$(3) \qquad\qquad q(\lambda I - T) < \infty \Rightarrow T^* \text{ has SVEP at } \lambda.$$

Denote by $H_0(T)$ the *quasi-nilpotent part* of $T \in L(X)$, i.e. is the set

$$(4) \qquad\qquad H_0(T) := \{x \in X : \lim_{n \to \infty} \|T^n x\|^{\frac{1}{n}} = 0\}.$$

It is easily seen that $\ker (T^m) \subseteq H_0(T)$ for every $m \in \mathbb{N}$. Moreover, T is quasi-nilpotent if and only if $H_0(T) = X$, see [1, Theorem 1.68], while if T is invertible $H_0(T) = \{0\}$. From [1, Theorem 2.31] we know that:

$$(5) \qquad\qquad H_0(\lambda I - T) \text{ closed} \Rightarrow T \text{ has SVEP at } \lambda.$$

Let $\sigma_a(T)$ denote the classical *approximate point spectrum* of T, i. e. the set

$$\sigma_a(T) := \{\lambda \in \mathbb{C} : \lambda I - T \text{ is not bounded below}\},$$

while *surjectivity spectrum* of $T \in L(X)$ is defined by

$$\sigma_s(T) := \{\lambda \in \mathbb{C} : \lambda I - T \text{ is not surjective}\}.$$

¿From definition of SVEP we also have

$$(6) \qquad\qquad \lambda \notin \text{acc } \sigma_a(T) \Rightarrow T \text{ has SVEP at } \lambda,$$

and dually,

$$(7) \qquad\qquad \lambda \notin \text{acc } \sigma_s(T) \Rightarrow T^* \text{ has SVEP at } \lambda.$$

2. Browder's theorem and a-Browders's theorem

We shall say that T satisfies *Browder's theorem* if

$$\sigma_{\mathrm{w}}(T) = \sigma_{\mathrm{b}}(T),$$

while, $T \in L(X)$ is said to satisfy *a-Browder's theorem* if

$$\sigma_{\mathrm{uw}}(T) = \sigma_{\mathrm{ub}}(T).$$

¿From the equalities (4) we have

T satisfies Browder's theorem $\Leftrightarrow$ T^* satisfies Browder's theorem.

Browder's theorem and a-Browders's theorem may be characterized by localized SVEP in the following way:

Theorem 2.1. [4], [5] *If $T \in L(X)$ the following equivalences hold:*

(i) *T satisfies Browder's theorem $\Leftrightarrow$ T has SVEP at every $\lambda \notin \sigma_{\mathrm{w}}(T)$;*

(ii) *T satisfies a-Browder's theorem $\Leftrightarrow$ T has SVEP at every $\lambda \notin \sigma_{\mathrm{uw}}(T)$.*

Moreover, the following statements hold:

(iii) *If T has SVEP at every $\lambda \notin \sigma_{\mathrm{lw}}(T)$ then a-Browder's theorem holds for T^*.*

(iv) *If T^* has SVEP at every $\lambda \notin \sigma_{\mathrm{uw}}(T)$ then a-Browder's theorem holds for T.*

Obviously,

$$a\text{-Browder's theorem} \Rightarrow \text{Browder's theorem}$$

and the converse is not true. Note the opposite implications of (iii) and (iv) in Theorem 2.1 in general do not hold, see [5]. By Theorem 2.1 we also have:

$$T, \text{ or } T^* \text{ has SVEP} \Rightarrow a\text{-Browder's theorem holds for both } T, T^*.$$

Let us consider the following set:

$$\Delta(T) := \sigma(T) \setminus \sigma_{\mathrm{w}}(T).$$

If $\lambda \in \Delta(T)$ then $\lambda I - T \in W(X)$ and since $\lambda \in \sigma(T)$ it then follows that $\alpha(\lambda I - T) = \beta(\lambda I - T) > 0$, so we can write

$$\Delta(T) = \{\lambda \in \mathbb{C} : \lambda I - T \in W(X),\, 0 < \alpha(\lambda I - T)\}.$$

The set $\Delta(T)$ has been recently studied in [26], where the points of $\Delta(T)$ are called *generalized Riesz points*. Note that $\Delta(T)$ may be empty. This for instance holds for every quasi-nilpotent operator. Define

$$p_{00}(T) := \sigma(T) \setminus \sigma_{\mathrm{b}}(T) = \{\lambda \in \sigma(T) : \lambda I - T \in B(X)\}.$$

Clearly, $p_{00}(T) \subseteq \Delta(T)$. The next result describes Browder's theorems by means of the spectral picture of the operator. Let write iso K for the set of all isolated points of $K \subseteq \mathbb{C}$.

Theorem 2.2. [4] *For an operator $T \in L(X)$ the following statements are equivalent:*

(i) *T satisfies Browder's theorem;*

(ii) *$p_{00}(T) = \Delta(T)$;*

(iii) *Every $\lambda \in \Delta(T)$ is an isolated point of $\sigma(T)$;*

(iv) *$\Delta(T) \subseteq \partial\sigma(T)$, $\partial\sigma(T)$ the topological boundary of σT);*

(v) *$int\,\Delta(T) = \emptyset$;*

(vi) *$\sigma(T) = \sigma_w(T) \cup iso\,\sigma(T)$.*

Let us consider the following set:

$$\Delta_a(T) := \sigma_a(T) \setminus \sigma_{uw}(T).$$

Since $\lambda I - T \in W_a(X)$ implies that $(\lambda I - T)(X)$ is closed, we can write

$$\Delta_a(T) = \{\lambda \in \mathbb{C} : \lambda I - T \in W_+(X),\, 0 < \alpha(\lambda I - T)\}.$$

It should be noted that the set $\Delta_a(T)$ may be empty. This is, for instance, the case of a right weighted shift on $\ell^2(\mathbb{N})$. Indeed, a right weighted shift has SVEP and iso $\sigma_a(T) = \emptyset$. Both these conditions entails that $\sigma_a(T) = \sigma_{aw}(T)$, see Corollary 2.10 of [5]. Define

$$p_{00}^a(T) := \sigma_a(T) \setminus \sigma_{ub}(T) = \{\lambda \in \sigma_a(T) : \lambda I - T \in B_+(X)\}.$$

It is easily seen that every $\lambda \in p_{00}^a(T)$ is an isolated point of $\sigma_a(T)$. Furthermore

$$p_{00}^a(T) \subseteq \Delta_a(T) \subseteq \sigma_a(T)$$

.

The following characterization of operators satisfying a-Browder's theorem is analogous to the result established in Theorem 2.2 for Browder's theorem.

Theorem 2.3. [5] *For a bounded operator $T \in L(X)$ the following statements are equivalent:*

(i) *T satisfies a-Browder's theorem;*

(ii) *$p_{00}^a(T) = \Delta_a(T)$;*

(iii) *$\Delta_a(T) \subseteq iso\,\sigma_a(T)$;*

(iv) *$\Delta_a(T) \subseteq \partial\sigma_a(T)$, $\partial\sigma_a(T)$ the topological boundary of $\sigma_a(T)$;*

(v) *$int\,\Delta_a(T) = \emptyset$;*

(vi) *$\sigma_a(T) = \sigma_{uw}(T) \cup iso\,\sigma_a(T)$.*

The next result shows that Browder's theorem and a-Browder's theorem for T are equivalent to the fact that $H_0(\lambda I - T)$ is closed as λ ranges in $\Delta(T)$, or $\Delta_a(T)$, respectively.

Theorem 2.4. [4],[5] *For a bounded operator $T \in L(X)$ the following statements are equivalent:*

(i) *Browder's theorem holds for T;*

(ii) *$H_0(\lambda I - T)$ is finite-dimensional for every $\lambda \in \Delta(T)$;*

(iii) *$H_0(\lambda I - T)$ is closed for all $\lambda \in \Delta(T)$.*

Analogously, for a bounded operator $T \in L(X)$ the following statements are equivalent:

(i) *a-Browder's theorem holds for T;*

(ii) $H_0(\lambda I - T)$ *is finite-dimensional for every* $\lambda \in \Delta_a(T)$;

(iii) $H_0(\lambda I - T)$ *is closed for every* $\lambda \in \Delta_a(T)$.

The *reduced minimum modulus* of a non-zero operator T is defined by

$$\gamma(T) := \inf_{x \notin \ker T} \frac{\|Tx\|}{\operatorname{dist}(x, \ker T)}.$$

It is well-known that $T(X)$ is closed if and only if $\gamma(T) > 0$, see for instance [1, Theorem 1.13].

Both Browder's theorem and a-Browder's theorem may be also characterized by means of the discontinuity of the mapping $\lambda \rightarrow \gamma(\lambda I - T)$ at the points of $\Delta(T)$ and $\Delta_a(T)$, respectively.

Theorem 2.5. [4],[5] *For a bounded operator $T \in L(X)$ the following statements are equivalent:*

(i) T *satisfies Browder's theorem;*

(ii) *the mapping* $\lambda \rightarrow \gamma(\lambda I - T)$ *is not continuous at every* $\lambda \in \Delta(T)$.

Analogously, following statements are equivalent:

(iv) T *satisfies a-Browder's theorem;*

(v) *the mapping* $\lambda \rightarrow \gamma(\lambda I - T)$ *is not continuous at every* $\lambda \in \Delta_a(T)$.

3. WEYL'S THEOREMS

For a bounded operator $T \in L(X)$ define

$$\pi_{00}(T) := \{\lambda \in \operatorname{iso} \sigma(T) : 0 < \alpha(\lambda I - T) < \infty\},$$

and

$$\pi_{00}^a(T) := \{\lambda \in \operatorname{iso} \sigma_a(T) : 0 < \alpha(\lambda I - T) < \infty\}.$$

Clearly, for every $T \in L(X)$ we have

(8) $p_{00}(T) \subseteq \pi_{00}(T) \subseteq \pi_{00}^a(T)$ and $p_{00}^a(T) \subseteq \pi_{00}^a(T)$.

According to Coburn [12], a bounded operator $T \in L(X)$ is said to satisfy *Weyl's theorem* if

$$\Delta(T) = \sigma(T) \setminus \sigma_w(T) = \pi_{00}(T).$$

We have

Weyl's theorem holds for $T \Rightarrow p_{00}(T) = \pi_{00}(T) = \Delta(T)$.

In fact, if T satisfies Weyl's theorem then $\Delta(T) = \pi_{00}(T)$. We show the equality $p_{00}(T) = \pi_{00}(T)$. It suffices to prove the inclusion $\pi_{00}(T) \subseteq p_{00}(T)$. Let λ be an arbitrary point of $\pi_{00}(T)$. Since λ is isolated in $\sigma(T)$ then both T and T^* have SVEP at λ and from the equality $\pi_{00}(T) = \sigma(T) \setminus \sigma_w(T)$ we know that $\lambda I - T \in W(X)$. The SVEP for both T and T^* at λ by Corollary 2.7 of [6] implies that $p(\lambda I - T) = q(\lambda I - T) < \infty$, so $\lambda \in p_{00}(T)$.

The following result shows the relationships between Browder's theorem and Weyl's theorem:

Theorem 3.1. [2] *For every $T \in L(X)$ the following statements are equivalent:*

(i) T *satisfies Weyl's theorem;*

(ii) T *satisfies Browder's theorem and $p_{00}(T) = \pi_{00}(T)$.*

The conditions $p_{00}(T) = \pi_{00}(T)$ is equivalent to several other conditions, see [1, Theorem 3.84]. In particular,

(9) $p_{00}(T) = \pi_{00}(T) \Leftrightarrow H_0(\lambda I - T)$ is finite-dimensional for all $\lambda \in \pi_{00}(T)$.

According to Rakočević [32], a bounded operator $T \in L(X)$ is said to satisfy *a-Weyl's theorem* if

(10) $$\Delta_a(T) = \sigma_a(T) \setminus \sigma_{\mathrm{uw}}(T) = \pi_{00}^a(T).$$

It is not difficult to prove the implication

$$a\text{-Weyl's theorem holds for } T \Rightarrow p_{00}^a(T) = \pi_{00}^a(T) = \Delta_a(T).$$

In fact, suppose that T satisfies a-Weyl's theorem. By definition then $\Delta_a(T) = \pi_{00}^a(T)$. We show now the equality $p_{00}^a(T) = \pi_{00}^a(T)$. It suffices to prove the inclusion $\pi_{00}^a(T) \subseteq p_{00}^a(T)$. Let λ be an arbitrary point of $\pi_{00}^a(T)$. Since λ is isolated in $\sigma_a(T)$ then T has SVEP at λ and from the equality $\pi_{00}^a(T) = \sigma_a(T) \setminus \sigma_{\mathrm{uw}}(T)$ we know that $\lambda I - T \in W_+(X)$. Hence $\lambda I - T \in \Phi_+(X)$ and the SVEP at λ by Corollary 2.7 of [6] implies that $p(\lambda I - T) < \infty$, so $\lambda \in p_{00}^a(T)$.

Theorem 3.1 has a companion for a-Weyl's theorem. In fact, a-Browder's theorem and a-Weyl's theorem are related by the following result:

Theorem 3.2. [2] *If $T \in L(X)$ the following statements are equivalent:*

(i) T *satisfies a-Weyl's theorem;*

(ii) T *satisfies a-Browder's theorem and $p_{00}^a(T) = \pi_{00}^a(T)$.*

It easily seen that the conditions $p_{00}^a(T) = \pi_{00}^a(T)$ holds if and only if $\sigma_{\mathrm{ub}}(T) \cap \pi_{00}^a(T) = \emptyset$, and the last equality is equivalent to several other conditions, see [1, Theorem 3.105]. In particular, $p_{00}^a(T) = \pi_{00}^a(T) \Leftrightarrow H_0(\lambda I - T)$ is finite-dimensional for all $\lambda \in \pi_{00}(T)$ and $(\lambda I - T)(X)$ is closed for all $\lambda \in \pi_{00}^a(T) \setminus \pi_{00}(T)$. From Theorem 3.2 it immediately follows that

$$a\text{-Weyl's theorem} \Rightarrow a\text{-Browder's theorem.}$$

It is not difficult to find an example of operator satisfying a-Browder's theorem but not a-Weyl's theorem. For instance, if $T \in L(\ell^2)$ is defined by

$$T(x_0, x_1, \dots) := (\frac{1}{2}x_1, \frac{1}{3}x_2, \dots) \quad \text{for all } (x_n) \in \ell^2,$$

then T is quasi-nilpotent, so has SVEP and consequently satisfies a-Browder's theorem. On the other hand T does not satisfy a-Weyl's theorem, since $\sigma_a(T) = \sigma_{\mathrm{uw}}(T) = \{0\}$ and $\pi_{00}^a(T) = \{0\}$. Note that the condition $\Delta_a(T) = \emptyset$

does not ensure that a-Weyl's theorem holds.

Let us define

$$\Delta_{00}(T) := \Delta(T) \cup \pi_{00}(T),$$

and

$$\Delta_{00}^a(T) := \Delta_a(T) \cup \pi_{00}^a(T).$$

The following result provides a characterization of Weyl's theorem and a-Weyl's theorem by means of the discontinuity of the map $\lambda \to \gamma(\lambda I - T)$.

Theorem 3.3. [4], [5] $T \in L(X)$ satisfies Weyl's theorem precisely when the mapping $\lambda \to \gamma(\lambda I - T)$ is discontinuous at every $\lambda \in \Delta_{00}(T)$. Analogously, $T \in L(X)$ satisfies a-Weyl's theorem precisely when the mapping $\lambda \to \gamma(\lambda I - T)$ is discontinuous at every $\lambda \in \Delta_{00}^a(T)$.

Comparing Theorem 3.3 with Theorem 2.5 we see that Browder's theorem (respectively, a-Browder's theorem) and Weyl's theorem (respectively, a-Weyl's theorem) correspond to the discontinuity of the map $\lambda \to \gamma(\lambda I - T)$ at the points of $\Delta(T)$ and $\Delta_{00}(T)$, with $\Delta(T) \subseteq \Delta_{00}(T)$ (respectively, at points of $\Delta_a(T)$ and $\Delta_{00}^a(T)$, with $\Delta_a(T) \subseteq \Delta_{00}^a(T)$).

Taking into account the equivalence (i) $\Leftrightarrow$ (iii) of Theorem 2.5, the equivalence (9), and the characterization established in Theorem 3.1 we easily deduce the following result:

Theorem 3.4. $T \in L(X)$ satisfies Weyl's theorem if and only if $H_0(\lambda I - T)$ is finite dimensional for all $\lambda \in \Delta_{00}(T)$.

The following implication holds:

(11) a-Weyl's theorem $\Rightarrow \Delta_{00}^a(T) \supseteq \Delta_{00}(T)$.

Indeed, suppose that a-Weyl's theorem holds for T. We claim that $\Delta(T) \subseteq \Delta_a(T)$. Let $\lambda \in \Delta(T) = \sigma(T) \setminus \sigma_w(T)$. From the inclusion $\sigma_{uw}(T) \subseteq \sigma_w(T)$ it follows that $\lambda \notin \sigma_{uw}(T)$, and since $\lambda I - T$ is Weyl we also have $0 < \alpha(\lambda I - T) < \infty$, otherwise we would have $0 = \alpha(\lambda I - T) = \beta(\lambda I - T)$ and hence $\lambda \notin \sigma(T)$. Therefore $\lambda \in \sigma_a(T) \setminus \sigma_{uw}(T) = \Delta_a(T)$.

On the other hand, by (8) we also have $\pi_{00}(T) \subseteq \pi_{00}^a(T)$, thus if T satisfies a-Weyl's theorem then

$$\Delta_{00}^a(T) = \Delta_a(T) \cup \pi_{00}^a(T) \supseteq \Delta(T) \cup \pi_{00}(T) = \Delta_{00}(T).$$

The implication (11), together with and Theorem 3.3, then yields:

a-Weyl's theorem $\Rightarrow$ Weyl's theorem.

Note that the reverse of this implication in general does not hold.

The next result shows that under SVEP both Weyl's theorem and a-Weyl's theorem are equivalent.

Theorem 3.5. [2] If $T \in L(X)$ has SVEP then the following statements are equivalent:

(i) Weyl's theorem holds for T^*;

(ii) *a-Weyl's theorem holds for T^*.*

Analogously, if the dual T^* of T has SVEP then the following statements are equivalent:

(iii) *Weyl's theorem holds for T;*

(iv) *a-Weyl's theorem holds for T.*

Let $\mathcal{P}_0(X)$, X a Banach space, denote the class of all operators $T \in L(X)$ such that there exists $p := p(\lambda) \in \mathbb{N}$ for which

(12) $$H_0(\lambda I - T) = \ker (\lambda I - T)^p \quad \text{for all } \lambda \in \pi_{00}(T).$$

Theorem 3.6. *$T \in \mathcal{P}_0(X)$ if and only if $p_{00}(T) = \pi_{00}(T)$.*

Proof. Suppose $T \in \mathcal{P}_0(X)$ and $\lambda \in \pi_{00}(T)$. Then there exists $p \in \mathbb{N}$ such that $H_0(\lambda I - T) = \ker (\lambda I - T)^p$. Since λ is isolated in $\sigma(T)$ we have

$$X = H_0(\lambda I - T) \oplus K(\lambda I - T) = \ker (\lambda I - T)^p \oplus K(\lambda I - T),$$

see [1, Theorem 3.74], from which we obtain

$$(\lambda I - T)^p(X) = (\lambda I - T)^p(K(\lambda I - T)) = K(\lambda I - T),$$

so $x = \ker (\lambda I - T)^p \oplus (\lambda I - T)^p(X)$ which implies by [1, Theorem 3.6] that $p(\lambda I - T) = q(\lambda I - T) \leq p$. By definition of $\pi_{00}(T)$ we know that $\alpha(\lambda I - T) < \infty$ and this implies by Theorem 3.4 of [1] that $\beta(\lambda I - T)$ is also finite. Therefore $\lambda \in p_{00}(T)$, from which we conclude that $p_{00}(T) = \pi_{00}(T)$.

Conversely, if $p_{00}(T) = \pi_{00}(T)$ and $\lambda \in \pi_{00}(T)$ then $p := p(\lambda I - T) = q(\lambda I - T) < \infty$. By Theorem 3.16 of [1] it then follows that $H_0(\lambda I - T) = \ker(\lambda I - T)^p$ for some $p := p(\lambda) \in \mathbb{N}$. ∎

¿From Theorem 3.1 and Theorem 3.6 we easily deduce that

(13) $$T \in \mathcal{P}_0(X) \text{ has SVEP} \Rightarrow \text{Weyl's theorem holds for } T.$$

A large number of the commonly considered operators on Banach spaces and Hilbert spaces have SVEP and belong to the class $\mathcal{P}_0(X)$.

(a) A bounded operator $T \in L(X)$ on a Banach space X is said *paranormal* if $\|Tx\| \leq \|T^2x\|\|x\|$ for all $x \in X$. $T \in L(X)$ is called *totally paranormal* if $\lambda I - T$ is paranormal for all $\lambda \in \mathbb{C}$. For every totally paranormal operator we have

(14) $$H_0(\lambda I - T) = \ker (\lambda I - T) \quad \text{for all } \lambda \in \mathbb{C},$$

see [28]. The class of totally paranormal operators includes all hyponormal operators on Hilbert spaces H. In the sequel denote by T' the Hilbert adjoint of $T \in L(H)$. The operator $T \in L(H)$ is said to be *hyponormal* if

$$\|T'x\| \leq \|Tx\| \quad \text{for all } x \in X.$$

A bounded operator $T \in L(H)$ is said to be *quasi-hyponormal* if

$$\|T'Tx\| \leq \|T^2x\| \quad \text{for all} \quad x \in H.$$

Also quasi-normal operators are totally paranormal, since these operators are hyponormal, see Conway [13].

An operator $T \in L(H)$ is said to be *-paranormal* if

$$\|T'x\|^2 \leq \|T^2x\|$$

holds for all unit vectors $x \in H$. $T \in L(H)$ is said to be *totally *-paranormal* if $\lambda I - T$ is *-paranormal for all $\lambda \in \mathbb{C}$. Every totally $*$-paranormal operator satisfies property (14), see [25].

(b) The condition (14) is also satisfied by every injective p-hyponormal operator, see [10], where an operator $T \in L(H)$ on a Hilbert space H is said to be *p-hyponormal*, with $0 < p \leq 1$, if $(T'T)^p \geq (TT')^p$, [10].

(c) An operator $T \in L(H)$ is said to be *log-hyponormal* if T is invertible and satisfies $\log (T'T) \geq \log (TT')$. Every log-hyponormal operator satisfies the condition (14), see [10].

(d) A bounded operator $T \in L(X)$ is said to be *transaloid* if the spectral radius $r(\lambda I - T)$ is equal to $\|\lambda I - T\|$ for every $\lambda \in \mathbb{C}$. Every transaloid operator satisfies the condition (14), see [14] .

(e) Given a Banach algebra A, a map $T : A \to A$ is said to be a *multiplier* if $(Tx)y = x(Ty)$ holds for all $x, y \in \mathcal{A}$. For a commutative semi-simple Banach algebra A, let $M(A)$ denote the commutative Banach algebra of all multipliers, [29]. If $T \in M(A)$, A a commutative semi-simple Banach algebra, then $T \in L(A)$ and the condition (14) is satisfied, see [6]. In particular, this condition holds for every convolution operator on the group algebra $L^1(G)$, where G is a locally compact abelian group.

(f) An operator $T \in L(X)$, X a Banach space, is said to be *generalized scalar* if there exists a continuous algebra homomorphism $\Psi : \mathcal{C}^\infty(\mathbb{C}) \to L(X)$ such that

$$\Psi(1) = I \quad \text{and} \quad \Psi(Z) = T,$$

where $\mathcal{C}^\infty(\mathbb{C})$ denote the Fréchet algebra of all infinitely differentiable complex-valued functions on $\mathbb{C}$, and Z denotes the identity function on $\mathbb{C}$. Every generalized scalar operator is decomposable and hence has SVEP, see [29]. An operator similar to the restriction of a generalized scalar operator to one of its closed invariant subspaces is called *subscalar*. The interested reader can find a well organized study of these operators in the Laursen and Neumann book [29]. Every generalized scalar operator satisfies the condition

(15) $$H_0(\lambda I - T) = \ker (\lambda I - T)^p \quad \text{for all } \lambda \in \mathbb{C}.$$

for some $p = p(\lambda) \in \mathbb{N}$, see [30].

(g) An operator $T \in L(H)$ on a Hilbert space H is said to be *M-hyponormal* if there is $M > 0$ for which $TT' \leq MT'T$. M-hyponormal operators, p-hyponormal operators, log-hyponormal operators, and algebraically hyponormal operators are generalized scalars, so they satisfy the condition (15), see [30]. Also *w-hyponormal operators* on Hilbert spaces are generalized scalars, see for definition and details [11].

(h) An operator $T \in L(X)$ for which there exists a complex nonconstant polynomial h such that $h(T)$ is paranormal is said to be *algebraically paranormal*. If $T \in L(H)$ is algebraically paranormal then T satisfies the condition (12), see [2], but in general the condition (15) is not satisfied by paranormal operators, (for an example see [7, Example 2.3]). Note that if T is paranormal then T has SVEP, see [7]. It should be noted that this implies that also every algebraically paranormal operator has SVEP, see Theorem 2.40 of [1] or [15].

(i) An operator $T \in L(X)$ is said to be *hereditarily normaloid* if every restriction $T|M$ to a closed subspace of X is normaloid, i.e. the spectral radius of $T|M$ coincides with the norm $\|T|M\|$. If, additionally, every invertible part of T is also normaloid then T is said to be *totally hereditarily normaloid*.

Let CHN denote the class of operators such that either T is totally hereditarily normaloid or $\lambda I - T$ is hereditarily normaloid for every $\lambda \in \mathbb{C}$. The class CHN is very large; it contains p-hyponormal operators, M-hyponormal operators, paranormal operators and w-hyponormal operators, see [23]. Also every *totally $*$-paranormal* operator belongs to the class CHN. Note that every $T \in CHN$ satisfies the condition (12) with $p(\lambda) = 1$ for all $\lambda \in \pi_{00}(T)$, see [19], so $T \in \mathcal{P}_0(X)$.

Theorem 3.7. *Suppose that T is any of the operators (a)-(i). Then Weyl's theorem holds for T.*

Proof. The condition (15), and in particular the condition (14), entails that T has SVEP. Weyl's theorem for the operators (a)-(g) then follows from the implication (13). Also, every algebraically paranormal operators on Hilbert spaces has SVEP, so Weyl's theorem for these operators follows again from the implication (13). Weyl's theorem for operators $T \in CHN$ follows from Theorem 3.1, since SVEP is satisfied at every $\lambda \notin \sigma_w(T)$ and hence T satisfies Browder's theorem, see Theorem 2.9 of [19]. ∎

The result of Theorem 3.7 may be improved as follows. Suppose that $T \in L(X)$ be *algebraic*, i.e. there exists a non-trivial polynomial q such that $q(T) = 0$. In [31] Oudghiri proved that if T has property (15) and K is an algebraic operators commuting with T then $T + K$ satisfies Weyl's theorem. This result has been extended in [7] to paranormal operators, i.e. if T is a paranormal operator on a Hilbert space and K is an algebraic operators commuting with T then $T + K$ satisfies Weyl's theorem. Note that Weyl's theorem is not generally transmitted to perturbation of operators satisfying Weyl's theorem.

It has already been observed that Browder's theorem for T and T^* are equivalent. In general Weyl's theorem for T in not transmitted to the dual T^*, see Example 1.1 of [3]. However, for most of the operators listed before Weyl's theorem holds also for T^* (or, in the case of Hilbert spaces operators, for T'). This is, for instance, the case of operators satisfying the condition (15), see [30]. Also for operators $T \in CHN$, and in particular for algebraically paranormal operators, Weyl's theorem holds for T', see [19, Corollary 2.16] and [2, Theorem 4.3]. In all these cases, if $\mathcal{H}(\sigma(T))$ denotes the set of all analytic functions defined on a neighborhood of $\sigma(T)$, then Weyl's theorem holds $f(T)$, where the operator $f(T)$, $f \in \mathcal{H}(\sigma(T))$, is defined by the classical functional calculus. The extension of Weyl's theorem from T to $f(T)$ depends essentially by the fact that all these operators are *isoloid*, i.e. every isolated point of the spectrum is an eigenvalue.

The following result follows from Theorem 3.5 and from the implication (13).

Theorem 3.8. *If $T \in L(X)$ satisfies the property (15) then a-Weyl's theorem holds for $f(T^*)$ for every $f \in \mathcal{H}(\sigma(T))$. Analogously, if T^* satisfies the property (15) then a-Weyl's holds for $f(T)$ for every $f \in \mathcal{H}(\sigma(T))$.*

In the case of an operator defined on a Hilbert space H, instead of the dual T^* it is more appropriate to consider the Hilbert adjoint T' of $T \in L(H)$.

Theorem 3.9. [2] *If $T \in L(H)$ and T' satisfies property (15) then a-Weyls theorem holds for $f(T)$ for all $f \in \mathcal{H}(\sigma(T))$.*

The previous theorem provides a general framework for a-Weyl's theorem, from which all the results listed in the sequel follow as special cases. In literature a-Weyl's theorem for these classes has been proved separately.

(i) *If T' is log-hyponormal or p-hyponormal then a-Weyl's theorem holds for $f(T)$* [17, Theorem 3.3], [18, Theorem 4.2].

(ii) *If T' is M-hyponormal then a-Weyl's theorem holds for $f(T)$*, [17, Theorem 3.6].

(iii) *If T' is totally *-paranormal then a-Weyl's theorem holds for $f(T)$*, [25, Theorem 2.10].

(iv) *If $T' \in L(H)$ is totally paranormal then a-Weyl's theorem holds for $f(T)$.*

Theorem 3.10. [2, Theorem 3.6] *Let $T \in L(H)$. Then the following statements hold:*

(i) *If $T \in L(H)$ is algebraically paranormal then Weyl's theorem holds for $f(T)$ for all $f \in \mathcal{H}(\sigma(T)$.*

(ii) *If T' is algebraically paranormal then a-Weyl's theorem holds for $f(T)$ for all $f \in \mathcal{H}(\sigma(T)$.*

Every quasi-hyponormal operator is paranormal [24], so from Theorem 3.9 we deduce the following result of S. V. Djordjević and D. S. Djordjević [16, Theorem 3.4].

Corollary 3.11. *If $T' \in L(H)$ is quasi-hyponormal then a-Weyl's holds for $f(T)$ for every $f \in \mathcal{H}(\sigma(T))$.*

REFERENCES

[1] P. Aiena *Fredholm and local spectral theory, with application to multipliers*. Kluwer Acad. Publ. (2004).

[2] P. Aiena *Classes of Operators Satisfying a-Weyl's theorem*, Studia Math.**169** (2005), 105-122.

[3] P. Aiena, M. T. Biondi *Weyl's theorems through local spectral theory*, Inter. Conference on Operator Theory and Operator Algebras, Suppl. Rend. Circ. Mat. di Palermo **73**, (2004), 143-167.

[4] P. Aiena, M. T. Biondi A. *Browder's theorem and localized SVEP*; Mediterranean Journ. of Math. **2** (2005), 137-151.

[5] P. Aiena, C. Carpintero, E. Rosas *Some characterization of operators satisfying a-Browder theorem*, J. Math. Anal. Appl. **311**, (2005), 530-544.

[6] P. Aiena, M. L. Colasante, M. Gonzalez *Operators which have a closed quasi-nilpotent part*, Proc. Amer. Math. Soc. **130**, (9) (2002), 2701-2710.

[7] P. Aiena, J. R. Guillen *Weyl's theorem for perturbations of paranormal operators*, (2005) To appear in Proc. Amer. Math. Soc.

[8] P. Aiena, T. L. Miller, M. M. Neumann *On a localized single valued extension property*, Math. Proc. Royal Irish Acad. **104A**,1,(2004), 17-34.

[9] P. Aiena, E. Rosas *The single valued extension property at the points of the approximate point spectrum*, J. Math. Anal. Appl. **279** (1), (2003), 180-188.

[10] P. Aiena, F. Villafãne *Weyl's theorem for some classes of operators.* (2005). To appear on Int. Eq. Oper. Theory.

[11] Lin Chen, Ruan Yingbin, Yan Zikun *w-hiponormal operators are subscalar.*, Int. Eq. Oper. Theory **50**, no. 2, 165-168.

[12] L. A. Coburn *Weyl's theorem for nonnormal operators* Michigan Math. J. **20** (1970), 529-544.

[13] J. B. Conway *Subnormal operators* Michigan Math. J. **20** (1970), 529-544.

[14] R. E. Curto, Y. M. Han *Weyl's theorem, a-Weyl's theorem, and local spectral theory*,(2002), J. London Math. Soc. (2) **67** (2003), 499-509.

[15] R. E. Curto, Y. M. Han *Weyl's theorem for algebraically paranormal operators*,(2003), To appear on Int. Eq. Oper. Theory.

[16] S. V. Djordjević, D. S. Djordjević *Weyl's theorems: continuity of the spectrum and quasi-hyponormal operators*. Acta Sci. Math (Szeged) **64**, no. 3 (1998), 259-269.

[17] S. V. Djordjević, Y. M. Han *A note on a-Weyl's theorem*. J. Math. Anal. Appl. **260**, no. 8, (2001), 200-213.

[18] B. P. Duggal, S. V. Djordjević *Weyl's theorems and continuity of spectra in the class of p-hyponormal operators*. Studia Math. **143** (2000), no.1, 23-32.

[19] Duggal B. P. *Hereditarily normaloid operators.* (2004), preprint.

[20] N. Dunford *Spectral theory I. Resolution of the identity* Pacific J. Math. **2** (1952), 559-614.

[21] N. Dunford *Spectral operators* Pacific J. Math. **4** (1954), 321-354.

[22] J. K. Finch *The single valued extension property on a Banach space* Pacific J. Math. **58** (1975), 61-69.

[23] T. Furuta, M. Ito, T. Yamazaki *A subclass of paranormal operators including class of log-hyponormal ans several related classes* Scientiae Mathematicae **1** (1998), 389-403.

[24] T. Furuta *Invitation to Linear Operators* Taylor and Francis, London (2001).

[25] Y. M. Han, A. H. Kim *A note on *-paranormal operators* Int. Eq. Oper. Theory. **49** (2004), 435-444.

[26] R. Harte, Woo Young Lee, L. L. Littlejohn *On generalized Riesz points* J. Operator Theory **47** (2002), 187-196.

[27] H. Heuser *Functional Analysis.* (1982), Wiley, New York.

[28] K. B. Laursen *Operators with finite ascent.* Pacific J. Math. **152** (1992), 323-36.

[29] K. B. Laursen, M. M. Neumann *Introduction to local spectral theory.*, Clarendon Press, Oxford 2000.

[30] M. Oudghiri *Weyl's and Browder's theorem for operators satisfying the SVEP* Studia Math. **163**, 1, (2004), 85-101.

[31] M. Oudghiri *Weyl's theorem and perturbations.* (2004). To appear on Int. Equat. and
      Oper. Theory.
[32] V. Rakočević *Operators obeying a-Weyl's theorem.* Rev. Roumaine Math. Pures Appl.
      34 (1989), no. 10, 915-919.

DIPARTIMENTO DI METODI E MODELLI MATEMATICI, UNIVERSITÀ DI PALERMO, VIALE
DELLE SCIENZE, I-90128 PALERMO (ITALY), E-MAIL PAIENA@UNIPA.IT

DEPARTAMENTO DE MATEMÁTICAS,, UNIVERSIDAD UCLA DE BARQUISIMETO (VENEZUELA)
E-mail address: mtbiondi2005@hotmail.com

MULTIPLICATIONS AND ELEMENTARY OPERATORS IN THE BANACH SPACE SETTING

EERO SAKSMAN AND HANS-OLAV TYLLI

1. INTRODUCTION

This expository survey is mainly dedicated to structural properties of the *elementary operators*

$$(1.1) \qquad \mathcal{E}_{A,B}; \ S \mapsto \sum_{j=1}^{n} A_j S B_j,$$

where $A = (A_1, \ldots, A_n), B = (B_1, \ldots, B_n) \in L(X)^n$ are fixed n-tuples of bounded operators on X and X is a (classical) Banach space. The simplest operators contained in (1.1) are the left and right multiplications on $L(X)$ defined by $L_U : S \mapsto US$ and $R_U : S \mapsto SU$ for $S \in L(X)$, where $U \in L(X)$ is fixed. Thus the operators in (1.1) can be written as

$$\mathcal{E}_{A,B} = \sum_{j=1}^{n} L_{A_j} R_{B_j}.$$

This concrete class includes many important operators on spaces of operators, such as the two-sided multiplications $L_A R_B$, the commutators (or inner derivations) $L_A - R_A$, and the intertwining maps (or generalized derivations) $L_A - R_B$ for given $A, B \in L(X)$. Elementary operators also induce bounded operators between operator ideals, as well as between quotient algebras such as the Calkin algebra $L(X)/K(X)$, where $K(X)$ are the compact operators on X. Note also that definition (1.1) makes sense in the more general framework of Banach algebras.

Elementary operators first appeared in a series of notes by Sylvester [Sy1884] in the 1880's, in which he computed the eigenvalues of the matrix operators corresponding to $\mathcal{E}_{A,B}$ on the $n \times n$-matrices. The term *elementary operator* was coined by Lumer and Rosenblum [LR59] in the late 50's. The literature related to elementary operators is by now very large, and there are many excellent surveys and expositions of certain aspects. Elementary operators on C*-algebras were extensively treated by Ara and Mathieu in [AM03, Chapter 5]. Curto [Cu92] gives an exhaustive overview of spectral properties of elementary operators, Fialkow [Fi92] comprehensively discusses their structural properties (with an emphasis on Hilbert space aspects and methods),

---

The first author was partly supported by the Academy of Finland project #201015 and the second author by the Academy of Finland projects #53893 and #210970.

and Bhatia and Rosenthal [BR97] deals with their applications to operator equations and linear algebra. Mathieu [Ma01b], [Ma01a] surveys some recent topics in the computation of the norm of elementary operators, and elementary operators on the Calkin algebra. These references also contain a number of applications, and we also note the survey by Carl and Schiebold [CS00], where they describe an intriguing approach to certain nonlinear equations from soliton physics which involves some elementary operators (among many other tools).

This survey concentrates on aspects of the theory of elementary operators that, roughly speaking, involves "Banach space techniques". By such methods we mean e.g. basic sequence techniques applied in X or in $K(X)$, facts about the structure of complemented subspaces of classical Banach spaces X, as well as useful special properties of the space X (such as approximation properties or the Dunford-Pettis property). The topics and results covered here are chosen to complement the existing surveys, though some overlap will be unavoidable. Our main motivation is to draw attention to the usefulness of Banach space methods in this setting. In fact, it turns out that Banach space techniques are helpful also when X is a Hilbert space.

This survey will roughly be divided as follows. In Section 2 we discuss various qualitative properties such as (weak) compactness or strict singularity of the basic two-sided multiplications $S \mapsto ASB$ for $A, B \in L(X)$. In Section 3 we concentrate on questions related to the norms and spectra of elementary operators in various settings. We include a quite detailed proof, using only elementary concepts, of the known formula $\sigma(\mathcal{E}_{A,B}) = \sigma_T(A) \circ \sigma_T(B)$ for the spectrum of $\mathcal{E}_{A,B}$ in terms of the Taylor joint spectra of the n-tuples A and B. We also describe the state of the art in computing the norm of elementary operators. Section 4 discusses properties of the induced elementary operators on the Calkin algebra $L(X)/K(X)$, such as a solution to the Fong-Sourour conjecture in the case where the Banach space X has an unconditional basis, and various rigidity properties of these operators. The results included here demonstrate that elementary operators have nicer properties on the Calkin algebra. There is some parallel research about tensor product operators $A \widehat{\otimes}_\alpha B$ for various tensor norms α and fixed operators A, B, which may be more familiar to readers with a background in Banach space theory.

The ideas and arguments will be sketched for a number of results that we highlight here, and several open problems will be stated. The topics selected for discussion have clearly been influenced by our personal preferences and it is not possible to attempt any exhaustive record of Banach space aspects of the theory of elementary operators in this exposition. Further interesting results and references can be found in the original papers and the surveys mentioned above.

Elementary operators occur in many circumstances, and they can be approached using several different techniques. This survey is also intended for non-experts in Banach space theory, and we have accordingly tried to ensure that it is as widely readable as possible by recalling many basic concepts.

Our notation will normally follow the references [LT77] and [Wo91], and we just recall a few basic ones here. We put $B_X = \{x \in X : \|x\| \leq 1\}$ and $S_X = \{x \in X : \|x\| = 1\}$ for the Banach space X. If $A \subset X$ is a given subset, then $[A]$ denotes the closed linear span of A in X. Moreover, $L(X,Y)$ will be the space of bounded linear operators $X \to Y$ and $K(X,Y)$, respectively, $W(X,Y)$ the closed subspaces of $L(X,Y)$ consisting of the compact, respectively the weakly compact operators. The class of finite rank operators $X \to Y$ is denoted by $\mathcal{F}(X,Y)$. We refer e.g. to [LT77], [DJT95], [JL03] or [Wo91] for more background and any unexplained terminology related to Banach spaces.

2. QUALITATIVE ASPECTS

In this section we focus on concrete qualitative properties of the basic two-sided multiplication operators $L_A R_B$ for fixed $A, B \in L(X)$, where

$$(2.1) \qquad L_A R_B(S) = ASB$$

for $S \in L(X)$ and X is a Banach space. The first qualitative result for $L_A R_B$ is probably due to Vala [Va64], who characterized the compact multiplication operators on the space of bounded operators. Recall that X has the approximation property if for every compact subset $D \subset X$ and $\varepsilon > 0$ there is a finite rank operator $U \colon X \to X$ so that $\sup_{x \in D} \|x - Ux\| < \varepsilon$.

Theorem 2.1. *Suppose that $A, B \in L(X)$ are non-zero bounded operators. Then $L_A R_B$ is a compact operator $L(X) \to L(X)$ if and only if $A \in K(X)$ and $B \in K(X)$.*

Proof. The necessity-part is a simple general fact which we postpone for a moment (see part (i) of Proposition 2.3). The following straightforward idea for the sufficiency-part comes from the proof of [ST94, Thm. 2], which dealt with weak compactness (see also [DiF76] for a similar idea for the ϵ-tensor product).

Suppose that $A \in K(X)$ and $B \in K(X)$. We first consider the situation where X has the approximation property. In this case there is a sequence $(A_n) \subset \mathcal{F}(X)$ of finite-rank operators satisfying $\|A - A_n\| \to 0$ as $n \to \infty$. Clearly

$$\|L_{A_n} R_B - L_A R_B\| \leq \|A_n - A\| \cdot \|B\| \to 0$$

so that $L_{A_n} R_B \to L_A R_B$ as $n \to \infty$, whence it is enough to prove the claim assuming that A is a rank-1 operator, that is, $A = x^* \otimes y$ for fixed $x^* \in X^*$ and $y \in X$. In this case one gets for $S \in L(X)$ and $z \in X$ that

$$(x^* \otimes y) \circ S \circ Bz = \langle x^*, SBz \rangle y = \langle B^* S^* x^*, z \rangle y,$$

that is, $L_A R_B(S) = B^* S^* x^* \otimes y$. Hence

$$(2.2) \qquad L_A R_B(B_{L(X)} \subset \Phi \circ B^*(B_{X^*}),$$

where $\Phi : X^* \to L(X,Y)$ is the bounded linear operator $\Phi(z^*) = \|x^*\|z^* \otimes y$ for $z^* \in X^*$. Since $\Phi \circ B^*$ is compact by assumption, the claim follows.

How should one proceed in the general situation? The main problem compared to the preceding argument is that it may not be possible to approximate the operator A by finite dimensional ones. For that end consider first the case where A is replaced by any rank-1 operator $C : X \to Y$ and Y is an arbitrary Banach space. Exactly the same argument as above applies and we obtain that $L_C R_B : L(X) \to L(X, Y)$ is compact. Also, if Y has the approximation property, then we *may* approximate any $C \in K(X, Y)$ by finite-rank operators and deduce that $L_C R_B$ is compact. Next let $J : X \to \ell^\infty(B_{X^*})$ be the isometric embedding defined by

$$Jx = (x^*(x))_{x^* \in B_{X^*}}, \quad x \in X,$$

and recall that $\ell^\infty(B_{X^*})$ has the approximation property. By choosing $C = JA$ in the previous reasoning we get that $L_C R_B$ is a compact operator. Finally, observe that $L_C R_B = L_J \circ (L_A R_B)$, where $L_J : L(X) \to L(X, \ell^\infty(B_{X^*}))$ is an isometric embedding, which forces $L_A R_B$ to be compact. $\qquad\square$

Vala's argument in [Va64] was quite different. He applied an Ascoli-Arzela type characterization of compact sets of compact operators, which was inspired by a symmetric version of the Ascoli-Arzela theorem used by Kakutani.

It is less straightforward to formulate satisfactory characterizations of arbitrary compact elementary operators $\mathcal{E}_{A,B}$, because of the lack of uniqueness in the representations of these operators (for instance, $L_{A-\lambda} - R_{A-\lambda} = L_A - R_A$ for every $A \in L(X)$ and every scalar λ). One possibility is to assume some linear independence among the representing operators. We next state a generalization of Theorem 2.1 of this type, due to Fong and Sourour [FS79] (cf. also [Fi92, Thm. 5.1]).

Theorem 2.2. *Let* $A = (A_1, \ldots, A_n), B = (B_1, \ldots, B_n) \in L(X)^n$, *where* X *is any Banach space. Then* $\mathcal{E}_{A,B}$ *is a compact operator* $L(X) \to L(X)$ *if and only if there are* $r \in \mathbb{N}$ *and compact operators* $C_1, \ldots, C_r$ *in the linear span of* $\{A_1, \ldots, A_n\}$ *and compact operators* $D_1, \ldots, D_r$ *in the linear span of* $\{B_1, \ldots, B_n\}$ *so that*

$$\mathcal{E}_{A,B} = \sum_{j=1}^{r} L_{C_j} R_{D_j}.$$

Vala's result (Theorem 2.1) raised the problem when the basic maps $L_A R_B$ are weakly compact, that is, when $L_A R_B(B_{L(X)})$ is a relatively weakly compact set. It is difficult in general to characterize the weakly compact subsets of $L(X)$, and the picture for weak compactness is much more complicated compared to Theorem 2.1. As the first step Akemann and Wright [AW80] characterized the weakly compact multipliers $L_A R_B$ in the case of Hilbert spaces H as follows: $L_A R_B$ *is weakly compact* $L(H) \to L(H)$ *if and only if* $A \in K(H)$ *or* $B \in K(H)$ (see Example 2.6 below). The weakly compact one-sided multipliers L_A and R_A on $L(H)$ were identified much earlier by Ogasawara [Og54] (cf. also [Y75] for some additional information).

Subsequently the weak compactness of $L_A R_B$ was studied more systematically by the authors [ST92], Racher [Ra92], and in a more general setting by

Lindström and Schlüchtermann [LSch99]. The following basic general facts were noticed in [ST92]. The original proof of part (ii) in [ST92] is somewhat cumbersome, and easier alternative arguments were given in [Ra92] and [ST94] (see also [LSch99]). The argument included below is arguably the simplest one conceptually.

Proposition 2.3. *Let X be any Banach space and let $A, B \in L(X)$.*
(i) If $L_A R_B$ is a weakly compact operator $L(X) \to L(X)$, and $A \neq 0 \neq B$, then $A \in W(X)$ and $B \in W(X)$.
(ii) If $A \in K(X)$ and $B \in W(X)$, or if $A \in W(X)$ and $B \in K(X)$, then $L_A R_B$ is weakly compact $L(X) \to L(X)$.

Proof. (i) The identity $L_A R_B(x^* \otimes x) = B^* x^* \otimes Ax$ for $x^* \in X^*, x \in X$, is the starting point. Fix $x \in S_X$ with $Ax \neq 0$, and note that

$$B^*(B_{X^*}) \otimes Ax \subset \{L_A R_B(x^* \otimes x) : x^* \in B_{X^*}\} \subset L_A R_B(B_{L(X)}),$$

where the right-hand set is relatively weakly compact in $L(X)$ by assumption. It follows that the adjoint B^* (and consequently also B) is a weakly compact operator. The fact that $A \in W(X)$ is seen analogously.

(ii) Suppose first that $A \in K(X)$ and $B \in W(X)$. In this situation the proof is quite analogous to the corresponding one of Theorem 2.1: if X has the approximation property then one notes that it is enough to consider the case where $A = x^* \otimes y$ for some $x^* \in X^*$ and $y \in X$. Here the inclusion (2.2) again yields the weak compactness of $L_A R_B$, since B^* is also weakly compact by Gantmacher's theorem. In the general case one again picks an isometric embedding $J : X \to \ell^\infty(B_{X^*})$ and one applies the approximation property of $\ell^\infty(B_{X^*})$ to obtain that

$$L_{JA} R_B = L_J \circ (L_A R_B) : L(X) \to L(X, Y)$$

is weakly compact. Recall next a useful fact: (relative) weak (non-)compactness is unchanged under isometries. Hence, as $L_J : L(X) \to L(X, Y)$ is an isometric embedding we obtain that $L_A R_B$ is weakly compact.

Next consider the case $A \in W(X)$ and $B \in K(X)$. From the preceding case applied to $L_{B^*} R_{A^*}$ we get that $G \equiv \{U^* : U \in L_A R_B(B_{L(X)})\}$ is a relatively weakly compact set, since obviously

$$G \subset L_{B^*} R_{A^*}(B_{L(X^*)}).$$

This implies that $L_A R_B(B_{L(X)})$ is also a relatively weakly compact set, since the map $U \to U^*$ is an isometric embedding $L(X) \to L(X^*)$.   □

For any given Banach space X the exact conditions for the weak compactness of $L_A R_B$ on $L(X)$ fall between the extremal conditions contained in (i) and (ii) of Proposition 2.3, and examples demonstrating a wide variety of different behaviour were included in [ST92] and [Ra92]. To get our hands on these examples we will need more efficient criteria for the weak compactness of $L_A R_B$, which can be obtained by restricting attention to suitable classes of Banach spaces. Before that we also observe that in the study of the maps $S \mapsto ASB$ one is naturally lead to consider (possibly differ-ent) Banach spaces X_1, X_2, X_3, X_4 and compatible operators $A \in L(X_3, X_4)$ and $B \in L(X_1, X_2)$. In this case (2.1) still defines a bounded linear op-erator $L_A R_B \colon L(X_2, X_3) \rightarrow L(X_1, X_4)$ (strictly speaking $L_A R_B$ is now a composition operator, but by a minor abuse of language we will still talk about multiplication operators). Above we restricted attention to the case $X_1 = X_2 = X_3 = X_4 = X$ for notational simplicity.

Remark 2.4. Theorem 2.1 and Proposition 2.3 remain valid in the general setting, with purely notational changes in the proofs. For instance, part (ii) of Proposition 2.3 can be stated as follows: *If $A \in K(X_3, X_4)$ and $B \in W(X_1, X_2)$, or if $A \in W(X_3, X_4)$ and $B \in K(X_1, X_2)$, then $L_A R_B$ is a weakly compact operator $L(X_2, X_3) \rightarrow L(X_1, X_4)$.*

The first type of examples belong to the class of Banach spaces, where the duals of the spaces $K(X, Y)$ of compact operators can be described using trace-duality. We refer to [DiU77] for the definition and the properties of the Radon-Nikodym property (RNP). It suffices to recall here that X has the RNP if X is reflexive or if X is a separable dual space, while X fails the RNP if X contains a linear isomorphic copy of c_0 or $L^1(0, 1)$. The following concrete range-inclusion criterion for the weak compactness of $L_A R_B$ is quite efficient. We again restrict our attention to the case $X_1 = \ldots = X_4 = X$, and the formulation below is far from optimal.

Proposition 2.5. *Let X be a Banach space having the approximation prob-lem, and suppose that X^* or X^{**} has the RNP. Let $A, B \in L(X)$ be non-zero operators. Then $L_A R_B$ is weakly compact $L(X) \rightarrow L(X)$ if and only if $A, B \in W(X)$ and*

$$(2.3) \qquad\qquad L_{A^{**}} R_{B^{**}}(L(X^{**})) \subset K(X^{**}).$$

If X is reflexive, then (2.3) reduces to

$$(2.4) \qquad\qquad \{ASB : S \in L(X)\} \subset K(X).$$

Condition (2.3) is based on the trace duality identifications $K(X)^* = N(X^*)$ and $N(X^*)^* = L(X^{**})$, where the first identification requires suit-able approximation properties and the RNP conditions (see e.g. section 2 of the survey [Ru84]). Recall here that $T = \sum_{n=1}^{\infty} y_n^* \otimes y_n$ is a nuclear operator on the Banach space Y, denoted $T \in N(Y)$, if the sequences $(y_n^*) \subset Y^*$ and $(y_n) \subset Y$ satisfy $\sum_{n=1}^{\infty} \|y_n^*\| \cdot \|y_n\| < \infty$. The nuclear norm $\|T\|_N$ is the infimum of $\sum_{n=1}^{\infty} \|y_n^*\| \cdot \|y_n\|$ over all such representations of T. Recall that trace duality is defined by

$$(2.5) \qquad\qquad \langle T, S \rangle \equiv trace(T^* S) = \sum_{n=1}^{\infty} S x_n^{**}(x_n^*).$$

for $S \in L(X^{**})$ and $T = \sum_{n=1}^{\infty} x_n^{**} \otimes x_n^* \in N(X^*)$. For $S \in K(X)$ and $T \in N(X^*)$ one has analogously that $\langle T, S \rangle = \sum_{n=1}^{\infty} S^{**} x_n^{**}(x_n^*)$. Thus one has $K(X)^{**} = L(X^{**})$ in this setting, where the canonical embedding $K(X) \subset$

$K(X)^{**}$ coincides with the natural isometry $S \mapsto S^{**}$ from $K(X)$ into $L(X^{**})$. One easily checks on the rank-1 operators that the (pre)adjoints of $L_A R_B$ satisfy

$$(2.6) \qquad (L_A R_B : K(X) \to K(X))^* = L_{A^*} R_{B^*} : N(X^*) \to N(X^*),$$

$$(2.7) \qquad (L_{A^*} R_{B^*} : N(X^*) \to N(X^*))^* = L_{A^{**}} R_{B^{**}} : L(X^{**}) \to L(X^{**}).$$

Hence $(L_A R_{B|K(X)})^{**} = L_{A^{**}} R_{B^{**}}$, where $L_A R_{B|K(X)}$ is the restricted operator $K(X) \to K(X)$, so that (2.3) reduces to a well-known general criterion (see [Wo91, Thm. 2.C.6(c)]) for the weak compactness of bounded operators.

The following examples from [ST92] demonstrate some typical applications of (2.3). For $p = 2$ this is the result of Akemann and Wright [AW80] cited above.

Example 2.6. *Let $1 < p < \infty$ and $A, B \in L(\ell^p)$. Then $L_A R_B$ is weakly compact on $L(\ell^p)$ if and only if $A \in K(\ell^p)$ or $B \in K(\ell^p)$.*

Proof. The implication "$\Leftarrow$" is included in Proposition 2.3.(ii). For the converse implication we first look at the simplest case where $p = 2$. Note that to apply (2.4) we must exhibit, for any given pair $A, B \notin K(\ell^2)$, an operator $S \in L(\ell^2)$ so that $ASB \notin K(\ell^2)$. This is easy to achieve. Indeed, by assumption there are closed infinite-dimensional subspaces $M_1, M_2 \subset \ell^2$, so that the restrictions $A_{|M_2}$ and $B_{|M_1}$ are bounded below. Pick constants $c_1, c_2 > 0$ such that

$$\|Bx\| \geq c_1\|x\| \text{ for } x \in M_1, \quad \|Ax\| \geq c_2\|x\| \text{ for } x \in M_2.$$

Define the bounded operator S on $\ell^2 = B(M_1) \oplus (BM_1)^\perp$ by requiring that S is an isometry from $B(M_1)$ onto M_2 and $S = 0$ on $(BM_1)^\perp$. Thus $\|ASBy\| \geq c_1\|S(By)\| \geq c_1 c_2\|y\|$ for $x \in M_1$, whence $ASB \notin K(\ell^2)$.

To argue as above for $p \neq 2$ one needs for any $U \notin K(\ell^p)$ to find a subspace $M \subset \ell^p$ so that M is isomorphic to ℓ^p, M and $U(M)$ are complemented in ℓ^p, and $U_{|M}$ is bounded below. This basic sequence argument is familiar to Banach space theorists, and we refer e.g. to [LT,Prop. 2.a.1 and 1.a.9] or [Pi80, 5.1.3] for the details. $\qquad \square$

A more serious refinement of these ideas yields the exact distance formula

$$dist(L_A R_B, W(L(\ell^p))) = dist(A, K(\ell^p)) \cdot dist(B, K(\ell^p))$$

for $A, B \in L(\ell^p)$ and $1 < p < \infty$, see [ST94, Thm. 2.(ii)].

Condition (2.4) gives, after some additional efforts, the following identification of the weakly compact maps $L_A R_B$ for the direct sum $X = \ell^p \oplus \ell^q$, see [ST92, Prop. 3.5]. Here we represent operators S on $\ell^p \oplus \ell^q$ as an operator matrix

$$S = \begin{pmatrix} S_{11} & S_{12} \\ S_{21} & S_{22} \end{pmatrix},$$

where $S_{11} \in L(\ell^p), S_{12} \in L(\ell^q, \ell^p), S_{21} \in L(\ell^p, \ell^q)$ and $S_{22} \in L(\ell^q)$.

Example 2.7. *Let* $1 < p < q < \infty$ *and* $A, B \in L(\ell^p \oplus \ell^q)$. *Then* $L_A R_B$ *is weakly compact on* $L(\ell^p \oplus \ell^q)$ *if and only if* $A \in K(\ell^p \oplus \ell^q)$, *or* $B \in K(\ell^p \oplus \ell^q)$, *or*

$$(2.8) \quad A \in \begin{pmatrix} K(\ell^p) & L(\ell^q, \ell^p) \\ L(\ell^p, \ell^q) & L(\ell^q) \end{pmatrix} \quad \text{and} \quad B \in \begin{pmatrix} L(\ell^p) & L(\ell^q, \ell^p) \\ L(\ell^p, \ell^q) & K(\ell^q) \end{pmatrix}.$$

Examples 2.6 and 2.7 provide ample motivation to consider the more delicate case $X = L^p(0,1)$ for $1 < p < \infty$ (recall that ℓ^p and $\ell^p \oplus \ell^2$ embed as complemented subspaces of $L^p(0,1)$). According to Proposition 2.3.(ii) the operator $L_A R_B$ is weakly compact on $L(L^p(0,1))$ if $A \in K(L^p(0,1))$ or $B \in K(L^p(0,1))$, but these conditions are far from being necessary. For this recall that $U \in L(X,Y)$ is a *strictly singular* operator, denoted $U \in S(X,Y)$, if the restriction $U_{|M}$ does not define an isomorphism $M \to U(M)$ for any closed infinite-dimensional subspaces $M \subset X$. It is known that $UV \in K(L^p(0,1))$ whenever $U, V \in S(L^p(0,1))$, see [Mi70, Teor. 7]. Hence condition (2.4) immediately yields the following fact:

- *If* $A, B \in S(L^p(0,1))$, *then* $L_A R_B$ *is weakly compact on* $L(L^p(0,1))$.

The preceding cases do not yet exhaust all the possibilities. In fact, note that (2.8) allows weakly compact multiplications $L_A R_B$ arising from non-strictly singular operators A, B on $\ell^p \oplus \ell^2$, and this example easily transfers to $L^p(0,1)$ by complementation. The following question remains unresolved, and it is also conceivable that there is no satisfactory answer.

Problem 2.8. *Let* $1 < p < \infty$ *and* $p \neq 2$. *Characterize those* $A, B \in L(L^p(0,1))$ *for which* $L_A R_B$ *is weakly compact on* $L(L^p(0,1))$.

For our second class of examples recall that X has the *Dunford-Pettis property* (DPP) if $\|U x_n\| \to 0$ as $n \to \infty$ for any $U \in W(X,Y)$, any weak-null sequence (x_n) of X and any Banach space Y. For instance, ℓ^1, $L^1(0,1)$, c_0, $C(0,1)$ and $L^\infty(0,1)$ have the DPP (in fact, more generally any $\mathcal{L}^1$- and $\mathcal{L}^\infty$-space X has the DPP). We refer to the survey [Di80] for further information about this property. Note that if X has the DPP, then $UV \in K(Y,Z)$ for all weakly compact operators $U \in W(X,Z)$ and $V \in W(Y,X)$. This fact and Proposition 2.5 suggest that there might be an analogue of Vala's theorem for weakly compact multipliers $L_A R_B$ on $L(X)$ if X has the DPP. It is an elegant result of Racher [Ra92] that this is indeed so (a more restricted version was contained in [ST92]). This fact provides plenty of non-trivial examples of weakly compact multiplications. The formulation included here of Racher's result is not the most comprehensive one.

Theorem 2.9. *Let* X *be a Banach space having the DPP, and suppose that* $A, B \in L(X)$ *are non-zero operators. Then* $L_A R_B$ *is weakly compact* $L(X) \to L(X)$ *if and only if* $A \in W(X)$ *and* $B \in W(X)$.

The proof of Racher's theorem is based on the following useful auxiliary fact.

Lemma 2.10. *Let X_1, X_2, X_3, X_4 be Banach spaces and suppose that $A = A_1 \circ A_2 \in W(X_3, X_4)$, $B = B_1 \circ B_2 \in W(X_1, X_2)$ factor through the reflexive spaces Z_1, respectively Z_2, so that*

$$(2.9) \qquad L_{A_2^{**}} R_{B_1}(L(X_2, X_3^{**})) \subset K(Z_2, Z_1).$$

Then $L_A R_B$ is weakly compact $L(X_2, X_3) \to L(X_1, X_4)$.

Proof. Let $(T_\gamma) \subset B_{L(X_2, X_3)}$ be an arbitrary net. It follows from Tychonoff's theorem and the w^*-compactness of $B_{X_3^{***}}$ that there is a subnet, still denoted by (T_γ), and an operator $S \in L(X_2, X_3^{**})$ so that

$$\langle y^*, K_3 T_\gamma x \rangle \to \langle y^*, Sx \rangle \quad \text{for all } x \in X_2, y^* \in X_3^*.$$

Here K_3 denotes the natural map $X_3 \to X_3^{**}$. Thus

$$\langle z^*, A_2^{**} K_3 T_\gamma B_1(z) \rangle \to \langle z^*, A_2^{**} S B_1(z) \rangle \quad \text{for all } z \in Z_2, z^* \in Z_1^*,$$

where $A_2^{**} S B_1 \in K(Z_2, Z_1)$ and $A_2^{**} K_3 T_\gamma B_1 \in K(Z_2, Z_1)$ for all γ by the assumption. A fundamental criterion for weak compactness in spaces of compact operators, due to Feder and Saphar [FeS75, Cor. 1.2], yields then that the net

$$A_2 T_\gamma B_1 = A_2^{**} K_3 T_\gamma B_1 \xrightarrow[\gamma]{w} A_2^{**} S B_1$$

in $K(Z_2, Z_1)$. Thus $L_{A_2} R_{B_1}$ is weakly compact $L(X_2, X_3) \to K(Z_2, Z_1)$, and so is $L_A R_B = L_{A_1} R_{B_2} \circ L_{A_2} R_{B_1}$. $\qquad\square$

The proof of Theorem 2.9 is immediate from Lemma 2.10. Recall first that the weakly compact operators $A, B \in W(X)$ factor as $A = A_1 \circ A_2$ and $B = B_1 \circ B_2$ through suitable reflexive spaces Z_1 and Z_2 by the well-known DFJP-construction, see [Wo91, Thm. II.C.5]. If X has the DPP, then $A_2 S B_1 \in K(Z_2, Z_1)$ for any $S \in L(X)$, so that (2.9) is satisfied.

It is also possible to prove analogues of Theorem 2.2 for the weak compactness of elementary operators $\mathcal{E}_{A,B}$. The following version of Theorem 2.9 is taken from [S95, Section 2]:
• *Suppose that X has the DPP. Then the elementary operator $\mathcal{E}_{A,B}$ is weakly compact on $L(X)$ if and only if there are m-tuples $U = (U_1, \ldots, U_m)$, $V = (V_1, \ldots, V_m) \in W(X)^m$ so that*

$$\mathcal{E}_{A,B} = \mathcal{E}_{U,V}.$$

It is possible to study multiplications and elementary operators in the general setting of Banach operator ideals in the sense of Pietsch [Pi80]. This extension is motivated by several reasons (one reason is the duality with the nuclear operators used in the proof of Proposition 2.5). Recall that $(I, \| \cdot \|_I)$ is a Banach operator ideal if $I(X, Y) \subset L(X, Y)$ is a linear subspace for any pair X, Y of Banach spaces, $\| \cdot \|_I$ is a complete norm on $I(X, Y)$ and

(i) $x^* \otimes y \in I(X, Y)$ and $\|x^* \otimes y\|_I = \|x^*\| \cdot \|y\|$ for $x^* \in X^*$ and $y \in Y$,
(ii) $ASB \in I(X_1, X_2)$ and $\|ASB\|_I \leq \|A\| \cdot \|B\| \cdot \|S\|_I$ whenever $S \in I(X, Y)$, $A \in L(Y, X_2)$ and $B \in L(X_1, X)$ are bounded operators.

There is a large variety of useful and interesting examples of Banach operator ideals. For instance, K, W, the nuclear operators $(N, \|\cdot\|_I)$ and the class Π_p of the p-summing operators are important examples (we refer to [Pi80] or [DJT95] for further examples). Conditions (i) and (ii) imply that the basic map $L_A R_B$ is bounded $I(X_2, X_3) \to I(X_1, X_4)$, and that in fact

$$(2.10) \qquad \|L_A R_B : I(X_2, X_3) \to I(X_1, X_4)\| = \|A\| \cdot \|B\|.$$

for any bounded operators $A \in L(X_3, X_4), B \in L(X_1, X_2)$.

The study of multiplication operators in the framework of Banach operator ideals was initiated by Lindström and Schlüchtermann in [LSch99]. Here one obviously meets the following general problem:

• Let $(I, \|\cdot\|_I)$ and $(J, \|\cdot\|_J)$ be Banach operator ideals. For which operators A and B does the map $L_A R_B : I(X_2, X_3) \to I(X_1, X_4)$ belong to J?

Note that Proposition 2.3 admits a more general version. Recall for this that $(I, \|\cdot\|_I)$ is a closed Banach operator ideal if the component $I(X,Y)$ is closed in $(L(X,Y), \|\cdot\|)$ for any pair X, Y. The ideal $(I, \|\cdot\|_I)$ is injective if $JS \in I(X, Z)$ for any isometry $J : Y \to Z$ yield that $S \in I(X,Y)$ and $\|JS\|_I = \|S\|_I$. Moreover, $(I, \|\cdot\|_I)$ is surjective if $SQ \in I(Z, Y)$ for any metric surjection $Q : Z \to X$ (that is, $\overline{QB_Z} = B_X$) imply that $S \in I(X,Y)$ and $\|SQ\|_I = \|S\|_I$. For instance, K and W are injective and surjective ideals.

Let I and J be Banach operator ideals, and let $A \in L(X_3, X_4)$, $B \in L(X_1, X_2)$, where $X_1, \ldots, X_4$ are any Banach spaces. The following general facts hold, see [LSch99, Section 2].

• If $A \neq 0 \neq B$ and the map $L_A R_B : I(X_2, X_3) \to I(X_1, X_4)$ belongs to J, then $A \in J(X_3, X_4)$ and $B^* \in J(X_2^*, X_1^*)$.
• Assume that I is injective, and that J is closed and injective. If $A \in K(X_3, X_4)$ and $B^* \in J(X_2^*, X_1^*)$, then the map $L_A R_B : I(X_2, X_3) \to I(X_1, X_4)$ belongs to J.
• Assume that I is surjective, and that J is closed and injective. If $A \in J(X_3, X_4)$ and $B \in K(X_1, X_2)$, then the map $L_A R_B : I(X_2, X_3) \to I(X_1, X_4)$ belongs to J.

Lindström and Schlüchtermann [LSch99] obtained several range inclusion results for the multiplications $L_A R_B$. We state two of their main results. Recall that the operator $U \in L(X,Y)$ is weakly conditionally compact if for every bounded sequence $(x_n) \subset X$ there is weakly Cauchy subsequence $(U x_{n_k})$. Clearly any weakly compact map is weakly conditionally compact. Note that part (ii) below provides a partial converse of Lemma 2.10. We refer to Section 4, or references such as [LT77] or [Wo91], for more background about unconditional bases.

Theorem 2.11. Let $X_1, \ldots, X_4$ be Banach spaces and $A \in L(X_3, X_4)$, $B \in L(X_1, X_2)$.
(i) Suppose that every $S \notin K(X_2, X_3)$ factors through a Banach space Z having an unconditional basis, and that X_3 does not contain any isomorphic copies of c_0. If the map $L_A R_B$ is weakly conditionally compact $L(X_2, X_3) \to$

$L(X_1, X_4)$, then

$$L_A R_B(L(X_2, X_3)) \subset K(X_1, X_4).$$

(ii) Suppose that every $S \notin K(X_2, X_3)$ factors through a Banach space Z having an unconditional basis, and that $L_A R_B$ is weakly compact $L(X_2, X_3) \to L(X_1, X_4)$. Then

$$A^{**} \circ \overline{B_{L(X_2, X_3)}}^{w^* OT} \circ B \subset K(X_1, X_4),$$

where $w^* OT$ denotes the w^*-operator topology in $L(X_2, X_3^{**})$.

The examples included in this section demonstrate that the conditions for $L_A R_B$ to belong to a given operator ideal I usually depend on geometric or structural properties of the Banach spaces involved. However, for suitable classical Banach spaces it is still possible to obtain complete descriptions. We next discuss some non-trivial results from [LST05] related to strict singularity. This class of operators is central for many purposes (such as in perturbation theory and the classification of Banach spaces). The main result of [LST05] completely characterizes the strictly singular multiplications $L_A R_B$ on $L(L^p(0, 1))$ for $1 < p < \infty$. The simple form of the characterization is rather unexpected, since the subspace structure of the algebras $L(X)$ is very complicated. The case $p = 2$ is essentially contained in Theorem 2.1, and it is excluded below.

Theorem 2.12. Let $1 < p < \infty, p \neq 2$, and suppose that $A, B \in L(L^p(0, 1))$ are non-zero. Then $L_A R_B$ is strictly singular $L(L^p(0, 1)) \to L(L^p(0, 1))$ if and only if $A \in S(L^p(0, 1))$ and $B \in S(L^p(0, 1))$.

In contrast to the simplicity of the statement above the proof of Theorem 2.12 is lengthy and quite delicate, and we are only able to indicate some of the main steps and difficulties here. The implication "$\Leftarrow$" is the non-trivial one (the converse implication follows from the generalities). As the starting point one notes that it suffices to treat the case $2 < p < \infty$, since the map $U \mapsto U^*$ preserves strict singularity on $L^p(0, 1)$, see [We77]. Assume to the contrary that there are operators $A, B \in S(L^p(0, 1))$ so that $L_A R_B$ is not strictly singular $L(L^p(0, 1)) \to L(L^p(0, 1))$. Hence there is an infinite-dimensional subspace $N \subset L(L^p(0, 1))$ so that $L_A R_B$ is bounded below on N. The first step consists of "modifying" N to obtain a block diagonal sequence $(S_k) \subset \mathcal{F}(L^p(0, 1))$, for which the restriction of $L_A R_B$ to $[S_k : k \in \mathbb{N}]$ is still bounded below and the image sequence $(AS_k B)$ is as close as we want to a block diagonal sequence $(U_k) \subset \mathcal{F}(L^p(0, 1))$. By a block diagonal sequence (S_k) is here meant that

$$S_k = (P_{m_{k+1}} - P_{m_k})S_k(P_{m_{k+1}} - P_{m_k}) \quad \text{for } k \in \mathbb{N},$$

where $(m_k) \subset \mathbb{N}$ is some increasing sequence and (P_r) is the sequence of basis projections associated to the Haar basis (h_n) of $L^p(0, 1)$. Note that $N \subset L(L^p(0, 1))$ so that this reduction cannot be achieved just by a straightforward approximation. In fact, the actual argument proceeds through several auxiliary results.

In the next step one invokes classical estimates on unconditional basic sequences in $L^p(0,1)$ to ensure that

$$(2.11) \qquad \qquad \| \sum_k c_k S_k \| \approx \|(c_k)\|_s \quad \text{for } (c_k) \in \ell^s,$$

where s satisfies $\frac{1}{2} = \frac{1}{p} + \frac{1}{s}$. The final challenge is to derive a contradiction from (2.11) by a subtle comparison with the Kadec-Pełczynski dichotomy. (This fundamental result [KP62] says that for any normalized basic sequence (f_n) in $L^p(0,1)$, where $2 < p < \infty$, there is a subsequence (f_{n_k}) so that $[f_{n_k} : k \in \mathbb{N}]$ is complemented in $L^p(0,1)$ and (f_{n_k}) is either equivalent to the unit vector basis of ℓ^p or ℓ^2).

Problem 2.13. *Find a simpler approach to Theorem 2.12.*

The delicacy above is further illustrated by the facts that Theorem 2.12 remains true for $X = \ell^p \oplus \ell^q$, see [LST05, Thm. 4.1], but not for $X = \ell^p \oplus \ell^q \oplus \ell^r$, where $1 < p < q < r < \infty$.

Example 2.14. *Let $1 < p < q < r < \infty$ and define $J_1, J_2 \in S(\ell^p \oplus \ell^q \oplus \ell^r)$ by*

$$J_1(x,y,z) = (0,0,j_1 y), \quad J_2(x,y,z) = (0,j_2 x,0), \quad (x,y,z) \in \ell^p \oplus \ell^q \oplus \ell^r,$$

where $j_1 : \ell^q \to \ell^r$ and $j_2 : \ell^p \to \ell^q$ are the natural inclusion maps. Then $L_{J_1} R_{J_2}$ is not strictly singular $L(\ell^p \oplus \ell^q \oplus \ell^r) \to L(\ell^p \oplus \ell^q \oplus \ell^r)$.

Indeed, by passing to complemented subspaces it is enough to check that the related composition map $L_{j_1} R_{j_2}$ is not strictly singular $L(\ell^q) \to L(\ell^p, \ell^r)$. This fact follows from the straightforward computation that

$$\| \sum_n a_n j_2^* e_n^* \otimes j_1 e_n \|_{\ell^p \to \ell^r} = \| \sum_n a_n e_n^* \otimes e_n \|_{\ell^q \to \ell^q} = \sup_n |a_n|$$

for $(a_n) \in c_0$, where $(e_n) \subset \ell^q$ is the unit vector basis and $(e_n^*) \subset \ell^{q'}$ is the biorthogonal sequence.

The reference [LST05] also characterizes the strictly singular multiplications $L_A R_B$ on $L(X)$ when X is a $\mathcal{L}^1$-space. This result is based on Theorem 2.9 and the non-trivial fact, essentially due to Bourgain [B81], that here $L(X^{**})$ has the DPP.

Qualitative results for the multiplication operators $L_A R_B$ are often helpful when studying other aspects of multiplication or elementary operators. We also mention an interesting application [BDL01, Section 5], where maps of the form $L_A R_B$ are used to linearize the analytic composition operators $C_\phi : f \mapsto f \circ \phi$ on certain vector-valued spaces of analytic functions. Here ϕ is an analytic self-map of the unit disc $D = \{z \in \mathbf{C} : |z| < 1\}$.

Other developments. There is a quite extensive theory of tensor norms of Banach spaces and tensor products of operators, which parallels the study of the multiplication operators $L_A R_B$. Recall that the norm α, defined on the algebraic tensor products $X \otimes Y$ for all Banach spaces X, Y, is called a *tensor norm* if

(iii) $\alpha(x \otimes y) = \|x\| \cdot \|y\|$ for $x \otimes y \in X \otimes Y$,

(iv) $\|A \otimes B : (X_1 \otimes Y_1, \alpha) \to (X \otimes Y, \alpha)\| \leq \|A\| \cdot \|B\|$ for any bounded operators $A \in L(X_1, X)$, $B \in L(Y_1, Y)$ and Banach spaces X_1, X, Y_1, Y.

The α-tensor product $X \widehat{\otimes}_\alpha Y$ is the completion of $(X \otimes Y, \alpha)$. Property (iv) states that any $A \in L(X_1, X)$ and $B \in L(Y_1, Y)$ induce a bounded linear operator $A \widehat{\otimes}_\alpha B : X_1 \widehat{\otimes}_\alpha Y_1 \to X \widehat{\otimes}_\alpha Y$. We refer to [DeF93] for a comprehensive account of tensor norms and tensor products of operators.

Tensor norms and Banach operator ideals are related to each other, but this correspondence is not complete. For instance, recall that $X^* \widehat{\otimes}_\epsilon Y = K(X, Y)$ if X^* or Y has the approximation property (see e.g. [DeF93, 5.3]), while $(X \widehat{\otimes}_\pi Y)^* = L(X^*, Y)$ for any pair X, Y. Given $A \in L(X)$ and $B \in L(Y)$ one may then identify, under appropriate conditions, the tensor product operator $A^* \widehat{\otimes}_\epsilon B$ with the map $L_{A^*} R_B$ and $(A \widehat{\otimes}_\pi B)^*$ with $L_A R_{B^*}$. There are many results which are more natural to state either in terms of multiplication operators or tensor products of operators. An example of this for tensor products is the following celebrated result of J. Holub [Ho70], [Ho74] (see also [DeF93, 34.5]):

- $A \widehat{\otimes}_\epsilon B$ is a p-summing operator whenever A and B are p-summing operators.

References such as e.g. [DiF76], [Pi87], [CDR89], [DeF93, Chapter 34], [Ra92] and [LSch99] contain qualitative results for tensor products of operators which resemble some of the results of this section for the multiplication operators. Since the elementary operators are the main objects of this survey we have not pursued this aspect.

It is not known whether Vala's result (Theorem 2.1) holds for arbitrary Banach operator ideals (alternatively, for arbitrary tensor norms).

Problem 2.15. Let $(I, \|\cdot\|_I)$ be an arbitrary Banach operator ideal. Is $L_A R_B$ a compact operator $I(X_2, X_3) \to I(X_1, X_4)$ whenever $A \in K(X_3, X_4)$ and $B \in K(X_1, X_2)$?

The tensor version of this problem was discussed by Carl, Defant and Ramanujan [CDR89], where one finds a number of partial positive results.

3. Norms and Spectra in Various Settings

This section discusses several results related to the computation of the operator norm and of various spectra of (classes of) elementary operators. It has turned out that computing the norm of reasonably general (classes of) elementary operators is a difficult problem. In fact, only very recently Timoney [Ti05] provided the first general formula for $\|\mathcal{E}_{A,B}\|$ on $L(\ell^2)$, see Theorem 3.10 below.

Recall as our starting point that

$$\|L_A R_B : L(X) \to L(X)\| = \|A\| \cdot \|B\|$$

by (2.10) for any Banach space X and any $A, B \in L(X)$. The first non-trivial results concern the norm of the *inner derivations* (or commutator maps)

$$L_A - R_A \colon L(X) \to L(X); \quad S \mapsto AS - SA,$$

determined by $A \in L(X)$. These concrete operators occur in many different contexts. Since $L_{A-\lambda} - R_{A-\lambda} = L_A - R_A$ for any scalar λ, we immediately get the general upper bound

$$(3.1) \qquad \qquad \|L_A - R_A\| \leq 2 \cdot \inf_{\lambda \in \mathbb{K}} \|A - \lambda\|,$$

which holds for any X (where the scalar field $\mathbb{K}$ is either $\mathbb{R}$ or $\mathbb{C}$). Stampfli [St70] showed that the preceding estimate is exact for the norm of the inner derivations on $L(\ell^2)$.

Theorem 3.1. *Let H be a complex Hilbert space and $A \in L(H)$. Then*

$$(3.2) \qquad \qquad \|L_A - R_A\| = 2 \cdot \inf\{\|A - \lambda\| : \lambda \in \mathbb{C}\}$$

Stampfli's elegant formula also holds in the case of real scalars. Stampfli [St70] extended it to the *generalized derivations* (or intertwining operators) $L_A - R_B$ on $L(\ell^2)$, where $S \mapsto AS - SB$.

Theorem 3.2. *Let H is a complex Hilbert space and $A, B \in L(H)$. Then*

$$(3.3) \qquad \qquad \|L_A - R_B\| = \inf\{\|A - \lambda\| + \|B - \lambda\| : \lambda \in \mathbb{C}\}$$

Later Fialkow [Fi79, Example 4.14] observed that the operator $A \in L(\ell^2)$, defined by $Ae_{2n} = e_{2n-1}$ and $Ae_{2n-1} = 0$ for $n \in \mathbb{N}$, satisfies

$$\|L_A - R_A \colon C_2 \to C_2\| < 2 \cdot \inf_{\lambda \in \mathbb{C}} \|A - \lambda\|.$$

Here (e_n) is the unit coordinate basis of ℓ^2 and $(C_2, \|\cdot\|_{HS})$ is the Banach ideal of $L(\ell^2)$ consisting of the Hilbert-Schmidt operators. Hence (3.2) fails for arbitrary restrictions $L_A - R_A \colon J \to J$, where J is a Banach ideal of $L(\ell^2)$.

Let H be a complex Hilbert space. Fialkow [Fi79] called the operator $A \in L(H)$ *S-universal* if $\|L_A - R_A \colon J \to J\| = 2 \cdot \inf\{\|A - \lambda\| : \lambda \in \mathbb{C}\}$ for all Banach ideals $J \subset L(H)$. Barraa and Boumazgour [BB01] obtained, in combination with earlier results of Fialkow, the following neat characterization of S-universality. Let $W(A) = \{(Ax, x) : x \in S_H\}$ be the numerical range of $A \in L(H)$.

Theorem 3.3. *Let H be a complex Hilbert space and $A \in L(H)$. Then the following conditions are equivalent.*
(i) A is S-universal,
(ii) $\|L_A - R_A \colon C_2 \to C_2\| = 2 \cdot \inf_{\lambda \in \mathbb{C}} \|A - \lambda\|$,
(iii) $\mathrm{diam}(W(A)) = 2 \cdot \inf_{\lambda \in \mathbb{C}} \|A - \lambda\|$,
(iv) $\mathrm{diam}(\sigma(A)) = 2 \cdot \inf_{\lambda \in \mathbb{C}} \|A - \lambda\|$. (Here $\sigma(A)$ is the spectrum of A.)

There are several proofs of Stampfli's formula (3.2), see e.g. [AM03, Section 4.1]. We briefly discuss one of the approaches from [St70] of this fundamental result in order to convey an impression of the tools involved here. The *maximal numerical range* of $A \in L(H)$ is

$$W_0(A) = \{\lambda \in \mathbb{C} \colon \lambda = \lim_n (Ax_n, x_n), \text{where } (x_n) \subset S_H \text{ and } \|Ax_n\| \to \|A\|\}.$$

Here $(\cdot, \cdot)$ is the inner product on H. The set $W_0(A)$ is known to be non-empty, closed and convex.

Sketch of the proof of Theorem 3.1. We first claim that if $\mu \in W_0(A)$ then

$$(3.4) \qquad \|L_A - R_A\| \geq 2(\|A\|^2 - |\mu|^2)^{1/2}.$$

Indeed, by assumption there is a sequence $(x_n) \subset S_H$ so that $\mu = \lim_n (Ax_n, x_n)$ and $\|A\| = \lim_n \|Ax_n\|$. Write $Ax_n = \alpha_n x_n + \beta_n y_n$ for $n \in \mathbb{N}$, where $y_n \in \{x_n\}^\perp$ and $\|y_n\| = 1$. Note that $\alpha_n = (Ax_n, x_n) \to \mu$ and $|\alpha_n|^2 + |\beta_n|^2 = \|Ax_n\|^2 \to \|A\|^2$ as $n \to \infty$. Define the rank-2 operators $V_n \in L(H)$ by

$$V_n = (x_n \otimes x_n - y_n \otimes y_n) \circ P_n,$$

where P_n is the orthogonal projection onto $[x_n, y_n]$. Here $(u \otimes v)x = (x, u)v$ for $u, v, x \in H$. Thus $\|V_n\| = 1$ for $n \in \mathbb{N}$. We obtain that

$$\lim_n \|AV_n x_n - V_n Ax_n\| = \lim_n \|\alpha_n x_n + \beta_n y_n - (\alpha_n x_n - \beta_n y_n)\|$$
$$= \lim_n 2|\beta_n| = 2(\|A\|^2 - |\mu|^2)^{1/2}$$

Since $\|L_A - R_A\| \geq \limsup_n \|AV_n x_n - V_n Ax_n\|$ it follows that (3.4) holds.

Observe next that if $0 \in W_0(A - \lambda_0)$ for some scalar $\lambda_0 \in \mathbb{C}$, then (3.4) yields the lower estimate

$$\|L_A - R_A\| = \|L_{A-\lambda_0} - R_{A-\lambda_0}\| \geq 2\|A - \lambda_0\| \geq 2 \inf_{\lambda \in \mathbb{C}} \|A - \lambda\|.$$

Hence it follows from (3.1) that (3.2) holds.

The non-trivial part of the argument is to find $\lambda_0 \in \mathbb{C}$ so that $0 \in W_0(A - \lambda_0)$. This part is quite well-documented in the literature so we just refer to Stampfli [St70] (who included two different approaches), [Fi92, Section 2] or [AM03, Thm. 4.1.17]). □

Stampfli asked whether (3.2) also holds for the inner derivations on $L(X)$, where X is an arbitrary Banach space. This was disproved by the following example of Johnson [J71].

Example 3.4. Let $1 < p < \infty$ and $p \neq 2$. Then there is a rank-1 operator $A \in L(\ell^p)$ for which

$$\|L_A - R_A\| < 2 \cdot \inf\{\|A - \lambda\| : \lambda \in \mathbb{C}\}.$$

Johnson [J71] also provided examples of spaces X where (3.2) does hold.

Example 3.5. Let $\ell_n^1(\mathbb{R}) = (\mathbb{R}^n, \|\cdot\|_1)$. Then

$$\|L_A - R_A\| = 2 \cdot \inf\{\|A - \lambda\| : \lambda \in \mathbb{R}\}$$

for any $A \in L(\ell_n^1(\mathbb{R}))$.

Above $\ell_n^1(\mathbb{R})$ is not uniformly convex. Subsequently Kyle [Ky77] obtained an elegant connection between Stampfli's formula and isometric characterizations of Hilbert spaces within the class of uniformly convex spaces.

Theorem 3.6. *Let X be a uniformly convex Banach space over the scalars $\mathbb{K}$. Then the following conditions are equivalent.*

(i) X is isometric to a Hilbert space,
(ii) $\|L_A - R_A\| = 2 \cdot \inf_{\lambda \in \mathbb{K}} \|A - \lambda\|$ holds for any $A \in L(X)$,
(iii) $\|L_A - R_A\| = 2 \cdot \inf_{\lambda \in \mathbb{K}} \|A - \lambda\|$ holds for any rank-1 operator $A \in L(X)$.

There has been much recent work concerning the computation of norms (of classes) of elementary operators. An optimal outcome would be a formula for

$$\| \sum_{j=1}^{n} L_{A_j} R_{B_j} : L(X) \to L(X) \|$$

which in some sense involves the coefficients $A = (A_1, \ldots, A_n) \in L(X)^n$, $B = (B_1, \ldots, B_n) \in L(X)^n$ of a given elementary operator $\mathcal{E}_{A,B}$ so that their contribution to the norm is clarified in some non-trivial sense (at least in the case where X is a Hilbert space). One obvious obstruction is the non-uniqueness of the representation of such operators. Runde [Run00] observed that $\|\mathcal{E}_{A,B}\|$ is not symmetric in the sense that the norms do not remain uniformly bounded in the flip correspondence $\mathcal{E}_{A,B} \to \mathcal{E}_{B,A}$ (note that this is not well-defined as a map). Timoney [Ti01] gave the following simplified version of Runde's instructive example.

Example 3.7. Suppose that X is a Banach space with a normalized basis (e_n) and biorthogonal sequence $(e_n^*) \subset X^*$. Put $A_j = e_j^* \otimes e_1$ and $B_j = e_1^* \otimes e_j$ for $j \in \mathbb{N}$. Then

$$\| \sum_{j=1}^{n} L_{A_j} R_{B_j} \| \geq n, \quad \| \sum_{j=1}^{n} L_{B_j} R_{A_j} \| \leq C, \quad n \in \mathbb{N},$$

where C is the basis constant of (e_n).

Proof. For $S \in L(X)$ one gets that

$$\sum_{j=1}^{n} L_{A_j} R_{B_j}(S) = \sum_{j=1}^{n} (e_j^* \otimes e_1) \circ S \circ (e_1^* \otimes e_j) = (\sum_{j=1}^{n} \langle e_j^*, Se_j \rangle) e_1^* \otimes e_1.$$

Thus $\| \sum_{j=1}^{n} L_{A_j} R_{B_j} \| \geq n \|e_1^* \otimes e_1\| = n$ by choosing $S = I_X$. Moreover, for $S \in L(X)$ and $x \in X$ one obtains that

$$\| \sum_{j=1}^{n} L_{B_j} R_{A_j}(S)x \| = \| \langle e_1^*, Se_1 \rangle \sum_{j=1}^{n} e_j^*(x) e_j \|$$

$$\leq \|S\| \cdot \|P_n\| \cdot \| \sum_{j=1}^{\infty} e_j^*(x) e_j \| \leq C \cdot \|S\| \cdot \|x\|.$$

Above $x = \sum_{j=1}^{\infty} e_j^*(x) e_j$ for $x \in X$, P_n denotes the natural basis projection $X \to [e_1, \ldots, e_n]$ and $C = \sup_m \|P_m\|$ is the basis constant. $\square$

By exercising somewhat more care in the argument (see [Ti01, Thm. 1]) it is enough above to assume just that the basic sequence (e_n) spans a complemented subspace of X.

Let X be a Banach space and $A, B \in L(X)$. Clearly the symmetrized elementary operator $L_A R_B + L_B R_A$, for which $S \mapsto ASB + BSA$ for $S \in L(X)$, satisfies

$$\|L_A R_B + L_B R_A : L(X) \to L(X)\| \le 2\|A\| \cdot \|B\|.$$

For these maps it is natural to look for lower bounds having the form

$$(3.5) \qquad \|L_A R_B + L_B R_A : L(X) \to L(X)\| \ge c_X \|A\| \cdot \|B\|,$$

for some constant $c_X > 0$, possibly depending on X. In particular, Mathieu [Ma89] conjectured that $c_H = 1$. Recently Mathieu's conjecture was independently solved, using different methods, by Timoney [Ti03a] and by Blanco, Boumazgour and Ransford [BBR04].

Theorem 3.8. *Let H be a Hilbert space. Then*

$$\|L_A R_B + L_B R_A : L(H) \to L(H)\| \ge \|A\| \cdot \|B\|$$

for any $A, B \in L(H)$.

Earlier Stacho and Zalar [SZ96] showed that Mathieu's conjecture holds for self-adjoint $A, B \in L(H)$, and that

$$\|L_A R_B + L_B R_A : L(H) \to L(H)\| \ge 2(\sqrt{2} - 1)\|A\| \cdot \|B\|$$

for any $A, B \in L(H)$. The bound $2(\sqrt{2} - 1)$ occurs naturally in the following result due Blanco, Boumazgour and Ransford [BBR04, Thm. 5.1 and Prop. 5.3].

Theorem 3.9. *In (3.5) one has that $c_X \ge 2(\sqrt{2} - 1)$ for any Banach space X. The bound $2(\sqrt{2} - 1)$ cannot be improved e.g. on $X = (\mathbb{R}^2, \|\cdot\|_\infty)$, ℓ^∞ or ℓ^1.*

The general estimate in Theorem 3.9 also holds for the norm of the restrictions $L_A R_B + L_B R_A : I(X) \to I(X)$, where I is a Banach operator ideal.

Very recently Timoney [Ti05], building on his earlier work [Ti03b], obtained a couple of general formulas for the norm $\|\mathcal{E}_{A,B}\|$ in the Hilbert space case. His work provides a solution of the norm problem which involves matrix numerical ranges and a notion of tracial geometric mean. We briefly describe his solution, though we are not able to include any details here. The *tracial geometric mean* of the positive (semi-definite) $n \times n$-matrices U, V is

$$tgm(U, V) = trace\sqrt{\sqrt{U}V\sqrt{U}} = \sum_{j=1}^{n} \sqrt{\lambda_j(UV)}.$$

Here $\sqrt{\cdot}$ denotes the positive square root, and $(\lambda_j(UV))$ are the eigenvalues of UV ordered in non-increasing order and counting multiplicities. For the

n-tuple $A = (A_1, \ldots, A_n) \in L(H)^n$ and $x \in H$ one introduces the scalar $n \times n$-matrix

$$Q(A, x) = ((A_i^* A_j x, x))_{i,j=1}^n = ((A_j x, A_i x))_{i,j=1}^n.$$

The first version [Ti05, Thm. 1.4] of Timoney's formula reads as follows.

Theorem 3.10. *For any* $A = (A_1, \ldots, A_n), B = (B_1, \ldots, B_n) \in L(H)^n$ *one has*

$$\|\mathcal{E}_{A,B}\colon L(H) \to L(H)\| = \sup\{tgm(Q(A^*, x), Q(B, y))\colon x, y \in S_H\},$$

where $A^* = (A_1^*, \ldots, A_n^*)$.

Next put

$$\|(x_1, \ldots, x_n)\|_{S1} = trace\sqrt{((x_i, x_j))_{i,j=1}^n}$$

for $(x_1, \ldots, x_n) \in H^n$. This defines a norm on H^n, see [Ti05, Lemma 1.7]. For $A = (A_1, \ldots, A_n) \in L(H)^n$ let $\|A\|_{S1}$ denote the norm of A considered as an operator $H \to (H^n, \|\cdot\|_{S1})$. One gets the following alternative formula [Ti05, Thm. 1.10] for the norm.

Theorem 3.11. *For any* $A = (A_1, \ldots, A_n), B = (B_1, \ldots, B_n) \in L(H)^n$ *one has*

$$\|\mathcal{E}_{A,B}\colon L(H) \to L(H)\| = \sup\{\|\sqrt{Q(B, y)^t} A^*\|_{S1}\colon y \in S_H\},$$

where $A^* = (A_1^*, \ldots, A_n^*)$ *and* $Q(B, y)^t$ *is the transpose of the matrix* $Q(B, y)$.

Timoney [Ti05] also established versions of Theorems 3.10 and 3.11 for the norm of elementary operators on arbitrary C*-algebras $\mathcal{A}$. Moreover, he characterized the compact elementary operators $\mathcal{A} \to \mathcal{A}$, thus providing a complete generalization of Theorem 2.2 to the setting of C*-algebras (see [AM03, 5.3.26] for earlier results of Mathieu in the case of prime algebras).

For further results about the norms of elementary operators on $L(H)$ or on classes of C*-algebras we refer to e.g. [AM03, Sections 4.1,4.2 and 5.4], [AST05], [Ti05] (and the references therein), as well as to Theorems 4.13 and 4.14 concerning elementary operators on the Calkin algebra.

Let X be a complex Banach space and $A, B \in L(X)$. It is an easy exercise to check that the spectrum $\sigma(L_A) = \sigma(A) = \sigma(R_A)$ for any operator A. Since $L_A R_B = R_B L_A$ it follows immediately from elementary Gelfand theory that the spectra of $L_A R_B$ and $L_A - R_B$ satisfy

$$\sigma(L_A R_B) \subset \sigma(A)\sigma(B), \quad \sigma(L_A - R_B) \subset \sigma(A) - \sigma(B),$$

where $\sigma(A) - \sigma(B) \equiv \{\alpha - \beta\colon \alpha \in \sigma(A), \beta \in \sigma(B)\}$. Lumer and Rosenblum [LR59] showed the exact formula

$$(3.6) \qquad \sigma(\sum_{j=1}^n L_{f_j(A)} R_{g_j(B)}) = \{\sum_{j=1}^n f_j(\alpha)g_j(\beta)\colon \alpha \in \sigma(A), \beta \in \sigma(B)\},$$

which holds whenever f_j is holomorphic in a neighborhood of $\sigma(A)$ and g_j is holomorphic in a neighborhood of $\sigma(B)$ for $j = 1, \ldots, n$. (The result itself is attributed to Kleinecke in [LR59].) Hence simple choices in (3.6) of the holomorphic functions imply that in fact

$$\sigma(L_A R_B) = \sigma(A)\sigma(B), \quad \sigma(L_A - R_B) = \sigma(A) - \sigma(B).$$

For a long time it remained a considerable challenge to compute the spectrum $\sigma(\mathcal{E}_{A,B})$ of general elementary operators. A satisfactory formula was eventually obtained by Curto [Cu83] for Hilbert spaces X, and this result was later substantially improved by Curto and Fialkow [CuF87] (again for Hilbert spaces) and Eschmeier [E88] (for arbitrary Banach spaces). (Some of these facts were announced by Fainshtein [F84].) Here one expresses the spectrum $\sigma(\mathcal{E}_{A,B})$ and the essential spectrum $\sigma_e(\mathcal{E}_{A,B})$ in terms of the Taylor joint spectrum and the Taylor joint essential spectrum of the n-tuples $(A_1, \ldots, A_n)$ and $B = (B_1, \ldots, B_n)$. We refer to the survey [Cu92] for further references to the numerous intermediary results (including those for tensor products of operators) that culminated in Theorems 3.12 and 3.13 below.

Let $A = (A_1, \ldots, A_n) \in L(X)^n$ be an n-tuple such that the set $\{A_1, \ldots, A_n\}$ commutes. The Taylor joint spectrum $\sigma_T(A)$ consists of $\lambda = (\lambda_1, \ldots, \lambda_n) \in \mathbb{C}^n$ so that the Koszul complex corresponding to $A - \lambda$ is not exact. Actually we will not require the precise homological definition here, and we refer e.g. to [Ta70], [Cu88] or [Mu02] for the details. The set $\sigma_T(A) \subset \mathbb{C}^n$ is compact and non-empty. We will use below the convenient notation

$$U \circ V \equiv \{\sum_{j=1}^{n} \alpha_j \beta_j : (\alpha_1, \ldots, \alpha_n) \in U, (\beta_1, \ldots, \beta_n) \in V\}$$

for subsets $U, V \subset \mathbb{C}^n$. It will also be convenient to discuss simultaneously the spectrum of the restrictions $\mathcal{E}_{A,B} : I \to I$, where $I \subset L(X)$ is any Banach ideal, since the results do not depend on I and the arguments are similar.

Theorem 3.12. *Let X be a complex Banach space and $I \subset L(X)$ be any Banach ideal. Suppose that $A = (A_1, \ldots, A_n), B = (B_1, \ldots, B_n) \in L(X)^n$ are n-tuples such that $\{A_1, \ldots, A_n\}$ and $\{B_1, \ldots, B_n\}$ are commuting sets. Then*

$$(3.7) \qquad \qquad \sigma(\mathcal{E}_{A,B} : I \to I) = \sigma_T(A) \circ \sigma_T(B).$$

The essential spectrum of $S \in L(X)$ is

$$\sigma_e(S) = \{\lambda \in \mathbb{C} : \lambda - S \text{ is not a Fredholm operator}\}.$$

Recall that $S \in L(X)$ is a Fredholm operator if there are operators $T \in L(X)$ and $K_1, K_2 \in K(X)$ so that $ST = I_X + K_1$ and $TS = I_X + K_2$. Thus $\sigma_e(S)$ is the spectrum of the quotient element $S + K(X)$ in the corresponding Calkin algebra $L(X)/K(X)$. The references [E88] and [CuF87] also compute the essential spectrum $\sigma_e(\mathcal{E}_{A,B})$ in the preceding setting. Let $A = (A_1, \ldots, A_n) \in L(X)^n$ be a commuting n-tuple. We recall that the Taylor joint essential spectrum $\sigma_{Te}(A)$ consists of the $\lambda \in \mathbb{C}^n$ for which the Koszul complex of $A - \lambda$ is not Fredholm (see again e.g. [E88] or [Cu92] for the precise definition).

Theorem 3.13. *Let X be a complex Banach space and let $I \subset L(X)$ be any Banach ideal. Suppose that $A = (A_1, \dots, A_n), B = (B_1, \dots, B_n) \in L(X)^n$ are n-tuples such that $\{A_1, \dots, A_n\}$ and $\{B_1, \dots, B_n\}$ are commuting sets. Then*

$$(3.8) \qquad \sigma_e(\mathcal{E}_{A,B} \colon I \to I) = \sigma_{Te}(A) \circ \sigma_T(B) \cup \sigma_T(A) \circ \sigma_{Te}(B).$$

Theorems 3.12 and 3.13 remain valid for the spectrum and the essential spectrum of the analogous tensor product operators $\sum_{j=1}^n A_j \otimes B_j$ with respect to any tensor norm (this is the explicit point of view in [E88]). The arguments in [E88] and [CuF87] apply multivariable spectral theory and some homological algebra (the requisite background is discussed e.g. in the surveys [Cu88] and [Cu92]). A central idea is to determine the complete spectral picture and compute, or in the Banach space case to suitably estimate, the Taylor joint (essential) spectra of the commuting $2n$-tuples

$$(A \otimes I_X, I_X \otimes B) = (A_1 \otimes I, \dots, A_n \otimes I, I \otimes B_1, \dots, I \otimes B_n)$$

or $(L_A, R_B) = (L_{A_1}, \dots, L_{A_n}, R_{B_1}, \dots, R_{B_n})$. Theorems 3.12 and 3.13 are then obtained by applying the polynomial spectral mapping property to $P \colon \mathbb{C}^n \times \mathbb{C}^n \to \mathbb{C}$, where $P(z, w) = \sum_{j=1}^n z_j w_j$ for $z, w \in \mathbb{C}^n$. The references [E88] and [CuF87] contain plenty of additional information related to other classical subsets of the spectrum as well as index formulas. By contrast we will provide below minimalist approaches to (3.7) and (3.8), which are based on ideas from [S95].

Before proving (3.7) we mention that it is even possible to identify the *weak essential spectrum* $\sigma_w(\mathcal{E}_{A,B})$ in several situations. Here

$$\sigma_w(S) = \{\lambda \in \mathbb{C} \colon \lambda - S \text{ is not invertible modulo } W(Y)\}$$

for $S \in L(Y)$. Recall that $S \in L(Y)$ is invertible modulo $W(Y)$ if there are $T \in L(Y)$ and $V_1, V_2 \in W(Y)$ so that $ST = I_Y + V_1$ and $TS = I_Y + V_2$. We refer e.g. to Section 2 and Corollary 4.2 for results about weakly compact elementary operators. The following results were obtained in [ST94].

Theorem 3.14. *(i) Let $1 < p < \infty$ and $A, B \in L(\ell^p)$. Then*

$$\sigma_w(L_A R_B) = \sigma_e(A)\sigma_e(B).$$

(ii) Let X be an arbitrary complex Banach space and $A, B \in L(X)$. Then

$$(3.9) \quad \sigma_w(A^*)\sigma(B) \cup \sigma(A)\sigma_w(B) \subset \sigma_w(L_A R_B) \subset \sigma_e(A)\sigma(B) \cup \sigma(A)\sigma_e(B).$$

For instance, if X^* has the Dunford-Pettis property, then one obtains from (3.9) that

$$\sigma_w(L_A R_B) = \sigma_e(L_A R_B) = \sigma_e(A)\sigma(B) \cup \sigma(A)\sigma_e(B).$$

For this equality one has to recall a few well-known facts. Firstly, if Y has the DPP and $V \in W(Y)$, then $I_Y + V$ is a Fredholm operator. In fact, by assumption $V^2 \in K(Y)$ so that $I_Y - V^2 = (I_Y + V)(I_Y - V) = (I_Y - V)(I_Y + V)$ is a Fredholm operator, whence $I_Y + V$ is also Fredholm. This yields that $\sigma_w(S) = \sigma_e(S)$ for any $S \in L(Y)$. Secondly, X has the DPP if X^* has this property. Finally, $\sigma_e(S^*) = \sigma_e(S)$ for any operator S.

Subsequently Saksman [S95] extended Theorem 3.14 to the weak essential spectra of elementary operators in certain cases. We state a couple of results from [S95].

Theorem 3.15. *Let X be a complex Banach space and suppose that $A = (A_1, \ldots, A_n)$, $B = (B_1, \ldots, B_n) \in L(X)^n$ are commuting n-tuples.*
(i) If $X = \ell^p$ and $1 < p < \infty$, then

$$\sigma_w(\mathcal{E}_{A,B}) = \sigma_{Te}(A) \circ \sigma_{Te}(B).$$

(ii) If X^* has the DPP, then

$$\sigma_w(\mathcal{E}_{A,B}) = \sigma_{Te}(A) \circ \sigma_T(B) \cup \sigma_T(A) \circ \sigma_{Te}(B).$$

The ideas underlying Theorem 3.15, and the earlier Theorem 3.14 for the case $L_A R_B$, yield "elementary" approaches to Theorems 3.12 and 3.13 that do not use any homological algebra (as was pointed out on [S95, p. 182]). In order to give the reader some impressions of the techniques involved in computing the spectra of elementary operators we present here a fairly detailed argument for Theorem 3.12 along these lines. We also point out the main additional ideas needed for a proof of Theorem 3.13 using elementary tools, see Remark 3.17 below.

We begin by recalling some classical concepts of joint spectra. Let X be a complex Banach space and let $A = (A_1, \ldots, A_n) \in L(X)^n$ be a n-tuple of commuting operators. The (joint) approximative point spectrum $\sigma_\pi(A)$ of $A = (A_1, \ldots, A_n)$ consists of the points $\lambda = (\lambda_1, \ldots, \lambda_n) \in \mathbb{C}^n$ for which

$$\inf_{x \in S_X} \sum_{j=1}^n \|(A_j - \lambda_j)x\| = 0.$$

The joint approximative spectrum of $A = (A_1, \ldots, A_n)$ is then

$$\sigma_a(A) = \sigma_\pi(A) \cup \sigma_\pi(A^*),$$

where $A^* = (A_1^*, \ldots, A_n^*)$. The left spectrum $\sigma_l(A)$ of $A = (A_1, \ldots, A_n)$ consists of $(\lambda_1, \ldots, \lambda_n) \in \mathbb{C}^n$ such that $\sum_{j=1}^n S_j(A_j - \lambda_j) \neq I_X$ for all n-tuples $(S_1, \ldots, S_n) \in L(X)^n$. Similarly, $(\lambda_1, \ldots, \lambda_n) \in \sigma_r(A)$ if $\sum_{j=1}^n (A_j - \lambda_j)S_j \neq I_X$ for $(S_1, \ldots, S_n) \in L(X)^n$. The Harte joint spectrum of $A = (A_1, \ldots, A_n)$ is

$$\sigma_H(A) = \sigma_l(A) \cup \sigma_r(A).$$

For a single operator $S \in L(X)$ (that is, the case $n = 1$) one has $\sigma_T(S) = \sigma_H(S) = \sigma_a(S) = \sigma(S)$. According to the polynomial spectral mapping property for the Taylor spectrum one has

$$\sigma(P(A_1, \ldots, A_n)) = P(\sigma_T(A_1, \ldots, A_n)).$$

for any commuting n-tuple $A = (A_1, \ldots, A_n) \in L(X)^n$ and for any scalar polynomial $P: \mathbb{C}^n \to \mathbb{C}$. This property also holds for the joint spectra $\sigma_H(\cdot)$ and $\sigma_a(\cdot)$. We refer e.g to [Cu88] for a further discussion of multivariable spectral theory.

The following technical observation will be crucial. This fact goes back to Curto [Cu86, Thm. 3.15] (see also [S95, Prop. 1]).

Lemma 3.16. *Let* $A = (A_1, \ldots, A_n), B = (B_1, \ldots, B_n) \in L(X)^n$ *be commuting n-tuples, and* $P \colon \mathbb{C}^n \times \mathbb{C}^n \to \mathbb{C}$ *be a polynomial. Then*

$$P(\sigma_T(A) \times \sigma_T(B)) = P(\sigma_H(A) \times \sigma_H(B)) = P(\sigma_a(A) \times \sigma_a(B)).$$

Proof. We verify the inclusion $P(\sigma_T(A) \times \sigma_T(B)) \subset P(\sigma_a(A) \times \sigma_a(B))$. The other inclusions are similar (actually, one could also apply the fact that $\sigma_a(A) \subset \sigma_H(A) \subset \sigma_T(A)$ for any commuting n-tuple A, see e.g. [Cu88]).

Let $P(\alpha_0, \beta_0) \in P(\sigma_T(A) \times \sigma_T(B))$, and consider the polynomial $Q_1 \colon \mathbb{C}^n \to \mathbb{C}$, where $Q_1(z) = P(z, \beta_0)$. It follows from the polynomial spectral mapping theorems, and the fact that these spectra coincide for a single operator, that

$$P(\alpha_0, \beta_0) = Q_1(\alpha_0) \in Q_1(\sigma_T(A)) = \sigma(Q_1(A)) = Q_1(\sigma_a(A)).$$

Hence there is $\alpha_1 \in \sigma_a(A)$ so that $P(\alpha_0, \beta_0) = Q_1(\alpha_1) = P(\alpha_1, \beta_0)$. By applying the same argument to $Q_2 \colon \mathbb{C}^n \to \mathbb{C}$, where $Q_2(w) = P(\alpha_1, w)$, we get $\beta_1 \in \sigma_a(B)$ so that

$$P(\alpha_0, \beta_0) = P(\alpha_1, \beta_0) = Q_2(\beta_0) = Q_2(\beta_1) = P(\alpha_1, \beta_1) \in P(\sigma_a(A) \times \sigma_a(B)).$$

$\square$

Proof of Theorem 3.12. The strategy is to prove the inclusions

$$(3.10) \qquad \sigma_H((L_A, R_B); L(I)) \subset \sigma_H(A) \times \sigma_H(B),$$

$$(3.11) \qquad P(\sigma_a(A) \times \sigma_a(B)) \subset \sigma(P(L_A, R_B) \colon I \to I),$$

for any polynomial $P \colon \mathbb{C}^n \times \mathbb{C}^n \to \mathbb{C}$. Above $\sigma_H((L_A, R_B); L(I))$ denotes the Harte spectrum of the commuting $2n$-tuple $(L_{A_1}, \ldots, L_{A_n}, R_{B_1}, \ldots, R_{B_n})$ in the algebra $L(I)$. By applying Lemma 3.16 with the choice $P(z, w) = \sum_{j=1}^n z_j w_j$ to (3.10) and (3.11) we get the desired identities

$$\sigma(\mathcal{E}_{A,B} \colon I \to I) = \sigma_T(A) \circ \sigma_T(B) = \sigma_H(A) \circ \sigma_H(B) = \sigma_a(A) \circ \sigma_a(B).$$

It is enough towards (3.10) to verify that

$$(3.12) \qquad \sigma_l((L_A, R_B); L(I)) \subset \sigma_l(A) \times \sigma_r(B),$$

$$(3.13) \qquad \sigma_r((L_A, R_B); L(I)) \subset \sigma_r(A) \times \sigma_l(B).$$

These inclusions were probably first noted by Harte. If $\lambda \notin \sigma_l(A)$ then there is an n-tuple $S = (S_1, \ldots, S_n) \in L(X)^n$ so that $\sum_{j=1}^n S_j(A_j - \lambda_j) = I_X$. It follows that

$$\sum_{j=1}^n L_{S_j} L_{A_j - \lambda_j} + \sum_{j=1}^n 0 \circ R_{B_j - \mu_j} = Id_I,$$

so that $(\lambda, \mu) \notin \sigma_l((L_A, R_B); L(I))$. If $\mu \notin \sigma_r(B)$ then there is an n-tuple $T = (T_1, \ldots, T_n) \in L(X)^n$ so that $\sum_{j=1}^n (B_j - \mu_j) T_j = I_X$. We get that

$$\sum_{j=1}^n 0 \circ L_{A_j - \lambda_j} + \sum_{j=1}^n R_{T_j} R_{B_j - \mu_j} = Id_I,$$

that is, $(\lambda, \mu) \notin \sigma_l((L_A, R_B); L(I))$. The verification of (3.13) is similar.

The proof of the lower inclusion (3.11) is the crucial step of the argument. Suppose that $(\lambda, \mu) \in \sigma_a(A) \times \sigma_a(B)$. We first factorize

$$P(z, w) - P(\lambda, \mu) = \sum_{j=1}^{n} G_j(z, w)(z_j - \lambda_j) + \sum_{j=1}^{n} H_j(z, w)(w_j - \mu_j),$$

where G_j and H_j are suitable polynomials for $j = 1, \ldots, n$, so that

$$\Phi = P(L_A, R_B) - P(\lambda, \mu) = \sum_{j=1}^{n} G_j(L_A, R_B) \circ L_{A_j - \lambda_j} + \sum_{j=1}^{n} H_j(L_A, R_B) \circ R_{B_j - \mu_j}$$

defines a bounded operator on I. Assume next to the contrary that $P(\lambda, \mu) \notin \sigma(P(L_A, R_B); I \to I)$, so that Φ is invertible $I \to I$. Since $\sigma_a(A) = \sigma_\pi(A) \cup \sigma_\pi(A^*)$, and similarly for $\sigma_a(B)$, we get four cases which are all handled somewhat differently.

Case 1. $\lambda \in \sigma_\pi(A)$, $\mu \in \sigma_\pi(B^*)$. There are sequences $(x_k) \subset S_X$ and $(x_k^*) \subset S_{X^*}$ so that $\|(A_j - \lambda_j)x_k\| \to 0$ and $\|(B_j^* - \mu_j)x_k^*\| \to 0$ as $k \to \infty$ for each $j = 1, \ldots, n$. Consider the rank-1 operator $x_k^* \otimes x_k \in I$, for which $\|x_k^* \otimes x_k\|_I = 1$ for $k \in \mathbb{N}$. Hence

$$\|(L_{A_j - \lambda_j})(x_k^* \otimes x_k)\|_I = \|x_k^*\| \cdot \|(A_j - \lambda_j)x_k\| \to 0,$$

$$\|(R_{B_j - \mu_j})(x_k^* \otimes x_k)\|_I = \|(B_j^* - \mu_j)x_k^*\| \cdot \|x_k\| \to 0$$

as $k \to \infty$ for $j = 1, \ldots, n$. Here $\| \cdot \|_I$ is the norm on the Banach ideal I. Deduce that

$$\begin{aligned}
\Phi(x_k^* \otimes x_k) &= \sum_{j=1}^{n} G_j(L_A, R_B)(L_{A_j - \lambda_j})(x_k^* \otimes x_k) + \\
&\quad + \sum_{j=1}^{n} H_j(L_A, R_B)(R_{B_j - \mu_j})(x_k^* \otimes x_k)
\end{aligned}$$

converges to 0 in I as $k \to \infty$, which contradicts the fact that Φ is invertible.

Case 2. $\lambda \in \sigma_\pi(A^*)$, $\mu \in \sigma_\pi(B)$. There are $(x_k) \subset S_X$ and $(x_k^*) \subset S_{X^*}$ so that $\|(A_j^* - \lambda_j)x_k^*\| \to 0$ and $\|(B_j - \mu_j)x_k\| \to 0$ as $k \to \infty$ for $j = 1, \ldots, n$. Define a linear functional ψ_k on I by

$$\psi_k(S) = \langle x_k^*, S x_k \rangle, \quad S \in I.$$

Thus $\psi_k \in I^*$ and $\|\psi_k\| = 1$ for $k \in \mathbb{N}$ since I is a Banach ideal. Since $G_j(L_A, R_B) \circ L_{A_j - \lambda_j} = L_{A_j - \lambda_j} \circ G_j(L_A, R_B)$ and $H_j(L_A, R_B) \circ R_{B_j - \mu_j} =$

$R_{B_j - \mu_j} \circ H_j(L_A, R_B)$ for $j = 1, \ldots, n$ we get for any $S \in I$ that

$$
\begin{aligned}
|\Phi^* \psi_k(S)| &\leq \sum_{j=1}^{n} |\langle (A_j^* - \lambda_j) x_k^*, (G_j(L_A, R_B)(S)) x_k \rangle| \\
&+ \sum_{j=1}^{n} |\langle x_k^*, (H_j(L_A, R_B)(S))((B_j - \mu_j) x_k) \rangle| \\
&\leq c \sum_{j=1}^{n} \|(A_j^* - \lambda_j) x_k^*\| \cdot \|S\|_I + d \sum_{j=1}^{n} \|(B_j - \mu_j) x_k\| \cdot \|S\|_I.
\end{aligned}
$$

Here $c = \max_{j \leq n} \|G_j(L_A, R_B)\|$ and $d = \max_{j \leq n} \|H_j(L_A, R_B)\|$ considered as operators $I \to I$. This implies that $\|\Phi^*(\psi_k)\| \to 0$ as $k \to \infty$, which again contradicts the invertibility of Φ.

Case 3. $\lambda \in \sigma_\pi(A)$, $\mu \in \sigma_\pi(B)$. There are $(x_k) \subset S_X$ and $(y_k) \subset S_X$ so that $\|(A_j - \lambda_j) x_k\| \to 0$ and $\|(B_j - \mu_j) y_k\| \to 0$ as $k \to \infty$ for $j = 1, \ldots, n$. Fix $x^*, y^* \in S_{X^*}$ and consider the normalized rank-1 operators $U_k = x^* \otimes x_k, V_k = y^* \otimes y_k \in I$ for $k \in \mathbb{N}$. Here $\|(A_j - \lambda_j) U_k\|_I = \|(A_j - \lambda_j) x_k\| \to 0$ and $\|(B_j - \mu_j) V_k\|_I \to 0$ as $k \to \infty$ for $j = 1, \ldots, n$. Note next that $\Phi \circ L_{A_j - \lambda_j} = L_{A_j - \lambda_j} \circ \Phi$ for each j, since $\{L_{A_1}, \ldots, L_{A_n}, R_{B_1}, \ldots, R_{B_n}\}$ commutes by assumption. Hence

$$
\Phi \circ L_{A_j - \lambda_j} \circ \Phi^{-1} \circ L_{U_k} = L_{(A_j - \lambda_j) U_k} \to 0
$$

as $k \to \infty$, considered as operators on I, for each j. This means that $L_{A_j - \lambda_j} \circ \Phi^{-1} \circ L_{U_k} \to 0$ as $k \to \infty$, since Φ is invertible on I by assumption. We get that

$$
\begin{aligned}
R_{V_k} L_{U_k} &= R_{V_k} \circ \Phi \circ \Phi^{-1} \circ L_{U_k} \\
&= \sum_{j=1}^{n} R_{V_k} \circ G_j(L_A, R_B) \circ L_{A_j - \lambda_j} \circ \Phi^{-1} \circ L_{U_k} \\
&+ \sum_{j=1}^{n} R_{V_k} \circ R_{B_j - \mu_j} \circ H_j(L_A, R_B) \circ \Phi^{-1} \circ L_{U_k} \to 0
\end{aligned}
$$

as $k \to \infty$. Above we also used the fact that $H_j(L_A, R_B)$ and $R_{B_j - \mu_j}$ commute for each j. This contradicts the fact that $\|R_{V_k} L_{U_k} : I \to I\| = \|U_k\| \cdot \|U_k\| = 1$ for $k \in \mathbb{N}$ by (2.10).

Case 4. $\lambda \in \sigma_\pi(A^*)$, $\mu \in \sigma_\pi(B^*)$. There are sequences $(x_k^*), (y_k^*) \subset S_{X^*}$ for which $\|(A_j^* - \lambda_j) x_k^*\| \to 0$ and $\|(B_j^* - \mu_j) y_k^*\| \to 0$ as $k \to \infty$ for $j = 1, \ldots, n$. Fix $x, y \in S_X$ and consider $U_k = x_k^* \otimes x, V_k = y_k^* \otimes y \in I$ for $k \in \mathbb{N}$. Thus $\|U_k(A_j - \lambda_j)\|_I \to 0$ and $\|V_k(B_j - \mu_j)\|_I \to 0$ as $k \to \infty$ for $j = 1, \ldots, n$. By arguing as in Case 3 we get that $L_{U_k} \circ \Phi^{-1} \circ L_{A_j - \lambda_j} \to 0$ as $k \to \infty$ for $= 1, \ldots, n$, from which we again deduce the contradiction that $\|L_{U_k} R_{V_k} : I \to I\| \to 0$ as $k \to \infty$. $\square$

Remark 3.17. Perhaps more interestingly, one may also recover an alternative proof of Theorem 3.13 by suitably modifying the arguments of [S95,

Prop. 10 and Thm. 11], see the Remark on [S95, p. 182]. However, the argument is more delicate than the previous one, and here we just indicate the main additional ideas. In this case the strategy is to show the inclusions

$$(3.14) \qquad \sigma_{He}((L_A, R_B); L(I)) \subset \sigma_{He}(A) \times \sigma_H(B) \cup \sigma_H(A) \times \sigma_{He}(B),$$

$$(3.15) \qquad \sigma_{He}(A) \times \sigma_a(B) \cup \sigma_a(A) \times \sigma_{He}(B^*) \subset \sigma''((L_A, R_B); \mathcal{C}(I)).$$

In (3.15) one meets a crucial observation, that is, the advantage of using the bicommutant spectrum $\sigma''((L_A, R_B); \mathcal{C}(I))$. Above $\sigma''((L_A, R_B); \mathcal{C}(I))$ is the algebraic joint spectrum of

$$(L_{A_1} + K(I), \ldots, L_{A_n} + K(I), R_{B_1} + K(I), \ldots, R_{B_n} + K(I))$$

in the (commutative) bicommutant subalgebra

$$\{L_{A_1} + K(I), \ldots, L_{A_n} + K(I), R_{B_1} + K(I), \ldots, R_{B_n} + K(I)\}''$$

of $\mathcal{C}(I) = L(I)/K(I)$ and $\sigma_{He}(A) = \sigma_{le}(A) \cup \sigma_{re}(A)$ is the Harte spectrum of $(L_{A_1+K(X)}, \ldots, L_{A_n+K(X)})$ computed on the Calkin algebra $\mathcal{C}(X) = L(X)/K(X)$. Another important tool in the proof is the construction of Fredholm inverses for operator n-tuples on X assuming their existence on the level of elementary operators on I.

The following approximation problem for the inverses of elementary operators has some practical interest. Note that the class $\mathcal{E}(L(X))$ of all elementary operators is a subalgebra of $L(L(X))$ for any Banach space X.

Problem 3.18. *Does the inverse $\mathcal{E}_{A,B}^{-1} \in \overline{\mathcal{E}(L(X))}$ whenever the elementary operator $\mathcal{E}_{A,B}$ is invertible on $L(X)$? Here $\overline{\mathcal{E}(L(X))}$ is the uniform closure of the subalgebra $\mathcal{E}(L(X))$ in $L(L(X))$.*

For instance, the above holds for invertible generalized derivations $L_A - R_B$, where $A, B \in L(X)$, or for invertible $\mathcal{E}_{A,B}$ if $A, B \in L(H)^n$ are commuting n-tuples of normal operators on a Hilbert space H. This is seen for $L_A - R_B$ by applying an integral representation of $(L_A - R_B)^{-1}$ due to Rosenblum [Ro56, Thm. 3.1]. One should mention here a striking approximation result due to Magajna [M93, Cor. 2.3]:

• *Let $\mathcal{A}$ be a C^*-algebra, and suppose that $\phi : \mathcal{A} \to \mathcal{A}$ is a bounded linear operator so that $\phi(I) \subset I$ for any closed 2-sided ideals $I \subset \mathcal{A}$. Then $\phi \in \overline{\mathcal{E}(\mathcal{A})}^{SOT}$, the closure in the strong operator topology of $L(\mathcal{A})$.*

Above $\mathcal{E}(\mathcal{A})$ is the class of elementary operators on $\mathcal{A}$. Magajna's result yields some information related to Problem 3.18. Suppose that $\mathcal{E}_{A,B}$ is invertible on $L(\ell^2)$ for the n-tuples $A, B \in L(\ell^2)^n$. By applying Magajna's result to $\phi = \mathcal{E}_{A,B}^{-1}$ we get that

$$\mathcal{E}_{A,B}^{-1} \in \overline{\mathcal{E}(L(X))}^{SOT}.$$

In fact, $K(\ell^2)$ is the only non-trivial ideal of $L(\ell^2)$ and $\mathcal{E}_{A,B}(K(\ell^2)) = K(\ell^2)$ since $(\mathcal{E}_{A,B|K(\ell^2)})^{**} = \mathcal{E}_{A,B}$ by (4.17) and (4.18).

Other developments. McIntosh, Pryde and Ricker [MPR88] estimate the growth of the norm $\|S\|$ of solutions to the elementary operator equation

$$\mathcal{E}_{A,B}(S) = Y$$

for commuting n-tuples $A, B \in L(X)^n$ consisting of generalized scalar operators. (Recall that $S \in L(X)$ is a generalized scalar operator if there is $s \geq 0$ and $C < \infty$ so that $\|exp(itS)\| \leq C(1 + |t|^s)$ for $t \in \mathbb{R}$.) Recently Shulman and Turowska [ShT05] have obtained an interesting approach to some operator equations that include those arising from elementary operators.

Moreover, we note that Arendt, Räbiger and Sourour [ARS94] discuss the spectrum of the map $S \mapsto AS + SB$ in the setting of unbounded operators A, B.

4. ELEMENTARY OPERATORS ON CALKIN ALGEBRAS

Let X be an arbitrary Banach space and let $\mathcal{E}_{A,B} = \sum_{j=1}^{n} L_{A_j} R_{B_j}$ be the elementary operator on $L(X)$ associated to the n-tuples $A = (A_1, \ldots, A_n)$, $B = (B_1, \ldots, B_n) \in L(X)^n$. Since $K(X)$ is a closed 2-sided ideal of $L(X)$ the operator $\mathcal{E}_{A,B}$ induces the related elementary operator

$$\mathcal{E}_{a,b}; \quad s \mapsto \sum_{j=1}^{n} a_j s b_j, \quad \mathcal{C}(X) \to \mathcal{C}(X),$$

on the Calkin algebra $\mathcal{C}(X) = L(X)/K(X)$, where we denote quotient elements by $s = S + K(X) \in \mathcal{C}(X)$ for $S \in L(X)$. The quotient norm $\|S\|_e \equiv dist(S, K(X))$ for $S \in L(X)$ is called the essential norm. (We will change here freely between these notations.)

In this section we will see that the operators $\mathcal{E}_{a,b}$ on the quotient algebra $\mathcal{C}(X)$ has several remarkable properties which are not shared by $\mathcal{E}_{A,B}$ on $L(X)$. Roughly speaking, $\mathcal{E}_{a,b}$ are quite "rigid" operators and this also tells something about $\mathcal{E}_{A,B}$. Let H be a Hilbert space. Fong and Sourour [FS79, p. 856] asked *whether the compactness of $\mathcal{E}_{a,b} : \mathcal{C}(H) \to \mathcal{C}(H)$ actually implies that $\mathcal{E}_{a,b} = 0$, that is, whether the Calkin algebra $\mathcal{C}(H)$ admits any non-trivial compact elementary operators $\mathcal{E}_{a,b}$.*

Clearly such a property does not hold for the elementary operators on $L(X)$, since already $L_A R_B$ is a non-zero finite rank operator on $L(X)$ whenever $A \neq 0 \neq B$ are finite rank operators on X. We next recall why the Fong-Sourour conjecture holds for the simplest operators $L_a R_b$ to get a feeling for the matter. In fact, if $A, B \notin K(H)$ then by a simple modification of the argument of Example 2.6 (for $p = 2$) one finds $A_1, A_2, B_1, B_2 \in L(H)$ so that $A_1 A A_2 = I_H = B_1 B B_2$. It follows that

$$I_{\mathcal{C}(H)} = L_{a_1} R_{b_2} \circ L_a R_b \circ L_{a_2} R_{b_1},$$

whence $L_a R_b$ is not weakly compact on $\mathcal{C}(H)$.

The Fong-Sourour conjecture was independently solved in the positive by Apostol and Fialkow [AF86], Magajna [M87] and Mathieu [Ma88]. Apostol and Fialkow [AF86, Thm. 4.1] established the stronger result that

(4.1) $$\|\mathcal{E}_{a,b} : \mathcal{C}(\ell^2) \to \mathcal{C}(\ell^2)\| = dist(\mathcal{E}_{a,b}, K(\mathcal{C}(\ell^2)))$$

for arbitrary n-tuples A, B of $L(\ell^2)$. Magajna's solution is algebraic in nature, while Mathieu used tools from C*-algebras. Actually, Mathieu established the somewhat stronger result that $\mathcal{C}(\ell^2)$ does not admit any non-zero weakly compact elementary operators $\mathcal{E}_{a,b}$.

It turns out that (4.1) is a particular case of a rigidity phenomenon for elementary operators $\mathcal{E}_{a,b}$ on the Calkin algebra $\mathcal{C}(X)$, where X is any Banach space having an unconditional Schauder basis. Recall that the normalized Schauder basis (e_j) of X is *unconditional* if $\sum_{j=1}^{\infty} \theta_j a_j e_j$ converges in X for any sequence $(\theta_j) \in \{-1, 1\}^{\mathbf{N}}$ of signs whenever $\sum_{j=1}^{\infty} a_j e_j$ converges in X. The unconditional basis constant of (e_j) is

$$C = \sup\{\|M_\theta\| : \theta = (\theta_j) \in \{-1, 1\}^{\mathbf{N}}\},$$

where the diagonal operators $M_\theta \in L(X)$ are given by $M_\theta(\sum_{j=1}^{\infty} a_j e_j) = \sum_{j=1}^{\infty} \theta_j a_j e_j$ for $\sum_{j=1}^{\infty} a_j e_j \in X$. The basis (e_j) is *1-unconditional* if $C = 1$. An important consequence of unconditionality is that X admits plenty of nice projections: for any non-empty subset $A \subset \mathbf{N}$

$$P_A(\sum_{j=1}^{\infty} a_j e_j) = \sum_{j \in A} a_j e_j \quad \text{for} \quad \sum_{j=1}^{\infty} a_j e_j \in X,$$

defines a projection $P_A \in L(X)$ such that $\|P_A\| \leq C$. Recall that the class of Banach spaces having an unconditional basis is substantial: it contains e.g. the sequence spaces ℓ^p $(1 \leq p < \infty)$ and c_0, as well as the function spaces $L^p(0,1)$ $(1 < p < \infty)$ and H^1. By contrast, $L^1(0,1)$ and $C(0,1)$ do not have any unconditional bases, cf. [LT77, 1.d.1] or [Wo91, II.D.10 and II.D.12].

The following result from [ST99, Thm. 3] extends the Apostol-Fialkow formula (4.1) to spaces X having an unconditional basis. If the unconditional constant $C > 1$ for X, then the identity (4.1) is here replaced by inequalities between $\|\mathcal{E}_{a,b}\|$ and $dist(\mathcal{E}_{a,b}, K(\mathcal{C}(X)))$ that involve C.

Theorem 4.1. *Suppose that X is a Banach space having an unconditional basis (e_j) with unconditional basis constant C. Let $A, B \in L(X)^n$ be arbitrary n-tuples. Then the elementary operators $\mathcal{E}_{A,B} : L(X) \to L(X)$ and $\mathcal{E}_{a,b} : \mathcal{C}(X) \to \mathcal{C}(X)$ satisfy*

(4.2) $$dist(\mathcal{E}_{a,b}, W(\mathcal{C}(X))) \geq C^{-4} \|\mathcal{E}_{a,b} : \mathcal{C}(X) \to \mathcal{C}(X)\|,$$

(4.3) $$dist(\mathcal{E}_{A,B}, W(L(X))) \geq C^{-4} \|\mathcal{E}_{a,b} : \mathcal{C}(X) \to \mathcal{C}(X)\|.$$

In particular, if (e_j) is a 1-unconditional basis, then

$$\|\mathcal{E}_{a,b} : \mathcal{C}(X) \to \mathcal{C}(X)\| = dist(\mathcal{E}_{a,b}, K(\mathcal{C}(X))) = dist(\mathcal{E}_{a,b}, W(\mathcal{C}(X))),$$

and

(4.4) $$\|\mathcal{E}_{a,b}\| \leq dist(\mathcal{E}_{A,B}, W(L(X))) \leq \|\mathcal{E}_{A,B} : L(X) \to L(X)\|.$$

Recall that e.g. ℓ^p ($1 \leq p < \infty$), c_0 and direct sums such as $\ell^p \oplus \ell^q$ ($1 \leq p < q < \infty$) have 1-unconditional bases, but the unconditional constant of any unconditional basis of $L^p(0,1)$ for $1 < p < \infty$, $p \neq 2$, is strictly greater that 1 (see e.g. [Wo91, II.D.13 and p. 68]).

In particular, Theorem 4.1 solves a generalized version of the Fong-Sourour conjecture for elementary operators on $\mathcal{C}(X)$ for this class of Banach spaces. Part (ii) should be compared with Proposition 2.5.

Corollary 4.2. *Suppose that X is a Banach space having an unconditional basis (e_j), and let $A, B \in L(X)^n$ be arbitrary n-tuples.*
(i) If $\mathcal{E}_{a,b}$ is weakly compact $\mathcal{C}(X) \to \mathcal{C}(X)$, then $\mathcal{E}_{a,b} = 0$.
(ii) If $\mathcal{E}_{A,B}$ is weakly compact $L(X) \to L(X)$, then $\mathcal{E}_{a,b} = 0$ (and consequently $\mathcal{E}_{A,B}(L(X)) \subset K(X)$).

Surprisingly enough, it is actually possible to improve the estimate (4.4) from Theorem 4.1 in the case $X = \ell^p$.

Theorem 4.3. *Let $1 < p < \infty$ and $A, B \in L(\ell^p)^n$ be arbitrary n-tuples. Then*

$$\|\mathcal{E}_{a,b} : \mathcal{C}(\ell^p) \to \mathcal{C}(\ell^p)\| = dist(\mathcal{E}_{A,B}, W(L(\ell^p))) = dist(\mathcal{E}_{A,B}, K(L(\ell^p))).$$

The proof of (4.1) in [AF86] is based on Voiculescu's non-commutative Weyl - von Neumann theorem (see e.g. [Da96, Section II.5] for a description of this result). By contrast, the proof of Theorem 4.1 is quite different and it draws on fundamental properties of unconditional bases. The following simple facts will also be used here.
- if $(x_n) \subset X$ is a normalized weak-null sequence, then

$$(4.5) \qquad \|S\|_e \geq \limsup_{n \to \infty} \|Sx_n\| \quad \text{for } S \in L(X)$$

(This holds since $\|Ux_n\| \to 0$ as $n \to \infty$ for any compact operator $U \in K(X)$.)
- Let (e_j) be a Schauder basis for X so that $\|Q_n\| = 1$ for all $n \in \mathbb{N}$, where Q_n is the natural basis projection $X \to [e_r : r \geq n+1]$ (this property holds e.g. if (e_j) is a 1-unconditional basis). Then

$$(4.6) \qquad \|S\|_e = \lim_{n \to \infty} \|Q_n S\| \quad \text{for } S \in L(X).$$

Proof of Theorem 4.1 (sketch). For notational simplicity we may assume that the basis (e_j) is 1-unconditional on X. Otherwise we just pass to the equivalent norm $|\cdot|$ on X, where

$$|x| = \sup\{\|M_\theta x\| : \theta = (\theta_j) \in \{-1,1\}^{\mathbb{N}}\}, \quad x \in X,$$

and (e_j) is an 1-unconditional basis in $(X, |\cdot|)$. Above $\|x\| \leq |x| \leq C\|x\|$ for $x \in X$, where C is the unconditional constant of the basis (e_j) in $(X, \|\cdot\|)$.

Suppose that after normalization one has $\|\mathcal{E}_{a,b} : \mathcal{C}(X) \to \mathcal{C}(X)\| = 1$ for the n-tuples $A = (A_1, \ldots, A_n), B = (B_1, \ldots, B_n) \in L(X)^n$. In order to show (4.2) we are required to verify that

$$(4.7) \qquad dist(\mathcal{E}_{a,b}, W(\mathcal{C}(X))) \geq 1.$$

The strategy of the argument is to construct, for any given $\varepsilon > 0$, an operator $T \in L(X)$ and a sequence $(S_j) \subset L(X)$ so that

$$(4.8) \qquad 1 = \|T\|_e \leq \|T\| < 1 + \varepsilon,$$

$$(4.9) \qquad 1 = \|S_j\|_e = \|S_j\|, \quad j \in \mathbb{N},$$

$$(4.10) \quad \sup_j |\lambda_j| \leq \|\sum_{j=1}^{\infty} \lambda_j S_j\|_e \leq \|\sum_{j=1}^{\infty} \lambda_j S_j\| \leq 2 \cdot \sup_j |\lambda_j| \quad \text{for all } (\lambda_j) \in c_0,$$

$$(4.11) \qquad dist(\mathcal{E}_{A,B}(TS_j), K(X)) \geq 1 - \varepsilon, \quad j \in \mathbb{N}.$$

To verify (4.7) from these conditions suppose that $\mathcal{V} \in W(\mathcal{C}(X))$ is arbitrary. By (4.10) the closed linear span $[s_j : j \in \mathbb{N}] \subset \mathcal{C}(X)$ is isomorphic to c_0, so that $s_j \xrightarrow{w} 0$ in $\mathcal{C}(X)$ as $n \to \infty$. Since c_0 has the DPP and $\mathcal{V} \circ L_t$ is weakly compact on $\mathcal{C}(X)$, it follows that $\|\mathcal{V}(ts_j)\| \to 0$ as $j \to \infty$. As $\|ts_j\| \leq \|T\|_e \cdot \|S_j\|_e \leq 1$ for $j \in \mathbb{N}$ we get that

$$\|\mathcal{E}_{a,b} - \mathcal{V}\| \geq \limsup_{j \to \infty} \|\mathcal{E}_{a,b}(ts_j) - \mathcal{V}(ts_j)\| \geq \limsup_{j \to \infty} \|\mathcal{E}_{a,b}(ts_j)\| \geq 1 - \varepsilon$$

from (4.11). The verification of (4.3) follows a similar outline, since $[S_j : j \in \mathbb{N}] \subset L(X)$ is also linearly isomorphic to c_0 by (4.10).

The heart of the argument lies in the construction of $T \in L(X)$ and $(S_j) \subset L(X)$ which satisfy (4.8) - (4.11). We indicate some of the ideas for completeness. Let $P_m \in L(X)$ be the natural basis projection of X onto $[e_1, \ldots, e_m]$, and $Q_m = I - P_m$ for $m \in \mathbb{N}$. Moreover, put $P_{(m,n]} = P_n - P_m$ for $m < n$, where $(m, n] = \{m + 1, \ldots, n\}$.

Since $\|\mathcal{E}_{a,b}\| = 1$ there is $T \in L(X)$ such that $1 = \|T\|_e \leq \|T\| < 1 + \varepsilon$ and $\|\mathcal{E}_{A,B}(T)\|_e \geq 1 - \frac{\varepsilon}{8}$. By an inductive process it is possible to choose increasing sequences $(m_k), (n_k) \subset \mathbb{N}$ and normalized sequences $(y_j) \subset S_Y$ and $(y_j^*) \subset S_{Y^*}$ so that the following properties are satisfied:

$$(4.12) \qquad 0 = m_1 < n_1 < m_2 < n_2 < \ldots,$$

$$(4.13) \qquad Q_k^* y_k^* = y_k^*, \quad k \in \mathbb{N},$$

$$(4.14) \qquad \langle y_k^*, \mathcal{E}_{A,B}(TU_k)y_k \rangle > 1 - \frac{\varepsilon}{2}, \quad k \in \mathbb{N},$$

$$(4.15) \qquad |\langle y_k^*, \mathcal{E}_{A,B}(TU_l)y_k \rangle| \leq \varepsilon \cdot 2^{-l-k}, \quad k, l \in \mathbb{N}, \ k \neq l.$$

Above we have denoted $U_k = P_{(m_k, n_k]}$ for $k \in \mathbb{N}$. The induction is fairly lengthy and delicate, and we are forced to refer to [ST99, pp. 8-9] for the actual details.

The operators (S_j) will be chosen as suitable disjointly supported basis projections on X that are obtained by "cut-and-paste" as follows. First fix a disjoint partition $\mathbb{N} = \cup_{i=1}^{\infty} N_i$ into infinite sets, and let $M_i = \cup_{k \in N_i} (m_k, n_k]$ for $i \in \mathbb{N}$. Put

$$S_j = P_{M_j} \quad \text{for } j \in \mathbb{N},$$

the basis projection on X related to $M_j \subset \mathbb{N}$. It follows from the 1-uncondition-ality of (e_j) that $\|M_\lambda\| = \sup_{j \geq 1} |\lambda_j|$ for any bounded real-valued $\lambda = (\lambda_j) \in \ell^\infty$, so that (by splitting into real and imaginary parts)

$$\sup_{j \in \mathbb{N}} |\lambda_j| \leq \|M_\lambda\| \leq 2 \cdot \sup_{j \in \mathbb{N}} |\lambda_j| \quad \text{for all } \lambda = (\lambda_j) \in \ell^\infty.$$

Since the sets M_j are pairwise disjoint this fact yields that

$$\left\| \sum_{j=1}^\infty \lambda_j S_j \right\|_e \leq \left\| \sum_{j=1}^\infty \lambda_j S_j \right\| \leq 2 \cdot \sup_j |\lambda_j|, \quad (\lambda_j) \in c_0.$$

To obtain the lower bound in (4.10) fix $k \in \mathbb{N}$ and enumerate $N_k = \{k(m) : m \in \mathbb{N}\}$. From (4.5) we get that $\| \sum_{j=1}^\infty \lambda_j S_j \|_e \geq \limsup_{m \to \infty} |\lambda_k| \cdot \|S_k e_{k(m)}\| = |\lambda_k|$. Thus (4.10) holds.

Finally, (4.13) - (4.15) are used to enforce (4.11). Put again $N_k = \{k(m) : m \in \mathbb{N}\}$ for $k \in \mathbb{N}$. Since $S_k = \sum_{m=1}^\infty U_{k(m)}$ in the strong operator topology in $L(X)$, we get from (4.13) - (4.15) that

$$\begin{aligned}
\|Q_{k(m)} \mathcal{E}_{A,B}(TS_k)\| &\geq |\langle y_{k(m)}^*, Q_{k(m)} \mathcal{E}_{A,B}(TS_k) y_{k(m)} \rangle| \\
&= |\langle y_{k(m)}^*, \mathcal{E}_{A,B}(TS_k) y_{k(m)} \rangle| \\
&\geq |\langle y_{k(m)}^*, \mathcal{E}_{A,B}(TU_{k(m)}) y_{k(m)} \rangle| - \sum_{l; l \neq m} |\langle y_{k(m)}^*, \mathcal{E}_{A,B}(TU_{k(l)}) y_{k(m)} \rangle| \\
&\geq 1 - \frac{\varepsilon}{2} - \varepsilon \sum_{l; l \neq m} 2^{-k(m)-k(l)} > 1 - \varepsilon.
\end{aligned}$$

Finally, since $\|Q_{k(m)}\| = 1$ for each m by the 1-unconditionality we get

$$dist(\mathcal{E}_{A,B}(TS_k), K(X)) \geq \limsup_{m \to \infty} \|Q_{k(m)} \mathcal{E}_{A,B}(TS_k)\| \geq 1 - \varepsilon.$$

from (4.6) and the preceding estimate. $\qquad\square$

Theorem 4.1 remains valid if X has an unconditional finite-dimensional Schauder decomposition (the Schatten classes C_p for $1 \leq p < \infty$ are concrete examples of spaces having this property, but failing to have an unconditional basis, cf. [DJT95, pp. 364-368]). The exact class of Banach spaces for which Theorem 4.1 holds remains unknown.

Problem 4.4. *Does Theorem 4.1 hold for $\mathcal{C}(X)$ if X is a (classical) Banach space that fails to have any unconditional structure, e.g. if $X = L^1(0,1)$ or $X = C(0,1)$? An obvious starting point is to check if $dist(L_a R_b, K(\mathcal{C}(X)))$ is comparable to $\|L_a R_b\|$.*

In [ST98, Thm. 7] one obtains a somewhat complementary rigidity result for non-zero elementary operators $\mathcal{E}_{a,b}$ on $\mathcal{C}(X)$, where X is a reflexive space having an unconditional basis. For such spaces one gets an alternative ap-proach to the Fong-Sourour type results in Corollary 4.2 by showing that the non-zero operators $\mathcal{E}_{a,b}$ are automatically non strictly singular (note that this yields yet another proof of the original conjecture for $\mathcal{C}(\ell^2)$). The argument does not give any information about $\|\mathcal{E}_{a,b}\|$, but it is a simpler variant of the ideas underlying Theorem 4.1.

For $S \in L(Y)$ let $R(S) \in L(Y^{**}/Y)$ be the operator defined by

$$R(S)(y^{**} + Y) = S^{**}y^{**} + Y, \quad y^{**} \in Y^{**}.$$

It is a general fact that if S is bounded below on the subspace $M \subset Y$ then $R(S)$ is bounded below on the subspace $M^{**}/M \subset Y^{**}/Y$, see e.g. [GST95, Prop. 1.4]. We will apply this observation to the restriction $\widetilde{\mathcal{E}}_{A,B} = \mathcal{E}_{A,B|K(X)}$, for which

$$\tag{4.16} R(\widetilde{\mathcal{E}}_{A,B}) = \mathcal{E}_{a,b}.$$

This holds in trace duality for any $A = (A_1, \ldots, A_n)$, $B = (B_1, \ldots, B_n) \in L(X)^n$, provided e.g. X is a reflexive Banach space having an unconditional basis. In fact, in this case $\mathcal{C}(X)^* = K(X)^\perp$, and from (2.6) - (2.7) we get that

$$\tag{4.17} (\mathcal{E}_{A,B} : K(X) \to K(X))^* = \mathcal{E}_{A^*,B^*} : N(X^*) \to N(X^*),$$

$$\tag{4.18} (\mathcal{E}_{A^*,B^*} : N(X^*) \to N(X^*))^* = \mathcal{E}_{A,B} : L(X) \to L(X)),$$

where $A^* = (A_1^*, \ldots, A_n^*), B^* = (B_1^*, \ldots, B_n^*)$ (see the discussion following Proposition 2.5).

Theorem 4.5. *Suppose that X is a reflexive Banach space having an unconditional basis (e_j), and let $A, B \in L(X)^n$ be arbitrary n-tuples. If $\mathcal{E}_{a,b} \neq 0$, then $\mathcal{E}_{a,b}$ fixes a copy of the non-separable quotient space ℓ^∞/c_0: there is a subspace $M \subset \mathcal{C}(X)$ so that M is isomorphic to ℓ^∞/c_0 and $\mathcal{E}_{a,b}$ is bounded below on M.*

Sketch. Assume without loss of generality that $\|\mathcal{E}_{a,b}\| = 1$. Let $\varepsilon > 0$ and pick $T \in L(X)$ so that $1 = \|T\|_e \le \|T\| < 1 + \varepsilon$ and $\|\mathcal{E}_{A,B}(T)\| > 1 - \varepsilon$.

By a gliding hump argument (see [ST98, p. 233] for the details) one obtains an increasing sequence $(m_k) \subset \mathbb{N}$ and a sequence $(y_k) \subset S_X$ so that

$$\tag{4.19} \|\mathcal{E}_{A,B}(TU_j)y_j\| > 1 - \varepsilon, \quad j \in \mathbb{N},$$

where $U_j = P_{(m_j, m_{j+1}]}$ for all j. (Here we retain the notation from Theorem 4.1.) Unconditionality implies that

$$\left\| \sum_{j=1}^\infty \lambda_j U_j \right\| \approx \sup_j |\lambda_j| \quad \text{for all } (\lambda_j) \in c_0.$$

Thus $U_j \xrightarrow{w} 0$ in $K(X)$ as $j \to \infty$, so that (TU_j) and $(\mathcal{E}_{A,B}(TU_j))$ are weak-null sequences of $K(X)$, but $(\mathcal{E}_{A,B}(TU_j))$ is not norm-null by (4.19). Hence the Bessaga-Pelczynski basic sequence selection principle [LT77, 1.a.12] produces a subsequence (U_{j_k}) so that both (TU_{j_k}) and $(\mathcal{E}_{A,B}(TU_{j_k}))$ are basic sequences in $K(X)$. This means that

$$\left\| \sum_{k=1}^\infty \lambda_j TU_{j_k} \right\| \approx \left\| \sum_{k=1}^\infty \lambda_j \mathcal{E}_{A,B}(TU_{j_k}) \right\| \approx \sup_k |\lambda_k| \quad \text{for all } (\lambda_k) \in c_0,$$

so that the restriction $\widetilde{\mathcal{E}}_{A,B} = \mathcal{E}_{A,B|K(X)}$ is bounded below on the subspace $N = [TU_{j_k} : k \in \mathbb{N}] \approx c_0$ in $K(X)$. It follows from the facts cited prior to this theorem that $\mathcal{E}_{a,b}$ is bounded below on the subspace $N^{**}/N \approx \ell^\infty/c_0$.   □

The circle of results (Theorems 4.1 and 4.5) related to the Fong-Sourour conjecture concern the non-existence of small non-zero elementary operators $\mathcal{E}_{a,b}$ on a natural class of Calkin algebras. We next discuss a different rigidity property from [ST98] for $\mathcal{E}_{a,b}$ on $C(X)$ for more special spaces X. These properties are related to the size of the kernel $Ker(\mathcal{E}_{a,b})$ and the cokernel $C(X)/\overline{Im(\mathcal{E}_{a,b})}$. We first recall some earlier results which served as motivation. Gravner [G86] noted the following surprising facts, which contain earlier results on $C(\ell^2)$ by Fialkow for the generalized derivation $L_a - R_b$ and by Weber for $L_a R_b$. The operators $\mathcal{E}_{A,B}$ do not enjoy such rigidity properties on $L(X)$ for any space X (just consider a left multiplication L_A).

Fact 4.6. Let $A = (A_1, \ldots, A_n), B = (B_1, \ldots, B_n) \in L(\ell^2)^n$ be n-tuples so that $\{A_1, \ldots, A_n\}$ or $\{B_1, \ldots, B_n\}$ are commuting sets in $L(\ell^2)$.

(i) If $\mathcal{E}_{a,b}$ is injective on $C(\ell^2)$, then $\mathcal{E}_{a,b}$ is actually bounded below on $C(\ell^2)$.

(ii) If the range $\mathcal{E}_{a,b}(C(\ell^2))$ is dense in $C(\ell^2)$, then $\mathcal{E}_{a,b}$ is a surjection $C(\ell^2) \to C(\ell^2)$.

Results from [LeS71] and [AT86] yield similar mapping results for the basic maps L_a, R_b and $L_a R_b$ on $C(X)$ for a number of classical Banach spaces X.

Fact 4.7. Suppose that $A, B \in L(X)$.

(i) Let $X = \ell^p$ $(1 \le p < \infty)$, c_0 or $L^p(0,1)$ $(1 \le p < \infty)$. Then L_a is injective $C(X) \to C(X)$ if and only if L_a is bounded below on $C(X)$.

(ii) Let $X = \ell^p$ $(1 < p \le \infty)$, c_0, or $L^p(0,1)$ $(1 < p \le \infty)$. Then R_b is injective $C(X) \to C(X)$ if and only if R_b is bounded below on $C(X)$.

(iii) Let $X = \ell^p$ $(1 < p < \infty)$, c_0, or $L^p(0,1)$ $(1 < p < \infty)$. Then $L_a R_b$ is injective $C(X) \to C(X)$ if and only if $L_a R_b$ is bounded below on $C(X)$.

Note that part (iii) follows easily from (i) and (ii), since $L_a R_b = R_b L_a$. In fact, if $L_a R_b$ is injective on $C(X)$, then L_a and R_b are both injective and consequently also bounded below on $C(X)$.

Fact 4.7 originated in studies of properties of the semi-Fredholm classes $\Phi_+(X)$ and $\Phi_-(X)$ in $C(X)$. Recall that the operator $S \in \Phi_+(X)$ if its range $Im(S)$ is closed in X and its kernel $Ker(S)$ is finite-dimensional, while $S \in \Phi_-(X)$ if $Im(S)$ has finite codimension in X (thus $\Phi(X) = \Phi_+(X) \cap \Phi_-(X)$ are the Fredholm operators). We refer e.g. to [LeS71], [AT86], [AT87] and [T94] for a more careful discussion and for further results about the classes $\Phi_\pm(X)$ modulo the compact operators.

Facts 4.6 and 4.7 suggest the question whether similar rigidity facts would hold for general elementary operators $\mathcal{E}_{a,b}$ on $C(X)$. The following unexpected example from [AT87] demonstrates that some limitations apply.

Example 4.8. *There is a reflexive Banach space X, which fails to have the compact approximation property, and an isometric embedding $J \in L(X)$ so that*

(i) *L_j is one-to-one but not bounded below on $C(X)$,*

(ii) *R_{j*} is one-to-one but not bounded below on $C(X^*)$.*

In contrast with Example 4.8 it was shown in [ST98, Thm. 3 and 4] that one has the following striking dichotomies for *arbitrary* elementary operators $\mathcal{E}_{a,b}$ on $C(\ell^p)$. We stress that the generality is much greater compared to Fact 4.6 also in the classical case $p = 2$, since there are no commutativity assumptions on the n-tuples.

Theorem 4.9. *Let $1 < p < \infty$ and $A = (A_1, \dots, A_n), B = (B_1, \dots, B_n) \in L(\ell^p)^n$ be arbitrary n-tuples. If $\mathcal{E}_{a,b}$ is not bounded below on $C(\ell^p)$, then its kernel $Ker(\mathcal{E}_{a,b})$ is non-separable. In particular, if $\mathcal{E}_{a,b}$ is injective on $C(\ell^p)$ then there is $c > 0$ so that*

$$\|\mathcal{E}_{A,B}(S)\|_e \geq c\|S\|_e, \quad S \in L(\ell^p).$$

Theorem 4.10. *Let $1 < p < \infty$ and $A = (A_1, \dots, A_n), B = (B_1, \dots, B_n) \in L(\ell^p)^n$ be arbitrary n-tuples. If the range $\mathcal{E}_{a,b}(C(\ell^p)) \neq C(\ell^p)$, then the quotient*

$$C(\ell^p)/\overline{Im(\mathcal{E}_{a,b})}$$

is non-separable. In particular, if $\mathcal{E}_{a,b}(C(\ell^p))$ is dense in $C(\ell^p)$, then $\mathcal{E}_{a,b}$ is a surjection.

Proof of Theorem 4.9 (sketch). Let $A = (A_1, \dots, A_n), B = (B_1, \dots, B_n) \in L(\ell^p)^n$ be n-tuples so that $\mathcal{E}_{a,b}$ is not bounded below on $C(\ell^p)$. Recall that by (4.16) we have $R(\widetilde{\mathcal{E}}_{A,B}) = \mathcal{E}_{a,b}$, where $\widetilde{\mathcal{E}}_{A,B} = \mathcal{E}_{A,B|K(\ell^p)}$ is the restriction to $K(\ell^p)$ and

$$R(\widetilde{\mathcal{E}}_{A,B})(S + K(\ell^p)) = \mathcal{E}_{A,B}(S) + K(\ell^p), \quad S + K(\ell^p) \in C(\ell^p).$$

By the crucial fact cited prior to Theorem 4.5 our assumption yields that $\widetilde{\mathcal{E}}_{A,B}$ is not bounded below on $K(\ell^p)$.

Let P_n and $Q_n = I - P_n$ again denote the natural basis projections with respect to the unit vector basis (e_k) in ℓ^p for $n \subset \mathbb{N}$.

Step 1. We claim that for any $\varepsilon > 0$ and $m \in \mathbb{N}$ there is $S \in K(\ell^p)$ so that

$$\|S\| = 1, \quad Q_m S Q_m = S, \quad \text{and} \quad \|\widetilde{\mathcal{E}}_{A,B}(S)\| < \varepsilon.$$

Fix $m \in \mathbf{N}$ and put $E_m = Q_m(\ell^p) = [e_s : s \geq m+1]$. The key observation here is that the related elementary operator

$$\mathcal{E}_{aj_m, q_m b} : C(E_m) \to C(\ell^p)$$

fails also to be bounded below. Here J_m denotes the inclusion $E_m \subset \ell^p$ and $AJ_m = (A_1 J_m, \dots, A_n J_m), Q_m B = (Q_m B_1, \dots, Q_m B_n)$. One should check that $\mathcal{E}_{aj_m, q_m b}$ is obtained from the restriction $\widetilde{\mathcal{E}}_{AJ_m, Q_m B} : K(E_m) \to K(\ell^p)$ in trace duality, that is

$$R(\widetilde{\mathcal{E}}_{AJ_m, Q_m B}) = \mathcal{E}_{aj_m, q_m b}.$$

Note that with only a minor loss of precision one may view the operators $\mathcal{E}_{aj_m,q_mb}$ and $\mathcal{E}_{a,b}$ as practically the same. This is so because $Q_m = I - P_m$, where P_m has finite rank and the codimension of E_m is finite – hence one expects that these differences are wiped away at the Calkin level. In particular, we thus know that $\mathcal{E}_{aj_m,q_mb}$ is not bounded below on $\mathcal{C}(E_m)$. From the above cited general fact we infer that $\mathcal{E}_{AJ_m,Q_mB}$ also fails to be bounded below on $K(E_m)$, and we may pick a normalized $S_0 \in K(E_m)$ so that

$$\|\mathcal{E}_{AJ_m,Q_mB}(S_0)\| = \|\widetilde{\mathcal{E}}_{A,B}(J_mS_0Q_m)\| < \varepsilon.$$

Finally, $S = J_mS_0Q_m \in K(\ell^p)$ is the desired operator. We refer the reader to [ST98, pp. 221-222] for a precise version of the above partly heuristic argument.

Step 2. By a gliding hump argument we next obtain an increasing sequence $(m_k) \subset \mathbb{N}$ and a normalized sequence $(S_k) \subset K(\ell^p)$ of finite rank operators so that

$$(4.20) \qquad\qquad P_{(m_k,m_{k+1}]}S_kP_{(m_k,m_{k+1}]} = S_k,$$

$$(4.21) \qquad\qquad \|\widetilde{\mathcal{E}}_{A,B}(S_k)\| < 1/2^k$$

for all $k \in \mathbb{N}$. Above $P_{(r,s]} \equiv P_s - P_r \in L(\ell^p)$ is the natural projection onto $[e_{r+1},\ldots,e_s]$ for $r,s \in \mathbb{N}$ and $r < s$. Here (4.20) states that (S_k) is a block-diagonal sequence on ℓ^p.

We outline the general step of the induction for completeness. Suppose that we have already found $m_1 = 1 < m_2 < \ldots < m_{n+1}$ and $S_1,\ldots,S_n \in K(\ell^p)$ that satisfy (4.20) - (4.21). From Step 1 we get $S \in K(\ell^p)$ so that

$$S = Q_{m_{n+1}}SQ_{m_{n+1}}, \quad \|S\| = 1 \text{ and } \|\widetilde{\mathcal{E}}_{A,B}(S)\| < 1/2^{n+2}.$$

For $r > m_{n+1}$ put $S = P_rSP_r + Z_r$, where $Z_r \equiv Q_rS + SQ_r - Q_rSQ_r \to 0$ as $r \to \infty$, since S is a compact operator on ℓ^p (and $1 < p < \infty$). By continuity we may then pick $r = m_{n+2} > m_{n+1}$ so that

$$\|P_{m_{n+2}}SP_{m_{n+2}}\| \approx 1 \quad \text{and} \quad \|\widetilde{\mathcal{E}}_{A,B}(P_{m_{n+2}}SP_{m_{n+2}})\| < 1/2^{n+2}.$$

Then $S_{n+1} = \|P_{m_{n+2}}SP_{m_{n+2}}\|^{-1}P_{m_{n+2}}SP_{m_{n+2}}$ satisfies (4.20) and (4.21).

Step 3. We next explain how to build an isometric copy of ℓ^∞ inside the kernel $\mathrm{Ker}(\mathcal{E}_{a,b})$. First fix a countable partition $\mathbb{N} = \bigcup_r N_r$ into infinite sets N_r and consider the operators $U_r = \sum_{k \in N_r} S_k \in L(\ell^p)$ for $r \in \mathbb{N}$ (the U_r are defined in the strong operator topology). One must verify the following facts for any $(c_r) \in \ell^\infty$:

$$(4.22) \qquad \|\sum_{r=1}^{\infty} c_rU_r\|_e = \|\sum_{r=1}^{\infty} c_rU_r\| = \sup_r |c_r|,$$

$$(4.23) \qquad \sum_{r=1}^{\infty} c_ru_r \in \mathrm{Ker}(\mathcal{E}_{a,b}).$$

The equality (4.22) is quite easy to verify on ℓ^p as one may essentially treat the disjoint normalized blocks S_k as diagonal elements (cf. the proof of (4.10)).

The inclusion (4.23) in turn follows by observing that $\sum_{r=1}^{\infty} c_r \mathcal{E}_{A,B}(U_r) \in K(\ell^p)$, since the sum can formally be rewritten as $\sum_{k=1}^{\infty} a_k \mathcal{E}_{A,B}(S_k)$, where $|a_k| \leq \|(c_r)\|_{\infty}$ for each k and where we have norm convergence thanks to (4.21). We leave the details to the reader. □

The strategy of the proof of Thm. 4.10 is to embed ℓ^{∞} isomorphically into the quotient $\mathcal{C}(\ell^p)/\overline{Im(\mathcal{E}_{a,b})}$ whenever $\mathcal{E}_{a,b}$ is not surjective on $\mathcal{C}(\ell^p)$. For this purpose one builds certain block diagonal sequences in the nuclear operators $N(\ell^{p'})$, where p' is the dual exponent of p, as well as a related sequence $(\phi_r) \subset K(\ell^p)^{\perp} = \mathcal{C}(\ell^p)^*$ of norm-1 functionals which are used to norm the desired ℓ^{∞}-copy. We refer to [ST98, Thm. 4] for the details.

To illustrate Theorem 4.10 in a simple special case recall that $1 = I_{\ell^2} + K(\ell^2) \notin Im(L_a - R_a)$ for any $A \in L(\ell^2)$ by a well-known commutator fact. It follows from Theorem 4.10 that the quotient $\mathcal{C}(\ell^2)/\overline{Im(L_a - R_a)}$ is non-separable, and so is $L(\ell^2)/\overline{Im(L_A - R_A)} + K(\ell^2)$. (Apparently Stampfli [St73] first noticed that $L(\ell^2)/\overline{Im(L_A - R_A)}$ is non-separable for any $A \in L(\ell^2)$). This fact should be contrasted with the following remarkable result of Anderson [A73], which is based on C*-algebraic tools.

Theorem 4.11. *There are* $A \in L(\ell^2)$ *for which*

$$I_{\ell^2} \in \overline{Im(L_A - R_A)},$$

that is, there is $(S_n) \subset L(\ell^2)$ *so that* $\|I_{\ell^2} - (AS_n - S_n A)\| \to 0$ *as* $n \to \infty$.

It remains unclear how far the techniques behind Theorems 4.9 and 4.10 can be pushed beyond ℓ^p. We state this as a problem, where the case $X = L^p(0,1)$ for $p \neq 2$ would be particularly interesting.

Problem 4.12. *Do Theorems 4.9 and 4.10 hold in the class of reflexive Banach spaces* X *having an unconditional basis?*

It was pointed out in [ST98, p. 226] that Theorem 4.9 remains valid for $X = \ell^p \oplus \ell^q$, where $1 < p < q < \infty$, and also for $X = c_0$. The technical obstruction in the argument of Theorem 4.9 is the following: given a block diagonal sequence (U_r) as in Step 3 above the sum $\sum_k c_k U_k$ does not always define a bounded operator on X for $(c_k) \in c_0$. (There is an analogous obstruction related to Theorem 4.10.) This phenomenon already occurs for $\ell^p \oplus \ell^q$, but in that case it can be circumvented.

Other developments. Magajna [M95] obtained a surprising formula for the norm of an arbitrary elementary operator $\mathcal{E}_{a,b}$ on the Calkin algebra $\mathcal{C}(\ell^2)$ in terms of the completely bounded norm. (Clearly this result is also closely related to Section 3.) We refer e.g. to [P03, Section 1] for a discussion of the cb-norm.

Theorem 4.13. *Let* $A, B \in L(\ell^2)^n$ *be arbitrary n-tuples. Then*

$$\|\mathcal{E}_{a,b}\colon \mathcal{C}(\ell^2) \to \mathcal{C}(\ell^2)\| = \|\mathcal{E}_{a,b}\|_{cb}.$$

Subsequently Archbold, Mathieu and Somerset [AMS99, Thm. 6] characterized the precise class of C*-algebras where the preceding identity holds. Recall for this that the C*-algebra $\mathcal{A}$ is *antiliminal* if no non-zero positive element $x \in \mathcal{A}$ generates an abelian hereditary C*-subalgebra. (The C*-subalgebra $\mathcal{B}$ of $\mathcal{A}$ is hereditary if $x \in \mathcal{B}$ whenever $y \in \mathcal{B}_+$, $0 \le x \le y$ and $x \in \mathcal{A}$. Here $\mathcal{B}_+$ is the positive part of $\mathcal{B}$.) The Calkin algebra $\mathcal{C}(\ell^2)$ is an antiliminal algebra, see [AM03, pp. 34-35]. The C*-algebra $\mathcal{A}$ is called *antiliminal-by-abelian*, if there are C*-algebras $\mathcal{J}$ and $\mathcal{B}$ so that $\mathcal{J}$ is abelian, $\mathcal{B}$ is antiliminal and

$$0 \longrightarrow \mathcal{J} \longrightarrow \mathcal{A} \longrightarrow \mathcal{B} \longrightarrow 0$$

is a short exact sequence. Finally, recall that the map $L_a R_b; s \mapsto asb$ is a bounded linear operator $\mathcal{A} \to \mathcal{A}$ for any $a, b \in M(\mathcal{A})$, the multiplier algebra of $\mathcal{A}$. (Roughly speaking, $M(A)$ is the maximal unital C*-algebra which contains $\mathcal{A}$ as a closed 2-sided ideal, cf. [AM03, pp. 27-28].) Thus $\mathcal{E}_{a,b} = \sum_{j=1}^{n} L_{a_j} R_{b_j}$ defines a bounded elementary operator $\mathcal{A} \to \mathcal{A}$ for any $a = (a_1, \ldots, a_n), b = (b_1, \ldots, b_n) \in M(\mathcal{A})^n$.

Theorem 4.14. *The following conditions are equivalent for any C*-algebra* $\mathcal{A}$:

(i) $\|\mathcal{E}_{a,b}\| = \|\mathcal{E}_{a,b}\|_{cb}$ *for any n-tuples* $a, b \in M(\mathcal{A})^n$,
(ii) $\mathcal{A}$ *is antiliminal-by-abelian*.

We refer to [M95] and [AMS99], or [AM03, Sect. 5.4] for the proofs of Theorems 4.13 and 4.14.

5. Concluding remarks

There is a rich and well-developed structural theory of special elementary operators on $L(\ell^2)$ to which we have paid less attention because of the restraints of this survey. A good introduction is found in [Fi92].

Recall that if $A, B \in L(X)$, then the commutator $AB - BA$ cannot have the form $\lambda I_X + K$ for $\lambda \ne 0$ and $K \in K(X)$. A classical result of Brown and Pearcy completely identifies the set of commutators on $L(\ell^2)$:

$$\{AB - BA : A, B \in L(\ell^2)\} = \{S \in L(\ell^2) : S \ne \lambda I_X + K \text{ for } K \in K(\ell^2), \lambda \ne 0\}.$$

This characterization was subsequently extended by Apostol [Ap72], [Ap73] to the case of $X = \ell^p$ for $1 \le p < \infty$ and $X = c_0$, but the picture for other classical Banach spaces remains incomplete.

References

[AW80] C. Akemann and S. Wright, *Compact actions on C*-algebras*, Glasgow Math. J. **21** (1980), 143-149.

[A73] J.H. Anderson, *Derivation ranges and the identity*, Bull. Amer. Math. Soc. **79** (1973), 705-708.

[Ap72] C. Apostol, *Commutators on ℓ^p-spaces*, Rev. Roum. Math. Pures Appl. **42** (1972), 1513-1534.

[Ap73] C. Apostol, *Commutators on c_0-spaces and ℓ^∞-spaces*, Rev. Roum. Math. Pures Appl. **43** (1973), 1025-1032.

[AF86] C. Apostol and L. Fialkow, *Structural properties of elementary operators*, Canad. J. Math. **38** (1986), 1485-1524.

[AM03] P. Ara and M. Mathieu, *Local multipliers of C*-algebras*, Springer Monographs in Mathematics (Springer-Verlag, 2003).

[AMS99] R.J. Archbold, M. Mathieu and D.W.B Somerset, *Elementary operators on antiliminal C*-algebras*, Math. Ann. **313** (1999), 609-616.

[AST05] R.J. Archbold, D.W.B Somerset and R.M. Timoney, *On the central Haagerup tensor product and completely bounded mappings on a C*-algebra*, J. Funct. Anal. **226** (2005), 406-428.

[ARS94] W. Arendt, F. Räbiger and A. Sourour, *Spectral properties of the operator equation* $AX + XB = Y$, Quart. J. Math. Oxford **45** (1994), 133-149.

[AT86] K. Astala and H.-O. Tylli, *On semiFredholm operators and the Calkin algebra*, J. London Math. Soc. **34** (1986), 541-551.

[AT87] K. Astala and H.-O. Tylli, *On the bounded compact approximation property and measures of noncompactness*, J. Funct. Anal. **70** (1987), 388-401.

[BB01] M. Barraa and M. Boumazgour, *Inner derivations and norm equality*, Proc. Amer. Math. Soc. **130** (2001), 471-476.

[BR97] R. Bhatia and P. Rosenthal, *How and why to solve the operator equation* $AX - XB = Y$, Bull. London Math. Soc. **29** (1997), 1-21.

[BBR04] A. Blanco, M. Boumazgour and T.J. Ransford, *On the norm of elementary operators*, J. London Math. Soc. **70** (2004), 479-498.

[BDL01] J. Bonet, P. Domański and M. Lindström, *Weakly compact composition operators on analytic vector-valued function spaces*, Ann. Acad. Sci. Fennica Math. **26** (2001), 233-248.

[B81] J. Bourgain, *On the Dunford-Pettis property*, Proc. Amer. Math. Soc. **81** (1981), 265-272.

[CDR89] B. Carl, A. Defant and M. Ramanujan, *On tensor stable operator ideals*, Michigan J. Math. **36** (1989), 63-75.

[CS00] B. Carl and C. Schiebold, *Ein direkter Ansatz zur Untersuchung von Solitongleichungen*, Jahresber. Deutsch. Math.-.Ver. **102** (2000), 102-148. [English translation: *A direct approach to the study of soliton equations*, Jenaer Schriften zur Mathematik und Informatik Math/Inf/18/01 (Friedrich-Schiller-Universität Jena, 2001)]

[Cu83] R.E. Curto, *The spectra of elementary operators*, Indiana Univ. J. Math. **32** (1983), 193-197.

[Cu86] R.E. Curto, *Connections between Harte and Taylor spectra*, Rev. Roum. Math. Pures Appl. **32** (1986), 203-215.

[Cu88] R.E. Curto, *Applications of several complex variables to multiparameter spectral theory*, in *Surveys of some recent results in operator theory, vol. II* (J.B. Conway and B.B. Morrel, eds.), pp. 25-90. Pitman Research Notes in Mathematics vol. 192 (Longman, 1988).

[Cu92] R.E. Curto, *Spectral theory of elementary operators*, in *Elementary operators and Applications* (M. Mathieu, ed.) (Proc. Int. Conf. Blaubeuren 1991), pp. 3-52. (World Scientific, 1992)

[CuF87] R.E. Curto and L.A. Fialkow, *The spectral picture of* $[L_A, R_B]$, J. Funct. Anal. **71** (1987), 371-392.

[Da96] K.R. Davidson, *C*-algebras by Example*, Fields Institute Monographs vol. 6 (American Mathematical Society, 1996).

[DeF93] A. Defant and K. Floret, *Tensor norms and operator ideals*, (North-Holland, 1993).

[Di80] J. Diestel, *A survey of results related to the Dunford-Pettis property*, Contemp. Math. **2** (1980), 15-60.

[DiF76] J. Diestel and B. Faires, *Remarks on the classical Banach operator ideals*, Proc. Amer. Math. Soc. **58** (1976), 189-196.

[DJT95] J. Diestel, H. Jarchow and A. Tonge, *Absolutely summing operators*. Cambridge Studies in advanced mathematics vol. **43** (Cambridge University Press, 1995).

[DiU77] J. Diestel and J.J. Uhl, jr., *Vector measures*. Mathematical Surveys **15** (American Mathematical Society, 1977).

[E88] J. Eschmeier, *Tensor products and elementary operators*, J. reine angew. Math. **390** (1988), 47-66.

[F84] A.S. Fainshtein, *The Fredholm property and the index of a function of left and right multiplication operators*, Akad. Nauk Azerbaidzhan. SSR Dokl. **40** (1984), 3-7. [in Russian]

[FeS75] M. Feder and P. Saphar, *Spaces of compact operators and their dual spaces*, Israel J. Math. **21** (1975), 38-49.

[Fi79] L. A. Fialkow, *A note on norm ideals and the operator $X \to AX - XB$*, Israel J. Math. **32** (1979), 331-348.

[Fi92] L. A. Fialkow, *Structural properties of elementary operators*, in *Elementary operators and Applications* (M. Mathieu, ed.) (Proc. Int. Conf. Blaubeuren 1991), pp. 55-113. (World Scientific, 1992).

[FS79] C.K. Fong and A. Sourour, *On the operator identity $\sum A_k X B_k = 0$*, Canad. J. Math. **31** (1979), 845-857.

[GST95] M. González, E. Saksman and H.-O. Tylli, *Representing non-weakly compact operators*, Studia Math. **113** (1995), 265-282.

[G86] J. Gravner, *A note on elementary operators on the Calkin algebra*, Proc. Amer. Math. Soc. **97** (1986), 79-86.

[Ho70] J. Holub, *Tensor product mappings*, Math. Ann. **188** (1970), 1-12.

[Ho74] J. Holub, *Tensor product mappings. II*, Proc. Amer. Math. Soc. **42** (1974), 437-441.

[J71] B.E. Johnson, *Norms of derivations on $L(X)$*, Pacific J. Math. **38** (1971), 465-469.

[JL03] W.B. Johnson and J. Lindenstrauss, *Basic concepts in the geometry of Banach spaces*, in *Handbook in the geometry of Banach spaces, vol. 1* (W.B. Johnson and J. Lindenstrauss, eds.), pp. 1- 83 (North-Holland, 2003).

[KP62] M.I. Kadec and A. Pełczynski, *Bases, lacunary sequences and complemented subspaces in the spaces L_p*, Studia Math. **21** (1962), 161-176.

[Ky77] J. Kyle, *Norms of derivations*, J. London Math. Soc. **16** (1977), 297-312.

[LeS71] A. Lebow and M. Schechter, *Measures of noncompactness and semigroups of operators*, J. Funct. Anal. **7** (1971), 1 - 26.

[LT77] J. Lindenstrauss and L. Tzafriri, *Classical Banach spaces I. Sequence spaces*, Ergebnisse der Mathematik vol. **92** (Springer-Verlag, 1977).

[LST05] M. Lindström, E. Saksman and H.-O. Tylli, *Strictly singular and cosingular multiplications*, Canad. J. Math. 57 (2005), 1249-1278.

[LSch99] M. Lindström and G. Schlüchtermann, *Composition of operator ideals*, Math. Scand. **84** (1999), 284-296.

[LR59] G. Lumer and M. Rosenblum, *Linear operator equations*, Proc. Amer. Math. Soc. **10** (1959), 32-41.

[M87] B. Magajna, *A system of operator equations*, Canad. Math. Bull. **30** (1987), 200-209.

[M93] B. Magajna, *A transitivity theorem for algebras of elementary operators*, Proc. Amer. Math. Soc. **118** (1993), 119-127.

[M95] B. Magajna, *The Haagerup norm on the tensor product of operator modules*, J. Funct. Anal. **129** (1995), 325-348.

[Ma88] M. Mathieu, *Elementary operators on prime C^*-algebras, II*, Glasgow Math. J. **30** (1988), 275-284.

[Ma89] M. Mathieu, *Properties of the product of two derivations of a C^*-algebra*, Canad. Math. Bull. **32** (1989), 490-497.

[Ma01a] M. Mathieu, *The norm problem for elementary operators*, in *Recent progress in functional analysis (Valencia, 2000)*, (K.D. Bierstedt, J. Bonét, M. Maestre, J. Schmets, eds.) pp. 363-368 (North-Holland, 2001).

[Ma01b] M. Mathieu, *Elementary operators on Calkin algebras*, Irish Math. Soc. Bull. **46** (2001), 33-42.

[MPR88] A. McIntosh, A. Pryde and W. Ricker, *Systems of operator equations and perturbation of spectral subspaces of commuting operators*, Michigan Math. J. **35** (1988), 43-65.

[Mi70] V.D. Milman, *Operators of class C_0 and C_0^**, Funkcii Funkcional. Anal. i Prilozen. **10** (1970), 15-26. [in Russian]

[Mu02] V. Müller, *On the Taylor functional calculus*, Studia Math. **159** (2002), 79-97.

[Og54] T. Ogasawara, *Finite-dimensionality of certain Banach algebras*, J. Sci. Hiroshima Univ. Ser. A **17** (1954), 359-364.

[Pi80] A. Pietsch, *Operator ideals*, (North-Holland, 1980).

[Pi87] A. Pietsch, *Eigenvalues and s-numbers*, (Geest and Portig K-G; Cambridge University Press, 1987).

[P03] G. Pisier, *Introduction to Operator Space Theory*, London Mathematical Society Lecture Notes vol. **294** (Cambridge University Press, 2003).

[Ra92] G. Racher, *On the tensor product of weakly compact operators*, Math. Ann. **294** (1992), 267-275.

[Ro56] M. Rosenblum, *On the operator equation $BX - XA = Q$*, Duke Math. J. **23** (1956), 263-269.

[Ru84] W. Ruess, *Duality and geometry of spaces of compact operators*, in *Functional Analysis: Surveys and recent results III* (North-Holland, 1984) pp. 59-79.

[Run00] V. Runde, *The flip is often discontinuous*, J. Operator Theory **48** (2002), 447-451.

[S95] E. Saksman, *Weak compactness and weak essential spectra of elementary operators*, Indiana Univ. Math. J. **44** (1995), 165-188.

[ST92] E. Saksman and H.-O. Tylli, *Weak compactness of multiplication operators on spaces of bounded linear operators*, Math. Scand. **70** (1992), 91-111.

[ST94] E. Saksman and H.-O. Tylli, *Weak essential spectra of multiplication operators on spaces of bounded linear operators*, Math. Ann. **299** (1994), 299-309.

[ST98] E. Saksman and H.-O. Tylli, *Rigidity of commutators and elementary operators on Calkin algebras*, Israel J. Math. **108** (1998), 217-236.

[ST99] E. Saksman and H.-O. Tylli, *The Apostol-Fialkow formula for elementary operators on Banach spaces*, J. Funct. Anal. **161** (1999), 1-26.

[ShT05] V. Shulman and L. Turowska, *Operator synthesis II: Individual synthesis and linear operator equations*, J. reine angew. Math. (to appear).

[SZ96] L.L. Stacho and B. Zalar, *On the norm of Jordan elementary operators in standard operator algebras*, Publ. Math. Debrecen **49** (1996), 127-134.

[St70] J. Stampfli, *The norm of a derivation*, Pacific J. Math. **33** (1970), 737-747.

[St73] J. Stampfli, *Derivations on $B(H)$: The range*, Illinois J. Math. **17** (1973), 518-524.

[Sy1884] J. Sylvester, C.R. Acad. Sci Paris **99** (1884), pp. 67-71, 115-118, 409-412, 432-436, 527-529.

[Ta70] J.L. Taylor, *A joint spectrum for several operators*, J. Funct. Anal. **6** (1970), 172-191.

[Ti01] R. M. Timoney, *Norms of clementary operators*, Irish Math. Soc. Bulletin **46** (2001), 13-17.

[Ti03a] R. M. Timoney, *Norms and CB norms of Jordan elementary operators*, Bull. Sci. Math. **127** (2003), 597-609.

[Ti03b] R. M. Timoney, *Computing the norm of elementary operators*, Illinois J. Math. **47** (2003), 1207-1226.

[Ti05] R. M. Timoney, *Some formulae for norms of elementary operators*, J. Operator Theory (to appear).

[T94] H.-O. Tylli, *Lifting non-topological divisors of zero modulo the compact operators*, J. Funct. Anal. **125** (1994), 389-415.

[Va64] K. Vala, *On compact sets of compact operators*, Ann. Acad. Sci. Fenn. A I Math. **351** (1964).

[We77] L. Weis, *On perturbations of Fredholm operators in $L_p(\nu)$ spaces*, Proc. Amer. Math. Soc. **67** (1977), 287-292.

[Wo91] P. Wojtaszczyk, *Banach spaces for analysts*, Cambridge studies in advanced mathematics vol. **25** (Cambridge University Press, 1991).

[Y75] K. Ylinen, *Weakly completely continuous elements of C^*-algebras*, Proc. Amer. Math. Soc. **52** (1975), 323-326.

DEPARTMENT OF MATHEMATICS AND STATISTICS, P.O. BOX 35 (MAD), FIN-40014 UNIVERSITY OF JYVÄSKYLÄ, FINLAND

E-mail address: saksman@maths.jyu.fi

DEPARTMENT OF MATHEMATICS AND STATISTICS, P.O. BOX 68 (GUSTAF HÄLLSTRÖMIN KATU 2B), FIN-0014 UNIVERSITY OF HELSINKI, FINLAND

E-mail address: hojtylli@cc.helsinki.fi

INTERPOLATION OF BANACH SPACES BY THE γ-METHOD

JESÚS SUÁREZ AND LUTZ WEIS

ABSTRACT. In this note we study some basic properties of the γ-interpolation method introduced in [6] e.g. interpolation of Bochner spaces and interpolation of analytic operator valued functions in the sense of Stein. As applications we consider the interpolation of almost summing operators (in particular γ-radonifying and Hilbert-Schmidt operators) and of γ-Sobolev spaces as introduced in [6]. We also compare this new interpolation method with the real interpolation method.

1. INTRODUCTION

In [6] a new method of interpolation was introduced, which in contrast to the classical complex and real interpolation method is based on almost summing sequences and gaussian averages. This becomes necessary in order to characterize the boundedness of the holomorphic functional calculus of sectorial operators in terms of the interpolation of the domains of its fractional powers (see [6, 7]). This is not possible using the real or complex interpolation method. Some information on this new method can be found in [6, 7]. In particular, it has equivalent formulations modelled after the complex, the real and discrete interpolation methods. Furthermore, in a Hilbert space, it agrees with the complex method.

In this note we provide some basic information concerning the γ-interpolation method. We show that interpolating Bochner spaces we obtain the 'right' result (see theorem 3.1) and that Stein's interpolation schema for analytic families of operators is still applicable (see theorem 4.2). With these tools we interpolate spaces $\gamma(H_i, X_i)$ of almost summing operators (or radonifying operators), with respect to the complex and the γ-method, where H_i are Hilbert spaces and X_i are B-convex Banach spaces. As a corollary we obtain an improvement of a result of Cobos and García-Davía [3] on the interpolation of Hilbert-Schmidt operators (see corollary 5.1) and show that the γ-Sobolev spaces introduced in [6] form an interpolation scale. Finally we compare the γ-method with the real interpolation method: the possible inclusions between

The work of the first author was supported in part by a Marie Curie grant HPMT-GH-01-00286-04 at Karlsruhe University under the direction of Prof. L. Weis and in part during a visit to the IMUB at Barcelona University.

Keywords: Interpolation of spaces of vector valued functions, γ-radonifying operators.
AMS-classification: 46B70, 47B07, 46E40.

293

interpolation spaces depend on the geometry of the Banach spaces, namely their type and cotype properties.

2. DEFINITIONS AND BASIC PROPERTIES

We will always work with an infinite dimensional separable Hilbert space and will reserve the letter H for this. In what follows, $\mathfrak{F}(Y, X)$ will denote the space of all finite rank operators from Y into X. Given (e_j) an orthonormal basis in H, we define the γ-norm of a finite rank operator $u \in \mathfrak{F}(H, X)$ given by $u = \sum_{j=1}^n x_j \otimes e_j$ and acting as $u(h) = \sum_{j=1}^n (h, e_j) x_j$, in the form

$$\|u\|_\gamma := \mathbb{E}\|\sum_{j=1}^n g_j x_j\|_X = \left(\int_\Omega \|\sum_{j=1}^n g_j(\omega) x_j\|_X^2 dP(\omega) \right)^{\frac{1}{2}},$$

where $\{g_n\}$ is a sequence of standard Gaussian variables on a fixed probability space (Ω, σ, P).

To extend this notation to operators of possibly infinite rank we proceed as follows. If $v : H \to X$ is any bounded operator, we define

$$\|v\|_\gamma := \sup \{\|vu\|_\gamma : u \in \mathfrak{F}(H, H), \|u\| \leq 1\}$$

and then form the spaces

$$\gamma_+(H, X) := \{v \in \mathcal{L}(H, X) : \|v\|_\gamma < \infty\}$$

and

$$\gamma(H, X) := \overline{\mathfrak{F}(H, X)}^\gamma$$

We call $\gamma(H, X)$ the space of radonifying operators. The following result can be found in [6]:

Proposition 2.1. Let be X a Banach space, then $\gamma(H, X) = \gamma_+(H, X)$ if and only if X contains no copy of c_0.

Now we sketch the basic ideas of the γ-method as developed in [6]. Assume that $X = (X_0, X_1)$ is a compatible pair of Banach spaces.

First, we introduce admissible classes $\mathcal{A}$ and $\mathcal{A}_+$ of operators. Let $u : L_2(dt) + L_2(e^{-2t} dt) \to X_0 + X_1$ be a bounded operator, we will say $u \in \mathcal{A}$ iff $u \in \gamma(L_2(e^{-2jt} dt), X_j)$ for $j = 0, 1$, and set

$$\|u\|_{\mathcal{A}} := \max_{j=0,1} \|u_j\|_{\gamma(L_2(e^{-2jt} dt), X_j)}.$$

We will say $u \in \mathcal{A}_+$ iff $u \in \gamma_+(L_2(e^{-2jt} dt), X_j)$ for $j = 0, 1$, and set

$$\|u\|_{\mathcal{A}_+} := \max_{j=0,1} \|u_j\|_{\gamma_+(L_2(e^{-2jt} dt), X_j)},$$

where in all cases u_j denotes the operator u on $L_2(e^{-2jt} dt)$ for $j = 0, 1$. These spaces are complete under their corresponding norm.

As in [6] we define the interpolation space $(X_0, X_1)_\theta^\gamma$, which we will briefly call X_θ^γ, as follows :
if $x \in X_0 + X_1$ we introduce the norm

$$\|x\|_\theta^\gamma = \inf \left\{ \|u\|_{\mathcal{A}} : u \in \mathcal{A}, u(e_\theta) = x \right\},$$

where e_θ denotes the function $e_\theta(t) = e^{\theta t}$. Now $(X_0, X_1)_\theta^\gamma$ is the space of all $x \in X_0 + X_1$ such that $\|x\|_\theta^\gamma < \infty$ equipped with $\|\cdot\|_\theta^\gamma$.
The same definitions work by replacing γ by γ_+ and by interchanging the roles of $\mathcal{A}$ and $\mathcal{A}_+$, respectively. The resulting interpolation space we denote by $X_\theta^{\gamma_+}$.

Proposition 2.2. *Suppose that (X_0, X_1) and (Y_0, Y_1) are compatible pairs of Banach spaces. Suppose that $S : X_0 + X_1 \to Y_0 + Y_1$ is a bounded operator such that $S(X_0) \subset Y_0$ and $S(X_1) \subset Y_1$. Then $S : (X_0, X_1)_\theta^\gamma \to (Y_0, Y_1)_\theta^\gamma$ is a bounded operator with*

$$\|S\|_{X_\theta^\gamma \to Y_\theta^\gamma} \leq \|S\|_{X_0}^{1-\theta} \|S\|_{X_1}^\theta.$$

The same holds true replacing γ by γ_+. See [6] for a detailed proof.

It is important for us that the interpolation we have described has an alternative formulation as a complex method. Denote by $\mathcal{S}$ the strip $\{z : 0 < \Re(z) < 1\}$ and consider the space $\mathbb{A}(X_0, X_1)$ of all analytic functions $\mathbf{F} : \mathcal{S} \to X_0 + X_1$ which are of the form $\mathbf{F}(z) = u(e^{zt})$ where $u \in \mathcal{A}$. We define a norm on this space by $\|\mathbf{F}\|_{\mathbb{A}} = \|u\|_{\mathcal{A}}$. Then we have the formula

$$\|x\|_\theta^\gamma = \inf \left\{ \|\mathbf{F}\|_{\mathbb{A}} : \mathbf{F}(\theta) = x \right\}.$$

Most of the proofs involved in the paper use density arguments, so it is interesting to observe that:

Lemma 2.1. *The set of finite-rank operators is dense in $\mathcal{A}$.*

As a consequence we get:

Corollary 2.1. *For $\theta \in (0, 1)$, $X_0 \cap X_1$ is dense in X_θ^γ.*

See a proof of this facts in [6].
Closer to the "complex spirit" is the following lemma:

Lemma 2.2. *Suppose $\mathbf{F} : \mathcal{S} \to X_0 + X_1$ is a bounded analytic function such that the boundary values $\mathbf{F}(j + it) = \lim_{\xi \to j} \mathbf{F}(\xi + it)$ exists t-a.e. and are in X_j for $j = 0, 1$. Suppose the functions $\mathbf{F}_j(it) := \mathbf{F}(j + it)$ are X_j strongly measurable and $\mathbf{F}_j \in \gamma_j(X_j) := \gamma(L_2(e^{-2jt}dt), X_j)$ for $j = 0, 1$. Then $\mathbf{F} \in \mathbb{A}$ and*

$$\|\mathbf{F}\|_{\mathbb{A}} = (2\pi)^{\frac{1}{2}} \max_{j=0,1} \|\mathbf{F}_j\|_{\gamma_j}.$$

By $\mathbf{F}_j \in \gamma_j(X_j)$ we mean that the operator $u_j : L_2(e^{-2jt}dt) \to X_j$ defined by

$$u_j(h) := \int_{\mathbb{R}} h(t)\mathbf{F}_j(t)dt$$

for $h \in L_2(e^{-2jt}dt)$ belongs to $\gamma_j(X_j)$ and $\|\mathbf{F}_j\|_{\gamma_j} = \|u_j\|_{\gamma_j}$. See [6] for a detailed proof.

Lemma 2.1 guarantees that the space of functions

$$\mathbb{A}_0(X_0, X_1) := \{\mathbf{F} \in \mathbb{A}(X_0, X_1) : \mathbf{F}(z) = u(e^{zt}), u \in \mathcal{A}, rank(u) < \infty\}$$

is dense in $\mathbb{A}(X_0, X_1)$. This is also true for

$$\mathbb{A}_{00}(X_0, X_1) := \{\mathbf{F} \in \mathbb{A}(X_0, X_1) : Range(\mathbf{F}) \subseteq X_0 \cap X_1\}.$$

3. Interpolation of L_p spaces

The interpolation of L_p spaces by the γ-method follows basically the arguments of the proof for the complex method. Two technical lemmas are needed.

Lemma 3.1. *For $1 \le p < \infty$, $\gamma(H, L_p(\Omega, A))$ and $L_p(\Omega, \gamma(H, A))$ are isomorphic Banach spaces.*

Proof. A family of finite dimensional maps $T(\omega) = \sum_{j=1}^{n} f_j(\omega) \otimes e_n$ with $f_j \in L_p(\Omega, A)$ can be thought of as an element of both spaces, $L_p(\Omega, \gamma(H, A))$ and $\gamma(H, L_p(\Omega, A))$. With Kahane's inequalities [4] we obtain:

$$
\begin{aligned}
\|T\|^p_{L_p(\Omega, \gamma(H,A))} &= \int_\Omega \|T\omega\|^p_{\gamma(H,A)} d\mu(\omega) \\
&\sim \int_\Omega \left(\int_S \left\| \sum T\omega(e_n) g_n(s) \right\|^p_A dP(s) \right) d\mu(\omega) \\
&= \int_S \left(\int_\Omega \left\| \sum T\omega(e_n) g_n(s) \right\|^p_A d\mu(\omega) \right) dP(s) \\
&= \int_S \left\| \sum T\omega(e_n) g_n(s) \right\|^p_{L_p(\Omega, A)} dP(s) \\
&\sim \|T\|^p_{\gamma(H, L_p(\Omega, A))},
\end{aligned}
$$

where $\sim$ stands for "equivalent norms". Since such elements T are dense in both spaces, the claim follows. $\square$

The second lemma is essentially Hadamard's three line lemma.

Lemma 3.2. *For $\mathbf{F} \in \mathbb{A}_0(X_0, X_1)$, we have*

$$\|\mathbf{F}(\theta)\|^\gamma_\theta \le (2\pi)^{\frac{1}{2}} \|\mathbf{F}_0\|^{1-\theta}_{\gamma_0} \|\mathbf{F}_1\|^\theta_{\gamma_1}.$$

Proof. Note that the analytic function

$$\Phi(z) = (\log \|\mathbf{F}_0\|_{\gamma_0})(1-z) + (\log \|\mathbf{F}_1\|_{\gamma_1})z$$

has real part $(1 - \theta) \log \|\mathbf{F}_0\|_{\gamma_0} + \theta log \|\mathbf{F}_1\|_{\gamma_1}$ for $z = \theta + it$. Let $\theta \in (0,1)$, then by lemma 2.2 $e^{-\Phi(z)}\mathbf{F}(z)$ is a function in $\mathbb{A}(X_0, X_1)$ and

$$
\begin{aligned}
(2\pi)^{-\frac{1}{2}} \|e^{-\Phi(\theta)}\mathbf{F}(\theta)\|^\gamma_\theta &\le \max_{j=0,1} \left[\|e^{-\log \|\mathbf{F}_j\|_{\gamma_j}} \mathbf{F}_j\|_{\gamma_j} \right] \\
&\le \max_{j=0,1} \left[e^{-\log \|\mathbf{F}_j\|_{\gamma_j}} \|\mathbf{F}_j\|_{\gamma_j} \right] \le 1.
\end{aligned}
$$

With all this we have

$$\log \|(2\pi)^{-\frac{1}{2}}\mathbf{F}(\theta)\|_\theta^\gamma \le \Re\Phi(\theta) \le (1-\theta)\log\|\mathbf{F}_0\|_{\gamma_0} + \theta\log\|\mathbf{F}_1\|_{\gamma_1}.$$

This gives the desired inequality. $\qquad\square$

Corollary 3.1. *For* $x \in X_0 \cap X_1$,

$$\|x\|_\theta^\gamma = (2\pi)^{\frac{1}{2}}\inf\{\|\mathbf{F}_0\|_{\gamma_0}^{1-\theta}\|\mathbf{F}_1\|_{\gamma_1}^\theta : \mathbf{F}(\theta) = x\}.$$

Theorem 3.1. *Assume that* A_0 *and* A_1 *are Banach spaces and that* $1 \le p_0 < \infty$, $1 \le p_1 < \infty$, $\theta \in (0,1)$, *if* $\frac{1}{p} = \frac{1-\theta}{p_0} + \frac{\theta}{p_1}$ *then*

$$(L_{p_0}(\Omega, A_0), L_{p_1}(\Omega, A_1))_\theta^\gamma = L_p(\Omega, (A_0, A_1)_\theta^\gamma)$$

with equivalent norms.

Proof. If we denote S the space of simple functions with values in $A_0 \cap A_1$, then S is dense in $L_{p_0}(\Omega, A_0) \cap L_{p_1}(\Omega, A_1)$ and thus in $(L_{p_0}(\Omega, A_0), L_{p_1}(\Omega, A_1))_\theta^\gamma$ and then also in $L_p(\Omega, (A_0, A_1)_\theta^\gamma)$ simply because $A_0 \cap A_1$ is dense in $(A_0, A_1)_\theta^\gamma$. So it is enough to consider functions a in S.

Let us see that $\|a\|_{(L_{p_0}(\Omega, A_0), L_{p_1}(\Omega, A_1))_\theta^\gamma} \le (2\pi)^{\frac{1}{2}}K\|a\|_{L_p(\Omega, A_\theta^\gamma)}$ where the constant K comes from Kahane's inequality [4]. Given a simple function, $a \in S$, and given $\varepsilon > 0$ there are functions $g(\cdot, \omega) \in \mathbb{A}_0(A_0, A_1)$ for $\omega \in \Omega$ (which are also steps functions with respect to ω) such that $\|g(\cdot, \omega)\|_{\mathbb{A}_0} \le (1 + \varepsilon)\|a(\omega)\|_{A_\theta^\gamma}$ with $x \in \Omega$ and $g(\theta, \omega) = a(\omega)$.

We set

$$\mathbf{F}(z, \omega) := g(z, \omega)\left(\frac{\|a(\omega)\|_{A_\theta^\gamma}}{\|a\|_{L_p(\Omega, A_\theta^\gamma)}}\right)^{p(\frac{1}{p_0} - \frac{1}{p_1})(\theta - z)}.$$

Using Lemma 3.1, we have that

$$
\begin{aligned}
\|\mathbf{F}(i\cdot, \cdot)\|_{\gamma(L_{p_0}(\Omega, A_0))}^{p_0} &\sim \int_\Omega \|\mathbf{F}(i\cdot, \omega)\|_{\gamma(A_0)}^{p_0} d\mu(\omega) \\
&\le \int_\Omega \|g(i\cdot, \omega)\|_{\gamma(A_0)}^{p_0} A(\omega) d\mu(\omega) \\
&\le (1 + \varepsilon)^{p_0}\|a\|_{L_p(\Omega, A_\theta^\gamma)}^{p_0},
\end{aligned}
$$

where $A(\omega) = \|a(\omega)\|_{A_\theta^\gamma}^{p_0 p(\frac{1}{p_0} - \frac{1}{p_1})\theta}\|a\|_{L_p(\Omega, A_\theta^\gamma)}^{-p_0 p(\frac{1}{p_0} - \frac{1}{p_1})\theta}$.

The same idea works to show that

$$\|\mathbf{F}(1 + i\cdot, \cdot)\|_{\gamma(L_{p_1}(\Omega, A_1))} \le K(1 + \varepsilon)\|a\|_{L_p(\Omega, A_\theta^\gamma)};$$

thus, it follows that:

$$\|a\|_{(L_{p_0}(\Omega, A_0), L_{p_1}(\Omega, A_1))_\theta^\gamma} \le (2\pi)^{\frac{1}{2}}K\|a\|_{L_p(\Omega, A_\theta^\gamma)}.$$

Conversely, since $a \in S$ there are functions $\mathbf{F}(\cdot, \omega) \in \mathbb{A}_0(A_0, A_1)$ with $\mathbf{F}(\theta, \omega) = a(\omega)$ for $\omega \in \Omega$ (which are step functions with respect to ω) and

$$\|a\|_{L_p(\Omega, A_\theta^\gamma)} = \left(\int_\Omega \|a(\omega)\|_{A_\theta^\gamma}^p d\mu(\omega) \right)^{\frac{1}{p}} \leq \left(\int_\Omega \|\mathbf{F}(\theta, \omega)\|_{A_\theta^\gamma}^p d\mu(\omega) \right)^{\frac{1}{p}}$$

Using Lemma 3.2 and Hölder's inequality with $\frac{1-\theta}{pp_0} + \frac{\theta}{pp_1} = 1$, we obtain:

$$\leq (2\pi)^{\frac{1}{2}} \left(\int_\Omega \|\mathbf{F}(\cdot, \omega)\|_{\gamma(A_0)}^{(1-\theta)p} \|\mathbf{F}(\cdot, \omega)\|_{\gamma(A_1)}^{\theta p} d\mu(\omega) \right)^{\frac{1}{p}}$$

$$\leq (2\pi)^{\frac{1}{2}} \left(\int_\Omega \|\mathbf{F}(\cdot, \omega)\|_{\gamma(A_0)}^{p_0} d\mu(\omega) \right)^{\frac{1-\theta}{p_0}} \left(\int_\Omega \|\mathbf{F}(\cdot, \omega)\|_{\gamma(A_1)}^{p_1} d\mu(\omega) \right)^{\frac{\theta}{p_1}}$$

$$\leq (2\pi)^{\frac{1}{2}} K \|\mathbf{F}\|_{\gamma(L_{p_0}(\Omega, A_0))}^{1-\theta} \|\mathbf{F}\|_{\gamma(L_{p_1}(\Omega, A_1))}^{\theta}$$

$$\leq (2\pi)^{\frac{1}{2}} K \|\mathbf{F}\|_{\mathbb{A}(L_{p_0}(\Omega, A_0) L_{p_1}(\Omega, A_1))},$$

where we applied Lemma 3.2 to the operators in $\gamma(L_2(e^{-2jt}dt), L_{p_j}(\Omega, A_j))$ defined by $\mathbf{F}$. This finishes the proof. $\qquad\square$

Following [4], we call by $Rad(X)$ the completion of the space of finite sequences, $(x_n) \subseteq X$ under the norm

$$\|(x_n)\| := \mathbb{E}\| \sum_{n=1}^\infty r_n x_n \|_X = \left(\int_0^1 \| \sum_{n=1}^\infty r_n(t) x_n \|_X^2 dt \right)^{\frac{1}{2}}.$$

As usually $r_n(\cdot)$ denotes the nth-Rademacher function.

Corollary 3.2. *If X_i is B-convex for $i = 0, 1$, then*

$$(Rad(X_0), Rad(X_1))_\theta^\gamma = Rad(X_\theta^\gamma)$$

with equivalent norms.

Proof. B-convexity implies that the canonical projections $\mathbf{P} : L_2([0, 1], X_i) \longrightarrow Rad(X_i)$ are bounded for $i = 0, 1$, by a result of G. Pisier (see [4]) for dyadic simple functions $f : [0, 1] \to X_0 \cap X_1$ the projection $\mathbf{P}$ is defined by

$$\mathbf{P}f := \sum_n \left(\int_0^1 r_n(u) f(u) du \right) r_n.$$

By [13, Th. 1.2.4], $\mathbf{P}$ defines a projection of $L = (L_2([0, 1], X_0), L_2([0, 1], X_1))_\theta^\gamma$ onto the closed span R of the finite Rademacher sequences in L. Furthermore, R is isomorph to $(Rad(X_0), Rad(X_1))_\theta^\gamma$. By theorem 3.1 the norms of L and $L_2([0, 1], X_\theta^\gamma)$ are equivalent on R and therefore R is isomorphic to $Rad(X_\theta^\gamma)$. $\qquad\square$

Corollary 3.3. *If X_i has type $p_i > 1$ and cotype q_i for $i = 0, 1$ and $\frac{1}{p} = \frac{1-\theta}{p_0} + \frac{\theta}{p_1}, \frac{1}{q} = \frac{1-\theta}{q_0} + \frac{\theta}{q_1}$ then $(X_0, X_1)_\theta^\gamma$ has type p and cotype q.*

Proof. That X_i has type p_i means that the natural mapping $l_{p_i}(X_i) \longrightarrow Rad(X_i)$ is bounded, so we have that the following operator is bounded too:

$$l_p(X_\theta^\gamma) = (l_{p_0}(X_0), l_{p_1}(X_1))_\theta^\gamma \longrightarrow (Rad(X_0), Rad(X_1))_\theta^\gamma = Rad(X_\theta^\gamma)$$

A similar argument works for the cotype. $\qquad\square$

Theorem 3.1 admits the following generalization:

Theorem 3.2. *Let A_0, A_1 be Banach spaces and let $1 \le p_0 < \infty$, $1 \le p_1 < \infty$, $\theta \in (0, 1)$, if $\frac{1}{p} = \frac{1-\theta}{p_0} + \frac{\theta}{p_1}$ and $\mu = \mu_0^{\frac{(1-\theta)p}{p_0}} \mu_1^{\frac{\theta p}{p_1}}$ then*

$$(L_{p_0}(\Omega(\mu_0), A_0), L_{p_1}(\Omega(\mu_1), A_1))_\theta^\gamma = L_p(\Omega(\mu), (A_0, A_1)_\theta^\gamma)$$

with equivalent norms.

Proof. For a given $\mathbf{F} \in \mathbb{A}(L_{p_0}(\Omega(\mu_0), A_0), L_{p_1}(\Omega(\mu_1), A_1))$ we set

$$\widetilde{\mathbf{F}}(z, \omega) = \mu_0(\omega)^{\frac{(1-z)p}{p_0}} \mu_1(\omega)^{\frac{zp}{p_1}} \mathbf{F}(z, \omega).$$

The mapping $\mathbf{F} \to \widetilde{\mathbf{F}}$ is clearly an isomorphism between $\mathbb{A}(L_{p_0}(\Omega(\mu_0), A_0), L_{p_1}(\Omega(\mu_1), A_1))$ and $\mathbb{A}(L_{p_0}(\Omega(\mu), A_0), L_{p_1}(\Omega(\mu), A_1))$. Now the arguments in the preceding theorem with obvious modifications go through. $\qquad\square$

4. ABSTRACT STEIN INTERPOLATION

In this section (X_0, X_1) and (Y_0, Y_1) are interpolation couples. Let us recall the following theorem that can be found in [14]:

Theorem 4.1. *Let $\{T_z : z \in \overline{S}\}$ be a family of linear mappings $T_z : X_0 \cap X_1 \to Y_0 + Y_1$ with the following properties:*

(1) *$\forall x \in X_0 \cap X_1$ the function $T_{(\cdot)}x : \overline{S} \to Y_0 + Y_1$ is continuous, bounded, and analytic on S.*

(2) *For $j = 0, 1$, $x \in X_0 \cap X_1$, the function $\mathbb{R} \ni s \to T_{j+is}x \in Y_j$ is continuous and*

$$M_j := sup\{\|T_{j+is}x\|_{Y_j} : s \in \mathbb{R}, x \in X_0 \cap X_1, \|x\|_{X_j} \le 1\} < \infty.$$

Then for all $t \in [0, 1]$, $T_t(X_0 \cap X_1) \subset Y_{[t]}$ and

$$\|T_t x\|_{[t]} \le M_0^{1-t} M_1^t \|x\|_{[t]}$$

for all $x \in X_0 \cap X_1$.

In order to formulate the analogous Stein interpolation result for the γ-method, we need the notion of γ-boundedness. A set of operators $\tau \subseteq B(X, Y)$

is γ-bounded, if there is a constant C such that for all $T_1, ..., T_n \in \tau$ and $x_1, ..., x_n \in X$ we have:

$$\mathbb{E}\| \sum_{k=1}^{n} g_k T_k x_k \| \leq C \mathbb{E}\| \sum_{k=1}^{n} g_k x_k \|.$$

The smallest constant C possible in this inequality is denoted by $R(\tau)$.

Theorem 4.2. *Let $\{T_z : z \in \overline{S}\}$ be a family of linear mappings $T_z : X_0 \cap X_1 \to Y_0 + Y_1$ with the following properties:*

(1) $\forall x \in X_0 \cap X_1$, *the function* $T_{(\cdot)}x : \overline{S} \to Y_0 + Y_1$ *is continuous, bounded and analytic on S.*

(2) *For $j = 0, 1$, $x \in X_0 \cap X_1$, the function $s \mapsto T_{j+is}x \in Y_j$ is strongly measurable, continuous and the operators $T_{j+is} \in B(X_j, Y_j)$ thereby defined on Y_j satisfy for $j = 0, 1$:*

$$M_j = R\left(\{T_{j+it} : t \in \mathbb{R}\}\right) < \infty.$$

Then, for every $\theta \in (0, 1)$, we have:

(1) $T_\theta (X_0 \cap X_1) \subseteq Y_\theta^\gamma.$

(2) $\|T_\theta x\|_{Y_\theta^\gamma} \leq M_0^{1-\theta} M_1^\theta \|x\|_{X_\theta^\gamma}$ *for all $x \in X_0 \cap X_1$.*

Proof. It is enough to consider $M_0 = M_1 = 1$, otherwise replacing T by $M_0^{z-1} M_1^{-z} T(z)$. We define $T_\theta : \mathbb{A}_{00}(X_0, X_1) \to \mathbb{A}(Y_0, Y_1)$ by $\mathbf{F} \mapsto T_{(\cdot)}\mathbf{F}(\cdot)$. Indeed , by the γ-boundedness of $\{T_{j+it} : t \in \mathbb{R}\}$ and an estimate in [6] we have

$$\|T_{j+i\cdot}\mathbf{F}(j + i\cdot)\|_{\mathbb{A}(Y_0, Y_1)} \leq \|\mathbf{F}(j + i\cdot)\|_{\mathbb{A}_{00}(X_0, X_1)}$$

for $j = 0, 1$. The set $\{\mathbf{F} \in \mathbb{A}_{00}(X_0, X_1) : \mathbf{F}(\theta) = 0\}$ is dense in $\{\mathbf{F} \in \mathbb{A}(X_0, X_1) : \mathbf{F}(\theta) = 0\}$ (compare [14, Th. 1.1]) and $T_{(\cdot)}$ maps the first set into $\{\mathbf{F} \in \mathbb{A}(Y_0, Y_1) : \mathbf{F}(\theta) = 0\}$. Therefore for $x \in X_0 \cap X_1$ and $\mathbf{F} \in \mathbb{A}_{00}(X_0, X_1)$ with $\mathbf{F}(\theta) = x$ we have:

$$\|T(\theta)x\|_{Y_\theta^\gamma} \leq \|T_{(\cdot)}\mathbf{F}(\cdot)\|_{\mathbb{A}(Y_0, Y_1)} \leq \|\mathbf{F}\|_{\mathbb{A}_{00}(X_0, X_1)}$$

and taking infimum over such all $\mathbf{F}$ gives the result. $\square$

Remark: Let us state a special case where the γ-boundedness assumption of theorem 4.2 can be checked directly.

If the functions $s \to T_{j+is} \in B(X_j, Y_j)$ are differentiable and satisfy

$$\begin{aligned} \widetilde{M_j} &= \|T_j\| \\ &+ e^j \int_{\mathbb{R}} \|T_{j+is}\|_{B(X_j, Y_j)} \sqrt{j + 4s^2} e^{-s^2} ds \\ &+ e^j \int_{\mathbb{R}} \|\frac{d}{ds} T_{j+is}\|_{B(X_j, Y_j)} e^{-s^2} ds < \infty, \end{aligned}$$

for $j = 0, 1$ then, for $x \in X_0 \cap X_1$ we still can conclude:

$$\|T_\theta x\|_{Y_\theta^\gamma} \leq \widetilde{M_0}^{1-\theta} \widetilde{M_1}^\theta \|x\|_{X_\theta^\gamma}.$$

Proof. Consider the analytic function $S_z = e^{-\theta^2} e^{z^2} T_z$ with $S_\theta = T_\theta$. Our assumption implies that

$$\|S_j\| + \int_{\mathbb{R}} \|\frac{d}{ds} T_{j+is}\|_{B(X_j, Y_j)} ds \leq \widetilde{M}_j.$$

By [8, Sect. 3] we conclude that $R(\{S_{j+it} : t \in \mathbb{R}\}) \leq \widetilde{M}_j$ and we may apply the theorem. $\square$

5. INTERPOLATION OF $\gamma(H, X)$

We are ready to obtain our stability results for the class of γ-radonifying operators.

Theorem 5.1. *Let H_0, H_1 be separable Hilbert spaces with $H_0 \cap H_1$ dense in H_i for $i = 0, 1$. If X_0, X_1 are B-convex, then*

$$(\gamma(H_0, X_0), \gamma(H_1, X_1))_{[\theta]} = \gamma((H_0, H_1)_{[\theta]}, (X_0, X_1)_{[\theta]})$$

with equivalent norms.

Proof. Step 1: Representation of $(H_0, H_1)_{[\theta]}$.

Let (e_n) be an orthonormal basis of $H_0 \cap H_1$ with respect to the inner product $(h, g) = (h, g)_0 + (h, g)_1$ where $(\cdot, \cdot)_j$ is the inner product of H_j so that (e_n) is also a complete orthogonal system in H_j for $j = 0, 1$. To construct such (e_n) inductively, assume that $e_1, \ldots e_N$ are already chosen. Then pick some h_0 in the kernel of the map $h \in H_0 \cap H_1 \to \{(h, e_n)_j\}_{n=1,\ldots N; j=0,1} \in \mathbb{C}^{2N}$ and normalize h_0 with respect to $(\cdot, \cdot)$ to obtain e_{N+1}.
Then clearly

$$H_j = \left\{ \sum_n \beta_n e_n : \sum_n |\beta_n|^2 \alpha_{j,n}^2 < \infty \right\},$$

where $\alpha_{j,n} = \|e_n\|_{H_j}$ and we get

$$(H_0, H_1)_{[\theta]} = \left\{ \sum_n \beta_n e_n : \sum_n |\beta_n|^2 \alpha_{0,n}^{2(1-\theta)} \alpha_{1,n}^{2\theta} < \infty \right\}$$

with orthonormal basis $(\alpha_{0,n}^{(1-\theta)} \alpha_{1,n}^\theta)^{-1} e_n$, $n \in \mathbb{N}$. To see this, identify $H_0 \cap H_1$ with l_2 via the maps $\beta_n \to \sum \beta_n e_n$ and use e.g. [1, Sect. 5.6].

Step 2: Operators of dimension m.

By H_0^m we denote the subspace of H_0^m spanned by the $e_1, \ldots, e_m$. Let $Rad^m(X_j)$ be the span of the $\sum_{n=1}^m r_n x_n$ with $x_1, \ldots, x_m \in X_j$. Note that $\gamma(H_0, X_0) \cap \gamma(H_1, X_1)$ and $\gamma(H_0, X_0) + \gamma(H_1, X_1)$ can be seen as continuously embedded subspaces of $\gamma(H_0 \cap H_1, X_0 + X_1)$ (due to the operator ideal property of γ) and $Rad(X_0) \cap Rad(X_1)$ and $Rad(X_0) + Rad(X_1)$ are naturally embedded into $L_2([0,1], X_0 + X_1)$. For a fixed $m \in \mathbb{N}$ and all $z \in \overline{S}$ we define maps

$$T_z^m : \gamma(H_0^m, X_0) \cap \gamma(H_1^m, X_1) \longrightarrow Rad^m(X_0) + Rad^m(X_1)$$

by $T_z^m(S) := \left(\alpha_{0,n}^{-1+z}\alpha_{1,n}^{-z}S(e_n)\right)_{n=1}^m$. For $z = j + it$ with $j \in [0,1]$ and $t \in \mathbb{R}$ and all $S \in \gamma(H_j^m, X_j)$ we obtain by the contraction principle

$$\|T_{j+it}^m(S)\|_{Rad^m(X_j)} \quad \leq \mathbb{E}\|\sum_{n=1}^m r_n\alpha_{0,n}^{-1+j+it}\alpha_{1,n}^{-j+it}S(e_n)\|_{X_j}$$

$$\leq 2\mathbb{E}\|\sum_{n=1}^m r_n\alpha_{0,n}^{-1+j}\alpha_{1,n}^{-j}S(e_n)\|_{X_j} = 2\|S\|_{\gamma(H_j^m, X_j)}.$$

The remaining assumptions of theorem 4.2 follow directly from the special form of T_z^m. Hence for $\theta \in (0,1)$ there is a map

$$T_\theta^m : (\gamma(H_0^m, X_0), \gamma(H_1^m, X_1))_{[\theta]} \to (Rad^m(X_0), Rad^m(X_1))_{[\theta]}$$

with $\|T_\theta^m\| \leq 2$. By interpolating in a similar way the 'inverse' functions

$$(T_z^m)^{-1} : Rad^m(X_0) \cap Rad^m(X_1) \to \gamma(H_0^m, X_0) + \gamma(H_1^m, X_1)$$

which assign to $(x_1, ..., x_m)$ the operator $S = (T_z^m)^{-1}(x_n)$ given by $S(e_n) = \alpha_{0,n}^{1-z}\alpha_{1,n}^z x_n$ with $n = 1, ..., m$ it follows that T_θ^m is an isomorphism with $\|(T_\theta^m)^{-1}\| \leq 2$ for all $\theta \in (0,1)$.

Step 3: Infinite dimensional operators.

Since X_0 and X_1 are B-convex Banach spaces, we have

$$(Rad(X_0), Rad(X_1))_{[\theta]} = Rad(X_{[\theta]})$$

with equivalent norms (see the argument in the proof of Corollary 3.2). Notice that we have consistent retractions

$$P_j : Rad(X_j) \to Rad^m(X_j)$$

given by $(x_n)_{n\in\mathbb{N}} \to (x_n)_{n=1}^m$ with $\|P_j\| = 1$ and hence by [13, Sect. 1.2.4] there is a constant M which depends on X but not on m or θ so that for $x_1, ..., x_m \in X_0 \cap X_1$

$$M^{-1}\|(x_n)\|_{Rad^m(X_{[\theta]})} \leq \|(x_n)\|_{(Rad^m(X_0), Rad^m(X_1))_{[\theta]}} \leq M\|(x_n)\|_{Rad^m(X_{[\theta]})}.$$

Next we observe that the consistent retractions $Q_j : \gamma(H_j, X_j) \to \gamma(H_j^m, X_j)$ given by $S \to S_{|span[e_1,...,e_m]}$ have norm 1 and, again by [13, Sect. 1.2.4], ensure that the inclusion $(\gamma(H_0^m, X_0), \gamma(H_1^m, X_1))_{[\theta]} \subseteq (\gamma(H_0, X_0), \gamma(H_1, X_1))_{[\theta]}$ becomes an isometric embedding.
For $x_1, ..., x_m \in X_0 \cap X_1$ and the associated operator $S = (T_\theta^m)^{-1}(x_n)$ given by $S(e_n) = \alpha_{0,n}^{1-\theta}\alpha_{1,n}^\theta x_n$ with $x_k = 0$ for $k = m+1, m+2, ...$ we obtain

$$\|S\|_{(\gamma(H_0, X_0), \gamma(H_1, X_1))_{[\theta]}} \quad = \|S\|_{(\gamma(H_0^m, X_0), \gamma(H_1^m, X_1))_{[\theta]}}$$

$$\sim^M \|(x_n)\|_{Rad^m(X_{[\theta]})} = \mathbb{E}\|\sum_{n=1}^m r_n x_n\|_{X_{[\theta]}}$$

$$= \mathbb{E}\|\sum_{n=1}^m r_n\alpha_{0,n}^{\theta-1}\alpha_{1,n}^{-\theta}S(e_n)\|_{X_{[\theta]}} = \|S\|_{\gamma(H_{[\theta]}, X_{[\theta]})}.$$

Hence the norms of $\gamma(H_{[\theta]}, X_{[\theta]})$ and $(\gamma(H_0, X_0), \gamma(H_1, X_1))_{[\theta]}$ are equivalent on all finite dimensional operators $H_0 \cap H_1 \to X_0 \cap X_1$. Since these operators are dense in both spaces the result follows. $\qquad\square$

For the γ-method this results holds under an additional assumption:

Definition 1. *A Banach space has Pisier's property (α) e.g. [8] if for all $\alpha_{i,j} \in \mathbb{C}$ we have*

$$\mathbb{E}'\mathbb{E}\| \sum_{i,j} \alpha_{ij} r_i' r_j x_{ij}\| \leq \sup_{i,j} |\alpha_{ij}| \mathbb{E}'\mathbb{E}\| \sum_{i,j} r_i' r_j x_{ij}\|$$

where (r_i') and (r_j) are two independent Bernoulli sequences.

Theorem 5.2. *Let H_0, H_1 be separable Hilbert spaces with $H_0 \cap H_1$ dense in H_i for $i = 0, 1$. If X_0 and X_1 are B-convex and have property (α) then*

$$(\gamma(H_0, X_0), \gamma(H_1, X_1))_\theta^\gamma = \gamma((H_0, H_1)_\theta^\gamma, (X_0, X_1)_\theta^\gamma)$$

with equivalent norms.

Proof. First we recall that $(H_0, H_1)_\theta^\gamma = (H_0, H_1)_{[\theta]}$ with equivalent norms for Hilbert spaces H_j, $j = 0, 1$. Using theorem 4.2 in place of theorem 4.1 we can now repeat the proof of theorem 5.1. Property (α) comes in since we have to check the γ-boundedness of the sets $\{T_{j+it}^m : t \in \mathbb{R}\}$ in $B(\gamma(H_j^m, X_j), Rad^m(X_j))$ with a constant independent of m. Indeed for $S_1, ..., S_N \in \gamma(H_j^m, X_j)$ and $t_1, ..., t_N \in \mathbb{R}$ we have

$$
\begin{aligned}
\mathbb{E}\| \sum_{n=1}^N r_n T_{j+it_n}(S_n)\|_{Rad^m(X_j)} &= \mathbb{E}\mathbb{E}'\| \sum_{l=1}^m \sum_{n=1}^N r_l' r_n \alpha_{0,n}^{-1+j+it_n} \alpha_{1,n}^{-j-it_n} S_n(e_l)\| \\
&\leq C\mathbb{E}\mathbb{E}'\| \sum_{l=1}^m \sum_{n=1}^N r_l' r_n \alpha_{0,n}^{-1+j} \alpha_{1,n}^{-j} S_n(e_l)\| \\
&\leq C\mathbb{E}\| \sum_{n=1}^N r_n S_n\|_{\gamma(H_j^m, X_j)},
\end{aligned}
$$

where the constant C only depends on the property (α) constant of X [8, Section 4]. $\qquad\square$

Corollary 5.1. *Let H_0, H_1 be separable Hilbert spaces with $H_0 \cap H_1$ dense in H_i for $i = 0, 1$. If X_0, X_1 are Hilbert spaces and S_2 denotes the Hilbert-Schmidt class, then:*

$$(S_2(H_0, X_0), S_2(H_1, X_1))_{[\theta]} = S_2((H_0, H_1)_{[\theta]}, (X_0, X_1)_{[\theta]}),$$

$$(S_2(H_0, X_0), S_2(H_1, X_1))_\theta^\gamma = S_2((H_0, H_1)_\theta^\gamma, (X_0, X_1)_\theta^\gamma).$$

This corollary improves a result of [3, Prop. 5.1.] which states:

Let $H_1 \subset H_0$, $X_1 \subset X_0$ be Hilbert spaces with H_1 dense in H_0, X_1 dense in X_0 and the inclusions being compact, then

$$(\mathcal{S}_2(H_0, X_0), \mathcal{S}_2(H_1, X_1))_{[\theta]} = \mathcal{S}_2\left((H_0, H_1)_{[\theta]}, (X_0, X_1)_{[\theta]}\right).$$

with equal norms.

We recall that we obtain isometries in Corollary 5.1 since all the constants appearing in the proofs of theorems 5.1 and 5.2 derive from the B-convexity constants and Kahane's contraction principle and will be 1 in the Hilbert case. For a more direct proof of Corollary 5.1 see [12].

In [7] "Sobolev spaces" of the following kind were introduced:

$$\gamma^s(\mathbb{R}^n, X) := \gamma(H^s, X)$$

where H^s is the completion of $\mathcal{S}(\mathbb{R}^n)$, the Schwartz class, in the norm

$$\|f\|_s = \|\mathcal{F}^{-1}[(1 + |\cdot|^2)^{\frac{s}{2}} \widehat{f}(\cdot)]\|_{L_2(\mathbb{R}^n)}.$$

Theorems 5.1 and 5.2 allow us to extend the well known interpolation results for H^s to $\gamma^s(\mathbb{R}^n, X)$:

Corollary 5.2. Let X_0, X_1 be B-convex Banach spaces. If $s = (1-\theta)s_0 + \theta s_1$ for $s_j \in \mathbb{R}$, $j = 0, 1$, then

$$(\gamma^{s_0}(\mathbb{R}^n, X_0), \gamma^{s_1}(\mathbb{R}^n, X_1))_{[\theta]} = \gamma^s(\mathbb{R}^n, X_{[\theta]})$$

with equivalent norms. The same is true for γ-interpolation method if X_0 and X_1 have in addition property (α).

The same is of course true if we replace the H^s in the definition of $\gamma^s(\mathbb{R}^n, X)$ by another scale of Sobolev spaces, e.g. the scale defined by Riesz potentials instead of Bessel potentials.

6. Comparison with the Real Interpolation Method

If (X_0, X_1) is an interpolation couple so that X_0 and X_1 have finite cotype, then the interpolation space X_θ^γ has also a discrete description (cf. [6, 7]): X_θ^γ consists of all $x \in X_0 + X_1$ which can be represented as a sum $x = \sum_{n \in \mathbb{Z}} x_n$ convergent in $X_0 + X_1$ with $x_n \in X_0 \cap X_1$ satisfying

$$\|x\|_\theta = \inf \left\{ \|(x_n)\| : x = \sum_{n \in \mathbb{Z}} x_n, \ x_n \in X_0 \cap X_1 \right\} < \infty$$

where

$$\|(x_n)\| = \max_{j=0,1} \mathbb{E} \| \sum_{n \in \mathbb{Z}} r_n 2^{n(j-\theta)} x_n \|_{X_j} \qquad (*)$$

defines an equivalent norm on X_θ^γ.
If we replace in $(*)$ the Rademacher averages with:

$$\max_{j=0,1} \left(\sum_{n \in \mathbb{Z}} \|2^{n(j-\theta)} x_n\|_{X_j}^{p_j} \right)^{\frac{1}{p_j}}$$

we obtain a well known characterization of the Lions-Peetre interpolation space $(X_0, X_1)_{\theta, p_0, p_1}$, [9, Chapitre 2]. By using the well known equality $(X_0, X_1)_{\theta, p_0, p_1} = (X_0, X_1)_{\theta, p}$ with $\frac{1}{p} = \frac{1-\theta}{p_0} + \frac{\theta}{p_1}$, see [11], and the definition of type and cotype we obtain immediately the following observation:

Theorem 6.1. *Let (X_0, X_1) be an interpolation couple so that X_j has Rademacher type $p_j \geq 1$ and cotype $q_j < \infty$ for $j = 0, 1$, then for $\frac{1}{p} = \frac{1-\theta}{p_0} + \frac{\theta}{p_1}$, $\frac{1}{q} = \frac{1-\theta}{q_0} + \frac{\theta}{q_1}$ with $\theta \in (0, 1)$ we have continuous embeddings*

$$(X_0, X_1)_{\theta, p} \subset (X_0, X_1)_\theta^\gamma \subset (X_0, X_1)_{\theta, q}.$$

A corresponding result for the complex method making the stronger assumption of Fourier type is due to Peetre [10].

Acknowledgement: The first author want to express his sincere gratitude to his colleagues in Karlsruhe, specially to B. Haak and J. Zimmerschied for endless hours of mathematics and help. To L. Weis for the opportunity to stay in Karlsruhe and particularly for sharing his ideas with me. To all of them thanks for such a nice atmosphere where to work. To M.J. Carro for some fruitful comments about the text. Finally, special thanks to my advisor J.M.F. Castillo for helping me with some mistakes and encourage me to finish it.

References

[1] J. BERGH and J. LÖFSTRÖM, *Interpolation spaces. An introduction*, Grundlehren der mathematischen Wissenschaften **223**, Springer-Verlag.

[2] A.P. CALDERÓN, Intermediate spaces and interpolation, the complex method, *Studia Math.* **24** (1964), 113-190.

[3] F. COBOS and M.A. GARCÍA-DAVÍA, Remarks on Interpolation properties of Schatten classes, *Bull. London Math. Soc.* **26** (1994), 465-471.

[4] J. DIESTEL, H. JARCHOW and A. TONGE, *Absolutely Summing Operators*, Cambridge Studies in advanced mathematics **43**, Cambridge Univ. Press. (1995).

[5] N.J. KALTON and L. WEIS, The H^∞-calculus and sums of closed operators. Math. Ann. **321**, No.2, 319-345 (2001).

[6] N.J. KALTON and L. WEIS, Euclidean structures and their application to spectral theory, in preparation.

[7] P. KUNSTMANN, N. KALTON and L. WEIS, Perturbation and interpolation theorems for the H^∞-calculus with applications to differential operators, to appear in Math. Annalen 2006.

[8] P. KUNSTMANN and L. WEIS, Maximal L_p-regularity for parabolic equations, Fourier multiplier theorems and H^∞-calculus, in Functional Analytic Methods for Evolution Equations (eds. M. Iannelli, R. Nagel and S. Piazzera), Springer Lecture Notes **1855**, 65-311 (2004).

[9] J.L. LIONS, and J. PEETRE, Sur une classe d'espaces d'interpolation, Publ. Math., Inst. Hautes Étud. Sci. **19**, 5-68 (1964).

[10] J. PEETRE, Sur la transformation de Fourier des fonctions à valeurs vectorielles. Rend. Sem. Mat. Univ. Padova **42**, 15-26 (1969).

[11] J. PEETRE, Sur le nombre de paramètres dans la définition de certains espaces d'interpolation, Ric. Mat. **12**, 248-261 (1963).

[12] J. SUÁREZ, A note on interpolation of Hilbert-Schmidt operators, in preparation.

[13] H. TRIEBEL *Interpolation theory, function spaces, differential operators.* North-Holland Mathematical Library, **18** (1978).

[14] J. VOIGT, Abstract Stein interpolation. Math. Nachr. **157** (1992), 197–199.

DEPARTAMENTO DE MATEMÁTICAS, UNIVERSIDAD DE EXTREMADURA, AVENIDA DE ELVAS S/N, 06071 BADAJOZ, SPAIN.
MATHEMATISCHES INSTITUT I, UNIVERSITÄT KARLSRUHE, 76128 KARLSRUHE, GERMANY.

JESUS.SUAREZ@MATH.UNI-KARLSRUHE.DE
LUTZ.WEIS@MATH.UNI-KARLSRUHE.DE

SOLVABILITY OF AN INTEGRAL EQUATION IN BC($\mathbb{R}_+$)

J. CABALLERO, B. LÓPEZ AND K. SADARANGANI

ABSTRACT. The aim of this paper is to prove the existence of solutions of an integral equation in the space of continuous and bounded functions on $\mathbb{R}_+$. The main tool used in our considerations is the technique associated with measures of noncompactness.

1. INTRODUCTION

Integral equations comprise a very important and significant part of mathematical analysis and their applications to real world problems ([1, 3, 10, 12, 13], among others). The theory of integral equations is now well developed with the help of several tools of functional analysis, topology and fixed-point theory. In this paper we study the existence of solutions of an integral equation in the space of continuous and bounded functions on $\mathbb{R}_+$. The main tool used in our study is associated with the technique of measures of noncompactness which has already been successfully applied in the solvability of some integral equations [8, 9, 10].

2. NOTATION AND AUXILIARY FACTS

Assume E is a real Banach space with norm $\| \cdot \|$ and zero element 0. Denote by $B(x, r)$ the closed ball centered at x and with radius r and by B_r the ball $B(0, r)$. If X is a nonempty subset of E we denote by $\overline{X}$, $ConvX$ the closure and the closed convex closure of X, respectively. The symbols λX and $X + Y$ denote the usual algebraic operations on sets. Finally, let us denote by $\mathfrak{M}_E$ the family of nonempty bounded subsets of E and by $\mathfrak{N}_E$ its subfamily consisting of all relatively compact sets.

Throughout this paper, we will also adopt the following definition of measure of noncompactness [7].

Definition 1. . A function $\mu : \mathfrak{M}_E \longrightarrow \mathbb{R}_+ = [0, \infty)$ is said to be a *measure of noncompactness* in the space E if it satisfies the following conditions:

(1) The family $ker\mu = \{X \in \mathfrak{M}_E : \mu(X) = 0\}$ is nonempty and $ker\mu \subset \mathfrak{N}_E$.
(2) $X \subset Y \Rightarrow \mu(X) \leq \mu(Y)$.
(3) $\mu(\overline{X}) = \mu(ConvX) = \mu(X)$.
(4) $\mu(\lambda X + (1 - \lambda)Y) \leq \lambda\mu(X) + (1 - \lambda)\mu(Y)$ for $\lambda \in [0, 1]$.

1991 *Mathematics Subject Classification.* (2000) Primary: 45M99. Secondary: 47H09.

(5) If $\{X_n\}_n$ is a sequence of nonempty, bounded and closed subsets of E such that $X_n \supset X_{n+1}$ ($n = 1, 2, ...$) and $\lim_{n \to \infty} \mu(X_n) = 0$ then the set $X_\infty = \bigcap_{n=1}^\infty X_n$ is nonempty.

Now let us assume that Ω is a nonempty subset of a Banach space E and $T : \Omega \to \Omega$ is a continuous operator transforming bounded subsets of Ω to bounded ones. We say that T satisfies the Darbo condition with constant $k \geq 0$ with respect to a measure of noncompactness μ if

$$\mu(TX) \leq k\mu(X)$$

for each $X \in \mathfrak{M}_E$ such that $X \subset \Omega$.

If $k < 1$ then T is called a contraction with respect to μ.

In what follows we will need the following fixed point theorem which is a version of the classical fixed point theorem for strict contractions in the context of measures of noncompactness [7].

Theorem 1. *Let Ω be a nonempty bounded closed and convex subset of E, μ a measure of noncompactness in E and $T : \Omega \longrightarrow \Omega$ a contraction with respect to μ. Then T has at least one fixed point in Ω.*

For further facts concerning measures of noncompactness and fixed point theory we refer to [7, 4, 11, 14, 15].

In the sequel, we will work in the space $BC(\mathbb{R}_+)$ consisting of all real functions which are bounded and continuous on $\mathbb{R}_+$. The space $BC(\mathbb{R}_+)$ is equipped with the standard norm $\|x\| = \sup\{|x(t)| : t \geq 0\}$.

In our considerations we will use a measure of noncompactness which appears in [7] and is defined by

$$\mu(X) = w_0(X) + \lim_{t \to \infty} \sup diam X(t)$$

where

$$diam X(t) = sup\{|x(t) - y(t)| : x, y \in X\},$$
$$w(x, \varepsilon) = sup \{| x(t) - x(s) | : t, s \in \mathbb{R}_+, | t - s | \leq \varepsilon\},$$
$$w(X, \varepsilon) = sup \{w(x, \varepsilon) : x \in X\}$$
$$w_0(X) = \lim_{\varepsilon \to 0^+} w(X, \varepsilon).$$

3. EXISTENCE THEOREM

In this section we will study the solvability of the following integral equation

(1) $$x(t) = h(t) + (Tx)(t) \int_0^t f(\phi(t, s)) \max_{[0,s]} |x(\tau)| ds$$

for $t \in \mathbb{R}_+$. Equations of such kind have been studied in other papers ([2, 5, 6]).

We will assume that the functions involved in equation (1) satisfy the following conditions.

(i) $h \in BC(\mathbb{R}_+)$ and is uniformly continuous.

(ii) The operator $T : BC(\mathbb{R}_+) \to BC(\mathbb{R}_+)$ is continuous and satisfies the Darbo condition with respect to the measure of noncompactness μ (defined in section 2) and with constant Q.

(iii) The function $\phi : \mathbb{R}_+ \times \mathbb{R}_+ \to \mathbb{R}$ is continuous.

(iv) The function $f : Im\phi \to \mathbb{R}$ is continuous.

(v) There exist nonnegative constants c and d such that

$$\|Tx\| \leq c + d\|x\|,$$

for each $x \in BC(\mathbb{R}_+)$.

(vi) There exist continuous functions $a, b : \mathbb{R}_+ \to \mathbb{R}_+$ such that
- $\lim\limits_{t \to \infty} a(t) = 0$,
- b is a bounded function and $b \in L^1(\mathbb{R}_+)$, and
- $|f(\phi(t, s))| \leq a(t) \cdot b(s)$.

(vii) There exists a function $\varphi : \mathbb{R}_+ \to \mathbb{R}_+$, $\varphi \in L^1(\mathbb{R}_+)$ such that

$$|f(\phi(t, s)) - f(\phi(t', s))| \leq \varphi(s) \cdot |t - t'|,$$

for $t, t', s \in \mathbb{R}_+$.

(viii) There exists $r_0 > 0$ such that $\|h\| + (c + d \cdot r_0) \cdot \|a\| \cdot \|b\|_1 \cdot r_0 \leq r_0$ and $(\|a\| \cdot \|b\|_1 \cdot r_0) \cdot Q < 1$.

Before we formulate our main result we present some remarks and preliminary results.

Remark 1. *Note that assumption (vi) implies that the function a is bounded and by $\|a\|$ we denote the supremum of the function a on $\mathbb{R}_+$.*

Remark 2. *As $a(t)$ and $b(s)$ are bounded, by assumption (vi) we get*

$$|f(\phi(t, s))| \leq \|a\| \cdot \|b\|.$$

Now we will prove the following lemmas which will be needed further on.

Lemma 1. *Suppose that $x \in BC(\mathbb{R}_+)$ and $\varepsilon > 0$. Then*

$$w(x, \varepsilon) = \sup_{L > 0} w^L(x, \varepsilon)$$

where $w^L(x, \varepsilon) = \sup\{|x(t) - x(s)| : t, s \in [0, L], |t - s| \leq \varepsilon\}$.

The proof is rather straighforward.

Lemma 2. *Suppose that $x \in BC(\mathbb{R}_+)$ and we define*

$$(Gx)(t) = \max_{[0,t]} |x(\tau)| \quad \text{for} \quad t \in \mathbb{R}_+.$$

Then $Gx \in BC(\mathbb{R}_+)$.

Proof. Without loss of generality we can assume that $x \geq 0$ and let L be a positive number. Now, we will prove that for $\varepsilon > 0$

$$w^L(Gx, \varepsilon) \leq w^L(x, \varepsilon).$$

Suppose the contrary. This means that there exist $t_1, t_2 \in [0, L]$, $t_1 \leq t_2$, $t_2 - t_1 \leq \varepsilon$ such that

$$(2) \qquad w(x, \varepsilon) < |(Gx)(t_2) - (Gx)(t_1)|,$$

and, as Gx is an increasing function, we have

$$(3) \qquad 0 < (Gx)(t_2) - (Gx)(t_1).$$

By the continuity of x we can find $0 \leq \tau_2 \leq t_2$ with the property $(Gx)(t_2) = x(\tau_2)$. Taking into account the inequality (3) and that $t_1 \leq \tau_2$ (if $\tau_2 < t_1$, $(Gx)(t_1) = (Gx)(t_2)$ and this contradicts (3)) we get

$$(Gx)(t_2) - (Gx)(t_1) = x(\tau_2) - (Gx)(t_1) \leq x(\tau_2) - x(t_1)$$

and as, $\tau_2 - t_1 \leq t_2 - t_1 \leq \varepsilon$

$$(Gx)(t_2) - (Gx)(t_1) \leq x(\tau_2) - x(t_1) \leq w^L(x, \varepsilon),$$

which is a contradiction.

Thus, for $\varepsilon > 0$

$$(4) \qquad w^L(Gx, \varepsilon) \leq w^L(x, \varepsilon).$$

As x is continuous on $\mathbb{R}_+$, in particular x is a continuous function on $[0, L]$ for every $L > 0$. From (4), we can deduce that Gx is a continuous function on $[0, L]$, for each $L > 0$. Finally, as L is arbitrary, we obtain that Gx is continuous on $\mathbb{R}_+$.

On the other hand, it is obvious that if x is bounded on $\mathbb{R}_+$, then Gx is also bounded. This completes the proof.

$\square$

Lemma 3. *Let $(x_n), x \in BC(\mathbb{R}_+)$. Suppose that $x_n \to x$ in $BC(\mathbb{R}_+)$. Then $Gx_n \to Gx$ in $BC(\mathbb{R}_+)$.*

Proof. Note that for $t \in \mathbb{R}_+$ and $y \in BC(\mathbb{R}_+)$

$$(Gy)(t) = \|y_{|[0,t]}\|$$

where $y_{|[0,t]}$ denotes the restriction of the function y on the interval $[0, t]$ and the norm is considered in the space $C([0, t])$. In view of this fact we can deduce

$$
\begin{aligned}
\|Gx_n - Gx\| &= \sup_{t \in \mathbb{R}_+} |(Gx_n)(t) - (Gx)(t)| = \\
&= \sup_{t \in \mathbb{R}_+} \left| \|x_{n|[0,t]}\| - \|x_{|[0,t]}\| \right| \leq \\
&\leq \sup_{t \in \mathbb{R}_+} \|(x_n - x)_{|[0,t]}\| \leq \|x_n - x\|.
\end{aligned}
$$

As $x_n \to x$ in $BC(\mathbb{R}_+)$ we obtain the desired result. $\square$

Now, we present our existence result

Theorem 2. *Under assumptions* $(i) - (viii)$, *equation (1) has at least one solution in* $BC(\mathbb{R}_+)$.

Proof. Let us define the operator A on the space $BC(\mathbb{R}_+)$ by

$$(Ax)(t) = h(t) + (Tx)(t) \int_0^t f(\phi(t,s)) \max_{[0,s]} |x(\tau)| ds.$$

Now we will prove that if $x \in BC(\mathbb{R}_+)$ then $Ax \in BC(\mathbb{R}_+)$.
To do this, it is sufficient to prove that if $x \in BC(\mathbb{R}_+)$ then $Bx \in BC(\mathbb{R}_+)$, where

$$(Bx)(t) = \int_0^t f(\phi(t,s)) \max_{[0,s]} |x(\tau)| ds.$$

In fact, let $t_0 \in \mathbb{R}_+$ and $\varepsilon > 0$. Then, there exist two possibilities:

   a) $t_0 \neq 0$.
   b) $t_0 = 0$.

   a) Let us assume $\|x\| \neq 0$ (in contrary case $(Bx)(t) \equiv 0$).

   Let $\delta < \min \left(\dfrac{\varepsilon}{2\|x\|\|\varphi\|_1}, \dfrac{\varepsilon}{2\|a\|\|b\|} \right)$ and $|t - t_0| < \delta$ (we can consider $t > t_0$), then

$$|(Bx)(t) - (Bx)(t_0)| =$$

$$= \left| \int_0^t f(\phi(t,s)) \max_{[0,s]} |x(\tau)| ds - \int_0^{t_0} f(\phi(t_0,s)) \max_{[0,s]} |x(\tau)| ds \right| \le$$

$$\le \left| \int_0^t f(\phi(t,s)) \max_{[0,s]} |x(\tau)| ds - \int_0^t f(\phi(t_0,s)) \max_{[0,s]} |x(\tau)| ds \right| +$$

$$+ \left| \int_0^t f(\phi(t_0,s)) \max_{[0,s]} |x(\tau)| ds - \int_0^{t_0} f(\phi(t_0,s)) \max_{[0,s]} |x(\tau)| ds \right| \le$$

$$\le \int_0^t |f(\phi(t,s)) - f(\phi(t_0,s))| \max_{[0,s]} |x(\tau)| ds +$$

$$+ \int_{t_0}^t |f(\phi(t_0,s))| \max_{[0,s]} |x(\tau)| ds \le$$

$$\le \|x\| \int_0^t |t - t_0| \varphi(s) ds + \|a\|\|b\|\|x\|(t - t_0) \le$$

$$\le \|x\|(t - t_0)\|\varphi\|_1 + \|a\|\|b\|\|x\|(t - t_0) \le$$

$$\le \|x\|\|\varphi\|_1 \delta + \|a\|\|b\|\|x\|\delta < \varepsilon.$$

   b) Let $t_0 = 0$, then $(Bx)(0) = 0$ and let consider $\delta < \dfrac{\varepsilon}{\|a\|\|b\|\|x\|}$. Then, if $t < \delta$

$$|(Bx)(t) - (Bx)(t_0)| = |(Bx)(t)| \le \int_0^t |f(\phi(t,s))| \max_{[0,s]} |x(\tau)| ds \le$$

$$\le \|a\|\|b\|\|x\|t < \varepsilon.$$

In the sequel we show that Ax is a bounded function for $x \in BC(\mathbb{R}_+)$. In fact, for $t \in \mathbb{R}_+$ and taking into account our assumptions we can get

$$|(Ax)(t)| \leq \|h\| + (c+d\|x\|) \cdot \|a\| \cdot \|x\| \int_0^t b(s)ds \leq \|h\| + (c+d\|x\|) \cdot \|a\| \cdot \|x\| \cdot \|b\|_1.$$

Using assumptions (i) and (vi), we get that Ax is bounded for $x \in BC(\mathbb{R}_+)$. Moreover, by $(viii)$ we obtain that A transforms the ball B_{r_0} into itself.

Now, we prove that the operator A is continuous on B_{r_0}. Let $\{x_n\}$ be a sequence in B_{r_0}, such that $x_n \to x$, we will prove that $Ax_n \to Ax$. For each $t \in I$ we have

$$|(Ax_n)(t) - (Ax)(t)| \leq$$

$$\leq \left| (Tx_n)(t) \int_0^t f(\phi(t,s)) \max_{[0,s]} |x_n(\tau)| ds - \right.$$

$$\left. -(Tx)(t) \int_0^t f(\phi(t,s)) \max_{[0,s]} |x_n(\tau)| ds \right| +$$

$$+ |(Tx)(t)| \int_0^t |f(\phi(t,s))| \left| \max_{[0,s]} |x_n(\tau)| - \max_{[0,s]} |x(\tau)| \right| ds \leq$$

$$\leq \|Tx_n - Tx\| \cdot \|a\| \cdot r_0 \cdot \|b\|_1 + (c + d \cdot r_0) \cdot \|a\| \cdot \|b\|_1 \cdot \|x_n - x\|$$

As T is continuous, $\|Tx_n - Tx\| \to 0$ as $n \to \infty$, so $\|Ax_n - Ax\| \to 0$ as $n \to \infty$.

In the sequel, we prove that the operator A satisfies the Darbo condition with respect to the measure of noncompactness μ introduced in section 2.

Let X be a nonempty subset of B_{r_0}. Fix $\varepsilon > 0$, $x \in X$, $L > 0$ and $t_1, t_2 \in [0, L]$ such that $t_2 - t_1 \leq \varepsilon$ and $t_1 < t_2$. Then we obtain

$$|(Ax)(t_2) - (Ax)(t_1)| \leq$$

$$\leq |h(t_2) - h(t_1)| + |(Tx)(t_2) - (Tx)(t_1)| \int_0^{t_2} a(t_2) \cdot b(s) \cdot r_0 ds +$$

$$+ |(Tx)(t_1)| \int_0^{t_2} |f(\phi(t_2, s)) - f(\phi(t_1, s))| \cdot r_0 ds +$$

$$+ |(Tx)(t_1)| \int_{t_1}^{t_2} |f(\phi(t_1, s))| \cdot r_0 ds \leq w^L(h, \varepsilon) + w^L(Tx, \varepsilon) \cdot \|a\| \|b\|_1 \cdot r_0 +$$

$$+ (c + d \cdot r_0) \|\varphi\|_1 (t_2 - t_1) + (c + d \cdot r_0) \cdot \|a\| \cdot \|b\| \cdot r_0 \cdot (t_2 - t_1) \leq$$

$$\leq w^L(h, \varepsilon) + (\|a\| \cdot \|b\|_1 \cdot r_0) w^L(Tx, \varepsilon) + \varepsilon \cdot (c + dr_0) \cdot \|\varphi\|_1 +$$

$$\varepsilon \|a\| \cdot \|b\| \cdot r_0 \cdot (c + dr_0).$$

Thus we have,

$$w^L(Ax, \varepsilon) \leq w^L(h, \varepsilon) + (\|a\| \cdot \|b\|_1 \cdot r_0) w^L(Tx, \varepsilon) + \varepsilon \cdot (c + dr_0) \cdot \|\varphi\|_1 +$$

$$+ \varepsilon \|a\| \cdot \|b\| \cdot r_0 \cdot (c + dr_0)$$

and taking the supremum over $L > 0$, we obtain by Lemma 1

$$w(Ax, \varepsilon) \leq w(h, \varepsilon) + (\|a\| \cdot \|b\|_1 \cdot r_0) w(Tx, \varepsilon) + \varepsilon \cdot [(c + dr_0) \cdot \|\varphi\|_1 +$$

$$+ \|a\| \cdot \|b\| \cdot r_0 \cdot (c + dr_0)]$$

Consequently, applying supremum in $x \in X$ and applying limit when $\varepsilon \to 0$

$$w_0(AX) \leq (\|a\| \cdot \|b\|_1 \cdot r_0)w_0(TX)$$

and by assumption (ii), as the operator T satisfies the Darbo condition with constant Q, then

$$(5) \qquad\qquad w_0(AX) \leq (\|a\| \cdot \|b\|_1 \cdot r_0) \cdot Q \cdot w_0(X).$$

In the sequel, we study the term related to the diameter which appears in the expression of the measure μ.

Let us take a nonempty subset X of B_{r_0}, $x, y \in X$ and $t \in \mathbb{R}_+$. Then

$$|(Ax)(t) - (Ay)(t)| =$$

$$= \left| (Tx)(t) \int_0^t f(\phi(t,s)) \max_{[0,s]} |x(\tau)| ds - (Ty)(t) \int_0^t f(\phi(t,s)) \max_{[0,s]} |y(\tau)| ds \right| \leq$$

$$\leq \left| (Tx)(t) \int_0^t f(\phi(t,s)) \max_{[0,s]} |x(\tau)| ds - (Ty)(t) \int_0^t f(\phi(t,s)) \max_{[0,s]} |x(\tau)| ds \right| +$$

$$+ \left| (Ty)(t) \int_0^t f(\phi(t,s)) \max_{[0,s]} |x(\tau)| ds - (Ty)(t) \int_0^t f(\phi(t,s)) \max_{[0,s]} |y(\tau)| ds \right| \leq$$

$$\leq |(Tx)(t) - (Ty)(t)| \cdot \int_0^t |f(\phi(t,s))| \max_{[0,s]} |x(\tau)| ds +$$

$$+ |(Ty)(t)| \cdot \int_0^t |f(\phi(t,s))| \cdot \left| \max_{[0,s]} |x(\tau)| - \max_{[0,s]} |y(\tau)| \right| ds \leq$$

$$\leq |(Tx)(t) - (Ty)(t)| \cdot r_0 \cdot a(t) \cdot \|b\|_1 + (c + dr_0) \cdot a(t) \cdot \|b\|_1 \cdot \|x - y\| \leq$$

$$\leq |(Tx)(t) - (Ty)(t)| \cdot r_0 \cdot a(t) \cdot \|b\|_1 + (c + dr_0) \cdot a(t) \cdot \|b\|_1 \cdot 2 \cdot r_0.$$

Applying supremum in X we obtain

$$diam(AX)(t) \leq diam(TX)(t) \cdot r_0 \cdot a(t) \|b\|_1 + (c + dr_0)a(t) \|b\|_1 2r_0.$$

Taking the upper limit when $t \to \infty$, and taking into account that $a(t) \to 0$ as $t \to \infty$, we obtain

$$(6) \qquad\qquad \limsup_{t \to \infty} diam(AX)(t) = 0.$$

Now, combining (5) and (6) we obtain

$$\mu(AX) \leq (\|a\| \cdot \|b\|_1 \cdot r_0) \cdot Q \cdot \mu(X).$$

Finally, by hypothesis (viii) and applying Theorem 1 we complete the proof. $\qquad \square$

Remark 3. *If we replace the assumptions (i), (ii) and (viii) of the previous Theorem with:*

(i') $h \in BC(\mathbb{R}_+)$

(ii') *The operator $T : BC(\mathbb{R}_+) \longrightarrow BC(\mathbb{R}_+)$ is continuous and satisfies that for each ball B_r there exists a constant k_r such that*

$$\|Tx - Ty\| \leq k_r \|x - y\|, \forall x, y \in B_r \subset BC(\mathbb{R}_+)$$

(viii') *There exists $r_0 > 0$ such that $\|h\| + (c + d \cdot r_0) \cdot \|a\| \cdot \|b\|_1 \cdot r_0 \leq r_0$ and $\|a\| \|b\|_1 (c + (d + k_{r_0})r_0) < 1$*

we can prove, in the same way that the proof of Theorem 2, that the operator A is a contraction in B_{r_0}. Consequently, we guarantee the uniqueness of the solution of the equation (1) in this ball.

4. SOME REMARKS

In this section, we are going to give some examples which show the relevance of the hypotheses of theorem 2.

Example 1. Let $h(t) = 1$. This function verifies assumption (i). Take $(Tx)(t) = \frac{1}{2}$. It is continuous and satisfies the Darbo condition with $Q = 0$, also this operator satisfies (v) with $c = \frac{1}{2}$ and $d = 0$. Let $\phi(t, s) = s$ which is a continuous function. Finally take $f(u) = \frac{1}{u}$. This function is not continuous at $u = 0$, so it doesn't verify assumption (iv). Nor does it satisfy assumption (vi), because $b(s) = \frac{1}{s}$ is not a bounded function. If we take $\varphi(s) \equiv 0$, assumption (vii) is satisfied.

In this case, our equation is:

$$x(t) = 1 + \frac{1}{2} \int_0^t \frac{1}{s} \max_{[0,s]} |x(\tau)| ds.$$

We have that $x(0) = 1$ and

$$x(t) \geq 1 + \frac{1}{2}|x(0)| \int_0^t \frac{1}{s} ds = 1 + \frac{1}{2}|x(0)|(\ln t - \ln k),$$

we deduce that $x(t) \to \infty$ as $k \to 0$. So $x(t) \notin BC(\mathbb{R}_+)$.

Example 2. Consider $h(t) = 1$ which satisfies assumption (i). Let $(Tx)(t) = \frac{1}{2}$. It is continuous and satisfies the Darbo condition with $Q = 0$, also this operator satisfies (v) with $c = \frac{1}{2}$ and $d = 0$. Take $\phi(t, s) = t + s$ and $f(u) = u$, they do not satisfy assumptions (vi) and (vii).

In this case, our equation is:

$$x(t) = 1 + \frac{1}{2} \int_0^t (t + s) \max_{[0,s]} |x(\tau)| ds.$$

We deduce that $x(0) = 1$ and $x(t) \geq 1 + \frac{1}{2}|x(0)| \int_0^t (t + s) ds = 1 + \frac{1}{2}|x(0)| \frac{3t^2}{2}$

so $x(t)$ is not a bounded solution.

In what follows we present an example where existence can be established by using theorem 2.

Example 3. Consider the integral equation

$$(7) \qquad x(t) = \frac{1}{2} \cos t + \frac{1}{2} \int_0^t \frac{e^{-s} \cdot \cos s}{1 + t} \max_{[0,s]} |x(\tau)| ds.$$

Observe that in this case, we can consider the function ϕ defined by

$$\phi(t,s) = \frac{e^{-s} \cdot \cos s}{1+t}$$

and $f(y) = y$ for each $y \in Im\phi$.

It is easy to prove that $f \circ \phi$ satisfies assumptions (iii) and (iv) with constant $M = 1$. Moreover, $(f \circ \phi)(t,s) = \frac{1}{t+1} \cdot (e^{-s} \cdot \cos s)$, so $f \circ \phi$ satisfies assumption (vi) with $a(t) = \frac{1}{t+1}$ and $b(s) = (e^{-s} \cdot \cos s)$. Obviously, $f \circ \phi$ satisfies assumption (vii) with $\varphi(s) = b(s)$.

The function h is given by the expression

$$h(t) = \frac{1}{2} \cos t$$

and satisfies assumption (i).

Observe that operator T is given by $(Tx)(t) = \frac{1}{2}$ which satisfies (ii) and (v) with $c = \frac{1}{2}$, $d = 0$ and $Q = \frac{1}{2}$.

Finally, inequalities of $(viii)$ are satisfied with $r_0 = 1$.

Theorem 2 tells us that our equation (7) has a solution in $BC(\mathbb{R}_+)$ which belongs to the set $B_1 = B(0,1)$.

REFERENCES

[1] R.P. Agarwal, D. O'Regan and P.J.Y. Wong, *Positive solutions of differential, difference and integral equations*, (Kluwer Academic Publishers, Dordrecht, 1999).

[2] V.G. Angelov and D.D. Bainov, On the functional differential equations with "maximums", *Appl. Anal.* **16** (1983), 187-194.

[3] I.K. Argyros, Quadratic equations and applications to Chandrasekhar's and related equations, *Bull. Austral. Math. Soc.*, **32** (1985), 275-292.

[4] J.M. Ayerbe Toledano, T. Dominguez and G. López Acedo, *Measures of Noncompactness in Metric Fixed Point Theory*, (Birkhauser, Basel, 1997).

[5] D.D. Bainov, S.D. Milusheva and J.J. Nieto, Partially multiplicative averaging for impulsive differential equations with supremum, *Proc. A. Razdmadze Math. Institute*, **110** (1994), 1-18.

[6] D.D. Bainov and A.I. Zahariev, Oscillatiiong and asymptotic properties of a class of functional differential equations with maxima, *Czechoslovak Math. J.* **34** (1984), 247-251.

[7] J. Banás and K. Goebel, *Measures of Noncompactness in Banach Spaces*, (Marcel Dekker, New York and Basel, 1980).

[8] J. Banás, J. Rocha and K. Sadarangani, Solvability of a nonlinear integral equation of Volterra type, *Journal of Computational and Applied Mathematics*, **157**, (2003), 31-48.

[9] J. Banás and K. Sadarangani, Solutions of some functional-integral equations in Banach algebra, *Mathematical and Computer Modelling*, **38**, (2003), 245-250.

[10] J. Caballero, J. Rocha and K. Sadarangani, Solvability of a Volterra integral equation of convolution type in the class of monotonic functions, *Intern. Math. Journal*, **4**, n.1,(2001), 69-77.

[11] T. Dominguez, M. A. Japon and G. López, *Metric fixed point results concerning measures of noncompactness*, Handbook of metric fixed point theory, Kluwer, Dordrecht,(2001), 239-268.

[12] S. Hu, M. Khavanin and W. Zhuang, Integral equations arising in the kinetic theory of gases, *Appl. Analysis*, **34**, (1989), 261-266.

[13] D. O'Regan and M M. Meehan,*Existence Theory for Nonlinear Integral and Integro-differential Equations*, (Kluwer Academic Publishers, Dordrecht, 1998).

[14] S. Reich, Fixed points in locally convex spaces, *Math. Z.* **125**, (1972), 17-31.

[15] S. Reich, Fixed points of condensing functions, *J. Math. Anal. Appl.* **41**, (1973), 460-467.

DEPARTAMENTO DE MATEMÁTICAS, UNIVERSIDAD DE LAS PALMAS DE GRAN CANARIA, CAMPUS DE TAFIRA BAJA, 35017, LAS PALMAS DE GRAN CANARIA, SPAIN.

E-mail address: `fefi@dma.ulpgc.es`, `blopez@dma.ulpgc.es`, `ksadaran@dma.ulpgc.es`

HARALD BOHR MEETS STEFAN BANACH

ANDREAS DEFANT AND CHRISTOPHER PRENGEL

ABSTRACT. We relate the early work of Harald Bohr on Dirichlet series with modern Banach space theory, in particular with questions on unconditionality for m-homogeneous polynomials or holomorphic functions on Banach spaces.

1. INTRODUCTION

We use various methods from analysis (local Banach space theory, combinatorial methods, probability theory, complex analysis in one and several variables, the theory of infinite dimensional holomorphy and number theory) in order to relate and to discuss

- Bohr's study of vertical strips in the complex plane on which Dirichlet series $\sum a_n \frac{1}{n^s}$ converge uniformly but not absolutely,

- the study of multidimensional variants of Bohr's power series theorem (originally a theorem on power series in one complex variable),

- the study of "Dineen's problem" which asked for the existence of infinite dimensional Banach spaces X and $m \geq 2$ such that the Banach space of m-homogeneous polynomials on X has an unconditional basis,

- the study of maximal subsets of a given Reinhardt domain R located in a Banach sequence space X on which each holomorphic function f has a convergent monomial expansion.

This article surveys on recent research on each of these topics, and in particular on joint work of the authors with J. C. Diaz, D. Garcia, L. Frerick, N. Kalton and M. Maestre. Eight problems are posed.

2. PRELIMINARIES

Standard notation and notions from Banach space theory are used, as presented e.g. in [32], [45] and [18]. See [26], and [30] for all needed background on polynomials and holomorphic functions on Banach spaces.

All considered Banach spaces X are assumed to be complex; the open unit ball is denoted by B_X. As usual ℓ_p^n, $1 \leq p \leq \infty$ and $n \in \mathbb{N}$, stands for $\mathbb{C}^n$ together with the p-norm $\|z\|_p := (\sum_{k=1}^n |z_k|^p)^{\frac{1}{p}}$ (with the obvious modification whenever $p = \infty$), and ℓ_p for the infinite dimensional version of these spaces.

Recall that the Banach-Mazur distance of two n-dimensional Banach spaces X and Y is given by $d(X, Y) := \inf \|R\| \|R^{-1}\|$, the infimum taken over all linear bijections $R : X \longrightarrow Y$ (see e.g. [45]).

As usual a sequence (x_k) of a Banach space X is said to be a (Schauder) basis if each $x \in X$ has a unique series representation $x = \sum_{k=1}^{\infty} \mu_k x_k$ (in this case, the coefficient functionals x_k^* defined by $x_k^*(x) = \mu_k$ are continuous). A basis (x_k) of a Banach space X is said to be unconditional if each series representation $x = \sum_{k=1}^{\infty} \mu_k x_k$ converges unconditionally, or equivalently, if there is a constant $c > 0$ such that for each choice of finitely many $\mu_1, \ldots, \mu_n \in \mathbb{C}$

$$\left\| \sum_{k=1}^{n} |\mu_k| x_k \right\| \leq c \left\| \sum_{k=1}^{n} \mu_k x_k \right\| ;$$

in this case the best constant c is denoted by $\chi((x_k))$ and called unconditional basis constant of (x_k). We say that (x_k) is a 1-unconditional basis whenever $\chi((x_k)) = 1$. Moreover, the unconditional basis constant of X is defined to be $\chi(X) := \inf \chi((x_k)) \in [1, \infty]$, the infimum taken over all unconditional bases (x_k) of X.

By a Banach sequence space X we mean a Banach space $X \subset \mathbb{C}^{\mathbb{N}}$ of sequences in $\mathbb{C}$ such that the closed unit ball $\overline{B}_X$ is closed in the topology of pointwise convergence and the canonical unit vectors $e_n = (\delta_{nk})_k$ form a 1-unconditional basis in X. The n-dimensional space $X_n := \text{span}\{e_1, \cdots, e_n\}$ is then said to be the nth section of X. A Banach sequence space X is called symmetric if every $x \in \mathbb{C}^{\mathbb{N}}$ belongs to X if and only if its decreasing rearrangement x^* belongs to X, and in this case they have equal norm. For $1 \leq p < \infty$ a Banach sequence space X is said to be p-convex if there exists a constant $c > 0$ such that for each choice of finitely many elements $x_1, \cdots, x_n \in X$

$$\left\| \left(\sum_{i=1}^{n} |x_i|^p \right)^{1/p} \right\|_X \leq c \left(\sum_{i=1}^{n} \|x_i\|_X^p \right)^{1/p} ;$$

X has non-trivial convexity if it is p-convex for some $1 \leq p < \infty$.

Finally, we recall that $\lambda_X(n) := \|\sum_{i=1}^{n} e_i\|_X$ stands for the fundamental function of X, and

$$X \cdot Y := \{xy \mid x \in X, \ y \in Y\}$$

defines the product space of two Banach sequence spaces X and Y; here the product of two sequences $x, y \in \mathbb{C}^{\mathbb{N}}$ is meant coordinatewise.

3. Dirichlet series and power series in infinitely many variables

An (ordinary) Dirichlet series is a series of the form

$$(3.1) \qquad \sum_{n=1}^{\infty} \frac{a_n}{n^s}$$

with coefficients $a_n \in \mathbb{C}$ and s a complex variable. Series of this type form an important tool in analytic number theory, and their structure theory helps understanding the Riemann zeta function given by

$$(3.2) \qquad \zeta(s) = \sum_{n=1}^{\infty} \frac{1}{n^s}.$$

The maximal domains where Dirichlet series converge absolutely, uniformly or conditionally are half planes $[\operatorname{Re} s > \sigma]$ where $\sigma = a, u$ or c is called the abscissa of absolute, uniform or conditional convergence, respectively. More precisely, define $\sigma := \inf \sigma_0$, the infimum taken over all σ_0 such that on $[\operatorname{Re} s > \sigma_0]$ we have convergence of the requested type.

We have $c \leq u \leq a$, but contrary to the situation for power series these values may differ from each other. For example for the series in (3.2) we have $c = a = 1$ whereas the alternating series $\sum (-1)^{n+1}/n^s$ has conditional convergence abscissa $c = 0$ and absolute convergence abscissa $a = 1$. In both cases we have $u = a$.

In general $a - c \leq 1$ and by the example above this estimate is optimal, i. e. $\sup a - c = 1$, the least upper bound taken over all Dirichlet series as in (3.1).

Dirichlet series were studied extensively by Harald Bohr during his lifetime and one of his major problems was to determine the value

$$(3.3) \qquad T := \sup a - u .$$

This problem arised from the so called "absolute convergence problem" for Dirichlet series, i. e. to derive the position of the absolute convergence abscissa from simple analytic properties of the function defined by

$$f(s) = \sum_{n=1}^{\infty} \frac{a_n}{n^s} .$$

In a rather ingenious fashion Bohr shifted the problem of finding the precise value of T into a problem formulated entirely in terms of power series in infinitely many variables. Using the fundamental theorem of arithmetic he established a one to one correspondence between Dirichlet series and power series in infinitely many variables, given by

$$(3.4) \qquad \sum_{\alpha \in \mathbb{N}_0^{(\mathbb{N})}} c_\alpha z^\alpha \quad \leftrightsquigarrow \quad \sum_n \frac{a_n}{n^s} \quad , \quad c_\alpha = a_{p^\alpha}$$

where $p = (p_n)$ is the sequence of primes. Then he defined

$$(3.5) \qquad S := \inf q ,$$

the infimum taken over all $q > 0$ such that $\sum |c_\alpha| \varepsilon^\alpha < \infty$ for every $\varepsilon \in \ell_q \cap \,]0,1[\,^{\mathbb{N}}$ and every power series $\sum c_\alpha z^\alpha$ bounded in the domain $[|z_i| \leq 1]$, i. e. with

$$(3.6) \qquad \sup_n \sup_{|z_i| \leq 1} \Big| \sum_{\alpha \in \mathbb{N}_0^n} c_\alpha z^\alpha \Big| < \infty ,$$

and used the prime number theorem to prove that

(3.7)
$$T = \frac{1}{S}.$$

Now the problem was to find the exact value of S and Bohr [11] proved the following theorem.

Theorem 3.1. *Let $\sum c_\alpha z^\alpha$ be a power series bounded in the domain $[|z_i| \leq 1]$ and $\varepsilon \in \ell_2 \cap]0,1[^{\mathbb{N}}$. Then*

$$\sum_{\alpha \in \mathbb{N}_0^{(\mathbb{N})}} |c_\alpha| \varepsilon^\alpha < \infty,$$

in other words: $S \geq 2$.

Bohr did not know if 2 was the exact value of S, not even if $S < \infty$. In 1915 Toeplitz [44] proved $S \leq 4$ by constructing for every $\delta > 0$ an $\varepsilon \in \ell_{4+\delta} \cap]0,1[^{\mathbb{N}}$ and a 2-homogeneous polynomial $\sum_{|\alpha|=2} c_\alpha z^\alpha$ bounded on $[|z_i| \leq 1]$ with $\sum_{|\alpha|=2} |c_\alpha| \varepsilon^\alpha = \infty$. Then after some time Bohnenblust and Hille [10] proved in 1931 that 2 is in fact the exact value of S. They showed that the upper bound 4 of Toeplitz is best possible if one just considers 2-homogeneous polynomials instead of arbitrary power series, and they constructed examples of m-homogeneous polynomials for arbitrary m to get the result.

Thus the final conclusion is that the maximal possible width of the strip of uniform but non-absolute convergence of a Dirichlet series $\sum \frac{a_n}{n^s}$ is

$$T = \frac{1}{2}$$

The inequality $S \leq 2$ was reproved by Dineen and Timoney [28] in 1989; this will be explained in section 7.

4. MULTIDIMENSIONAL VERSIONS OF BOHR'S POWER SERIES THEOREM

Bohr's problem to find an upper bound for S lead him to investigate the relation between the absolute value of a power series in one variable and the sum of the absolute values of the individual terms. In 1913 he published the following result, now known as Bohr's power series theorem [12].

Theorem 4.1. *Let $\sum c_n z^n$ be a power series in $\mathbb{C}$ with*

(4.1)
$$\left| \sum_{n=1}^{\infty} c_n z^n \right| \leq 1$$

for all $|z| < 1$. Then for all $|z| < 1/3$

$$\sum_{n=1}^{\infty} |c_n z^n| \leq 1,$$

and the value $1/3$ is optimal, i. e. for every $r > 1/3$ there is a power series $\sum c_n z^n$ satisfying (4.1) and $\sum_{n=1}^{\infty} |c_n| r^n > 1$.

How can this theorem be extended to power series $\sum_{\alpha \in \mathbb{N}_0^n} c_\alpha z^\alpha$ in n complex variables?

The domains of convergence of such power series are Reinhardt domains. By a Reinhardt domain $R \subset \mathbb{C}^n$ we mean a non empty open set which satisfies the following property: If $z \in R$ and $u \in \mathbb{C}^n$ with $|u| \leq |z|$ then $u \in R$. In the literature one usually only assumes R to be open and n-circular (i. e. $z \in R$ and $u \in \mathbb{C}^n$ with $|u| = |z|$ implies $u \in R$) and the additional assumption that $\lambda R \subset R$ for every $\lambda \in \mathbb{C}$ with $|\lambda| \leq 1$ is called completeness of R. Thus our Reinhardt domains are always complete.

Examples of Reinhardt domains are unit balls of ℓ_p^n-spaces, or more generally of finite dimensional Banach spaces $(\mathbb{C}^n, \|\cdot\|)$ such that the canonical unit vectors form a 1-unconditional basis (see section 5).

Bohr's power series theorem motivates the following definition due to Boas and Khavinson [9].

Definition 4.2. The Bohr radius $K(R)$ of a Reinhardt domain $R \subset \mathbb{C}^n$ is defined to be the least upper bound of all $r \geq 0$ such that every power series $\sum_{\alpha \in \mathbb{N}_0^n} c_\alpha z^\alpha$ with

$$\left| \sum_{\alpha \in \mathbb{N}_0^n} c_\alpha z^\alpha \right| \leq 1$$

for all $z \in R$ satisfies

$$\sum_{\alpha \in \mathbb{N}_0^n} |c_\alpha z^\alpha| \leq 1$$

for all z in the scaled domain rR. If only m-homogeneous polynomials $\sum_{|\alpha|=m} c_\alpha z^\alpha$ are considered instead of arbitrary power series, then we write $K_m(R)$, the mth Bohr radius of R.

Now Bohr's power series theorem in this terminology reads as follows:

$$(4.2) \qquad\qquad K(\mathbb{D}) = 1/3,$$

$\mathbb{D}$ as usual the open unit disc in $\mathbb{C}$.

Obviously, $K_m(\mathbb{D}) = 1$ for all m, hence the following result from [16, Theorem 2.2] is a sort of abstract multi dimensional variant of Bohr's power series theorem.

Proposition 4.3. *Let R be a bounded Reinhardt domain in $\mathbb{C}^n$. Then*

$$\frac{1}{3} \inf_m K_m(R) \leq K(R) \leq \min \left\{ \frac{1}{3}, \inf_m K_m(R) \right\}.$$

Clearly, the upper estimate is trivial from what was explained so far, so that the following problem arises.

Problem 4.4. *Decide whether or not*

$$K(R) = \frac{1}{3} \inf_m K_m(R).$$

The first multidimensional extension of Bohr's theorem for the n-dimensional polydisc was obtained by Dineen and Timoney [27] in 1989. They studied the existence of absolute monomial bases for spaces of holomorphic functions on locally convex spaces and saw a possibility to solve their basis problem with the help of a polydisc version of Bohr's theorem. Their result can be stated as follows: For all $\varepsilon > 0$ there is a constant $c_\varepsilon > 0$ such that for all n

$$(4.3) \qquad K(B_{\ell_\infty^n}) \leq c_\varepsilon \sqrt{\frac{n^\varepsilon}{n}}.$$

In 1997 Boas and Khavinson [9] improved this estmate by showing that

$$(4.4) \qquad \frac{1}{c}\frac{1}{\sqrt{n}} \leq K(B_{\ell_\infty^n}) \leq c\sqrt{\frac{\log n}{n}},$$

and Aizenberg [2] established the ℓ_1-case

$$(4.5) \qquad \frac{1}{c} \leq K(B_{\ell_1^n}) \leq c.$$

So, essentially the sequence $(K(B_{\ell_\infty^n}))_n$ of Bohr radii of the n-dimensional polydiscs $B_{\ell_\infty^n}$ decreases to 0 like $\frac{1}{\sqrt{n}}$ whereas for the n-dimensional hypercones $B_{\ell_1^n}$ this sequence remains strictly larger than 0. In [8] Boas was then able to give the following asymptotic estimates for the scale of all finite dimensional ℓ_p-spaces:

$$(4.6) \qquad \frac{1}{c}\left(\frac{1}{n}\right)^{1-\frac{1}{\min\{p,2\}}} \leq K(B_{\ell_p^n}) \leq c\left(\frac{\log n}{n}\right)^{1-\frac{1}{\min\{p,2\}}},$$

$c > 0$ again a constant independent of n.

The proofs for the lower bounds use techniques which can be traced back to the original papers of Bohr, and are based on clever estimates for the moduli of the coefficients of the power series considered as a holomorphic function.

So far all known non trivial upper estimates for multi dimensional Bohr radii use probabilistic methods, and the log-term in the preceding estimates is a consequence of these methods. Boas in [8, p. 239] conjectured that *presumably this logarithmic factor, an artifact of the proof, should not really be present* (see also [8, section 7, Problem 1] and the discussion in [9]).

We will see in section 6 that contrary to this commonly held opinion the lower estimate in (4.6) can be improved by a combination of Bohr's methods and methods from local Banach space theory.

We now discuss some upper and lower estimates for $K(R)$ for some more general classes of n-dimensional Reinhardt domains R. Note first that since each Reinhardt domain is the union of polydiscs (with varying radii) the Bohr radius of the n-dimensional polydisc is an extreme case in the following sense (see [9, Theorem 3]): For each n-dimensional Reinhardt domain R

$$(4.7) \qquad K(B_{\ell_\infty^n}) \leq K(R).$$

In order to improve this estimate we need a further notation which allows to compare the Bohr radii of two Reinhardt domains R_1 and R_2 in $\mathbb{C}^n$:

$$S(R_1, R_2) := \inf\{b > 0 : R_1 \subset bR_2\}.$$

To see an example, if R is a bounded Reinhardt domain in $\mathbb{C}^n$, then we have

$$(4.8) \qquad S(R, B_{\ell_p^n}) = \sup_{z \in R} \Big(\sum_{k=1}^{n} |z_k|^p\Big)^{1/p}, \ 1 \le p \le \infty$$

With this terminology at hands the following useful device can be formulated:

$$(4.9) \qquad K(R_1) \le S(R_1, R_2)S(R_2, R_1)K(R_2),$$

R_1 and R_2 again two Reinhardt domains in $\mathbb{C}^n$.

Now from (4.7) and (4.9) the following general lower bound is immediate –a result which in almost all concrete cases at least up to a log-term in n seems to lead to optimal estimates.

Corollary 4.5. *There is a constant $c > 0$ such that for each bounded Reinhardt domain R in $\mathbb{C}^n$*

$$\max\left\{K(B_{\ell_\infty^n}), \frac{c}{S(R, B_{\ell_1^n})S(B_{\ell_1^n}, R)}\right\} \le K(R).$$

Reasonable upper bounds for Bohr radii of general n-dimensional Reinhardt domains are technically more delicate. The following result is from [15, Theorem 4.2] and [16, Theorem 2.8].

Theorem 4.6. *For each bounded Reinhardt domain R in $\mathbb{C}^n$ we have*

$$K(R) \le e^3 2^{3/2} \sqrt{\log n}\ \frac{S(R, B_{\ell_2^n})}{S(R, B_{\ell_1^n})}.$$

Clearly, a combination of Corollary 4.5, Theorem 4.6 and (4.8) leads to the asymptotic of Boas given in (4.6) (with a slightly worse log-term).

In the 1-dimensional case the proof of the upper estimate, $K(\mathbb{D}) \le 1/3$, is simple, and uses Moebius transforms $\frac{z-a}{1-az}$, $0 < a < 1$ (see e.g. [8, section 2]). But in higher dimensions special power series which lead to reasonable upper estimates for $K(R)$, so far can only be generated by probabilistic arguments. This idea originates in [8], [9], and [27].

An m-homogeneous polynomial $\sum_{|\alpha|=m} c_\alpha z^\alpha$ is said to be unimodular whenever the moduli of all coefficients c_α equal 1. It is well-known (see e.g. [17, Theorem 2.3]) that each unimodular m-homogeneous polynomial satisfies the following lower norm estimate

$$c_m n^{\frac{m+1}{2}} \le \sup_{z \in B_{\ell_\infty^n}} \Big| \sum_{|\alpha|=m} c_\alpha z^\alpha \Big|,$$

c_m a constant only depending on the degree m and not on the number of variables n. The famous Kahane-Salem-Zygmund Theorem then states that this estimate is optimal in the following sense: For each n there exist signs ε_α such that

$$\sup_{z \in B_{\ell_\infty^n}} | \sum_{|\alpha|=m} \varepsilon_\alpha z^\alpha | \leq c_m n^{\frac{m+1}{2}},$$

and it is this type of polynomials which is needed to establish the estimates from (4.4) and (4.6) (in [8, 5.2] Boas uses instead of the Kahane-Salem-Zygmund Theorem a closely related result proved by probabilistic techniques due to Mantero and Tonge from [38]).

The main tool for the proof of Theorem 4.6 is of independent interest–it is an upper estimate for the expectation of the sup norm of a Gaussian random polynomial, the sup norm taken on bounded circled domains in $\mathbb{C}^n$. For its proof which is based on Slepian's lemma for Gaussian random processes, see [15] and [17].

Theorem 4.7. *Let U be a bounded circled set in $\mathbb{C}^n$ and $(g_\alpha)_{|\alpha|=m}$ and $(g_i)_{i=1}^n$ two families of independent standard Gaussian random variables on a probability space (Ω, μ). Then for each choice of coefficients c_α, $|\alpha| = m$*

$$\int \sup_{z \in U} | \sum_{|\alpha|=m} c_\alpha g_\alpha z^\alpha | d\mu$$

$$\leq C^m \sup_{|\alpha|=m} \left\{ |c_\alpha| \sqrt{\frac{\alpha!}{m!}} \right\} \sup_{z \in U} \left(\sum_{i=1}^n |z_i|^2 \right)^{\frac{m-1}{2}} \int \sup_{z \in U} | \sum_{i=1}^n g_i z_i | d\mu$$

where $C > 0$ is an absolut constant.

Note that since Gaussian averages dominate Rademacher averages this result for $U = B_{\ell_\infty^n}$ gives the above mentioned (special case of the) Kahane-Salem-Zygmund Theorem.

Parts of this result cannot be improved. For example the term $\sqrt{\alpha!/m!}$ cannot be replaced by $\alpha!/m!$; if yes, then the log-term in (4.4) would be superfluous (see the proof of Theorem 4.6 in [15]), but this is not the case by Theorem 6.3.

We finish this section illustrating that our results for large classes of X lead to almost optimal estimates. The following corollary is an immediate consequence of Corollary 4.5 and Theorem 4.6.

Corollary 4.8. *Let R be a bounded Reinhardt domain in $\mathbb{C}^n$ such that $B_{\ell_1^n} \subset R \subset B_{\ell_2^n}$. Then we have*

$$\frac{1}{c} \frac{1}{\sup_{z \in R} \sum_{k=1}^n |z_k|} \leq K(R) \leq c \sqrt{\log n} \frac{1}{\sup_{z \in R} \sum_{k=1}^n |z_k|},$$

$c > 0$ an absolute constant.

For "up to a log-term in the dimension" optimal estimates of the Bohr radius for concrete n-dimensional (convex and non convex) Reinhardt domains

R see again [15] and [16]. Recently much work has been done about multi-dimensional Bohr radii; for more information see [2], [3], [4], [5], [7], [8], [9], [13], [15], [16], [27], [29], [39].

5. UNCONDITIONALITY IN SPACES OF m-HOMOGENEOUS POLYNOMIALS

Let X be a Banach space and $m \in \mathbb{N}$. A function $p : X \to \mathbb{C}$ is said to be an m-homogeneous polynomial if there is an m-linear form $\varphi : \prod_{k=1}^{m} X \to \mathbb{C}$ such that for all $x \in X$

$$p(x) = \varphi(x, \ldots, x).$$

As usual we denote by $\mathcal{P}(^m X)$ the vector space of all m-homogeneous continuous polynomials p on X which together with the norm

$$\|p\| := \sup_{\|x\| \leq 1} |p(x)|$$

forms a Banach space. For m functionals $x_1^*, \ldots, x_m^* \in X^*$ the definition

$$x_1^* \otimes \cdots \otimes x_m^*(x) := \prod_{k=1}^{m} x_k^*(x), \ x \in X$$

gives the prototype of an m-homogeneous continuous polynomial on X. The closure of all linear combinations of such monomials leads to $\mathcal{P}_{appr}(^m X)$, the subspace of all so called approximable m-homogeneous polynomials on X.

The above given basis free definition of an m-homogeneous polynomial on infinite dimensional Banach spaces dates from 1932 and was given independently by Michal and Banach. Indeed, Banach intended to write a second volume of his famous book devoted to the non linear theory with this definition in its center. Dineen in his recent book [26, 2.6 Notes] claims that the development of the theory of polynomials on Banach spaces can be devided into two periods. The first period starts in the thirties of the last century and, although mainly motivated externally through holomorphic and differential functions on infinite dimensional spaces, it identified interesting concepts and results on polynomials in infinitely many variables. The second period begins in the eighties. It did not represent a break in the scientific sense but was more of a psychological change of attitude notable for the appearance of polynomials as the main object of study.

For a finite dimensional Banach space $X = (\mathbb{C}^n, \| \cdot \|)$ we obviously have that $\mathcal{P}(^m X)$ coincides with all polynomials of the form $\sum_{|\alpha|=m} c_\alpha z^\alpha$ in n complex variables z_k, and the monomials z^α, $\alpha \in \mathbb{N}_0^n$ and $|\alpha| = m$, clearly form a linear basis.

In infinite dimensions the question whether $\mathcal{P}(^m X)$ allows a Schauder basis is more involved. Let X be a Banach space, (x_k^*) a sequence of functionals in its dual X^*, and $\alpha = (\alpha_1, \ldots, \alpha_n) \in \mathbb{N}_0^{(\mathbb{N})}$ a multi index. Then we call the function

$$x_\alpha^*(x) := x_{\alpha_1}^* \otimes \cdots \otimes x_{\alpha_n}^*(x), \ x \in X$$

a monomial. If the order $|\alpha|$ equals m, then $x_\alpha^* \in \mathcal{P}(^m X)$. When do these monomials form a basis of $\mathcal{P}(^m X)$?

The following result was 1980 stated in Ryan [42] (although its proof in [42] contains a gap which was recently closed in [25], [31] and [41]).

Proposition 5.1. *Let X be a Banach space such that X^* has a basis (x_k^*). Then the monomials $(x_\alpha^*)_{\alpha \in \mathbb{N}_0^{(\mathbb{N})}}$ under an appropriate order of $\mathbb{N}_0^{(\mathbb{N})}$ form a basis of $\mathcal{P}_{appr}(^m X)$.*

Such an order can be defined inductively as follows: For $\alpha \in \mathbb{N}_0^{(\mathbb{N})}$ define $n(\alpha) := \max\{k \mid \alpha_k \neq 0\}$ and $\overline{\alpha}$ by

$$\overline{\alpha}_n := \begin{cases} \alpha_n - 1 & n = n(\alpha) \\ \alpha_n & \text{otherwise} . \end{cases}$$

Then for multiindices α, β of order m

$$\alpha \leq \beta \quad :\Longleftrightarrow \quad (1) \ n(\alpha) < n(\beta)$$
$$\text{or} \quad (2) \ n(\alpha) = n(\beta) \ \text{and} \ \overline{\alpha} \leq \overline{\beta} .$$

This order was used in [25], [31] and [41]. It is established as the socalled *square order* of the monomials, though it is different from the order defined by Ryan in [42] for which he originally used this name (that even the latter order works is more difficult to see and was also proved in [41]).

Hence, the monomials form a basis for the whole space $\mathcal{P}(^m X)$ provided $\mathcal{P}_{appr}(^m X) = \mathcal{P}(^m X)$. For reflexive spaces X even the converse holds as was shown by Alencar [1].

Proposition 5.2. *Let X be a reflexive Banach space such that X^* has a basis (x_k^*). Then the following are equivalent:*

(1) *The monomials (x_α^*) form a basis of $\mathcal{P}(^m X)$.*
(2) *$\mathcal{P}_{appr}(^m X) = \mathcal{P}(^m X)$.*
(3) *$\mathcal{P}(^m X)$ is reflexive.*

See [26] for a collection of results on the reflexivity of spaces of m-homogeneous polynomials on Banach spaces; for example, a result of Pełczyński [40] from 1957 states that $\mathcal{P}(^m \ell_p)$ is reflexive if and only if $m < p$. As a consequence, the monomials (square order) form a basis of $\mathcal{P}(^m \ell_p)$ if and only if $m < p$.

The following questions became to be known as **Dineen's problem:** *Let (x_k) be an unconditional basis of the infinite dimensional Banach space X, and let (x_k^*) be the sequence of its coefficient functionals in X^*. Can the monomials with respect to the coefficient functionals form an unconditional basis of $\mathcal{P}(^m X)$? Can $\mathcal{P}(^m X)$ at all have an unconditional basis?*

Dineen writes in [26, p. 305]: *... however, it is rarely, and perhaps never, the case that the monomials form an unconditional basis for $\mathcal{P}(^m X)$ when X is an infinite dimensional Banach space ....*

The following theorem is the main result of [19].

Theorem 5.3. *Let X be a Banach space with an unconditional basis and $m \geq 2$. Then the Banach space $\mathcal{P}(^m X)$ of all m-homogeneous polynomials on X has an unconditional basis if and only if X is finite dimensional.*

The proof of this result consists of three steps of independent interest which we briefly sketch:

The **first step** towards the solution of Dineen's problem was established in [14]. Following a program originally initiated by Gordon and Lewis it is proved that for each Banach space X which has a dual X^* with an unconditional basis (x_k^*), the approximable polynomials $\mathcal{P}_{appr}(^m X)$ have an unconditional basis if and only if the monomial basis with respect to (x_k^*) is an unconditional basis.

Let us give this equivalence a more quantitative formulation. A Banach space invariant closely related to $\chi(X)$ is the Gordon-Lewis constant invented in the classical paper [33]. A Banach space X is said to be a Gordon-Lewis space (or to have the Gordon-Lewis property) if every 1-summing operator $T : X \longrightarrow \ell_2$, i.e.

$$\pi_1(T) := \sup\left\{\sum_{i=1}^n \|Tx_i\| \mid \|\sum_{i=1}^n \lambda_i x_i\| \le 1, n \in \mathbb{N}, |\lambda_i| \le 1\right\} < \infty,$$

allows a factorization $T : X \xrightarrow{R} L_1(\mu) \xrightarrow{S} \ell_2$ (μ some measure, R and S bounded operators). In this case, there is a constant $c \ge 0$ such that for all $T : X \longrightarrow \ell_2$

$$\gamma_1(T) := \inf \|R\|\|S\| \le c\,\pi_1(T),$$

and the best such c is called the Gordon-Lewis constant of X and denoted by $gl(X) \in [0, \infty]$.

A fundamental tool for the study of unconditionality in Banach spaces is the so called Gordon-Lewis inequality from [33] (see also [24, 17.7]):

(5.1) $$gl(X) \le 2\chi(X).$$

Recall that $\chi(\mathcal{P}_{appr}(^m X))$ stands for the unconditional basis constant of $\mathcal{P}_{appr}(^m X)$, and write $\chi_{mon}(\mathcal{P}_{appr}(^m X))$ for the unconditional basis constant of the monomials with respect to (x_k^*). The following inequality was proved in [14, Theorem 1]); for our special situation it is a sort of converse of (5.1).

Proposition 5.4. *Let X be a Banach space and (x_k^*) a 1-unconditional basis of X^*. Then for all m*

$$\chi_{mon}(\mathcal{P}_{appr}(^m X)) \le c_m\, gl(\mathcal{P}_{appr}(^m X)) \le 2c_m\, \chi(\mathcal{P}_{appr}(^m X)),$$

where $c_m \le (m^{4m}/m!)2^m$.

In the **second step** of the proof of Theorem 5.3 we use this result to derive asymptotically optimal estimates for the unconditional basis constants of spaces of m-homogeneous polynomials on finite dimensional sections of Banach sequence spaces ([15, 6.2] and [14, Theorem 3]).

Proposition 5.5. *Let X be a Banach sequence space and let X_n be the linear span of $\{e_k \mid 1 \le k \le n\}$, $n \in \mathbb{N}$. Then*

$$\chi(\mathcal{P}(^m X_n)) \overset{m}{\asymp} \chi_{mon}(\mathcal{P}(^m X_n)) \overset{m}{\asymp} d(X, \ell_1^n)^{m-1}$$

provided X satisfies one of the following two conditions:

(1) *X is a subset of ℓ_2 and has non-trivial convexity.*

(2) X *is symmetric and 2-convex.*

In particular, for each $1 \leq p \leq \infty$

$$\chi(\mathcal{P}(^m\ell_p^n)) \overset{m}{\asymp} \chi_{mon}(\mathcal{P}(^m\ell_p^n)) \overset{m}{\asymp} n^{1-\frac{1}{\min(p,2)}}. \tag{5.2}$$

Here $\overset{m}{\asymp}$ *means that the right and the left side are equivalent up to constants only depending on* m.

A well known result of Tzafriri [46] states that each infinite dimensional Banach space X with an unconditional basis contains uniformly complemented ℓ_p^n's for some $p \in \{1, 2, \infty\}$.

Since the Gordon-Lewis constant is invariant under complemented subspaces, Proposition 5.3 combined with (5.2) assures that X can neither contain all ℓ_2^n's nor all ℓ_∞^n's uniformly complemented whenever we assume that $\chi(\mathcal{P}(^mX)) < \infty$.

But this contradicts the following independently interesting result from [19, Proposition 3.2] which is the **third step** completing the proof of Theorem 5.3.

Proposition 5.6. *Suppose* X *has an unconditional basis and* $m \geq 2$. *If* $\mathcal{P}(^mX)$ *is separable, then either* X *contains uniformly complemented* ℓ_2^n's *or* X *contains uniformly complemented* ℓ_∞^n's.

Let us finish with some remarks on the proof of this result which relies on a refinement of Tzafriri's theorem (just mentioned) using recent results on so called greedy bases.

A normalized basic sequence $(x_k)_{k=1}^\infty$ in a Banach space X is called *democratic* if there is a constant C such that if A, B are finite subsets of $\mathbb{N}$ with $|A| \leq |B|$ then

$$\left\| \sum_{k \in A} x_k \right\| \leq C \left\| \sum_{k \in B} x_k \right\|.$$

A basic sequence which is both unconditional and democratic is called *greedy*. In fact, greedy bases were originally defined in terms of approximation rates, and it is a theorem of Konyagin and Temlyakov [34] that this is equivalent to our definition. We refer to [22] and [23] for more information on greedy bases.

If $(x_k)_{k=1}^\infty$ is a greedy basic sequence then for its fundamental function we have that

$$\lambda_X(n) = \sup\{\|\sum_{j \in A} x_k\| : |A| \leq n\}.$$

An important principle needed for the proof of the preceding theorem is the following special case of [22, Proposition 5.3] (proved by use of Ramsey theory): *Suppose* X *is a Banach space with non-trivial cotype and* $(x_k)_{k=1}^\infty$ *is an unconditional basis of* X. *Then* $(x_k)_{k=1}^\infty$ *has a subsequence* $(x_{k_n})_{n=1}^\infty$ *which is greedy.*

Combined with this general principle the following modification of Tzafriri's result mentioned above easily implies Proposition 5.6 ([19, Theorem 3.1]).

Proposition 5.7. *Suppose X has a greedy basis $(x_k)_{k=1}^{\infty}$. Suppose X has non-trivial cotype $q < \infty$ and that for some $p > 1$ we have*

$$\liminf_{n \to \infty} n^{-\frac{1}{p}} \lambda_X(n) = 0.$$

Then X contains uniformly complemented ℓ_2^n's.

Two problems remain open.

Problem 5.8. *Let X be an infinite dimensional Banach space with an unconditional basis and $m \geq 2$. Can the monomials with respect to the coefficient functionals of this basis form an unconditional basis of $\mathcal{P}_{appr}(^m X)$?*

The discussion of this section suggests that the answer is no. It is interesting to remark that the monomials can form an unconditional *basic sequence* in $\mathcal{P}_{appr}(^m X)$ (or, equivalently, in $\mathcal{P}^m X$)). It can be seen easily that this happens for $X = \ell_1$, but as shown in [19, Theorem 4.3] there are Lorentz spaces $d(\omega, 1)$ different from ℓ_1 which never the less fulfill this property.

In this context the following equivalence from [19, Proposition 4.1] seems to be interesting, its proof has as its main ingredient again Proposition 5.4.

Proposition 5.9. *Let X be a Banach space with an unconditional basis (x_k) and coefficient functionals (x_k^*). Then for each m the following are equivalent:*

(1) *The monomials (x_α^*) form an unconditional basic sequence in $\mathcal{P}(^m X)$.*
(2) *$\mathcal{P}(^m X)$ is isomorphic to a Banach lattice.*

Problem 5.10. *Characterize those Banach spaces X with an unconditional basis for which $\mathcal{P}(^m X)$ is isomorphic to a Banach lattice.*

6. BOHR'S POWER SERIES THEOREM AND LOCAL BANACH SPACE THEORY

In this section we link the preceding cycle of ideas on the unconditionality of spaces of m-homogeneous polynomials on Banach spaces X with our study of multidimensional variants of Bohr's power series theorem from section 4.

In order to motivate our basic link between these topics recall from (4.6) that $K(B_{\ell_p^n})$ up to an absolute constant and up to a log-term in n equals

$$\frac{1}{n^{1 - \frac{1}{\min\{p,2\}}}},$$

whereas $\chi_{mon}(\mathcal{P}(^m \ell_p^n))$ up to an absolute constant by (5.2) equals

$$(n^{1 - \frac{1}{\min\{p,2\}}})^{m-1}.$$

Hence, for large degrees m the mth root of $\chi_{mon}(\mathcal{P}(^m \ell_p^n))$ approximates $K(B_{\ell_p^n})^{-1}$. It will turn out that this even holds for every Banach space $X = (\mathbb{C}^n, \| \cdot \|)$ for which the canonical basis vectors e_k form a 1-unconditional basis. Again we start with a definition.

Definition 6.1. Let $X = (\mathbb{C}^n, \|\cdot\|)$ be a Banach space such that the e_k's form a 1-unconditional basis (i. e., B_X is a Reinhardt domain). Then we define

$$r(X) := \sup_m \chi_{mon}(\mathcal{P}(^m X))^{\frac{1}{m}} .$$

A straight forward caculation shows that for each m

$$K_m(B_X) = \frac{1}{\sqrt[m]{\chi_{mon}(\mathcal{P}(^m X))}}$$

(see [15, Lemma 2.1]), hence we obtain as an immediate consequence of Proposition 4.3 the following basic link between Bohr radii and radii of unconditionality (see [15, Theorem 2.2]).

Theorem 6.2. *Let $X = (\mathbb{C}^n, \|\cdot\|)$ be a Banach space for which the e_k's form a 1-unconditional basis. Then we have*

$$\frac{1}{3}\frac{1}{r(X)} \le K(B_X) \le \min\left\{\frac{1}{3}, \frac{1}{r(X)}\right\}.$$

This "bridge" allows to use methods from Banach space theory, in particular the local theory, to study Dirichlet series, Bohr radii or related topics, e.g. from infinite dimensional holomorphy, and vice versa, it helps to use results from complex analysis in order to study unconditionality in spaces of m-homogeneous polynomials or holomorphic functions on Banach spaces.

We give two examples which stress this opinion.

The following lower estimate for the Bohr radii for the scale of Reinhardt domains $B_{\ell_p^n}$, $1 \le p \le \infty$, was given in [21]. It shows that–at least up to a log log-term in the dimension n–the log-terms in the estimates from (4.4), (4.6), Theorem 4.6 and Corollary 4.8 are not superfluous.

Theorem 6.3. *There is a constant $c > 0$ such that for each $1 \le p \le \infty$ and all n*

$$\frac{1}{c}\left(\frac{\log n / \log\log n}{n}\right)^{1 - \frac{1}{\min(p,2)}} \le K(B_{\ell_p^n}).$$

The proof of this result uses complex analysis (in particular, ideas of Bohr) to establish the estimate

$$\chi_{mon}(\mathcal{P}(^m \ell_p^n))^{\frac{1}{m}} \le c\left(1 + \frac{n}{m}\right)^{1 - \frac{1}{\min(p,2)}},$$

and alternatively, results from the local Banach space theory to show that

$$\chi_{mon}(\mathcal{P}(^m \ell_p^n))^{\frac{1}{m}} \le 4e\, n^{1 - \frac{1}{\min(p,2)}} \frac{1}{\left(n^{\frac{1}{m}}\right)^{1 - \frac{1}{\min(p,2)}}}.$$

A synthesis of both estimates together with the "bridge" from Theorem 6.2 leads to Theorem 6.3.

Problem 6.4. *Decide whether the term $\log\log n$ in Theorem 6.3 is superfluous.*

Aizenberg's result from (4.5) shows that among the scale of all domains $B_{\ell_p^n}$ the unit ball of ℓ_1^n plays a special role, it is the only one for which the sequence $(K(B_{\ell_1^n}))_n$ of Bohr radii does not converge to 0. Moreover, recall that

$$d(\ell_p^n, \ell_1^n) \asymp n^{1 - \frac{1}{\min(p,2)}}$$

(see e.g. [45]). Hence, (4.6) and its improvement Theorem 6.3, but also the results from Theorem 6.2 combined with Proposition 5.5, suggest that for each Banach space $X = (\mathbb{C}^n, \| \cdot \|)$ (such that the e_k's form a 1-unconditional basis) the following "asymptotic equality"

$$(6.1) \qquad K(B_X) \asymp \frac{1}{d(X, \ell_1^n)},$$

might be true (up to absolute constants and a log-term in the dimension n). The next theorem collects some results in this direction.

Theorem 6.5.

(1) *There is an absolute constant $c > 0$ such that for all Banach spaces $X = (\mathbb{C}^n, \| \cdot \|)$ for which the e_k's form a 1-unconditional basis, the following estimates hold:*

$$\frac{1}{c} \frac{1}{d(X, \ell_1^n)} \leq K(B_X) \leq c \sqrt{\log n} \, \frac{\sup_{\|z\|_X \leq 1} (\sum_{k=1}^n |z_k|^2)^{1/2}}{\sup_{\|z\|_X \leq 1} \sum_{k=1}^n |z_k|}.$$

(2) *There is $c > 0$ such that for all X as in (1)*

$$\frac{1}{c} \frac{1}{d(X, \ell_1^n)} \leq K(B_X) \leq c \sqrt{\log n} \, \frac{1}{d(X, \ell_1^n)^{1/2}}.$$

(3) *There is $c > 0$ such that for all X as in (1) which satisfy $B_X \subset B_{\ell_2^n}$ or are the nth section of a symmetric and 2-convex Banach sequence space, we have*

$$\frac{1}{c} \frac{1}{d(X, \ell_1^n)} \leq K(B_X) \leq c \sqrt{\log n} \, \frac{1}{d(X, \ell_1^n)}.$$

The lower estimate from (1) was proved in [21, 4.1], the upper estimate is an obvious reformulation of Theorem 4.6 and was proved in [15, 4.2]. The results from (2) and (3) are consequences of (1), see [15, 5.1, 5.2, 5.3].

Problem 6.6. *Decide whether or not (6.1) holds up to absolute constants and a log-term in the dimension n.*

Finally, we mention a problem which is very much related with Problem 5.10 but also with the content of the forthcoming section.

Problem 6.7. *Let X be a Banach sequence space, and denote by X_n its nth section. Does X equal ℓ_1 if $\inf_n K(B_{X_n}) > 0$? Equivalently, is $X = \ell_1$ whenever we have a constant c such that $\chi_{mon}(\mathcal{P}(^m X_n)) \leq c^m$ for all n, m?*

7. DOMAINS OF CONVERGENCE IN INFINITE DIMENSIONAL HOLOMORPHY

Most of the results of this section are taken from the forthcoming PhD-thesis [41] of the second named author–parts of the results of this thesis will also be contained in [20].

Let R be a Reinhardt domain in $\mathbb{C}^n$. Then every holomorphic function f on R has a power series expansion which converges to f in every point of R. More precisely, for every $f \in H(R)$ we can find a unique family of scalars $(c_\alpha)_{\alpha \in \mathbb{N}_0^n}$ such that

$$f(z) = \sum_{\alpha \in \mathbb{N}_0^n} c_\alpha z^\alpha$$

for every $z \in R$, and the coefficients can be calculated by

(7.1) $$c_\alpha = \frac{\partial^\alpha f(0)}{\alpha!} = \left(\frac{1}{2\pi i}\right)^n \int_{[|z|=r]} \frac{f(z)}{z^{\alpha+1}} dz,$$

where $r \in \mathbb{R}_{>0}^n$ is such that the polydisc $[|z| \leq r]$ is contained in R.

Now we consider the infinite dimensional situation. As in finite dimensions a Reinhardt domain R in a Banach sequence space X is a non empty open set for which $u \in R$ whenever $u \in \mathbb{C}^\mathbb{N}$ and $z \in R$ with $|u| \leq |z|$. Note that if $R \subset X$ is a Reinhardt domain, then all its finite dimensional sections $R_n = R \cap X_n$ are Reinhardt domains in $\mathbb{C}^n$.

Let f be a holomorphic function on a Reinhardt domain R in a Banach sequence space X. Then f has a power series expansion $\sum_{\alpha \in \mathbb{N}_0^n} c_\alpha^{(n)} z^\alpha$ on every finite dimensional section R_n of R, and for example from the Cauchy formula (7.1) we can see that $c_\alpha^{(n)} = c_\alpha^{(n+1)}$ for $\alpha \in \mathbb{N}_0^n \subset \mathbb{N}_0^{n+1}$. Thus we can find a unique family $(c_\alpha)_{\alpha \in \mathbb{N}_0^{(\mathbb{N})}}$ such that

(7.2) $$f(z) = \sum_{\alpha \in \mathbb{N}_0^{(\mathbb{N})}} c_\alpha z^\alpha$$

for all $z \in R_n$ and all $n \in \mathbb{N}$. The power series $\sum c_\alpha z^\alpha$ is called the *monomial expansion of f*, and $c_\alpha = c_\alpha(f)$ are the monomial coefficients of f; they satisfy (7.1) whenever $\alpha \in \mathbb{N}_0^n$.

Moreover, the set of all points $z \in R$ where the monomial expansion of f converges, is said to be the domain of convergence of f, notation

$$\mathrm{dom}(f).$$

The monomial expansion converges to $f(z)$ for all $z \in \mathrm{dom}(f)$, since the e_n's form a basis of X. Different from the finite dimensional situation we in general have that $\mathrm{dom}(f) \neq R$.

In this section we discribe the domain of convergence $\mathrm{dom}\,\mathcal{F}(R)$ for subsets $\mathcal{F}(R) \subset H(R)$ of holomorphic functions, that means the set

$$\mathrm{dom}\,\mathcal{F}(R) := \bigcap_{f \in \mathcal{F}(R)} \mathrm{dom}(f)$$

of all elements $z \in R$ for which the monomial expansion of each function $f \in \mathcal{F}(R)$ converges, and our main interest lies in determining the special sets

- dom $H_\infty(R)$, $H_\infty(R)$ all bounded holomorphic functions on R,
- dom $\mathcal{P}(X)$, $\mathcal{P}(X)$ all polynomials on X, and
- dom $\mathcal{P}(^mX)$, $\mathcal{P}(^mX)$ all m-homogeneous polynomials on X.

Note that convergence (of the net of finite sums) in (7.2) of course means unconditional = absolute convergence, hence the "size" of dom $\mathcal{F}(R)$ somewhat measures the "remaining unconditionality" of the function space $\mathcal{F}(R)$. Recall from Theorem 5.3 that $\mathcal{P}(^mX)$ has no unconditional basis unless X is finite dimensional.

Observe that the Bohr radius $K(\mathbb{D})$ of the open unit disc (which by Bohr's power series theorem equals $1/3$, Theorem 4.1) easily can be written in terms of holomorphic functions:

$$K(\mathbb{D}) = \sup\{r > 0 \mid \forall f \in H_\infty(\mathbb{D}) : \sum_{n=1}^\infty |c_n(f)|r^n \le \|f\|_\mathbb{D}\},$$

and from this point of view the following definition gives another natural multi dimensional Bohr radius for Reinhardt domains R in $\mathbb{C}^n$; it will be our main tool in order to "estimate" individual sequences $z \in$ dom $\mathcal{F}(R)$ (see 7.5).

Definition 7.1. Let R be a Reinhardt domain in $\mathbb{C}^n$ and $\lambda \ge 1$. Then we define

$$A_\lambda(R) := \sup\{\frac{1}{n}\sum_{i=1}^n r_i \mid r \in \mathbb{R}_{\ge 0}^n, \forall f \in H_\infty(R) :$$

$$\sum_{\alpha \in \mathbb{N}_0^n} |c_\alpha(f)|r^\alpha \le \lambda \|f\|_R\},$$

and call $A(R) := A_1(R)$ the *arithmetic Bohr radius* of R.

Obiously, we have

(7.3) $$K(B_{\ell_\infty^n}) \le A(B_{\ell_\infty^n});$$

in fact the result from (4.3) originally was more an estimate for $A(B_{\ell_\infty^n})$ than for $K(B_{\ell_\infty^n})$. In [28] Dineen and Timoney used their polydisc version of Bohr's theorem to reprove the estimate $S \le 2$ in the way that Bohr originally might have thought of; they showed that if $\varepsilon \in\,]0, 1[^{\,\mathbb{N}}$ is such that $\sum |c_\alpha|\varepsilon^\alpha < \infty$ for all power series $\sum c_\alpha z^\alpha$ bounded on $[|z_i| \le 1]$, then $\varepsilon \in \ell_{2+\delta}$ for all $\delta > 0$ (a sort of converse of Theorem 3.1). In order to prove (4.3) Dineen and Timoney used a probabilistic method from [38] which goes back to the mid 70's (see the discussion in the preceding section) and which of course was not accessible to Bohr.

The relation (7.3) can be extended to arbitrary bounded Reinhardt domains.

Proposition 7.2. Let R be a bounded Reinhardt domain in $\mathbb{C}^n$. Then

$$A(R) \ge \frac{S(R, B_{\ell_1^n})}{n} K(R).$$

Recall Bohr's definition of the number S from (3.5); in terms of bounded holomorphic functions it is not difficult to see that S can be rewritten as

$$S = \sup\{q > 0 \mid \ell_p \cap B_{c_0} \subset \operatorname{dom} H_\infty(B_{c_0})\}\,.$$

Thus with our notation the results of Theorem 3.1 and Dineen–Timoney from [28] (see also the end of section 3) can be stated as follows:

For all $\varepsilon > 0$

$$(7.4) \qquad \ell_2 \cap B_{c_0} \subset \operatorname{dom} H_\infty(B_{c_0}) \subset \ell_{2+\varepsilon} \cap B_{c_0}\,.$$

In 1999 Lempert [36] extended results of Ryan [43] from 1987 on the dual case; they proved that the monomial expansion of a holomorphic function on B_{ℓ_1} (more generally on rB_{ℓ_1}, $0 < r \leq \infty$) converges uniformly absolutely on compact subsets of B_{ℓ_1}, hence in particular the following equality holds:

$$(7.5) \qquad \operatorname{dom} H_\infty(B_{\ell_1}) = B_{\ell_1}\,.$$

It will turn out that (7.4) and (7.5) reflect extreme cases of more general (and up to some point precise) descriptions for domains of convergence for bounded Reinhardt domains R located in certain symmetric Banach sequence spaces X.

First we have the following general lower inclusion.

Theorem 7.3. *Let R be a Reinhardt domain in a Banach sequence space X. Then*

$$(\ell_1 \cup (X \cdot \ell_2)) \cap R \subset \operatorname{dom} H_\infty(R)\,.$$

For the proof of this theorem we use a careful analysis of the above cited Ryan-Lempert results for ℓ_1 and Bohr's Theorem 3.1 for c_0.

The next theorem shows that in many situations the preceding lower inclusions are optimal "up to an ε" .

Theorem 7.4. *Let R be a bounded Reinhardt domain in a symmetric Banach sequence space X and $\varepsilon > 0$.*

(1) *If $X \subset \ell_2$, then*

$$\ell_1 \cap R \subset \operatorname{dom} H_\infty(R) \subset \ell_{1+\varepsilon} \cap R\,.$$

(2) *If X is 2-convex and satisfies*

$$\left(\frac{1}{\lambda_X(n)\, n^\delta}\right)_n \in X \quad \text{for every } \delta > 0,$$

then

$$(X \cdot \ell_2) \cap R \subset \operatorname{dom} H_\infty(R) \subset (X \cdot \ell_{2+\varepsilon}) \cap R\,.$$

In particular, if R is a bounded Reinhardt domain in ℓ_p, $1 \leq p < \infty$ or c_0 ($p = \infty$), then

$$\ell_1 \cap R \subset \operatorname{dom} H_\infty(R) \subset \ell_{1+\varepsilon} \cap R \qquad \text{if } p \leq 2$$

$$\ell_q \cap R \subset \operatorname{dom} H_\infty(R) \subset \ell_{q+\varepsilon} \cap R \qquad \text{if } p \geq 2,\ \frac{1}{q} = \frac{1}{2} + \frac{1}{p}\,.$$

For the proof of the upper inclusions in the preceding theorem we use the arithmetic Bohr radius, the probabilistic arguments collected in Theorem 4.7, and apply the technique of Dineen and Timoney from [28] to our more general situation. The key is the following link between the elements of dom $H_\infty(R)$ and the arithmetic Bohr radius from 7.1 which can be proved with a Baire argument.

Lemma 7.5. *Let R be a Reinhardt domain in a Banach sequence space X and $R_n := R \cap X_n$ its n-dimensional section. Then for every $z \in$ dom $H_\infty(R)$ we can find a $\lambda \geq 1$ such that for all $n \in \mathbb{N}$*

$$\frac{1}{n} \sum_{i=1}^{n} |z_i| \leq A_\lambda(R_n).$$

Now for $A_\lambda(R_n)$ we have the following estimates.

Proposition 7.6. *Let $R_n \subset \mathbb{C}^n$ be a bounded Reinhardt domain and $\lambda \geq 1$. Then*

$$\max\left\{\frac{1}{3n}\frac{1}{s_{1,n}}, \frac{s_{n,1}}{n} K(R_n)\right\} \leq A_\lambda(R_n) \leq 17\,\lambda^{2/\log n}\frac{\sqrt{\log n}}{n}\,s_{n,2}$$

where

$$s_{p,n} = S(B_{\ell_p^n}, R_n) \quad , \quad s_{n,p} = S(R_n, B_{\ell_p^n}).$$

¿From this and Lemma 7.5 we can deduce an asymptotic upper bound for the decreasing rearrangement of the elements of dom $H_\infty(R)$.

Proposition 7.7. *Let R be a bounded Reinhardt domain in a symmetric Banach sequence space X. Then for every $z \in$ dom $H_\infty(R)$ there is a constant $c > 0$ such that for all n*

$$z_n^* \leq c\frac{\sqrt{\log n}}{n} \|\mathrm{id} : X_n \longrightarrow \ell_2^n\|.$$

This immediately implies the upper inclusion in Theorem 7.4 whenever $X \subset \ell_2$ since in this case the inclusion mapping is automatically continuous. For the upper inclusion in (2) some properties of the product space $X \cdot \ell_p$ have to be investigated.

In [36] Lempert treated the problem of solving the $\overline{\partial}$-equation in an open subset of a Banach space. He was able to prove a result for $(0,1)$-forms on balls rB_{ℓ_1} in ℓ_1. A crucial step in his study is that, as already mentioned, each bounded holomorphic function on rB_{ℓ_1} has a pointwise convergent monomial expansion, i.e., dom $H_\infty(rB_{\ell_1}) = rB_{\ell_1}$. Some years earlier monomial expansions of holomorphic functions in infinitely many variables had already been investigated by Ryan [43], and as in Lempert's work the sequence space ℓ_1 plays a special role.

Lempert in [36] asks "Why ℓ_1 of all Banach spaces?", and points out how from the technical point of view his approach depends on the structure of ℓ_1. Our (probabilistic) techniques permit to state the following partial answer.

Theorem 7.8. *Let X be a symmetric Banach sequence space. If X contains a bounded Reinhardt domain R such that $\operatorname{dom} H_\infty(R)$ is absorbant, in particular if $\operatorname{dom} H_\infty(R) = R$, then*

$$\ell_1 \subset X \subset \ell_{1+\varepsilon}$$

for all $\varepsilon > 0$.

Proof. Factoring through ℓ_∞^n we get from Corollary 7.7 that

$$z_n^* \leq c\frac{\sqrt{\log n}}{n}n^{1/2} = c\frac{\sqrt{\log n}}{n^{1/2}}$$

for all $z \in \operatorname{dom} H_\infty(R)$. This implies $\operatorname{dom} H_\infty(R) \subset \ell_{2+\varepsilon}$, and thus $X \subset \ell_{2+\varepsilon}$ for all $\varepsilon > 0$ since by the assumption for all $z \in X$ there is a $\lambda > 0$ such that $\lambda z \in \operatorname{dom} H_\infty(R)$. But now since the embedding $X \hookrightarrow \ell_{2+\varepsilon}$ is automatically continuous, factorization through $\ell_{2+\varepsilon}^n$ gives

$$\|\operatorname{id} : X_n \longrightarrow \ell_2^n\| \leq n^{1/2-1/(2+\varepsilon)}$$

for all $\varepsilon > 0$. Thus the estimate in Proposition 7.7 implies that $\operatorname{dom} H_\infty(R)$ and hence X is contained in $\ell_{1+\varepsilon}$ for all $\varepsilon > 0$. $\square$

This raises the following question.

Problem 7.9. *Does a symmetric Banach sequence space X equal ℓ_1 provided $\operatorname{dom} H_\infty(B_X) = B_X$? Or more generally, is it possible to take $\varepsilon = 0$ in Theorem 7.8 ?*

Surprisingly, the following equivalences show that this is a reformulation of Problem 6.7.

Proposition 7.10. *Let R be a bounded Reinhardt domain in a symmetric Banach sequence space X and $\mathcal{F}(R)$ a closed subspace of $H_\infty(R)$ which contains the space $\mathcal{P}(X)$ of all polynomials on X. Equivalent are:*

(1) $tB_X \subset \operatorname{dom} \mathcal{F}(R)$ *for some $t > 0$.*

(2) $\operatorname{dom} \mathcal{F}(R)$ *is absorbant.*

(3) *There is an $0 < r < 1$ and a constant $\lambda > 0$ such that*

$$\sup_{z \in rB_X} \sum_{\alpha \in \mathbb{N}_0^{(N)}} |c_\alpha(f)z^\alpha| \leq \lambda\|f\|_R$$

for all $f \in \mathcal{F}(R)$.

(4) *There is a constant $c > 0$ such that for all m*

$$\sup_n \chi_{mon}(\mathcal{P}(^m X_n)) \leq c^m.$$

Note that for $\operatorname{dom} \mathcal{P}(X)$ a description can be given which is completely analoguous to Theorem 7.4. But in this case it is quite astonishing that there exists X different from ℓ_1 for which $\operatorname{dom} \mathcal{P}(X) = X$ holds (this is the space mentioned after Problem 5.8).

Similar descriptions as for $\operatorname{dom} H_\infty(R)$ and $\operatorname{dom} \mathcal{P}(X)$ can be proved for the domain of convergence $\operatorname{dom} \mathcal{P}(^m X)$ of all m-homogeneous polynomials, which in this case of course depend on m.

For $X = \ell_p$ we have the following result. Note that in accordance with Theorem 5.3 the upper inclusions again show that the monomials in $\mathcal{P}(^m \ell_p)$ do not form an unconditional basis.

Theorem 7.11. *Let* $m \geq 2$ *and* $1 \leq p \leq \infty$. *Define* q_m *by* $\frac{1}{q_m} = \frac{1}{p} + \frac{1}{2}(1 - \frac{1}{m})$. *Then*

(1) *For* $p = \infty$:

$$\ell_{2(1-\frac{1}{m})-1} \subset \operatorname{dom} \mathcal{P}(^m c_0) \subset \ell_{2(1-\frac{1}{m})-1+\varepsilon}$$

(2) *For* $2 \leq p < \infty$:

$$\ell_{q_m} \subset \operatorname{dom} \mathcal{P}(^m \ell_p) \subset \ell_{q_m, \infty}$$

(3) *For* $1 \leq p \leq 2$:

$$\ell_{\max\{1, q_m\}} \subset \operatorname{dom} \mathcal{P}(^m \ell_p) \subset \ell_{(mp')', \infty}$$

For case (3) we conjecture the following.

Problem 7.12. *Show that for* $1 < p < 2$

$$\ell_{(mp')'} \subset \operatorname{dom} \mathcal{P}(^m \ell_p).$$

Again this is closely connected with the topic from section 5, unconditionality in spaces of m-homogeneous polynomials. The conjecture is equivalent to the fact that

$$(7.6) \qquad \sup_n \chi_{mon}(\operatorname{id} : \mathcal{P}(^m \ell_p^n) \longrightarrow \mathcal{P}(^m \ell_{(mp')'}^n)) < \infty,$$

where $\chi_{mon}(\operatorname{id} : \mathcal{P}(^m X_n) \longrightarrow \mathcal{P}(^m Y_n))$ for Banach sequence spaces X and Y is defined to be the best constant $c > 0$ such that for all $\mu_\alpha, \varepsilon_\alpha \in \mathbb{K}$ with $|\varepsilon_\alpha| \leq 1$

$$\Big\| \sum_{|\alpha|=m} \varepsilon_\alpha \mu_\alpha z^\alpha \Big\|_{\mathcal{P}(^m Y_n)} \leq \Big\| \sum_{|\alpha|=m} \mu_\alpha z^\alpha \Big\|_{\mathcal{P}(^m X_n)}.$$

Recall that $\mathcal{P}(^m X_n)$ is isometrically isomorphic to the symmetric injective tensor product $\otimes_{\varepsilon_s}^m X_n'$. It can be seen that (7.6) for $2 < p < \infty$ is equivalent to

$$\sup_n \chi_{mon}(\operatorname{id} : \otimes_\varepsilon^m \ell_p^n \longrightarrow \otimes_\varepsilon^m \ell_{mp}^n) < \infty,$$

with $\chi_{mon}(\operatorname{id} : \otimes_\varepsilon^m X_n \longrightarrow \otimes_\varepsilon^m Y_n)$ the best constant $c > 0$ satisfying the inequality

$$\Big\| \sum_{i \in I} \varepsilon_i \mu_i e_i \Big\|_{\otimes_\varepsilon^m Y_n} \leq c \Big\| \sum_{i \in I} \mu_i e_i \Big\|_{\otimes_\varepsilon^m X_n}$$

for all $\mu_i, \varepsilon_i \in \mathbb{K}$ with $|\varepsilon_i| \leq 1$, $i \in I := \{1, \cdots, n\}^m$; here $e_i := e_{i_1} \otimes \cdots \otimes e_{i_m}$, $i \in I$.

References

[1] R. Alencar, On reflexivity and basis of $\mathcal{P}(^mX)$, *Proc. R. Ir. Acad.* **85** (1985), 131-138.

[2] L. Aizenberg, Multidimensional analogues of Bohr's theorem on power series, *Proc. Amer. Math. Soc.* **128** (1999), no. 4, 1147–1155.

[3] ———, Generalization of Caratheodory's inequality and the Bohr radius for multidimensional power series, *University of Ljubljana, IMFM, Preprint series* **39** (2001).

[4] L. Aizenberg, A. Aytuna, and P. Djakov, Generalization of a theorem of Bohr for bases in spaces of holomorphic functions of several complex variables, *J. Math. Anal. Appl.* **258** (2001), 429–447.

[5] L.Aizenberg, E.Liflyand, and A.Vidras, Multidimensional analogue of van der Corput-Visser inequality and its application to the estimation of Bohr radius, *Ann. Pol. Math.* **80** (2003), 47–54.

[6] R. Alencar, R.M. Aron, and G. Fricke, Tensor products of Tsirelson's space. *Illinois J. of Math.* **31**(1) (1987), 17-23.

[7] C. Bénéteau, A. Dahlner, and D. Khavinson, Remarks on the Bohr phenomenon, *Comput. Methods Funct. Theory* **4** (2004), 1–19.

[8] H.P. Boas, Majorant series, *J. Korean Math. Soc.* **37** (2000), no. 2, 321–337.

[9] H.P. Boas and D. Khavinson, Bohr's power series theorem in several variables, *Proc. Amer. Math. Soc.* **125**, 10 (1997), 2975-2979.

[10] H.F. Bohnenblust and E. Hille, On the absolute convergence of Dirichlet series, *Ann of Math.* (2) **32** (1934), 600-622.

[11] H. Bohr, Über die Bedeutung der Potenzreihen unendlich vieler Variabeln in der Theorie der Dirichletschen Reihen $\sum \frac{a_n}{n^s}$, *Nachrichten von der Königlichen Gesellschaft der Wissenschaften zu Göttingen, Mathematisch-Physikalische Klasse* (1913), 441–488.

[12] H. Bohr, A theorem concerning power series, *Proc. London Math. Soc.* (2) **13** (1914), 1-5.

[13] E. Bombieri and J. Bourgain, A remark on Bohr's inequality, *International Math. Research Notices* **80** (2004), 4308-4330.

[14] A. Defant, J.C. Díaz, D. García, and M. Maestre, Unconditional basis and Gordon-Lewis constants for spaces of polynomials, *Journal of Funct. Anal.* **181** (2001), 119-145.

[15] A. Defant, D. García, and M. Maestre, Bohr's power series theorem and local Banach space theory, *J. reine angew. Math.* **557** (2003), 173-197.

[16] A. Defant, D. García, and M. Maestre, Asymptotic estimates for the first and second Bohr radii of Reinhardt domains, to appear in J. Approx. Theory 2005.

[17] A. Defant, D. García, and M. Maestre, Maximum moduli of unimodular polynomials, J. Korean Math. Soc. **41** (2004), 209-230.

[18] A. Defant and K. Floret, "Tensor Norms and Operator Ideals", North–Holland Math. Studies, **176**, 1993.

[19] A. Defant and N. Kalton, Unconditionality in spaces of m-homogeneous polynomials on Banach spaces; to appear in Quarterly J. Math. (2005).

[20] A. Defant, M. Maestre, and C. Prengel, Domains of convergence in infinite dimensional holomorphy, in preparation.

[21] A. Defant and L. Frerick, A logarithmical lower bound for multi dimensional Bohr radii, to appear in Israel J. Math. (2005).

[22] S.J. Dilworth, N.J. Kalton, and D. Kutzarova, On the existence of almost greedy bases in Banach spaces, Studia Math. 159 (2003) 67-101.

[23] S.J. Dilworth, N.J. Kalton, D. Kutzarova, and V.N. Temlyakov, The thresholding greedy algorithm, greedy bases and duality, Constr. Approx. 19 (2003), 575-597.

[24] J. Diestel, H. Jarchow, and A. Tonge, "Absolutely Summing Operators", Cambridge Stud. Adv. Math. **43**, 1995.

[25] V. Dimant and S. Dineen, Banach subspaces of holomorphic functions and related topics, *Math. Scand.*, **83** (1998), 142-160.

[26] S. Dineen, "Complex Analysis on Infinite Dimensional Banach Spaces", Springer-Verlag. Springer Monographs in Mathematics, Springer-Verlag, London, 1999.

[27] S. Dineen and R.M. Timoney, Absolute bases, tensor products and a theorem of Bohr, *Studia Math.* **94** (1989), 227-234.

[28] S. Dineen and R.M. Timoney, On a problem of H. Bohr, *Bull. Soc. Roy. Sci. Liège* **60**, 6 (1991), 401-404.

[29] C. Finet, H. Queffélec, and A. Volberg, Compactness of Composition Operators on a Hilbert space of Dirichlet Series, *J. Funct. Anal.* **211** (2004), 271-287.

[30] K. Floret, Natural norms on symmetric tensor products of normed spaces, *Note di Mat.* **17** (1997), 153-188.

[31] B. Grecu and R. Ryan, Schauder bases for symmetric tensor products, to appear.

[32] J. Lindenstrauss and L. Tzafriri, "Classical Banach spaces I and II". Springer-Verlag, 1977, 1979.

[33] Y. Gordon and D.R. Lewis, Absolutely summing operators and local unconditional structures, *Acta Math.* **133** (1974), 27-47.

[34] S.V. Konyagin and V.N. Temlyakov, A remark on greedy approximation in Banach spaces, *East J. Approx.* **5** (1999), 365-379.

[35] L. Lempert, Approximation de fonctions holomorphes d'un nombre infini de variables, *Ann. Inst. Fourier* **49** (1999), 1293-1304.

[36] _____, The Dolbeault complex in infinite dimensions II, *J. Am. Math. Soc.* **12** (1999), 775-793.

[37] _____, Approximation of holomorphic functions of infinitely many variables II, *Ann. Inst. Fourier* **50** (2000), 423-442.

[38] A. M. Mantero and A. Tonge, The Schur multiplication in Banach algebras, *Studia Math.* **68** (1980), 1-24.

[39] V. I. Paulsen, G. Popescu, and D. Singh, On Bohr's inequality, *Proc. London Math. Soc.* **85** (2002), 493-512.

[40] A. Pełczyński, A property of multilinear operations, *Studia Math.*, **16** (1957), 173-182.

[41] C. Prengel, Domains of convergence in infinite dimensional holomorphy, Ph. D. Thesis, Unversität Oldenburg, to appear 2005.

[42] R. Ryan, Applications of Topological Tensor Products to Infinite Dimensional Holomorphy. Ph. D. Thesis. Trinity College of Dublin, 1982.

[43] R. Ryan, Holomorphic mappings on ℓ_1, *Trans. Amer. Math. Soc.* **302** (1987), 797-811.

[44] O. Toeplitz, Über eine bei den Dirichletschen Reihen auftretende Aufgabe aus der Theorie der Potenzreihen von unendlichvielen Veränderlichen, *Nachrichten von der Königlichen Gesellschaft der Wissenschaften zu Göttingen, Mathematisch-Physikalische Klasse* (1913), 417-432.

[45] N. Tomczak–Jaegermann, "Banach–Mazur Distances and Finite–Dimensional Operator Ideals". Longman Scientific & Technical, 1989.

[46] L. Tzafriri, On Banach spaces with unconditional basis, *Israel J. Math.* **17** (1974), 84–93.

INSTITUT OF MATHEMATICS, CARL VON OSSIETZKY UNIVERSITY, D–26111, OLDENBURG, GERMANY

E-mail address: `defant@mathematik.uni-oldenburg.de`

INSTITUT OF MATHEMATICS, CARL VON OSSIETZKY UNIVERSITY, D–26111, OLDENBURG, GERMANY

E-mail address: `christopher.prengel@mail.uni-oldenburg.de`

To Joram and Lior with appreciation of their
contribution to the theory of Banach spaces

SELECTED PROBLEMS ON THE STRUCTURE OF COMPLEMENTED SUBSPACES OF BANACH SPACES

ALEKSANDER PEŁCZYŃSKI

ABSTRACT. We discuss several old and more recent problems on bases, complemented subspaces and approximation properties of Banach spaces

INTRODUCTION AND NOTATION

This paper is based on author's lecture at the conference "Contemporary Ramifications of Banach Space Theory" in honor of Joram Lindenstrauss and Lior Tzafriri, (Jerusalem, June 19-24, 2005). We present several open problems, some more than 20 years old, some recent. They mainly concern properties of various classes of Banach spaces expressed in terms of the structure of complemented subspaces of the spaces, bases and approximation by finite rank operators. The paper consists of 5 sections. In Section 1 we outline the proof of the classical Complemented Subspaces Theorem due to Lindenstrauss and Tzafriri [LT1] which characterizes Hilbert spaces among Banach spaces. We follow the presentation by Kadec and Mityagin [KaMy] emphasizing some points which are omitted in their approach. Section 2 is devoted to some problems which naturally arise from the Complemented Subspace Theorem. In Section 3 we discuss spaces which have few projections; their properties are on the other extreme from the assumption of the Complemented Subspace Theorem. Section 4 is devoted to approximation properties of some special non-separable function spaces and spaces of operators. Section 5 has a different character; it concerns bases for $L^p(G)$ consisting of characters where G is either a Torus group, or the Dyadic group.

We employ standard Banach space notation and terminology (cf. [LT1]). In particular, "subspace" means "closed linear subspace", "operator" means "bounded linear operator", "basis" means "Schauder basis", "projection' means "bounded linear idempotent" and "c-complemented subspace" means "the range of projection of norm at most c". The symbols $\mathbb{R}, \mathbb{C}, \mathbb{Z}, \mathbb{N}$ stand for real numbers, complex numbers, integers and positive integers, respectively. By $\#A$ we denote the number of elements of a finite set A. The symbol $\mathcal{F}(X)$

Keywords: Banach and Hilbert spaces; complemented subspaces, projections, bases, approximation property, finite rank operator, character of compact Abelian group.

2000 *Mathematics Subject Classification*: 46B)3, 46B07, 46B15, 46B25, 46B26, 46B28, 42A16, 42C10.

denotes the family of all finite-dimensional subspaces of a Banach space X; $d(E, F)$ stands for the multiplicative Banach-Mazur distance between the Banach spaces E and F.

1. Complemented subspace theorem (CST)

Solving the problem which had been open more than 40 years Lindenstrauss and Tzafriri proved in 1971 [LT1]

Theorem 1.1. *(CST). A Banach space X is isomorphic to a Hilbert space if and only if*

(1.1) *every subspace of X is complemented.*

Outline of the proof. Without loss of generality we assume that $\dim X = \infty$.

Step I: By a standard construction of finite dimensional Schauder decompositions we get (1.1) $\Longrightarrow$ (1.2), where

(1.2) $\quad \exists K \geq 1 : \forall E \in \mathcal{F}(X) \quad \exists$ projection $P : X \xrightarrow{onto} E$ with $||P|| \leq K$.

Step II: Given $E \in \mathcal{F}(X)$ and $\varepsilon > 0$ there exist $F \in \mathcal{F}(X)$ and projections $p : E + F \xrightarrow{onto} E$ and $q : E + F \xrightarrow{onto} F$ such that

$$\max(||p||, ||q||) \leq 1 + \varepsilon, \quad pq = qp = 0, \quad d(F, \ell^2_{\dim E}) < 1 + \varepsilon.$$

Using first a standard trick which goes back to S. Mazur we construct a subspace X_1, which is the intersection of the kernels of finitely many linear functionals on X each attaining its norm on points of an $\varepsilon/2-$net of the unit sphere of E, to get an infinite dimensional subspace almost orthogonal to a given finite dimensional subspace. Let $p_1 : E + X_1 \to X_1$ be defined by $p_1(e + x) = x$ for $e \in E$ and $x \in X_1$. The projection p is the restriction of p_1 to a subspace $E + F$ of $E + X_1$ where F is a suitable $n-$dimensional subspace of $E + X_1$; we put $q = Id - p$.

The construction of F is more sophisticated. It is an "almost hilbertian" subspace of X_1 with the property that the restriction $h_{|F}$ of the quotient map $h : E + X_1 \xrightarrow{onto} (E + X_1)/E$ is an "almost isometry". The construction uses the Dvoretzky Theorem on spherical sections (cf.[DV]) and the Krasnosielski-Krein-Milman lemma [KKrM] which is an easy consequence of the Borsuk-Ulam antipodal theorem; we refer to [P1, Added in proof]. The first proof of Step II is due to Figiel (cf.[P1]) and is based upon V. Milman's concentration of measure spectrum theorem [Mi].

By an inessential change of norm one may assume

(1.3) $\qquad d(F, \ell^2_{\dim E}) = 1 \quad \text{and} \quad ||p|| = ||q|| = 1.$

Step III:

Main Lemma. *Let E, F satisfy (1.3). Let $\dim E = n$. Let $T : E \to \ell^2_n$ be an isomorphism. Let $s : F \to \ell^2_n$ be an isometry. Put $d := d(E, \ell^2_n)$.*

Let K be the common upper bound of the norms of minimal projections from $G := E \oplus F$ onto its subspaces. Then

$$d \leq 4K^2.$$

Proof. (after Kadec and Mityagin [KaMy] (1973)). Let $D = \{e + s^{-1}Te : e \in E\} \subset G$. Let $R : G \xrightarrow{onto} D$ be a projection with $||R|| \leq K$. Define the map $\tilde{T} : D \to \ell^2_{2n} = \ell^2_n \oplus \ell^2_n$ by

$$\tilde{T}e = (Te, sq\sqrt{\mu}Re),$$

where $\mu > 0$ will be chosen later. Clearly

$$||\tilde{T}e|| = \sqrt{||Te||^2 + \mu||sqRe||^2} \leq \sqrt{||T||^2 + \mu K^2}||e||.$$

To estimate $||\tilde{T}e||$ from below use the identity

$$(1.4) \qquad Te = sqRe + TpRs^{-1}Te \quad \text{for} \quad e \in E.$$

Then

$$
\begin{aligned}
||\tilde{T}e|| &= \sqrt{||Te||^2 + \mu||Te - TpRs^{-1}Te||^2} \\
&\geq \sqrt{\frac{||pRs^{-1}Te||^2}{||pRs^{-1}T||^2} + \frac{||e - pRs^{-1}Te||^2}{\mu^{-1}||T^{-1}||^2}} \\
&\geq \frac{||pRs^{-1}Te|| + ||e - pRs^{-1}Te||}{\sqrt{||pRs^{-1}||^2 + \mu^{-1}||T^{-1}||^2}} \quad \text{(by Cauchy-Schwarz inequality)} \\
&\geq \frac{||e||}{\sqrt{K^2 + \mu^{-1}||T^{-1}||^2}}.
\end{aligned}
$$

Thus $\tilde{T}$ is invertible. A subspace of a Hilbert space is itself a Hilbert space; hence

$$d \leq ||\tilde{T}|| ||\tilde{T}^{-1}|| \leq \sqrt{(||T||^2 + K^2\mu)(K^2 + \mu^{-1}||T^{-1}||^2)}.$$

Choosing T so that $||T|| ||T^{-1}|| = d$ and $||T|| = ||T^{-1}||$ we get

$$d \leq \sqrt{(d + K^2\mu)(K^2 + \mu^{-1}d)}.$$

Putting $\mu = d/K^2$ which minimizes the right hand side of the latter inequality we get $d \leq 2K\sqrt{d}$.

It remains to verify (1.4). By the definition of R there are e_0 and e_1 in E such that

$$(1.5) \qquad Re = e_0 + s^{-1}Te_0; \quad Rs^{-1}Te = e_1 + s^{-1}Te_1.$$

Since R is a projection onto D, adding by sides the identities (1.5) we get

$$e + s^{-1}Te = R(e + s^{-1}Te) = (e_0 + e_1) + s^{-1}T(e_0 + e_1),$$

which implies the equality of the "coordinates"

(1.6) $$e = e_0 + e_1; \quad Te = T(e_0 + e_1).$$

Since $qe_0 = 0, qs^{-1} = s^{-1}, pe_1 = e_1, ps^{-1} = 0$, it follows from (1.5) that

$$sqRe = sqe_0 + sqs^{-1}Te_0 = Te_0; \quad TpRs^{-1}Te = Tpe_1 + Tps^{-1}Te_1 = Te_1.$$

Adding by sides these identities and invoking (1.6) we get (1.4). $\qquad\square$

Step IV: Proof of (CST). It follows from the previous steps that if X satisfies (1.2) and $\dim X = \infty$ then every finite dimensional subspace of X is within Banach-Mazur distance $4K^2$ from a Hilbert space. This yields $d(X, H) \leq 4K^2$, where H is a Hilbert space of suitable density character (cf. e.g. [L] (1963)). $\qquad\square$

2. Open problems related to (CST)

2.1. Lindestrauss-Tzafriri vs. Kakutani-Bohnenblust.

The proof of (CST) presented here gives for $\dim X = \infty$ that if X satisfies (1.2) for some $K \geq 1$ then $d(X, H) \leq 4K^2$. Kakutani (1939) [Kk], for the real case, and Bohnenblust (1941) [Bo], for the complex case, proved:

Theorem 2.1. *If $\dim X \geq 3$ and every $2-$dimensional subspace of X is $1-$complemented then $d(X, H) = 1$.*

However, a transparent proof in the complex case would be welcome. We put

$$\Phi_{\dim X}(K) = d(X, \ell^2_{\dim X}) \quad \text{if } X \text{ satisfies (1.2) for given } K \geq 1.$$

Kadec and Mityagin [KaMy] essentially asked

Problem 2.1. *Describe the function $K \to \Phi_{\dim X}(K)$.*

Very recently Kalton [Kal] showed that the function $K \to \Phi_{\dim X}(K)$ is continuous at 1. In particular he fully described the function for $\dim X = \infty$ when K is close to 1. Precisely he proved

Theorem 2.2. *There exists constants $0 < c < C < \infty$ so that*

$$c\sqrt{K-1} \leq \Phi_\infty(K) \leq C\sqrt{K-1} \quad \text{for } 1 \leq K \leq 2.$$

2.2. Operator version of (CST).

Definition 2.3. *Let X, Y be Banach spaces. An operator $T : X \to Y$ is said to be fully decomposable, in symbols $T \in \mathcal{FD}$, provided that*

$$\forall E \in \mathcal{F}(X) \; \exists T_E : X \to T(E) \text{ such that } T_E \text{ extends the restriction } T_{|E} : E \to T(E)$$

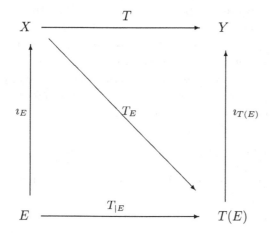

Here $\iota_E, \iota_{T(E)}$ denote the set theoretical inclusions.
For $K \geq 1$ let

$$\mathcal{FD}_K \;=\; \Big\{ T : X \to Y \ (X, Y\text{-Banach spaces}) \text{ such that}$$

$$\forall E \in \mathcal{F}(X) \exists T_E : X \to T(E) \text{ extending } T_{|E} \text{ with } \|T_E\| \leq K \Big\}.$$

Then

$$T \in \mathcal{FD} \Leftrightarrow T \in \mathcal{FD}_K \text{ for some } K \geq 1.$$

For the implication "$\Rightarrow$" assume to the contrary the existence of a $T \in \mathcal{FD} \setminus \bigcup_{K \geq 1} \mathcal{FD}_K$. Similarly as in Step 1 (using Mazur's trick described in Step 2) one constructs a subspace E_∞ of X which admits a Schauder decomposition (cf. [LT2], vol.I p.47 for definition) $(E_n)_{n=1}^\infty$ such that $E_n \in \mathcal{F}(X)$ but $T_{|E_n}$ has no extension $T_E : X \to T(E_n)$ with $\|T_E\| \leq n$. Thus $T_{|E_\infty}$ does not extend to a bounded operator from X into $T(E_\infty)$; otherwise all $T_{|E_n}$ would have extensions with uniformly bounded norms. A contradiction.

The implication "$\Leftarrow$" is a routine application of a standard compactness argument and the following

Fact. *If $T \in \mathcal{FD}_K$ for some $K \geq 1$ then T is weakly compact.*

Outline of the proof. If there were a non weakly compact $T \in \mathcal{FD}_K$ for some $K \geq 1$ then, by [?, Theorem 8.1], $(\Sigma) \in \mathcal{FD}_C$ for some $C \geq 1$ where $(\Sigma) : \ell^1 \to \ell^\infty$ is defined by $(\Sigma)((a_j)_{j=1}^\infty) = (\sum_{j=1}^n a_j)_{n=1}^\infty$. Thus $(\int) \in \mathcal{FD}_{2C}$, where $(\int) : L^1(0,1) \to C(0,1)$ is defined by $(\int)(f)(t) = \int_0^t f(s)ds$ for $0 \leq t \leq 1$. Hence $V \in \mathcal{FD}_{4C}$, where $V : L^1(\mathbb{T}) \to C(\mathbb{T})$ ($\mathbb{T}$ denotes the circle group) is defined by $V(e^{int}) = \frac{1}{in}e^{int}$ for $n = \pm 1, \pm 2, \ldots$ and $V(1) = 0$.

We get a contradiction by showing that $V \notin \bigcup_{K \geq 1} \mathcal{FD}_K$. For $n = 1, 2, \ldots$ let H_n^1 (resp. A_n) denote the subspace of the space $L^1(\mathbb{T})$ (resp. $C(\mathbb{T})$) spanned by the characters e^{ijt} for $j = 1, 2, \ldots, n$. Let $V_n : L^1(\mathbb{T}) \to A_n$ be an extension of the restriction $V_{|H_n^1}$. Since $V_{|H_n^1}$ is translation invariant, by averaging we get the translation invariant extension, say $\widetilde{V}_n : L^1(\mathbb{T}) \to A_n$; moreover $||\widetilde{V}_n|| \leq ||V_n||$. It is easy to see that $\widetilde{V}_n$ is the operator of convolution by the function $g_n = \sum_{j=1}^{n} \frac{1}{ij} e^{ijt}$. Let F_n denote the n-th Fejer kernel. Then $||F_n||_{L^1(\mathbb{T})} = 1$. The divergent of the harmonic series yields

$$\sup_n ||\widetilde{V}_n(F_n)||_{C(\mathbb{T})} \geq \sup_n |\widetilde{V}_n(F_n)(0)| = +\infty.$$

Thus no sequence (V_n) of extensions $(V_{|H_n^1})$ has uniformly bounded norms. Hence $V \notin \bigcup_{K \geq 1} \mathcal{FD}_K$. $\qquad\square$

A similar argument gives

Proposition 2.4. *Fully decomposable operators are weakly compact.*

Proposition 2.4 was discovered by T. Figiel and myself; it was published with our permission by Mascioni [Ma] in 1990. He invented an alternative proof.

Recall that an operator $T : X \to Y$ is *hilbertian* if it factors through a Hilbert space, i.e. $T = BA$ for some Hilbert space H and operators $A : X \to H$ and $B : H \to Y$. Obviously if T is hilbertian then $T \in \mathcal{FD}$.

Problem 2.2. *Is every fully decomposable operator hibertian?*

A restatement of (CST) says that the identity on a Banach space is fully decomposable if and only if it is hilbertian. An analysis of the proof of (CST) shows that to get a positive solution of Problem 2.2 it is enough to establish that the sum of two fully decomposable operators is fully decomposable; it is also enough to show that there is a function $F : [1, \infty) \to [1, \infty)$ such that if $T : X \to Y$ is in $\mathcal{FD}_K$ then there is $F(K)$s such that

$$T \oplus I_n : X \oplus \ell_n^2 \to Y \oplus \ell_n^2 \in \mathcal{FD}_{F(K)} \quad \text{for } n = 1, 2, \ldots,$$

where $I_n =$ denotes the identity operator on ℓ_n^2 and "$\oplus$" stands for Cartesian product.

2.3. **(CST) and subspaces with bases.** A vague question. How "large" should the family of complemented subspaces of a Banach space X be to insure that X is isomorphic to a Hilbert space?

Problem 2.3. *Assume that*

(2.1)   *every subspace of X with a basis is complemented in X.*

Is X isomorphic to a Hilbert space?

Recall that the *basis constant* of a Banach space Y, in symbol $bc(Y)$, is the least upper bound of the constants L such that there is a basis (e_n) for Y satisfying

$$\|\sum_{j=1}^{n} t_j e_j\| \le L\|\sum_{j=1}^{n+m} t_j e_j\| \quad \text{for all scalars } t_1, t_2, \ldots, t_m \text{ and } n, m = 1, 2, \ldots.$$

If Y has no basis we set $bc(Y) = \infty$. A standard argument yields the equivalence (2.1) $\Longleftrightarrow$ (2.2), where

(2.2) $\exists C \ge 1$ such that every $E \in \mathcal{F}(X)$ is $C \cdot bc(E)$−complemented in X.

Using the Main Lemma one gets

Proposition 2.5. *If X satisfies (2.1) then*

$$d(E, \ell_{\dim E}^2) \le (2K bc(E))^2 \quad \text{for every } E \in \mathcal{F}(X).$$

Thus, Problem 2.3 reduces to

Conjecture. *There are $C \ge 1$ and a function $L \to \Psi(L)$ for $L \ge 1$ with* $\lim_{L \to \infty} \Psi(L) = \infty$ *such that if $\dim E < \infty$ and $d(E, \ell_{\dim E}^2) \ge L$ then there is a subspace $F \subset E$ with $bc(F) \le C$ and $d(F, \ell_{\dim F}^2) \ge \Psi(L)$.*

Roughly speaking the Conjecture says that *"A finite-dimensional space which is far from Hilbert spaces has a subspace far from Hilbert spaces with a "nice" basis.*

Remark 2.6. Gowers (1996) [G] using a Ramsey type argument (Gowers Dichotomy) combined with a contribution of Komorowski and Tomczak-Jaegermann (1995)[KoT] proved

Theorem 2.7. *A Banach space all of whose subspaces are isomorphic is isomorphic to a Hilbert space.*

In analogy with Problem 2.3 one asks

Problem 2.4. *Assume that all separable infinite-dimensional subspaces with bases of a Banach space X are isomorphic. Is X isomorphic to a Hilbert space?*

Problem 2.4 also easily reduces to the Conjecture. Recall that I asked in 1964 [P2] *"Assume that a Banach space X has the property that all its separable subspaces with bases are isomorphic to ℓ^2. Is X isomorphic to a Hilbert space?"* This problem remains still open. It also reduces to the Conjecture.

3. SPACES WITH FEW PROJECTIONS

It is convenient to employ the following notation. Let X be an infinite-dimensional Banach space. Put

$\mathcal{S}(X) =:$ the family of all closed subspaces of X;

$\mathcal{C}(X) =:$ the family of all proper complemented subspaces of X;

$\mathcal{F}(X) =:$ the family of all finite-dimensional subspaces of X.

Obviously

$$\mathcal{S}(X) \supset \mathcal{C}(X) \supset \mathcal{F}(X);$$

(CST) can be restated as follows:

$$\mathcal{S}(X) = \mathcal{C}(X) \Leftrightarrow X \text{ is isomorphic to a Hilbert space.}$$

In this section we consider problems related to Banach spaces with properties on the other extreme from the assumption of (CST).

A Banach space X is an $\mathcal{I}-$space ($=indecomposable$ space), provided that $\mathcal{C}(X) = \mathcal{F}(X)$. If every $Y \in \mathcal{S}(X) \setminus \mathcal{F}(X \in \mathcal{I})$ then X is called a $\mathcal{HI}$-space ($=hereditarily\ indecomposable$ space). A Banach space is *locally badly complemented* provided that there is an increasing sequence (c_n) with $\lim_n c_n = \infty$ such that every $n-$dimensional subspace of X is at least c_n-complemented in X. $\mathcal{HI}$-spaces were constructed in 1993 by Gowers and Maurey [GMau1, GMau2, Mau]. Badly locally complemented spaces were constructed in 1983 by Pisier [Pi].

Problem 3.1. *Construct a $\mathcal{HI}$-space which is locally badly complemented.*

To some extent, finite dimensional spaces (constructed by Gluskin [Gl] and Szarek [Sa]) with the property that their half-dimensional subspaces are "badly" complemented are finite dimensional counterparts of Pisier's space. Discussing the topic with Nicole Tomczak-Jaegermann we came to the conclusion that the full finite-dimensional analogue of Pisier's space seems to be unknown.

Problem 3.2. *Does there exists an increasing sequence (c_n) with $\lim_n c_n = \infty$ such that for $n = 1, 2, \ldots$ there is an $n-$dimensional Banach space F_n such that every subspace E of F_n is at least c_m-complemented where $m = \min(\dim E, \dim F_n/E)$?*

The spaces constructed by Gowers and Maurey have bases. It follows from general theory (cf. [LT2]) that every $\mathcal{HI}-$space contains a $\mathcal{HI}-$space with a basis and that there are $\mathcal{HI}-$spaces without bases. The following is open:

Problem 3.3. *Is every $\mathcal{HI}-$space isomorphic to a subspace of a $\mathcal{HI}-$space with a basis?*

Recently, Koszmider (2004) [Kos] proved

Theorem 3.1. *There exists a compact Hausdorff space $\mathbb{K}$ such that $C(\mathbb{K})$ is an $\mathcal{I}-$space.*

Of course if Q is an infinite compact Hausdorff space then $C(Q)$ is not a $\mathcal{HI}-$space because it contains c_0. However there exist $\mathcal{L}_\infty-$spaces which do not contain c_0 (cf. Bourgain and Delbaen 1980 [BrDl]).

Problem 3.4. *Does there exist an* $\mathcal{L}_\infty$*–space which is* $\mathcal{HI}$*?*

I do not see any formal reason which prevents the existence of such a space.

4. Approximation property of some non-separable spaces

We introduce convenient notation and recall some classical definitions. Let X be a Banach space. Denote by $\mathbf{B}(X)$, $\mathbf{K}(X)$, $\mathbf{F}(X)$ the spaces of all operators from X into X which are, respectively, bounded, compact, of finite rank. Here AP stands for "approximation property"; BAP stands for "bounded approximation property", UAP stands for "uniform approximation property" and MAP stands for "metric approximation property. Recall their definitions:

$$X \in AP \overset{df}{\equiv} \forall A \subset X, A - \text{compact } \forall \varepsilon > 0 \ \exists T \in \mathbf{F}(X) \text{ such that:}$$
$$\|Tx - x\| < \varepsilon \quad \text{for } x \in A.$$

$$X \in BAP_K \overset{df}{\equiv} \exists K > 1 \ \forall A \subset X, \#A < \infty \ \exists T \in \mathbf{F}(X) > 1 \text{ such that:}$$
$$Tx = x \quad \text{for } x \in A; \ \|T\| < K.$$

$$X \in UAP_K \overset{df}{\equiv} \exists K > 1 \ \exists \text{ function } n \to N(n) \text{ such that:}$$
$$\forall A \subset X, \#A < \infty \ \exists T \in \mathbf{F}(X) \text{ such that:}$$
$$Tx = x \quad \text{for } x \in A; \ \|T\| < K; \ \dim T(X) \le N(\#A).$$

$$X \in MAP \overset{df}{\equiv} X \in BAP_K \text{ for all } K > 1.$$

"BAP" means "BAP_K for some K". Clearly $UAP \Rightarrow BAP \Rightarrow AP$. The AP and MAP have been introduced by Grothendieck [Gr]; the BAP, essentially, by Banach [B]; and the UAP, by Pełczyński and Rosenthal [PR]. Our knowledge of the approximation properties of special important non-separable spaces seems to be unsatisfactory. The problem whether the Hardy space H^∞ has the AP has been open from more than 25 years. A formally stronger question is whether the Disc Algebra A has the UAP. A counterpart to these problems for spaces of smooth functions is

Problem 4.1. *Does* $L^\infty_m(\mathbb{R}^n)$ *have the AP? Does* $C^m_0(\mathbb{R}^n)$ *have the UAP?* ($n = 2, 3, \ldots; m = 1, 2, \ldots$)

Here $L^\infty_m(\mathbb{R}^n)$ dentes the Sobolev space of uniformly bounded functions on $\mathbb{R}^n$ vanishing together with its first m distributional derivatives at infinity; $C^m_0(\mathbb{R}^n)$ stands for the space of functions on $\mathbb{R}^n$ with continuous partial derivatives of order $\le m$, vanishing at infinity together with the first m derivatives.

The most important negative result concerns spaces of operators; equivalently, tensor products; it is:

Theorem 4.1. *(Szankowski (1981) [Sz1]) If H is an infinite-dimensional Hilbert space then $\mathbf{B}(H) \notin AP$.*

Since the UAP is preserved while passing to dual and predual spaces [H] and $(\mathbf{K}(H))^{**} = \mathbf{B}(H)$, Theorem 4.1 implies that $\mathbf{K}(H) \notin UAP$. Also $\mathbf{K}(H)^* = \mathbf{S}_1(H) \notin UAP$ where $\mathbf{S}_p$ denotes the Schatten-von Neumann trace class p for $1 \leq p < \infty$. Analyzing the proof of Theorem 4.1, Szankowski (1984)[Sz2] proved that $\mathbf{S}_p(H) \notin UAP$ for $\infty > p > 80$ and $80/79 > p \geq 1$. It is therefore natural to ask

Problem 4.2. *Does $\mathbf{S}_p \notin UAP$ for all $p \neq 2$?*

It follows from Theorem 4.1 that if two Banach spaces X and Y contain ℓ_n^1 uniformly complemented for $n = 1, 2, \ldots$ then the projective tensor product $X \widehat{\otimes} Y$ and the injective tensor product $X \widecheck{\otimes} Y$ fail UAP. In particular if $1 < p_1 \leq p_2 < \infty$ then $\ell_{p_1} \widehat{\otimes} \ell_{p_2}$ and $\ell_{p_1} \widecheck{\otimes} \ell_{p_2}$ fail UAP. The limit case when one of the spaces is replaced either by ℓ^1 for injective tensor product or by c_0 for projective tensor product seems to be open. In particular

Problem 4.3. *Do $c_0 \widecheck{\otimes} c_0$ and its dual $\ell^1 \widehat{\otimes} \ell^1$ fail UAP?*

We pass to function spaces in $L^1 - type$ norms. First we mention the deep result due to P. Jones (1985) [J] (for a detailed presentation see [Mu, Chapt. 5.3]).

Theorem 4.2. *The Hardy space $H^1 \in UAP$.*

Remembering that the UAP is preserved while passing to the dual and predual spaces we derive from Theorem 4.2 that the spaces of functions on the circle of Bounded Mean Oscillation and Vanishing Mean Oscillation have UAP. Another recent result [ACPPr] says

Theorem 4.3. *$BV(\mathbb{R}^n) \in BAP_K$ for some $K = K(n)$ $(n = 1, 2, \ldots)$.*

Here $BV(\mathbb{R}^n)$ denotes the space of functions with bounded variation on $\mathbb{R}^n$ which can be defined as the space of $f : \mathbb{R}^n \to \mathbb{C}^n$ such that

$$(4.1) \qquad \|f\|_{BV(\mathbb{R}^n)} = sup \left| \int_{\mathbb{R}^n} f \, \mathrm{d}iv \, \vec{\phi} \; dx \right| < \infty,$$

where the supremum extends over all vector-valued C^∞-functions $\vec{\phi} = (\phi_j)_1^n$ with compact support such that $sup_{x \in \mathbb{R}^n} \sum_j |\phi_j(x)|^2 \leq 1$. Analogously one defines spaces $BV(\Omega)$ for $\Omega \subset \mathbb{R}^n$ an open non-empty set. Theorem 4.3 extends to $BV(\Omega)$ provided that there is a linear extension operator from $BV(\Omega)$ into $BV(\mathbb{R}^n)$. For "pathological" sets Ω nothing is known. Spaces $BV(\Omega)$ are Sobolev type of spaces, closely related to classical Sobolev spaces $L_1^1(\Omega)$ of absolutely summable functions with absolutely summable first distributional partial derivatives.

Problem 4.4. *Does $BV(\mathbb{R}^n) \in UAP$, equivalently does $L_1^1(\mathbb{R}^n) \in UAP$ for $n = 2, 3, \ldots$?*

The next question is a result of discussions with T. Figiel and W. B. Johnson.

Problem 4.5. *Does $BV(\mathbb{R}^n) \in MAP$ for $n = 2, 3, \dots$*

Comment: $BV(\mathbb{R}^n)$ is in the norm (4.1) a dual Banach space space [PWj]. By a result of Grothendieck [Gr, LT2]) $AP \Rightarrow MAP$ for separable dual spaces. Whether this implication is true for nonseparable dual spaces is open. The space $BV(\mathbb{R}^n)$ as a Banach space is isomorphic with the space $BV(\mathbb{T}^n)$ of $n-$periodic functions with bounded variation. The latter is an example of a translation invariant vector-valued space on the torus $\mathbb{T}^n$. For scalar valued translation invariant spaces the analogues are the following spaces of measures: Let G be a compact Abelian group, Γ the space of characters of G; i.e., the dual group of G. Let $M(G)$ denote the space of all complex-valued measure on G with bounded variation. Let $A \subset \Gamma$. Let us put

$$M_{A^\perp}(G) = \{\mu \in M(G) : \int_G \overline{\gamma(x)}\mu(dx) = 0 \quad \text{for } \gamma \in A\}.$$

$M_{A^\perp}(G)$ under the norm of total variation of measure is a dual Banach space. Very little is known about approximation properties of $M_{A^\perp}(G)$. A special case is

Problem 4.6. *Let $A \subset \Gamma$ be an infinite Sidon set. Does $M_{A^\perp}(G) \in AP$?*

We do not know the answer to Problem 4.6 even in the simplest cases: for the Dyadic Group when A are the Rademacher functions, and for the Circle Group when A is the set of exponents $\{e^{i2^k t} : k = 0, 1, \dots\}$.

5. BASES OF CHARACTERS ON L^p SPACES

We are interested in bases in $L^p(G)$ consisting of characters where G is either the finite torus $\mathbb{T}^d$ for $d = 1, 2, \dots$ or the infinite torus $\mathbb{T}^\mathbb{N}$ or the Dyadic Group $\mathbb{D}$.

If (x_n) is a basis in a Banach space the X then the *basis constant* of (x_n) is defined by

$$bc((x_n); X) = \inf\{C : \|\sum_{j=1}^n a_j x_j\| \le C\|\sum_{j=1}^{n+m} a_j x_j\|,$$

where the infimum extends over all scalar sequences (t_j) and $n, m = 1, 2, \dots$. Two bases (x_n) in X and (y_n) in Y are said to be *equivalent* provided that the map $x_n \to y_n$ extends to an isomorphism from X onto Y. We order the set of characters of $\mathbb{T}$ in the sequence

$$(e_n) := (1, e^{it}, e^{-it}, e^{2it}, e^{-2it}, \dots).$$

The characters of $\mathbb{D}$ are identified with the Walsh orthonormal system. They are ordered in the sequence (w_n) according to so called *Paley order* [ScWaSi]. Classical results due to F. Riesz for the exponents [LT2, Vol 2,Chapt

II, c] and to Paley for Walshes [ScWaSi] state that (e_n) and (w_n) form bases for $L^p(\mathbb{T})$ and $L^p(\mathbb{D})$, respectively. Moreover

$$bc((e_n), L^p(\mathbb{T})) = O(p, \frac{p}{p-1}), \quad bc((w_n)), L^p(\mathbb{D})) = O(p, \frac{p}{p-1}), \quad (1 < p < \infty).$$

It follows from [GeGi], [DeF, p.158] that if A and B are compact metric Abelian groups and (a_n) and (b_n) are bases of characters for $L^p(A)$ and $L^p(B)$, respectively, then the sequence of tensors

$$(c_n) = (a_1 \otimes b_1, a_2 \otimes b_1, a_2 \otimes b_2, a_1 \otimes b_2, a_3 \otimes b_1, a_3 \otimes b_2, \dots)$$

taken in the rectangular order forms a basis for $L^p(A \times B)$; moreover

$$bc((c_n) : L^p(A \times B)) = bc((a_n), L^p(A))bc((b_n); L^p(B)) \quad (1 < p < \infty).$$

Thus if $d = 1, 2, \dots$ and $1 < p < \infty$ then there is a basis (c_n) of characters for $L^p(\mathbb{T}^d)$ such that $bc((c_n), L^p(\mathbb{T}^d)) = O((\max \frac{p}{p-1})^d)$.

Problem 5.1. Let $1 < p < \infty$. Does there exist a basis of characters for $L^p(\mathbb{T}^\mathbb{N})$ for some (for all) $p \neq 2$?

The analogous problem for $\mathbb{D}$ is trivial because the groups $\mathbb{D}^d$ and $\mathbb{D}^\mathbb{N}$ are isomorphic for $d = 1, 2, \dots$.

Problem 5.2. Let $d \in \mathbb{N}$ and p with $1 < p \neq 2 < \infty$ be fixed. Does there exist a constant $C = C(d, p) > 0$ such that if for some permutation $\pi : \mathbb{N} \xrightarrow{onto} \mathbb{N}$ the sequence $(c_{\pi(n)})$ is a basis for $L^p(\mathbb{T}^d)$ then $bc((c_{\pi(n)}); L^p(\mathbb{T}^d)) \geq C(\max(p, \frac{p}{p-1}))^d$?

A related question is

Problem 5.3. Let $1 < p \neq 2 < \infty$. Does there exist an isomorphism $T : L^p(\mathbb{T}) \xrightarrow{onto} L^p(\mathbb{D})$ which takes characters into characters?

Problem 5.3 is usually stated as follows: Are the exponents $\{e^{int} : n \in \mathbb{Z}\}$ and the Walsh system permutably equivalent? In the natural orders, i.e the order described by (5.1) for the exponents and the Paley order for the Walsh system, the exponents and the Walsh system are non equivalent (cf. [Wo], 1976,). Wojtaszczyk 2000 [W] and Hinrichs and Wenzel 2003 [HiWe] investigated the problem of non-equivalence of these systems in natural and some other orders. A stronger negative fact might be true that there is no constant $C = C(k, p) \geq 1$ such that for each k there is a bounded linear operator $T_k : L^p(\mathbb{Z}_2^k) \to L^p(\mathbb{T})$ which takes characters of $\mathbb{Z}_2^k$ to characters of $\mathbb{T}$ and satisfies

$$||f||_{L^p(\mathbb{Z}_2^k)} \leq ||T_k(f)||_{L^p(\mathbb{T})} \leq C||f||_{L^p(\mathbb{Z}_2^k)} \quad (f \in L^p(\mathbb{Z}_2^k)).$$

Here $\mathbb{Z}_2^k$ stands for the kth Cartesian power of the cyclic group $\mathbb{Z}_2$. On the other hand it follows from [D, Theoreme 4.1] that for every $d \in N$ and every finite set F of characters of $\mathbb{T}^d$ there is a linear operator $T : L_F^p(\mathbb{T}^d) \, to L^p(\mathbb{T})$ which takes characters into characters and satisfies

$$||f||_{L_F^p(\mathbb{T}^d)} \leq ||T(f)||_{L^p(\mathbb{T})} \leq K_p||f||_{L_F^p(\mathbb{T}^d)} \quad (f \in L_F^p(\mathbb{T}^d)),$$

where T is independent of p for $1 \leq p < \infty$, and $K_p > 1$ is independent of d and F. Here $L_A^p(\mathbb{T}^d)$ denotes the subspace of $L^p(\mathbb{T}^d)$ spanned by F. Bourgain,Rosenthal and Schechtman (1981) [BrRS] proved: *For $1 < p \neq 2 < \infty$ there are $\aleph_1$ mutually non-isomorphic subspaces of $L^p(\mathbb{D})$ spanned by characters of $\mathbb{D}$ which are $\mathcal{L}_p$−spaces.* It is unknown whether $\aleph_1$ can be replaced by continuum. The following problem is also open for more than 20 years:

Problem 5.4. *Does there exist a family of $\aleph_1$ mutually non isomorphic subspaces of $L^p(\mathbb{T}^n)$ spanned by characters of $\mathbb{T}^n$ with the same properties?*

REFERENCES

[ACPPr] G. Alberti, M. Csörnyei, A. Pełczyński, D. Preiss, *BV has the bounded approximation property*, J. Geometric Analysis **15** (2005), 1-7.

[B] S. Banach, *Théorie des opérations lin/'eaires*, Monografie Matematyczne **1**, Warszawa 1932.

[Bo] H. F. Bohnenblust, *A characterization of complex Hilbert spaces*, Potugaliae Math. **3** (1942), 103-109.

[BrDl] J. Bourgain and F. Delban, *A class of $\mathcal{L}_\infty$ spaces*, Acta Math. **145** (1981), 155-176.

[BrRS] J. Bourgain, H. P. Rosenthal, G. Schechtman, *An ordinal L^p−index for Banach spaces, with application to complemented subspaces of L^p*, Ann of Math. **114** (1981), 193-228.

[D] M. Déchamps, *Sous-espaces invariants de $L^p(G)$, G groupe abélien compact*, Publications Mathematiques d'Orsay 1980-1981, Expose **3**, Université de Paris-Sud, Orsay 1981.

[DaDnSi] W.J. Davis, D.W. Dean and I. Singer, *Complemented subspaces and Λ systems in Banach spaces*, Israel J. Math. **6** (1968), 303-309.

[DeF] A. Defant and K. Floret, *Tensor Norms and Operator Ideals*, Mathematics Study **176**. North-Holland, Amsterdam 1993.

[DV] A. Dvoretzky, *Some results on convex bodies and Banach spaces*, Proc. Symposium on Linear spaces. Jerusalem 1961, 123-160.

[G] W. T. Gowers, *A new dichotomy for Banach spaces*, Geom. Func. Anal. **6** (1996), 1083-1093.

[GMau1] W. T. Gowers and B. Maurey, *The unconditional basic sequence problem*, J. Amer. Math. Soc. **6** (1993), 851-874.

[GMau2] W. T. Gowers and B. Maurey, *Banach spaces with small spaces of operators*, Math. Ann. **307** 1997), 543-568.

[GeGi] B. Gelbaum and J. Gil de Lamadrid, *Bases of tensor products of Banach spaces*, Pacific J. Math. **11** (1961), 1281-1286.

[Gl] E. D. Gluskin, *Finite-dimensional analogues of spaces without basis*, Dokl. Akad. Nauk SSSR **216** (1981), 1046-1050.

[Gr] A. Grothendieck, *Produits tensoriels topologiques et espaces nucléaires*, Mem. Amer. Math. Soc. **16** 1955.

[H] S. Heinrich, *Ultraproducts in Banach space theory*, J. Reine Angew. Math. **313** (1980), 72-104.

[HiWe] A. Hinrichs and J. Wenzel, *On the non-equivalence of rearranged Walsh and trigonometric systems in L_p*, Studia Math. **159** (2003), 435-445.

[J] P. W. Jones, *BMO and the Banach space approximation problem*, Amer J. Math. **107** (1985), 853-893.

[KKrM] , M. A. Krasnosielskiĭ, M. G. Krein and D. P. Milman, *On defect numbers of linear operators in Banach spaces and on some geometric questions*, Sb. Trud. Mat. Inst. AN USSR **11** (1948), 97-112 (Russian).

[Kal] N. J. Kalton, *The complemented subspace problem revised*, to appear.

[KaMy] M. I. Kadets and B. S. Mityagin, *Complemented subspaces in Banach spaces*, Uspehi Mat. **28** (1973) 77-94 (Russian).

[Kk] S. Kakutani, *Some characterizations of Euclidean spaces*, Jap. J. Math. **16** (1939), 93-97.

[Kos] P. Koszmider, *Banach spaces of continuous functions with few operators*, Math. Ann. **320** (2004), 151-183.

[KoT] R. Komorowski and N. Tomczak-Jaegermann, *Banach spaces without local unconditional structure*, Israel J. Math. **89** (1995), 205-226; *Erratum to "Banach spaces without local unconditional structure"*, ibidem **105** (1998), 85-92.

[L] J. Lindenstrauss, *On the modulus of smoothness and divergent series in Banach spaces*, Michigan Math. J. **10** (1964), 241-252.

[LT1] J. Lindenstrauss and L. Tzafriri, *On the comlemented subspaces problem*, Israel J. Math. **9** (1971), 263-269.

[LT2] J. Lindenstrauss and L. Tzafriri, *Classical Banach Spaces, Vol I and II*, Ergebn. Math. Grenzgeb. **92** and **97** , Springer, 1977 and 1979.

[Ma] V. Mascioni, *Some remarks on the uniform approximation property of Banach spaces*, Studia Math. **96** (1990), 243-253.

[Mau] B. Maurey, *Banach spaces with few operators*, Handbook of the Geometry of Banach Spaces, Vol2, Eds. W. B. Johnson and J. Lindenstrauss, Elsevier, Amsterdam 2003.

[Mi] V. D. Milman, *Spectra of bounded continuous functions defined on the unit sphere of a B-space*, Funkcional Anal. Prilozhen. **3** (1969), 67-79 (Russian)

[Mu] P.F.X. Müler, *Isomorphisms between H^1 spaces*, Monografie Matematyczne **66**, Birkhäuser Verlag, Basel 2005.

[P1] A. Pełczyński, *All separable Banach spaces admit for every $\varepsilon > 0$ fundamental total and bounded by $1 + \varepsilon$ biorthogonal sequence*, Studia Math. **55** (1976), 295-304.

[P2] A. Pełczyński, *Some problems on bases in Banach and Frechet spaces*, Israel J. Math. **2** (1964), 132-138.

[PR] A. Pełczyński and H. P. Rosenthal, *Localisation techniques in L^p spaces*, Studia Math. **52** (1975), 263-289.

[PWj] A. Pełczyński and M. Wojciechowski, *Spaces of functions with bounded variation and Sobolev spaces without local unconditional structure*, J. Reine Angew. Math. **558** (2003), 109-159.

[Pi] G. Pisier, *Counterexamples to a conjecture of Grothendieck*, Acta Math. **151** (1983), 181-208.

[Pl] G. Plebanek, *A construction of a Banach space $C(K)$ with few operators*, Top. Appl. **143** (2004), 217-239.

[ScWaSi] , F. Schipp, W. R. Wade and P. Simon, *An introduction to dyadic harmonic Analysis*, Akademia Kiado, Budapest 1990.

[Sa] S. J. Szarek, *The finite-dimensional basis problem with an appendix on nets of Grassman manifolds*, Acta Math. **151** (1983), 153-179.

[Sz1] A. Szankowski, *B(H) does not have the approximation property*, Acta Math. **147** (1981), 89-108.

[Sz2] A. Szankowski, *On the uniform approximation property in Banach spaces*, Israel J. Math. **49** (1984), 343-359.

[W] P. Wojtaszczyk, *Non similarity of Walsh and Trygonometric systems*, Studia Math. **142** (2000), 171-185.

[Wo] Wo-Sung-Yang, *A note of Walsh-Fourier series*, Proc. Amer. Math. Soc. **59** (1976), 305-310.

Aleksander Pełczyński, Institute of Mathematics Polish Academy of Sciences, Śniadeckich 8 Ip; 00956 Warszawa; POLAND

E-mail: olek@impan.gov.pl

LIST OF PARTICIPANTS

Acosta, María Dolores; Granada, Spain. dacosta@ugr.es
Aiena, Pietro; Palermo, Italy. paiena@unipa.it
Anes Gallego, Sandra; Badajoz, Spain. sandra_ anes@hotmail.com
Angosto Hernández, Carlos; Murcia, Spain. angosto@um.es
Aqzzouz, Belmesnaoui; Kénitra, Morocco. bakzzouz@hotmail.com
Argyros, Spiros; Athens, Greece. sargyros@atlas.uoa.gr
Arias, Pablo; Cáceres, Spain. jparias@unex.es
Arjona Polo, Patricia; Extremadura, Spain. parjona@unex.es
Arranz, Francisco; Extremadura, Spain. farranz@unex.es
Arvanitakis, Alexander; Athens, Greece. aarva@math.ntua.gr
Avilés, Antonio; Murcia, Spain. avileslo@um.es
Bandyopadhyay, Pradipta; Kolkata, India. pradipta@isical.ac.in
Becerra, Julio; Granada, Spain. juliobg@ugr.es
Benabdellah, Houcine ; Marrakesh, Morocco. benabdellah@ucam.ac.ma
Benítez, Carlos; Extremadura, Spain. cabero@unex.es
Blasco, Fernando; Politécnica de Madrid, Spain. fblasco@montes.upm.es
Bombal, Fernando; Madrid, Spain. bombal@mat.ucm.es
Caballero, Josefa; Las Palmas de Gran Canaria, Spain. jmena@ull.es
Cabello Sánchez, Félix; Extremadura, Spain. fcabello@unex.es
Cabrera, Ignacio J.; Las Palmas de Gran Canaria, Spain. icabrera@dma.ulpgc.es
Calderón, M. del Carmen; Sevilla, Spain. mccm@us.es
Calvo, Cármen; Extremadura, Spain. ccalvo@unex.es
Campión, María Jesús; Navarra, Spain. mjesus.campion@unavarra.es
Candeal, Juan; Zaragoza, Spain. candeal@unizar.es
Cascales, Bernardo; Murcia, Spain. beca@um.es
Castillo, Jesús M. F.; Badajoz, Spain. castillo@unex.es
Cilla, Raffaella; Catania, Italy. cilia@dmi.unict.it
Cobos, Fernando; Madrid, Spain. cobos@mat.ucm.es
Defant, Andreas; Oldenburg, Germany. defant@mathematik.uni-oldenburg.de
Dodos, Pandelis; Athens, Greece. pdodos@math.ntua.gr
Domanski, Pawel; Poznan, Poland. domanski@amu.edu.pl
El Harti, Rachid; Settat, Morocco. relharti@hotmail.com
Fabian, Marián; Prague, Czech Republic. FABIAN@math.cas.cz
Fejo Herrera, María; Extremadura, Spain.
Fernández, Antonio; Sevilla, Spain. afernandez@esi.us.es
Fernández-Cabrera, Luz M.; Madrid, Spain. luz fernandez-c@mat.ucm.es
Fry, Robert; St. F.X., Nova Scotia, Canada. rfry@stfx.ca
Flores, Julio; Madrid, Spain. j.flores@escet.urjc.es
García del Amo, Alejandro; Madrid, Spain. alejandro@escet.urjc.es
García González, Ricardo; Extremadura, Spain. rgarcia@unex.es

García Herrera, Ruth; Extremadura, Spain. jagarcia@unex.es
García Muõoz, José Antonio; Extremadura, Spain.
Gavira, Beatriz; Sevilla, Spain. bgavira@us.es
Godefroy, Giles; Paris, France. gig@ccr.jussieu.fr
González, Manuel; Cantabria, Spain. gonzalen@unican.es
Guirao Sánchez, Antonio J.; Murcia, Spain. ajguirao@um.es
Hernández, Francisco L.; Madrid, Spain. pacoh@mat.ucm.es
Induráin, Esteban; Pamplona, Spain. steiner@si.unavarra.es
Johnson, W.B.; Texas A & M, USA. johnson@math.tamu.edu
Kalton, Nigel; Missouri, USA. nigel@math.missouri.edu
Kanellopoulos, Vassilis; Athens, Greece. bkanel@math.ntua.gr
Lajara, Sebastián; Albacete, Spain. sebastian.lajara@uclm.es
Lalaoui Rhali, My Hachem Marrakech, Morocco. hmlalaoui@ucam.ac.ma
López, Belén; Las Palmas de Gran Canaria, Spain. blopez@dma.ulpgc.es
Maestre, Manuel; Valencia, Spain. Manuel.maestre@uv.es
Martín, Miguel; Granada, Spain. mmartins@ugr.es
Martín, Pedro; Extremadura, Spain. pjimenez@unex.es
Martínez, Antonio; Vigo, Spain. antonmar@uvigo.es
Martínez, Juan Francisco; Valencia, Spain. J.Francisco.Martinez@uv.es
Martinón, Antonio; La Laguna, Spain. anmarce@ull.es
Mayoral, Fernando; Sevilla, Spain. mayoral@us.es
Michels, Carsten; Oldenburg, Germany michels@mathematik.uni-oldenburg.de
Moltó, Aníbal; Valencia, Spain. Anibal.Molto@uv.es
Montesinos, Vicente; Politécnica de Valencia, Spain. vmontesinos@mat.upv.es
Moraes, Luiza Amália; Rio de Janeiro, Brasil. luiza@im.ufrj.br
Moreno, José Pedro; Madrid, Spain. josepedro.moreno@uam.es
Moreno, Yolanda; Extremadura, Spain. ymoreno@unex.es
Muñoz, María; Cartagena, Spain. maria.mg@upct.es
Naranjo, Francisco José; Sevilla, Spain. naranjo@us.es
Navarro Olmo, Rosa; Extremadura, Spain. rnavarro@unex.es
Oncina, Luis; Murcia, Spain. luis@um.es
Orihuela, José; Murcia, Spain. joseori@um.es
Ortiz Caraballo, Carmen; Extremadura, Spain. carortiz@unex.es
Pallarés, Antonio José; Murcia, Spain. apall@um.es
Papini, Pier Luigi; Bologna, Italy. papini@dm.unibo.it
Pełczyński, Alexander; Warszawa, Poland, A.Pelczynski@impan.gov.pl
Pello, Javier; Madrid, Spain. jpello@escet.urjc.es
Pérez-García, David; Madrid, Spain. dperezg@escet.urjc.es
Phelps, Robert; Washington, USA. phelps@math.washington.edu
Prado, José Antonio; Autónoma de Madrid, Spain. joseantonio.prado@uam.es
Pulgarín, Antonio; Extremadura, Spain. aapulgar@unex.es
Raynaud, Yves; Paris, France. yr@ccr.jussieu.fr
Rodríguez Palacios; A. Granada, Spain. apalacios@ugr.es
Rodríguez, José; Murcia, Spain. joserr@um.es
Sadarangani, Kishin; Las Palmas de Gran Canaria, Spain. ksadaran@dma.ulpgc.e

Schmith, Richard; Oxford, England. richard.smith@christ-church.oxford.ac.uk

Signes, María Teresa; Murcia, Spain. tmsignes@um.es

Suárez Granero, Antonio; Madrid, Spain. AS granero@mat.ucm.es

Suárez Jesús; Extremadura, Spain. jesussf@telefonica.net

Tylli, Hans-Olav; Helsinki, Finland. hojtylli@cc.helsinki.fi

Villanueva, Ignacio; Madrid, Spain. ignaciov@mat.ucm.es

Wengenroth, J.; Trier, Germany. wengen@uni-trier.de

Yañez, Diego; Extremadura, Spain. dyanez@unex.es

Printed in the United States
by Baker & Taylor Publisher Services